BLACKWELL'S BOOKSHOP, OXFORD
FOUNDED 1879

THIS BOOK BELONGS TO

AF759266

Boundary Integral Equation Methods for Solids and Fluids

Boundary Integral Equation Methods for Solids and Fluids

Marc Bonnet
CNRS and École Polytechnique, Paris, France

JOHN WILEY & SONS LTD
Chichester • Weinheim • New York • Brisbane • Singapore • Toronto

Ouvrage publie avec l'aide du ministere francais charge de la Culture
(Published with the support of the French Ministry for the Arts)

Copyright © 1995 by Editions Eyrolles and CNRS Editions, Paris, France.

Published by John Wiley & Sons Ltd,
Baffins Lane, Chichester,
West Sussex PO19 1UD, England

National 01243 779777
International (+44) 1243 779777
e-mail (for orders and customer service enquiries): cs-books@wiley.co.uk

Visit our Home Page on http://www.wiley.co.uk
or http://www.wiley.com

All rights reserved. No part of this publication may be reproduced, stored in a retrieval system, or transmitted, in any form or by any means, electronic, mechanical, photocopying, recording, scanning or otherwise, except under the terms of the Copyright, Designs and Patents Act 1988 or under the terms of a licence issued by the Copyright Licensing Agency, 90 Tottenham Court Road, London, UK W1P 9 HE, without the permission in writing of the publisher.

Other Wiley Editorial Offices

John Wiley & Sons, Inc, 605 Third Avenue,
New York, NY 10158-0012, USA

Wiley-VCH Verlag GmbH, Pappelallee 3,
D-69469 Weinham, Germany

Jacaranda Wiley Ltd, 33 Park Road, Milton,
Queensland 4064, Australia

John Wiley & Sons (Asia) Pte Ltd, 2 Clementi Loop #02-01,
Jin Xing Distripark, Singapore 0512

John Wiley & Sons (Canada) Ltd, 22 Worcester Road,
Rexdale, Ontario M9W 1L1, Canada

British Library Cataloguing in Publication Data

A catalogue record for this book is available from the British Library

ISBN 0 471 97184 7

Produced from Latex files supplied by the author
Printed and bound in Great Britain by Biddles Ltd, Guildford and Kings Lynn
This book is printed on acid-free paper responsibly manufactured from sustainable forestry, in which at least two trees are planted for each one used for paper production.

Contents

Part III ADVANCED TOPICS

Preface

This book is the result of graduate courses given since 1990 at the Ecole Polytechnique, and also in 1993 at the Faculty of Mathematics of the University of Bucharest. It is a somewhat updated English version of the French edition (published by CNRS Editions and Eyrolles, Paris, France in 1995)

Although I do not claim to offer an exhaustive coverage of the subject (which would be far beyond the range of a single book), a broad array of application areas is covered, including

- Potential problems
- Elasticity
- Time-domain and frequency-domain scalar and elastic waves
- Fracture mechanics
- Small-strain elastoplasticity

Moreover, a systematic and modern approach to the subject is adopted. The regularization viewpoint is consistently used, in order to completely avoid formulations with non-integrable singular terms. This allows both rigorous exposition and application-oriented formulations. A substantial amount of space is devoted to the so-called hypersingular integral equations. Other modern features, which are currently research topics (to which I have contributed) not commonly found in textbooks, notably include

- Domain derivative and BIE formulations

- Use of geometrical symmetry
- Galerkin-type variational BIE formulations

The chosen style of presentation is concise, but at the same time this book is, I hope, self-sufficient. The mathematical prerequisites are familiarity with the field equations and boundary-value problems associated with continuum mechanics; a good working knowledge of classical differential geometry; multidimensional, surface and curvilinear integration; and classical tensor algebra. On the other hand, the use of advanced mathematics (and in particular of functional analysis) is avoided. Numerous bibliographical references are provided.

This book has been typeset by the author using the scientific document preparation system LaTeX and the style file `w-book.cls` provided by John Wiley and Sons.

Acknowledgements

I wish first to express my deep respect and appreciation for Huy Duong Bui, under whose guidance I accomplished my doctoral thesis. His initiation to research in general and boundary integral formulations in particular has been the best I could dream of.

I happily acknowledge my gratitude to Pierre Bérest, head of the Solid Mechanics research department at the Ecole Polytechnique, for the excellent working environment and freedom of research I have enjoyed here since 1988, and to Qoc Son Nguyen, Jean Salençon and André Zaoui for their constant encouragement in this undertaking. My greetings also go to all my colleagues and friends at the Ecole Polytechnique and elsewhere.

I wish to express my profound affection to my wife Bénédicte and my sons Pierre and Benoît. To undertake writing a book is certainly not the best news one could bring home!

This book is dedicated to the memory of Robert Bonnet.

Introduction

A crucial step in the modelling of many engineering problems is the solution of partial differential equations, together with appropriate boundary conditions and (for evolution problems) initial conditions, associated with solid or fluid mechanics, heat transfer, acoustics, electromagnetics ... The shape of the geometrical domain physically occupied by the system under study in most practical situations precludes the use of classical techniques like separation of variables or integral transformations and thus the possibility of obtaining exact, analytical solutions. Therefore, it is most often necessary to resort to numerical methods of solution. The majority of these are based on finite difference, finite element or boundary element techniques, although other approaches such as characteristics or spectral methods are also encountered.

The boundary element method (BEM) is based on the discretization of boundary integral equation (BIE) formulations. The study of the latter is more than one century old and constitutes a significant area of classical mathematical physics, namely the potential theory (see e.g. the books by Kellogg [8], Gunther [6]). For example, the Somigliana identity for elastostatics, which provides the groundwork for most developments of boundary element methods in solid mechanics, was published in 1886 [13]. The development of the boundary element method as a numerical solution technique is, however, later than that of finite element (FE) or finite difference (FD) methods. The first numerical implementations of the BEM date back to the 1960 decade: pioneering works include those of Shaw [12], Rizzo [11], Cruse [5].

The classical potential theory deals mainly with scalar field variables (temperature, electrostatic potential, acoustic waves, velocity potential for inviscid fluid flow ...) which are governed by partial differential equations involving the Laplacian operator: Poisson or Helmholtz equations ('elliptic'); scalar wave equation ('hyperbolic'), transient heat conduction ('parabolic') ... For instance, solutions of the Poisson equation $\Delta u + f = 0$ are representable as linear combinations of *potentials*, i.e. fields generated by sources distributions on a volume V or a surface S, such as Newtonian potentials:

$$u(\boldsymbol{x}) = \frac{1}{4\pi} \int_V f(\boldsymbol{y}) \frac{\mathrm{d}V_y}{|y - x|}$$

single-layer potentials:

$$u(\boldsymbol{x}) = \frac{1}{4\pi} \int_S \varphi(\boldsymbol{y}) \frac{\mathrm{d}S_y}{|y - x|}$$

or double-layer potentials:

$$u(\boldsymbol{x}) = \frac{1}{4\pi} \int_S \psi(\boldsymbol{y}) \frac{\partial}{\partial n} \frac{1}{|y - x|} \, \mathrm{d}S_y$$

The problem of solving the Poisson equation thus reduces to that of solving an integral equation for a source density φ or ψ defined on the boundary of the geometrical domain under study; the source distribution f: body force per unit volume, mass density, internal heat source ... is known in advance. It is important to note that the various potentials are expressed in terms of singular integrals for $\boldsymbol{x} \in V, S$. The question of whether the initial problem admits a unique solution for a given set of boundary conditions thus leads us to consider the existence and uniqueness of solutions to the relevant boundary integral equation. Answers to such general questions can thus be obtained through suitable extensions to singular multidimensional integral equations of the Fredholm theorems (see Tricomi [14]). This approach has also been conducted for the fundamental boundary-value problems of static and dynamic elasticity: strongly singular elastic potential fields similar to the scalar potentials shown above are introduced e.g. by Kupradze [9], Kupradze et al. [10], who then proceed to establish the appropriate boundary integral equations and prove using a regularization approach that the Fredholm theorems still hold in this context.

The integral equations coming from potential theory are often termed 'indirect' because the unknown function in them is a secondary variable, namely a real or fictitious source distribution. They are classically used to formulate fundamental Dirichlet, Neumann, Robin boundary-value problems. Other integral formulations, termed 'direct', relate the values taken on the boundary by the primary physical variables: potential and flux, displacement and traction vectors ... and their practical applicability is not restricted to specific classes of boundary conditions.

In the vast majority of cases, both direct and indirect types of integral equations can be solved in an approximate manner only; this motivates the concept of boundary element methods (BEMs). Their definition relies heavily on the transposition of field variable interpolation and geometry representation techniques initially created and developed for finite element methods (Bathe [1], Batoz and Dhatt [2], Ciarlet and Lions [3, 4] Hugues [7], Zienkiewicz [15]): meshing, shape functions ... The conceptual advantage of integral equation methods over other techniques like FEMs is the dimensionality gain: the geometrical support for the geometry modelling and unknown field variable discretization is not the whole domain but only its boundary.

Boundary elements vs. finite elements

The relative computational merits of finite element and boundary element methods applied to a given problem, although frequently discussed, is not necessarily a critical issue in our opinion. The finite element method has obviously a dominant status in the field of computational methods in engineering, mostly because of its greater flexibility and wider range of applicability; this book is therefore not intended to advocate the opposite viewpoint. Integral equation methods are superior for certain classes of problems, often (though not necessarily) featuring at least one of the following characteristics: linear governing equations in general, wave propagation, unbounded media, cracks, moving or unknown boundaries. Thus FEMs and BEMs should be considered as complementary rather that competitive. This complementarity is most apparent in mixed solution strategies based on FEM/BEM coupling, where some nonlinear or heterogeneous subregion is modelled using finite elements while the complementary, linear and homogeneous, domain is treated using boundary elements on the interface.

Overview of the book contents

In an introductory chapter, the main areas of application of boundary integral equation methods are discussed, and governing equations for the most common classes of problems amenable to boundary element treatment are briefly recalled.

The main body is divided into four main parts. In the first part, boundary integral formulations related to static problems are established, together with the fundamentals of boundary element discretization. Important concepts appear at this stage, notably because all issues related to the singular nature of the integral operators in general are contained in static problems. Finally, one chapter is devoted to a concise review of classical mathematical results of potential theory.

The second part is devoted to boundary integral equation methods for evolution problems: scalar or elastic wave equation in time or frequency domain, diffusion equation.

The third part addresses several more advanced issues, which are still currently under research: variational boundary integral formulations for mixed boundary-value problems; problem size reduction using geometrical symmetries; differentiation of boundary integral formulations with respect to domain shape perturbations.

Finally, the fourth and last part deals with two important topics in solid mechanics: crack problems; treatment of initial strains or stresses, with application to classical small-strain elastoplasticity. The latter constitute an important class of nonlinear problems amenable to boundary element solution techniques.

A rather abundant general bibliography is given, where books and papers selected for their importance or representativity are listed. Given the abundance of literature on the subject, no attempt at completeness is made. Some technical auxiliary topics are briefly treated in the appendices: tangential differential operators and integration by parts on surfaces, and usual or modified shape functions. Other appendices appear at the end of some chapters.

Notation

We define here a few conventions and notations that are consistently, and implicitly, used throughout this book. Any departure from the following definitions will be explicitly stated.

Indices, vectors, tensors. Usual typeface letters (f,x ...) denote scalar quantities while boldface letters ($\boldsymbol{x}$, $\boldsymbol{u}$, $\boldsymbol{\sigma}$, $\boldsymbol{U}$, $\boldsymbol{\Sigma}$ ...) denote vectors, matrices or tensors. Einstein summation convention is used throughout, unless when explicitly stated otherwise. The range of latin indices $i, j, k, \ell \ldots$ (resp. greek indices $\alpha, \beta, \gamma, \delta \ldots$) is $1,2,3$ (resp. $1,2$).

Geometry. Points in the Euclidean three-dimensional physical space $\mathbb{R}^3$ are denoted $\boldsymbol{x} = x_i \boldsymbol{e}_i$, $\boldsymbol{y} = y_i \boldsymbol{e}_i$, $\boldsymbol{z} = z_i \boldsymbol{e}_i$, where $\mathcal{R} = (\boldsymbol{O}, \boldsymbol{e}_1, \boldsymbol{e}_2, \boldsymbol{e}_3)$ is an orthonormal Cartesian reference frame. In fundamental solutions (i.e. fields generated by a unit point source), the field point is denoted by $\boldsymbol{y}$ and the fixed source (or observation, or collocation) point by $\boldsymbol{x}$ or $\boldsymbol{z}$. A unit vector normal to a surface is denoted $\boldsymbol{n}$; unless explicitly stated otherwise, we use the unit normal to the boundary of a given domain which is exterior to the domain.

Differential operators. The partial derivative is often denoted using a comma:

$$\frac{\partial f}{\partial x} \equiv f_{,x} \qquad \frac{\partial^2 f}{\partial x \partial y} \equiv f_{,xy} \qquad \frac{\partial f}{\partial y_i} \equiv f_{,i} \ldots$$

Moreover, for two-point scalar, vector or tensor functions, the following notation convention is used to distinguish Cartesian partial derivatives with respect to the first or second argument:

$$f_{,\bar{j}}(\boldsymbol{x},\boldsymbol{y}) \equiv \frac{\partial}{\partial x_j} f(\boldsymbol{x},\boldsymbol{y}) \qquad f_{,j}(\boldsymbol{x},\boldsymbol{y}) \equiv \frac{\partial}{\partial y_j} f(\boldsymbol{x},\boldsymbol{y})$$

The normal derivative is denoted $(\cdot)_{,n}$:

$$f_{,n}(\boldsymbol{y}) = f_{,i}(\boldsymbol{y})n_i(\boldsymbol{y}) \qquad f_{,n}(\boldsymbol{x},\boldsymbol{y}) = f_{,i}(\boldsymbol{x},\boldsymbol{y})n_i(\boldsymbol{y})$$

Moreover, the main differential operators are defined and denoted as follows:

Gradient: $\boldsymbol{\nabla}(\cdot) = (\cdot)_{,i} \otimes \boldsymbol{e}_i$.

Surface gradient: $\boldsymbol{\nabla}_S(\cdot) = \boldsymbol{\nabla}(\cdot) - (\cdot)_{,n} \otimes \boldsymbol{n}$. Using orthonormal Cartesian coordinates, one has in component notation:

$$\boldsymbol{\nabla}_S(\cdot) = D_i(\cdot) \otimes \boldsymbol{e}_i \qquad D_i(\cdot) = (\cdot)_{,i} - (\cdot)_{,n} n_i$$

This defines the new notation D_i : projection of the Cartesian partial derivative $\partial/\partial y_i$ onto the tangent plane.

Divergence : div $= \boldsymbol{\nabla}(\cdot) : \mathbf{1}$. Using orthonormal Cartesian coordinates: $\mathrm{div}\boldsymbol{u} = u_{i,i}$.

Surface divergence: $\mathrm{div}_S(\cdot) = \mathrm{div}(\cdot) - (\cdot)_{,n}.\boldsymbol{n} = D_i(\cdot).\boldsymbol{e}_i$.

REFERENCES

1. BATHE, K. J. *Finite element procedures in engineering analysis*. Prentice-Hall, Englewood Cliffs, New Jersey, USA (1982).
2. BATOZ, J. L., DHATT, G. *Modélisation des structures par éléments finis*. Hermes, Paris (1990).
3. CIARLET, P. G., LIONS, J. L. (eds). *Finite element methods (part 1)*. Handbook of numerical analysis. North-Holland (1991).
4. CIARLET, P. G., LIONS, J. L. (eds). *Finite element methods (part 2), numerical methods for solids (part 2)*. Handbook of numerical analysis. North-Holland (1996).
5. CRUSE, T. A. Numerical solutions in three-dimensional elastostatics. *Int. J. Solids Struct.*, **5**, 1259–1274 (1969).
6. GUNTHER, N. M. *La théorie du potentiel et ses applications aux problèmes fondamentaux de la physique mathématique*. Gauthier-Villars, Paris (1934).
7. HUGUES, T. J. R. *The finite element method – linear static and dynamic finite element analysis*. Prentice-Hall, Englewood Cliffs, New Jersey, USA (1987).
8. KELLOGG, O. D. *Fundations of potential theory*. Springer-Verlag, Berlin, Heidelberg (1967).
9. KUPRADZE, V. D. *Dynamical problems in elasticity*, vol. 3 of *Progress in solids mechanics*. North-Holland (1963).

10. Kupradze, V. D. (ed.). *Three-dimensional problems of the mathematical theory of elasticity and thermoelasticity*. North-Holland (1979).

11. Rizzo, F. J. An integral equation approach to boundary-value problems of classical elastostatics. *Quart. Appl. Math.*, **25**, 83–95 (1967).

12. Shaw, R. P. Diffraction of acoustic pulses by obstacles of arbitrary shape with a Robin condition – part A. *J. Acoust. Soc. Am.*, **41**, 855–859 (1962).

13. Somigliana, C. Sopra l'equilibrio di un corpo elastico isotrope. *Il Nuovo Ciemento, serie III*, **20**, 181–185 (1886).

14. Tricomi, F. G. *Integral equations*. Interscience Publishers, New York (1957).

15. Zienkiewicz, O. C. *The finite element method: basic concepts and linear applications*. McGraw-Hill, fourth edn. (1989).

1

Basic principles and fields of application

This introductory chapter begins with a brief exposition of the basic ideas underlying boundary integral equation (BIE) formulations (Section 1.1). Then, the main fields of application of BIE methods are discussed (Section 1.2). Next, a review of various physical situations with governing partial differential equations involving the Laplace operator is given (Sections 1.3 to 1.5), together with a brief summary of the fundamental equations of linear elasticity. This enumeration covers many (but not all) classes of problems amenable to BIE methods of solution. In particular, many problems arising from various areas of physics and engineering sciences are governed by Laplace or Poisson equations in either scalar or vector form, and thus fall within the range of applicability of the methods presented in Chapters 2 and 3.

1.1 BASIC PRINCIPLES

A boundary-value problem over a geometrical domain $\Omega \subset \mathbb{R}^n$ $(n = 1, 2, 3)$, having u as unknown, associated with a linear second-order partial differential operator $\mathcal{L}$ has the typical structure:

$$\begin{cases} \mathcal{L}u + f = 0 & \text{in } \Omega \\ u = g_1 & \text{on } S_u \\ T^n(u) = g_2 & \text{on } S_T \end{cases} \tag{1.1}$$

in which g_1, g_2 and the source f are given, $u \to T^n(u)$ denotes a first-order partial differential operator acting on u and linear with respect to $\boldsymbol{n}$, the unit normal to the boundary $\partial\Omega$ exterior to Ω. The surfaces S_u, S_T define a partition of $\partial\Omega$: $S_u \cup S_T = \partial\Omega$ and $S_u \cap S_T = \emptyset$. Moreover, the operators $\mathcal{L}$ and T^n are assumed to satisfy the following *reciprocity identity*:

$$\int_\Omega (\mathcal{L}u.v - \mathcal{L}v.u) \, \mathrm{d}V = \int_{\partial\Omega} (T^n(u).v - T^n(v).u) \, \mathrm{d}S \tag{1.2}$$

The main idea is then to replace v with a so-called *fundamental solution*, that is, a particular solution to the field equation $(1.1)_1$ for a unit point source $f(\boldsymbol{y}) = \delta(\boldsymbol{y} - \boldsymbol{x})$ applied at a given point $\boldsymbol{x}$ not located on the boundary of Ω ($\boldsymbol{x} \in \mathbb{R}^3 - \partial\Omega$); denote by $G(\boldsymbol{x}, \boldsymbol{y})$ any such fundamental solution, where $\boldsymbol{y}$ is the (variable) field point. Recall that the Dirac distribution $\delta(\boldsymbol{y} - \boldsymbol{x})$ is defined by the following property:

$$\int_\Omega \delta(\boldsymbol{y} - \boldsymbol{x}) u(\boldsymbol{y}) \, \mathrm{d}V_y = \kappa u(\boldsymbol{x}) \qquad \text{where} \qquad \begin{cases} \kappa = 1 & \boldsymbol{x} \in \Omega \\ \kappa = 0 & \boldsymbol{x} \notin \Omega \end{cases} \tag{1.3}$$

Then, substitution of v by G into the reciprocity identity (1.2) leads to the following result, valid for any source point $\boldsymbol{x} \in \mathbb{R}^3 - \partial\Omega$:

$$\kappa u(\boldsymbol{x}) = \int_\Omega f(\boldsymbol{y}) G(\boldsymbol{x}, \boldsymbol{y}) \, \mathrm{d}V_y + \int_{\partial\Omega} [u(\boldsymbol{y}) T^n G(\boldsymbol{x}, \boldsymbol{y}) - T^n u(\boldsymbol{y}) G(\boldsymbol{x}, \boldsymbol{y})] \, \mathrm{d}S_y \tag{1.4}$$

This *integral representation formula* gives explicitly the value of u at any point $\boldsymbol{x}$ interior to Ω as the sum of:

- a domain integral over Ω containing the given source f, which defines a particular solution to $(1.1)_1$ that in general does not satisfy the boundary conditions $(1.1)_{2,3}$;
- a boundary integral involving the values assumed by $u, T^n u$ over $\partial\Omega$, one of them being, at any given point of $\partial\Omega$, prescribed by the boundary conditions $(1.1)_{2,3}$ and the other unknown.

This procedure defines in some sense a 'quasi-inversion' of the field equation $(1.1)_1$, since the determination of u in Ω reduces to that of the values of $u, T^n u$ on $\partial\Omega$ which remain unknown after the boundary conditions $(1.1)_{2,3}$ have been taken into account. Thus, the solution procedure is shifted from the domain to its boundary. In practical terms, this saves one space dimension for the expected discretization of the geometry and unknowns.

Principle of the boundary integral equation method

The representation formula (1.4) holds only for $\boldsymbol{x}$ not on the boundary of Ω. The natural next step is to formulate a *boundary integral equation* (BIE) that relates u and $T^n u$ on the boundary. At first sight one needs only to apply

(1.4) with $\boldsymbol{x}$ chosen on the boundary. However, it turns out that, for most fundamental solutions $G(\boldsymbol{x},\boldsymbol{y})$ in two or three space dimensions, $T^n G(\boldsymbol{x},\boldsymbol{y})$ has a strong singularity for $\boldsymbol{y}=\boldsymbol{x}$; as a consequence, identity (1.4) cannot be used in its present form when $\boldsymbol{x}\in\partial\Omega$. The actual formulation of a boundary integral equation relies upon a limiting process in (1.4) in which the singular behaviour of the fundamental solution is adequately accounted for. Once the correct BIE is formulated, the *boundary integral equation method* can be applied to the original boundary-value problem (1.1) in the following way:

1. The boundary integral equation is solved; as a result, $(u, T^n u)$ are known over the whole boundary $\partial\Omega$.

2. The *integral representation formula* (1.4) then expresses in an explicit way the value of $u(\boldsymbol{x})$ at any interior point.

The boundary integral equation cannot in most cases be solved analytically, so that the above scheme primarily outlines a *numerical* solution strategy, applicable to arbitrary geometries and boundary conditions.

Notice that the definition of $G(\boldsymbol{x},\boldsymbol{y})$ makes no use of boundary conditions, and therefore is not unique. The simplest and most commonly used choice is the fundamental solution defined over the whole infinite space $\mathbb{R}^n$.

1.2 OVERVIEW OF FIELDS OF APPLICATION

The formulation of boundary integral equations is linked, through the introduction of fundamental solutions, to the presence of a linear partial differential operator in the field equations under consideration; besides, the so-called indirect BIE formulations are based upon the superposition principle. The boundary element methods (BEMs) therefore are primarily solution techniques for problems governed by linear partial differential equations: Laplace or Poisson equations (scalar potential problems, arising from various areas of physical sciences), wave equations in time or frequency domains (acoustic or electromagnetic wave propagation), Navier equations of linear elastostatics or elastodynamics, biharmonic equation (elastic plate bending), ... The media involved must usually possess homogeneous physical properties.

Mechanics of deformable bodies

In the solid mechanics area, BEM techniques are in particular used to solve elasticity problems. In this case, the relative computational efficiency (in terms of number of unknowns and computer resources) of BEM and FEM depend, among other things, upon the bulkiness or slenderness of the body under study. Many of the applications of BEM in solid mechanics concern cracked solids, in the linear fracture mechanics framework: the difficulties associated with three-dimensional finite element meshing for cracked solids, and especially with mesh refinement near the crack front, make BEM techniques appear advantageous, the more so when numerical simulation of crack front advance is concerned.

Unbounded media

The ability to deal with unbounded domains without the need to introduce artificial boundaries is among the main strengths of BIE-based methods. This is of particular importance in dynamical problems, where outgoing waves are non-physically reflected by the artificial boundaries needed using FEMs or FDMs. Moreover, the domain discretization needed in the latter methods is known to introduce a non-physical anisotropy of the propagation properties of the medium under consideration.[1] Since it involves neither domain discretization nor boundaries at infinity, BEM is thus very well suited to problems for which it is desirable to consider unbounded media, especially dynamical problems. This includes applications in acoustics (exterior or underwater acoustics, sound radiation); elastic wave problems in e.g. geophysics, seismology or earthquake engineering; electromagnetic wave problems; elastic fields created by inclusions or initial strain distributions; coupled problems where one component of the system occupies an unbounded domain, like in soil–structure, fluid–structure or sol–fluid–structure interaction.

Problems involving nonlinearities

Although the boundary integral equation formulations rely upon the use of fundamental solutions to linear partial differential operators, they can accommodate to some extent geometrical or material nonlinearity. The general idea is to reject all nonlinear terms in the right-hand side of the field equations, treating them as if they were body sources (for instance, they act as initial strains or stress distributions for elastic–plastic problems). In this case, the primary unknowns are still defined on the boundary, but the source-like terms appear in domain integrals over the part of the body (potentially) affected by the nonlinearity. In particular, BIE methods remain interesting for problems where the nonlinear part of the body is small in relative terms.

Unknown or variable domains

The boundary integral equation approach is also well suited to formulate and solve problems with variable or unknown geometrical domains: crack or damage front propagation; shape optimization; inverse problems with domain

[1]See e.g. BAMBERGER A., CHAVENT G., LAILLY CL.: *Etude de schémas numériques pour les équations de l'élastodynamique linéaire*, Institut National de Recherche en Informatique et en Automatique (France), internal report no. 41, 1980.

(partly) unknown such as crack, inclusion or other defect identification ...; problems with a free surface (e.g. groundwater flow, gravity waves). This area of application of BIE methods, still relatively new, is expected to witness many developments in the future.

1.3 HEAT CONDUCTION

The temperature distribution T in a medium Ω with a homogeneous but possibly anisotropic conductivity represented by a second-order symmetric tensor $\boldsymbol{k}$, following the Fourier law $\boldsymbol{q} = -\boldsymbol{k}.\boldsymbol{\nabla}T$ ($\boldsymbol{q}$: thermal flux), at thermal equilibrium with an internal heat source distribution r and under external influences (prescribed temperature or thermal flux, known heat transfer coefficient), solves the stationary heat conduction equation:

$$\text{div}(\boldsymbol{k}.\boldsymbol{\nabla}T) + r = 0 \tag{1.5}$$

For a medium with isotropic conductivity ($\boldsymbol{k} = k\boldsymbol{I}$), eq. (1.5) reduces to a Poisson equation:

$$k\Delta T + r = 0 \tag{1.6}$$

When transient heat conduction is considered, the temperature is a function of space and time and solves the diffusion equation:

$$\text{div}\boldsymbol{k}.\boldsymbol{\nabla}T - \rho c\frac{\partial T}{\partial t} + r = 0 \tag{1.7}$$

where c is the specific heat and ρ the mass density.

The BIE formulations associated with the Poisson equation (1.6) or its generalization (1.5) for anisotropic media are treated in Chapters 2, 3. A BIE formulation for the diffusion equation (1.7) is presented in Chapter 9.

Nonlinear heat conductivity

When the conductivity is temperature-dependent ($k = k(T)$), it is useful to resort to the *Kirchhoff transformation*, where the following auxiliary unknown ϕ is introduced:

$$\phi(\boldsymbol{y}) = -\int_{T_0}^{T(y)} k(\tau)\,\mathrm{d}\tau \tag{1.8}$$

As a result, the stationary heat conduction equation (1.6) reduces to the Laplace equation:

$$\Delta\phi = r$$

while the heat flux becomes:

$$\boldsymbol{q} = \boldsymbol{\nabla}\phi$$

Thus, the effect of the transformation (1.8) is to linearize the heat conduction equation (1.6) *and* the boundary conditions for prescribed heat flux. At a point where the temperature value is prescribed, the transformation (1.8) gives a well-defined value of ϕ, and the formulation of the heat conduction problem is linear with respect to ϕ.

1.4 FLUID FLOW

Potential fluid flow

The velocity field $\boldsymbol{U}(\boldsymbol{y},t)$ associated with an incompressible and irrotational flow of a perfect (i.e. non-viscous) fluid is expressed in terms of a scalar potential $\phi(\boldsymbol{y},t)$ by $\boldsymbol{U} = \boldsymbol{\nabla}\phi$. From the incompressibility condition div$\boldsymbol{U} = 0$, the potential is a harmonic function at any given time:

$$\Delta_y \phi(\boldsymbol{y},t) = 0$$

In order to define a well-posed boundary-value problem, this field equation is supplemented by boundary conditions (i) kinematical: impervious rigid walls or obstacles ($\phi_{,n} = \boldsymbol{\nabla}\phi.\boldsymbol{n} = 0$), prescribed stream velocity, ... , or (ii) dynamical: prescribed external pressure ... Even though the Laplace operator Δ acts only on space coordinates, the potential ϕ is time-dependent for unsteady inviscid fluid flow. Once ϕ has been computed (for instance using the boundary element method), the pressure $p(\boldsymbol{y},t)$ at any point of the fluid domain Ω and any time instant t is explicitly known from the generalized Bernoulli theorem (with g: gravity acceleration, ρ: mass density, C: a constant for the entire fluid region):

$$\frac{p}{\rho} + \frac{1}{2}|\boldsymbol{U}|^2 + g + \frac{\partial \phi}{\partial t} = C$$

The velocity field $\boldsymbol{U}$ is also expressible in terms of a vector potential: $\boldsymbol{U} = \mathbf{rot}\boldsymbol{\psi}$, where irrotationality implies that $\boldsymbol{\psi}$ solves the vector Laplace equation:

$$\Delta\boldsymbol{\psi} = \mathbf{0}$$

The vector potential reduces for two-dimensional fluid flow in a plane orthogonal to $\boldsymbol{e}_3$ to a scalar *stream function* ψ ($\boldsymbol{\psi} = \psi\boldsymbol{e}_3$) which is harmonic in the flow plane.

Groundwater flows

The groundwater flow of an inviscid incompressible fluid according to the Darcy law ($\boldsymbol{U} = -\boldsymbol{k}.\boldsymbol{\nabla}\phi$, where $\phi = p/\rho + gz$ is the static head and $\boldsymbol{k}$

the hydraulic conductivity tensor constitutes another example of a physical problem governed by the Laplace equation:

$$k\Delta\phi = 0 \quad \text{(isotropic case)} \qquad \text{div}(\boldsymbol{k}.\boldsymbol{\nabla}\phi) = 0 \quad \text{(anisotropic case)}$$

Acoustics

The velocity potential ϕ for small irrotational perturbations, around an undisturbed reference state $(\boldsymbol{U}(\boldsymbol{y},t) = 0,\ p(\boldsymbol{y},t) = p_0,\ \rho(\boldsymbol{y},t) = \rho_0)$, of a compressible inviscid fluid is governed by the scalar wave equation:

$$\Delta\phi - \frac{1}{c^2}\frac{\partial^2\phi}{\partial t^2} = 0 \qquad \frac{1}{c^2} = \frac{\mathrm{d}h}{\mathrm{d}\rho}(\rho_0)$$

where c is the acoustic wave velocity and $\rho = h(p)$ the constitutive equation for the fluid, i.e the pressure is a given function h of the mass density, h being a physical characteristic. The physical variables are then given in terms of the velocity potential ϕ by:

$$\begin{aligned} \boldsymbol{U}(\boldsymbol{y},t) &= \boldsymbol{\nabla}\phi(\boldsymbol{y},t) \\ p(\boldsymbol{y},t) - p_0 &= -\rho_0\frac{\partial\phi}{\partial t} \\ \rho(\boldsymbol{y},t) - \rho_0 &= \frac{1}{c^2}[p(\boldsymbol{y},t) - p_0] \end{aligned}$$

Boundary integral equation formulations for the wave equation are treated in Chapters 7 (time domain) and 8 (frequency domain).

Incompressible viscous fluid flow at moderate velocities

The velocity and pressure fields $\boldsymbol{U}, p$ for the flow of an incompressible viscous fluid (viscosity μ) solve the Navier–Stokes equation and the incompressibility equation (conservation of mass):

$$\begin{cases} -\boldsymbol{\nabla}p + \mu\Delta\boldsymbol{U} = \rho\left(\dfrac{\partial\boldsymbol{U}}{\partial t} + \boldsymbol{\nabla}\boldsymbol{U}.\boldsymbol{U}\right) - \rho g \boldsymbol{e}_3 \\ \text{div}\boldsymbol{U} = 0 \end{cases}$$

For steady-state flow at moderate velocity (i.e. low Reynolds number) and without body forces, the above equations reduce to the Stokes equations:

$$\begin{cases} -\boldsymbol{\nabla}p + \mu\Delta\boldsymbol{U} = 0 \\ \text{div}\boldsymbol{U} = 0 \end{cases} \tag{1.9}$$

The latter are supplemented with boundary conditions, either kinematical (e.g. no-slip condition ($\boldsymbol{U} = \boldsymbol{U}_{wall}$ along a possibly moving rigid wall), given stream velocity) or dynamical (e.g. equilibrium with a given external pressure). A boundary integral formulation for the Stokes system (1.9) is presented in Chapter 9.

1.5 ELECTROMAGNETISM

The electric $\boldsymbol{E}$ and magnetic $\boldsymbol{B}$ vector fields as well as the electric charge density ρ in a medium with linear, isotropic and homogeneous physical properties (electric conductivity σ, electric and magnetic susceptibilities χ_e, χ_m) are governed by the Maxwell equations:

$$\begin{cases} \text{div}\boldsymbol{E} = \dfrac{\rho}{\varepsilon_0(1+\chi_e)} \\ \text{div}\boldsymbol{B} = 0 \end{cases} \qquad \begin{cases} \textbf{rot}\boldsymbol{E} + \dfrac{\partial \boldsymbol{B}}{\partial t} = \mathbf{0} \\ \textbf{rot}\boldsymbol{B} - \dfrac{1}{c^2}\dfrac{\partial \boldsymbol{E}}{\partial t} = \mu_0(1+\chi_m)\sigma\boldsymbol{E} \end{cases} \tag{1.10}$$

where $c = (\varepsilon_0(1+\chi_e)\mu_0(1+\chi_m))^{-1/2}$ is the velocity of electromagnetic waves in the medium under consideration; ε_0 and μ_0 = are the permittivity and permeability of free space (with $c_0 = (\varepsilon_0\mu_0)^{-1/2}$ being the electromagnetic wave velocity in free space). Combining appropriately the above equations, one obtains either of the two following partial differential equations:

$$\begin{aligned} \Delta\boldsymbol{E} - \frac{1}{c^2}\frac{\partial^2\boldsymbol{E}}{\partial t^2} - \mu_0(1+\chi_m)\sigma\frac{\partial\boldsymbol{E}}{\partial t} &= \frac{1}{\varepsilon_0(1+\chi_e)}\boldsymbol{\nabla}\rho \\ \Delta\boldsymbol{B} - \frac{1}{c^2}\frac{\partial^2\boldsymbol{B}}{\partial t^2} - \mu_0(1+\chi_m)\sigma\frac{\partial\boldsymbol{B}}{\partial t} &= \mathbf{0} \end{aligned} \tag{1.11}$$

Electrostatics

The electric field is the gradient of a scalar potential: $\boldsymbol{E} = -\boldsymbol{\nabla}V$ which solves the Poisson equation:

$$\Delta V + \frac{\rho}{\varepsilon_0(1+\chi_e)} = 0 \tag{1.12}$$

The electrostatic field generated in a spatial region Ω by a distribution of electric charge is then obtained by solving eq. (1.12) supplemented by boundary conditions: prescribed potential $V = V^D$, insulation ($\boldsymbol{\nabla}V.\boldsymbol{n} = 0$) ...

Waves

When the physical properties of the medium are such that:

$$\sigma T/(\varepsilon_0(1+\chi_e)) \ll 1 \qquad \text{or} \sigma/(\varepsilon_0(1+\chi_e)\omega) \ll 1$$

(T being a characteristic time, or alternatively ω being a characteristic angular frequency), the propagation terms dominate the diffusion terms in eqs (1.11). Discarding the diffusion terms and assuming zero electric charge distribution, the fields $\boldsymbol{E}, \boldsymbol{B}$ thus solve the vector wave equation (Chapter 7):

$$\Delta\boldsymbol{H} - \frac{1}{c^2}\frac{\partial^2\boldsymbol{H}}{\partial t^2} = \mathbf{0} \qquad (\boldsymbol{H} = \boldsymbol{E}, \boldsymbol{B})$$

For time-harmonic phenomena ($\boldsymbol{H}(\boldsymbol{y},t) = \hat{\boldsymbol{H}}(\boldsymbol{y})e^{i\omega t}$), or using a Fourier transform with respect to time (with ω the transformed variable) the above equation becomes the Helmholtz (vector) equation (Chapter 8):

$$\{\Delta + \omega^2\}\hat{\boldsymbol{H}} = \boldsymbol{0} \qquad (\boldsymbol{H} = \boldsymbol{E}, \boldsymbol{B})$$

Induction

Conversely, when the physical properties of the medium are such that:

$$\sigma T/(\varepsilon_0(1+\chi_e)) \gg 1 \qquad \text{or} \sigma/(\varepsilon_0(1+\chi_e)\omega) \gg 1$$

the diffusion terms dominate the propagation terms in eqs (1.11). Discarding the propagation terms and still assuming zero electric charge distribution, the fields $\boldsymbol{E}, \boldsymbol{B}$ thus solve the vector diffusion equation (Chapter 9):

$$\Delta \boldsymbol{H} - \mu_0(1+\chi_m)\sigma\frac{\partial \boldsymbol{H}}{\partial t} = 0 \qquad (\boldsymbol{H} = \boldsymbol{E}, \boldsymbol{B})$$

The above equations govern *eddy currents* $\boldsymbol{j} = \sigma\boldsymbol{E}$ that are induced in a conducting body by the presence of an exterior, given, magnetic field $\boldsymbol{B}_0$. The heat generated by eddy currents through the Joule effect can be either desired or unwanted, depending on the particular technological situation considered.

Two-dimensional problems

The magnetic field can also be expressed in terms of a vector potential: $\boldsymbol{B} = \textbf{rot}\boldsymbol{A}$. In two dimensions, this reduces to a scalar potential analogous to the stream function of ideal fluid flow: $\boldsymbol{A} = A\boldsymbol{e}_3$. Then, for example, electromagnetic wave propagation or eddy current problems can be formulated using scalar wave or diffusion equations.

1.6 LINEAR ELASTOSTATICS

The displacement, strain and stress fields (denoted respectively $\boldsymbol{u}, \boldsymbol{\varepsilon}, \boldsymbol{\sigma}$) corresponding to an equilibrium state of a linearly elastic homogeneous body Ω, assuming infinitesimal transformations, are related at any point of Ω by the following local equations:

$$\varepsilon_{ij} = (u_{i,j} + u_{j,i})/2 \qquad (1.13)$$

$$\sigma_{ij} = C_{ijkl}\varepsilon_{kl} = C_{ijkl}u_{k,l} \qquad (1.14)$$

$$\sigma_{ij,j} + \rho F_i = 0 \qquad (1.15)$$

where the body force vector $\rho\boldsymbol{F}$ may have various origins: gravity, inertia, thermal effects (see Chapter 4), inelastic terms, initial strains or stresses (see

Chapter 14) ... The elastic constants of the material, which appear in the constitutive equation (1.14), define the fourth-order Hooke tensor $\boldsymbol{C}$. The following symmetry properties hold:

$$C_{ijkl} = C_{klij} = C_{jikl} = C_{ijlk} \tag{1.16}$$

so that any given elastic constitutive law is characterized by at most 21 independent coefficients. Any triplet $(\boldsymbol{u}, \boldsymbol{\sigma}, \boldsymbol{F})$ satisfying the field equations (1.13–1.15) will be termed an *elastostatic state* over Ω.

Any 'well-posed' problem must be supplemented by suitable boundary conditions over $\partial\Omega$: prescribed force distribution $\boldsymbol{t}^D$ applied on a part S_T of $\partial\Omega$:

$$\boldsymbol{T}^n(\boldsymbol{u}) \equiv \boldsymbol{\sigma}.\boldsymbol{n} = \boldsymbol{t}^D \qquad \text{over } S_T$$

or prescribed displacements $\boldsymbol{u}^D$ over the complementary part $S_u = \partial\Omega - S_T$:

$$\boldsymbol{u} = \boldsymbol{u}^D \qquad \text{over } S_u$$

Combining equations (1.13), (1.14), (1.15), one gets the Navier equation for the displacement field:

$$\boldsymbol{\Delta}^\star \boldsymbol{u} + \rho \boldsymbol{F} = 0 \qquad \text{where} \quad \boldsymbol{\Delta}^\star \boldsymbol{u} \equiv C_{ijkl} u_{k,lj} \boldsymbol{e}_i \tag{1.17}$$

Isotropic elasticity

The Hooke tensor $\boldsymbol{C}$ is then expressed in terms of only two independent elastic constants: λ, μ (Lamé constants) or E, ν (Young modulus and Poisson ratio), where:

$$\lambda = \frac{\nu E}{(1-2\nu)(1+\nu)} \qquad \mu = \frac{E}{2(1+\nu)}$$

The Hooke tensor is given by

$$\begin{aligned} C_{ijkl} &= \lambda \delta_{ij}\delta_{kl} + \mu(\delta_{ik}\delta_{jl} + \delta_{il}\delta_{jk}) \\ &= \mu\left(\frac{2\nu}{1-2\nu}\delta_{ij}\delta_{kl} + \delta_{ik}\delta_{jl} + \delta_{il}\delta_{jk}\right) \end{aligned} \tag{1.18}$$

The traction vector is then expressed in terms in the displacement by

$$\boldsymbol{T}^n(\boldsymbol{u}) = 2\mu \boldsymbol{u}_{,n} + \lambda \boldsymbol{n} \text{div} \boldsymbol{u} + \mu \boldsymbol{n} \wedge \textbf{rot} \boldsymbol{u} \tag{1.19}$$

The partial differential operator $\boldsymbol{\Delta}^\star$ of the Navier equation (1.17) becomes

$$\boldsymbol{\Delta}^\star = \mu \boldsymbol{\Delta} + (\lambda + \mu)\boldsymbol{\nabla}\text{div} \tag{1.20}$$

Finally, a classical result of elasticity theory states that any solution to the isotropic Navier equation (1.17)–(1.20) has the following representation:

$$\boldsymbol{u} = 2(1-\nu)\boldsymbol{\Delta G} - \boldsymbol{\nabla}(\text{div}\boldsymbol{G}) \tag{1.21}$$

where, upon substitution of (1.21) into (1.17), the *Galerkin* vector $\boldsymbol{G}$ solves

$$\boldsymbol{\Delta\Delta G} + \frac{\rho}{2\mu(1-\nu)}\boldsymbol{F} = 0$$

Antiplane elasticity

This is a particular class of elasticity problems where the domain Ω is invariant under translations along the $\boldsymbol{e}_3$-direction and the displacement field assumes the following form:

$$\boldsymbol{u}(\boldsymbol{y}) = u(y_1, y_2)\boldsymbol{e}_3$$

This implies that the only nonzero strain and stress components are:

$$2\varepsilon_{\alpha 3} = u_{,\alpha} \quad \sigma_{\alpha 3} = 2\mu\varepsilon_{\alpha 3} \qquad (\alpha = 1, 2)$$

and the Navier equation (1.17) reduces to a scalar two-dimensional Poisson equation:

$$\Delta_2 u + F = 0$$

Part I

Statics

2

Integral equations for the Poisson equation

This chapter addresses the integral equations and integral representation formulas associated with solutions of the Poisson equation:

$$\Delta u + b = 0 \qquad \boldsymbol{y} \in \Omega \tag{2.1}$$

where $\Omega \subset \mathbb{R}^3$ is a bounded domain, u is the unknown field variable, termed *potential* in this chapter, and b denotes a given source. The unknown u must also obey at any point of the boundary $\partial\Omega$ to a given boundary condition. The fundamental boundary-value problems to which equation (2.1) corresponds are the Neumann problem, where the flux is known on the whole boundary:

$$q(\boldsymbol{y}) \equiv \boldsymbol{\nabla} u(\boldsymbol{y}).\boldsymbol{n}(\boldsymbol{y}) = q^D(\boldsymbol{y}) \qquad (\boldsymbol{y} \in \partial\Omega)$$

and the Dirichlet problem, where the potential is known on the whole boundary:

$$u(\boldsymbol{y}) = u^D(\boldsymbol{y}) \qquad (\boldsymbol{y} \in \partial\Omega)$$

Mixed boundary conditions are also frequently considered: flux prescribed over some subset S_q of $\partial\Omega$ and potential prescribed over the complementary subset $S_u = \partial\Omega \backslash S_q$. More general boundary conditions are sometimes considered; for example some linear combination of potential and flux is prescribed.[1]

The exposition will focus mainly on three-dimensional situations, the two-dimensional problems being treated as special three-dimensional ones in Section 2.5.

[1] In some heat transfer problems, the heat flux across $\partial\Omega$ is proportional to the difference between the body surface and ambient external temperatures.

Geometrical hypotheses. We consider a bounded domain[2] $\Omega \subset \mathbb{R}^3$, which boundary $\partial\Omega$ is assumed to be piecewise twice continuously differentiable and may have a finite number of edges or corners[3] and is not necessarily connected.[4] Each connected component of $\partial\Omega$ should have an interior of nonzero measure, in order to preclude the possibility of discontinuity lines or surfaces, cracks, screens, etc. for which a distinct treatment is needed (Chapter 5). The unit normal to $\partial\Omega$ is conventionally, and consistently, oriented outward from the domain Ω.

2.1 INTEGRAL REPRESENTATION

The steps that were followed in Section 1.1 in order to establish an integral representation formula associated with an unspecified linear partial differential operator $\mathcal{L}$, are now applied again for $\mathcal{L} \equiv \Delta$ (Laplace operator) and $T^n \equiv \boldsymbol{\nabla}().\boldsymbol{n}$ (normal derivative).

Reciprocity identity

Taking the product of the Poisson equation with unknown u^1 by a function u^2 and integrating the result over Ω, one gets:

$$\int_{\partial\Omega} q^1 u^2 \, \mathrm{d}S + \int_{\Omega} (b^1 u^2 - \boldsymbol{\nabla} u^1 . \boldsymbol{\nabla} u^2) \, \mathrm{d}V = 0 \tag{2.2}$$

where the divergence formula

$$\int_{\Omega} \operatorname{div} \boldsymbol{v} \, \mathrm{d}V = \int_{\partial\Omega} (\boldsymbol{v}.\boldsymbol{n}) \, \mathrm{d}S$$

has been used for $\boldsymbol{v} = u^2 \boldsymbol{\nabla} u^1$. Let (u^1, u^2) denote two solutions to (2.1), with respective sources (b^1, b^2). Upon application of formula (2.2) to (u^1, q^1, b^1) for the test function u^2 and to (u^2, q^2, b^2) for the test function u^1, followed by a subtraction of the resulting identities, one obtains the reciprocity identity:

$$\int_{\partial\Omega} (q^1 u^2 - q^2 u^1) \, \mathrm{d}S = \int_{\Omega} (b^2 u^1 - b^1 u^2) \, \mathrm{d}V \tag{2.3}$$

Accounting for the fact that u_1, u_2 are solutions to (2.1), another equivalent formulation of the latter identity is the well-known *Green's third formula*:

$$\int_{\partial\Omega} (q^1 u^2 - q^2 u^1) \, \mathrm{d}S = \int_{\Omega} (u^2 \Delta u^1 - u^1 \Delta u^2) \, \mathrm{d}V \tag{2.4}$$

[2]The extension to unbounded domains is treated in Section 2.3.
[3]More generally, the presence of finitely many points with three or more tangent planes is allowed, but the domain must be on one side of the boundary.
[4]Think of a body with several cavities.

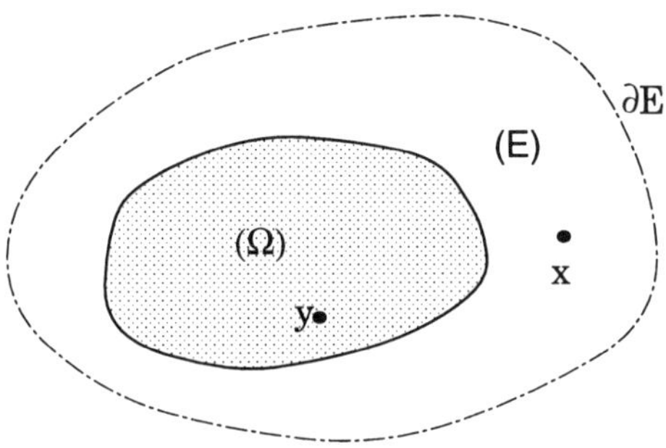

Figure 2.1 Domain Ω and auxiliary domain E used to define $G(x, y)$.

Fundamental solutions

Let E denote an open subset of $\mathbb{R}^3$. A *fundamental solution* is any solution in E to (2.1) with a unit point source $b(\boldsymbol{y}) = \delta(\boldsymbol{y} - \boldsymbol{x})$ applied at a given, fixed point $\boldsymbol{x} \in E$. Denote by $G(\boldsymbol{x}, \boldsymbol{y})$ any such solution, for which the following equation thus holds:

$$G_{,jj}(\boldsymbol{x}, \boldsymbol{y}) + \delta(\boldsymbol{y} - \boldsymbol{x}) = 0 \qquad (\boldsymbol{y} \in E) \tag{2.5}$$

In the above equation and throughout this book, the notation $f_{,j}$ denotes by convention the partial derivative of a two-point function $f(\boldsymbol{x}, \boldsymbol{y})$ with respect to the j-component of the field point $\boldsymbol{y}$. The fundamental solution $G(\boldsymbol{x}, \boldsymbol{y})$ is considered as a function of the field point $\boldsymbol{y}$, the fixed support $\{\boldsymbol{x}\}$ of the point source being treated as a parameter. Also, we will denote by $H(\boldsymbol{x}, \boldsymbol{y})$ the flux at $\boldsymbol{y}$ in the direction $\boldsymbol{n}$ associated with $G(\boldsymbol{x}, \boldsymbol{y})$:

$$H(\boldsymbol{x}, \boldsymbol{y}) = G_{,j}(\boldsymbol{x}, \boldsymbol{y}) n_j(\boldsymbol{y}) \tag{2.6}$$

In principle, one can define a fundamental solution for any given $E \subset \mathbb{R}^3$ and any set of boundary conditions ∂E.[5] However, the concept of fundamental solution is in practice restricted to cases where $G(\boldsymbol{x}, \boldsymbol{y})$ is known analytically.[6]

Global equilibrium. Let $G(\boldsymbol{x}, \boldsymbol{y})$ denote any fundamental solution satisfying (2.5). Then, for any given domain $\Omega \subseteq E$ and source point $\boldsymbol{x} \in E, \boldsymbol{x} \notin \partial\Omega$, integration of (2.5) over Ω using the divergence formula yields the following identity:

$$\kappa + \int_{\partial\Omega} H(\boldsymbol{x}, \boldsymbol{y}) \, \mathrm{d}S_y = 0 \qquad \text{with} \left\{ \begin{array}{ll} \kappa = 1 & \boldsymbol{x} \in \Omega \\ \kappa = 0 & \boldsymbol{x} \notin \Omega \end{array} \right. \tag{2.7}$$

[5]For time-independent problems like those described by the Poisson equation, boundary data of pure Neumann type must equilibrate the point source, eq. (2.7).
[6]Or at least semi-analytically, e.g. through some integral transform to be evaluated numerically.

which states that the flux H across $\partial\Omega$ equilibrates the unit point source applied at $\boldsymbol{x}$.

Fundamental solution for free space. The field G generated in the infinite space $E = \mathbb{R}^3$ by a unit point source located at $\boldsymbol{x}$ and its normal derivative (directional derivative at $\boldsymbol{y}$ along $\boldsymbol{n}$) H are given by:

$$G(\boldsymbol{x}, \boldsymbol{y}) = \frac{1}{4\pi r} \qquad H(\boldsymbol{x}, \boldsymbol{y}) = -\frac{1}{4\pi r^2} r_{,n} \tag{2.8}$$

where $r = |\boldsymbol{r}| = |\boldsymbol{y} - \boldsymbol{x}|$ is the Euclidean distance between the source $\boldsymbol{x}$ and the field point $\boldsymbol{y}$, and $r_{,n}$ is the normal derivative of r:

$$r_n = r_{,i} n_i = \frac{y_i - x_i}{r} n_i$$

The fundamental solution (2.8) is singular for $\boldsymbol{y} = \boldsymbol{x}$ $(r = 0)$. This was to be expected: a *concentrated* source of finite intensity necessarily generates, at its support, an infinite field. Besides, one has:

$$G(\boldsymbol{x}, \boldsymbol{y}) = G(\boldsymbol{y}, \boldsymbol{x}) \tag{2.9}$$

$$G_{,j}(\boldsymbol{x}, \boldsymbol{y}) = -G_{,\bar{j}}(\boldsymbol{x}, \boldsymbol{y}) \tag{2.10}$$

where the notation $f_{,\bar{j}}$ conventionally denotes the partial derivative of f with respect to the j-component of the source point $\boldsymbol{x}$.

Other fundamental solutions. There are some other geometrical configurations for which a fundamental solution is analytically known. When homogeneous Dirichlet boundary conditions over ∂E are considered, they are called *Green's functions* of the domain E.

- *Half-space:* E denotes the region $y_3 \leq 0$. Let $\bar{\boldsymbol{x}} = (x_1, x_2, -x_3)$ be the mirror image of a given source point $\boldsymbol{x} \in E$ with respect to the plane $y_3 = 0$ and put $\bar{r} = |\boldsymbol{y} - \bar{\boldsymbol{x}}|$ (Figure 2.2). Then, adequate superpositions of free-space fundamental solutions for sources at $\boldsymbol{x}$ and $\bar{\boldsymbol{x}}$ yield half-space fundamental solutions satisfying a homogeneous Dirichlet condition ($G(\boldsymbol{x}, \boldsymbol{y}) = 0$ on $y_3 = 0$):

$$G^D(\boldsymbol{x}, \boldsymbol{y}) = \frac{1}{4\pi}\left(\frac{1}{r} - \frac{1}{\bar{r}}\right) \tag{2.11}$$

  or a homogeneous Neumann condition ($H(\boldsymbol{x}, \boldsymbol{y}) = 0$ on $y_3 = 0$):

$$G^N(\boldsymbol{x}, \boldsymbol{y}) = \frac{1}{4\pi}\left(\frac{1}{r} + \frac{1}{\bar{r}}\right) \tag{2.12}$$

- *Sphere:* E is the sphere centred at the origin, of radius a, and the source $\boldsymbol{x}$ is at a distance $\rho \leq a$ from the centre. The fundamental solution

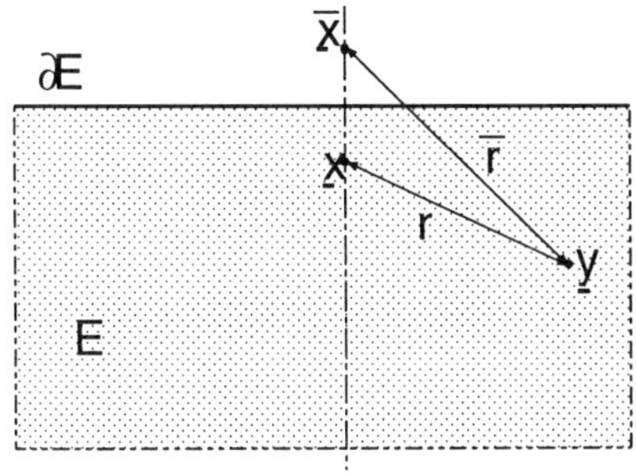

Figure 2.2 Half-space and the superposition method.

associated with Dirichlet boundary conditions on the sphere is given by

$$G(\boldsymbol{x},\boldsymbol{y}) = \frac{1}{4\pi}\left(\frac{1}{r} - \frac{a}{\rho}\frac{1}{\bar{r}}\right) \tag{2.13}$$

with $\bar{\boldsymbol{x}} = (a^2/\rho^2)\boldsymbol{x}$ (inverse of $\boldsymbol{x}$ with respect to the sphere), $\bar{r} = |\boldsymbol{y} - \bar{\boldsymbol{x}}|$.

General properties of the fundamental solutions. All fundamental solutions are singular at $\boldsymbol{y} = \boldsymbol{x}$:

$$G(\boldsymbol{x},\boldsymbol{y}) = O(r^{-1}) \qquad\qquad H(\boldsymbol{x},\boldsymbol{y}) = O(r^{-2}) \tag{2.14}$$

Moreover, the singular term is identical to that of the free-space fundamental solution if $\boldsymbol{x} \notin \partial E$, and generally different if $\boldsymbol{x} \in \partial E$ (for example, compare eq. (2.8) to eqs (2.11, 2.12)).

Besides, all fundamental solutions to the Poisson equation share the symmetry property (2.9). Indeed, consider two fundamental solutions for the same domain E and respective sources at $\boldsymbol{x}$ and $\boldsymbol{y}$. This defines two states, for which the reciprocity identity (2.3) gives:

$$\int_{\partial E} [H(\boldsymbol{x},\boldsymbol{z})G(\boldsymbol{y},\boldsymbol{z}) - H(\boldsymbol{y},\boldsymbol{z})G(\boldsymbol{x},\boldsymbol{z})]\,\mathrm{d}S_z = G(\boldsymbol{x},\boldsymbol{y}) - G(\boldsymbol{y},\boldsymbol{x})$$

The above identity implies that the symmetry property (2.9), i.e.:

$$G(\boldsymbol{x},\boldsymbol{y}) = G(\boldsymbol{y},\boldsymbol{x}) \qquad (\forall \boldsymbol{x},\boldsymbol{y},\ \boldsymbol{x} \neq \boldsymbol{y})$$

holds true for any fundamental solution with homogeneous Dirichlet or Neumann boundary data over ∂E. For the free-space solution ($E = \mathbb{R}^3$), since $\mathrm{d}S \sim r^2\,\mathrm{d}\omega$ (ω: solid angle) and $G(\cdot,\boldsymbol{z}) \sim r^{-1}$, $H(\cdot,\boldsymbol{z}) \sim r^{-2}$ for large r, the integral over ∂E in the above formula vanishes as ∂E goes to infinity.

Introducing small perturbations $\delta\boldsymbol{x}, \delta\boldsymbol{y}$ of the points $\boldsymbol{x}, \boldsymbol{y}$, the equality

$$G(\boldsymbol{x}+\delta\boldsymbol{x}, \boldsymbol{y}+\delta\boldsymbol{y}) = G(\boldsymbol{y}+\delta\boldsymbol{y}, \boldsymbol{x}+\delta\boldsymbol{x})$$

is true for any $\delta\boldsymbol{x},\delta\boldsymbol{y}$. Matching the first- and second-order terms in $\delta\boldsymbol{x},\delta\boldsymbol{y}$ on both sides of the equality, one obtains:

$$G_{,j}(\boldsymbol{x},\boldsymbol{y}) = G_{,\bar{j}}(\boldsymbol{y},\boldsymbol{x}) \tag{2.15}$$
$$G_{,\bar{i}j}(\boldsymbol{x},\boldsymbol{y}) = G_{,i\bar{j}}(\boldsymbol{y},\boldsymbol{x}) \tag{2.16}$$

In particular, eq. (2.15) generalizes property (2.10) to arbitrary fundamental solutions.

Integral representation formulas

The reciprocity identity (2.3) is applied to the unknown state (u,q,b) and a known fundamental solution defined by (2.5), the source point $\boldsymbol{x}$ being interior to either Ω or $\mathbb{R}^3 - \Omega$. This leads to:

$$\int_\Omega \delta(\boldsymbol{y}-\boldsymbol{x})u(\boldsymbol{y})\,\mathrm{d}V_y = \int_{\partial\Omega} \{q(\boldsymbol{y})G(\boldsymbol{x},\boldsymbol{y}) - u(\boldsymbol{y})H(\boldsymbol{x},\boldsymbol{y})\}\,\mathrm{d}S_y + \int_\Omega b(\boldsymbol{y})G(\boldsymbol{x},\boldsymbol{y})\,\mathrm{d}V_y$$

Moreover, the definition of the Dirac distribution implies:

$$\int_\Omega \delta(\boldsymbol{y}-\boldsymbol{x})u(\boldsymbol{y})\,\mathrm{d}V_y = \kappa u(\boldsymbol{x}) \qquad \text{with} \left\{ \begin{array}{ll} \kappa = 1 & \boldsymbol{x}\in\Omega \\ \kappa = 0 & \boldsymbol{x}\notin\Omega \end{array}\right. \tag{2.17}$$

As a consequence, the following results are obtained.

Integral representation formula for the potential. The value $u(\boldsymbol{x})$ at $\boldsymbol{x}$ of a solution to the Poisson equation is given in terms of the source b and the values of (u,q) on the boundary $\partial\Omega$ by the integral representation formula:

$$\kappa u(\boldsymbol{x}) = \int_{\partial\Omega} \{q(\boldsymbol{y})G(\boldsymbol{x},\boldsymbol{y}) - u(\boldsymbol{y})H(\boldsymbol{x},\boldsymbol{y})\}\,\mathrm{d}S_y + \int_\Omega b(\boldsymbol{y})G(\boldsymbol{x},\boldsymbol{y})\,\mathrm{d}V_y \tag{2.18}$$

where $\kappa = 1$ for $\boldsymbol{x}\in\Omega$ (interior representation formula) or $\kappa = 0$ for $\boldsymbol{x}\notin\Omega$ (exterior representation formula).

Integral representation formula for the gradient. Except when $\boldsymbol{x}\in\partial\Omega$, the integrals in (2.18) are differentiable functions of the variable $\boldsymbol{x}$. The integral representation formula for the gradient of u is thus obtained through differentiation of (2.18) under the integral sign:

$$\kappa u_{,i}(\boldsymbol{x}) = \int_{\partial\Omega} \{qG_{,\bar{i}}(\boldsymbol{x},\boldsymbol{y}) - u(\boldsymbol{y})n_j(\boldsymbol{y})G_{,\bar{i}j}(\boldsymbol{x},\boldsymbol{y})\}\,\mathrm{d}S_y + \int_\Omega b(\boldsymbol{y})G_{,\bar{i}}(\boldsymbol{x},\boldsymbol{y})\,\mathrm{d}V_y \tag{2.19}$$

Integral representation formula for the gradient: alternative formulation. In practice, the free-space fundamental solution (2.8) is often used in (2.19), and from property (2.9) the $\boldsymbol{x}$-derivatives can be recast into $\boldsymbol{y}$-derivatives:

$$\kappa u_{,i}(\boldsymbol{x}) = \int_{\partial\Omega} \{u(\boldsymbol{y})n_j(\boldsymbol{y})G_{,ij}(\boldsymbol{x},\boldsymbol{y}) - q(\boldsymbol{y})G_{,i}(\boldsymbol{x},\boldsymbol{y})\} \, \mathrm{d}S_y - \int_{\Omega} b(\boldsymbol{y})G_{,i}(\boldsymbol{x},\boldsymbol{y}) \, \mathrm{d}V_y$$

Then, an integration by parts can be performed: indeed, for any *closed* surface S which does not contain $\boldsymbol{x}$, one has:

$$\int_S u(\boldsymbol{y})G_{,ij}(\boldsymbol{x},\boldsymbol{y})n_j(\boldsymbol{y}) \, \mathrm{d}S_y = \int_S u(\boldsymbol{y})[D_{ji}G_{,j}(\boldsymbol{x},\boldsymbol{y}) + G_{,jj}(\boldsymbol{x},\boldsymbol{y})n_i(\boldsymbol{y})] \, \mathrm{d}S_y$$

The resulting expression involves the tangential differential operator $D_{ij}() = n_i()_{,j} - n_j()_{,i}$ introduced in Appendix A. Moreover, one has $G_{,jj}(\boldsymbol{x},\boldsymbol{y}) = 0$ on S since $\boldsymbol{x} \notin S$. Then, application of the integration by parts identity (A.8) yields the result:

$$\int_S u(\boldsymbol{y})D_{ji}G_{,j}(\boldsymbol{x},\boldsymbol{y}) \, \mathrm{d}S_y = \int_S D_{ij}u(\boldsymbol{y})G_{,j}(\boldsymbol{x},\boldsymbol{y}) \, \mathrm{d}S_y$$

This integration by parts technique appears for instance in Kupradze [3], Sladek and Sladek [4]. The potential gradient components are thus given by the following formula:

$$\kappa u_{,i}(\boldsymbol{x}) = \int_{\partial\Omega} \{D_{ij}u(\boldsymbol{y})G_{,j}(\boldsymbol{x},\boldsymbol{y}) - q(\boldsymbol{y})G_{,i}(\boldsymbol{x},\boldsymbol{y})\} \, \mathrm{d}S_y - \int_{\Omega} b(\boldsymbol{y})G_{,i}(\boldsymbol{x},\boldsymbol{y}) \, \mathrm{d}V_y \quad (2.20)$$

The above expression, equivalent to (2.19), involves integrals of $|\boldsymbol{y}-\boldsymbol{x}|^{-1,-2}$ terms, whereas the original formula (2.19) has $|\boldsymbol{y}-\boldsymbol{x}|^{-2,-3}$ terms. On the other hand, eq. (2.20) is limited to the free-space fundamental solution, while eq. (2.19) is general.

Necessity of an integral equation

For any observation point $\boldsymbol{x}$ interior to Ω, the value taken by the potential u and its gradient are known in terms of the source b and the values of (u, q) on $\partial\Omega$ from equations (2.18), (2.19). Considering a well-posed boundary-value problem, the value of either u or q is prescribed at any given point $\boldsymbol{x} \in \partial\Omega$, the remaining one being unknown. From equations (2.18), (2.19) one sees that solving a boundary-value problem for the Poisson equation essentially reduces to the determination at every point of the boundary of the value of (u, q) which is not prescribed by the boundary conditions. In other words, equations (2.18), (2.19) achieve a 'quasi-inversion' of the Poisson equation (2.1), the geometrical support of the unknowns being transferred from the domain to the boundary.

Let $G(\boldsymbol{x}, \boldsymbol{y})$ be the Green's function of Ω (i.e. $G(\boldsymbol{x}, \boldsymbol{y}) = 0$ on $\partial\Omega$ holds). The representation formulas (2.18), (2.19) then give explicitly the solution to the Dirichlet problem with data u^D:

$$u(\boldsymbol{x}) = -\int_{\partial\Omega} u^D(\boldsymbol{y}) H(\boldsymbol{x}, \boldsymbol{y}) \, \mathrm{d}S_y$$

However, the Green's function is known analytically in closed form only for some cases where Ω assume very simple shapes, so that in fact the above explicit formula has limited practical interest.

It is interesting to mention at this point that sometimes a sensible choice of fundamental solution allows us, for a given problem, to reduce the size of the geometrical support of boundary unknowns. For example, consider the domain $\Omega = E \setminus C$ where E is a half-space and C a bounded subset of E (cavity); see Figure 2.3. The geometrical support of unknowns reduces from $\partial E \cup \partial C$ to ∂C alone when the half-space fundamental solution (2.11) or (2.12) (depending on whether ∂E is a Dirichlet or Neumann boundary) is used in (2.18) instead of the free-space fundamental solution. For instance, if u is prescribed on ∂E and ∂C, one gets:

$$u(\boldsymbol{x}) = -\int_{\partial E} u^D(\boldsymbol{y}) H^D(\boldsymbol{x}, \boldsymbol{y}) \, \mathrm{d}S_y + \int_{\partial C} \{q(\boldsymbol{y}) G^D(\boldsymbol{x}, \boldsymbol{y}) - u^D(\boldsymbol{y}) H^D(\boldsymbol{x}, \boldsymbol{y})\} \, \mathrm{d}S_y$$

where the flux q on ∂C is the only remaining unknown.

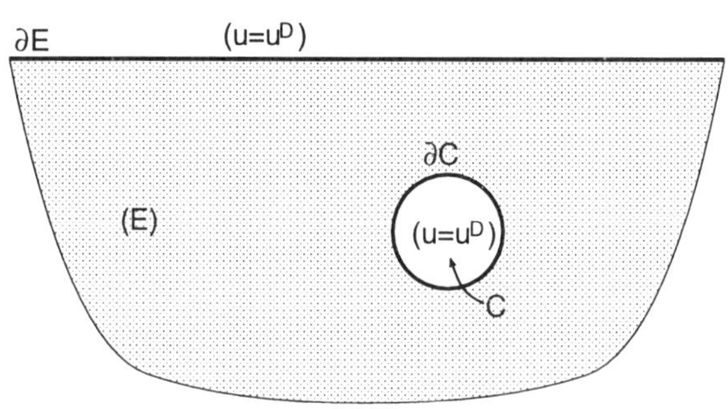

Figure 2.3 Half-space with cavity.

2.2 BOUNDARY INTEGRAL EQUATION

The remaining task is to define a *boundary integral equation*, i.e. a governing equation for the values of (u, q) over the boundary. A mere invocation of (2.18) or (2.19) for points $\boldsymbol{x}$ chosen on the boundary does not, however, provide a satisfactory answer. This is so because $H(\boldsymbol{x}, \boldsymbol{y})$ is singular like $|\boldsymbol{y} - \boldsymbol{x}|^{-2}$, so that the integral over $\partial\Omega$ may not be convergent for $\boldsymbol{x} \in \partial\Omega$. On the other hand, we stressed that eq. (2.17) does not hold for $\boldsymbol{x} \in \partial\Omega$. A proper formulation of a boundary integral equation must therefore invoke a *limiting process* designed to account properly for the strong singularity of the normal derivative of the fundamental solution.

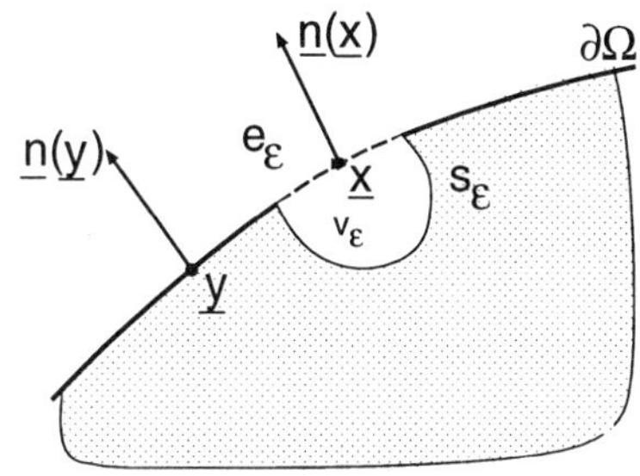

Figure 2.4 Exclusion neighbourhood $v_\varepsilon(x)$ and other geometrical notation used for the limiting process $\varepsilon \to 0$.

Limiting process

Let $\boldsymbol{x}$ denote a fixed point of the boundary surface $\partial\Omega$, and assume that $\partial\Omega$ is C^2-regular at $\boldsymbol{x}$. For a given (small) $\varepsilon > 0$, introduce a neighbourhood $v_\varepsilon(\boldsymbol{x})$ of $\boldsymbol{x}$, connected, its diameter being $\leq \varepsilon$, called an *exclusion neighbourhood* (these definitions and notations follow Guiggiani et al. [2]). The domain $\Omega_\varepsilon(\boldsymbol{x}) = \Omega - v_\varepsilon(\boldsymbol{x})$ obtained upon removing $v_\varepsilon(\boldsymbol{x})$ from Ω (Figure 2.4) is such that the point $\boldsymbol{x}$ is exterior to $\Omega_\varepsilon(\boldsymbol{x})$, while its boundary is $\partial\Omega_\varepsilon = (\partial\Omega - e_\varepsilon) + s_\varepsilon$, where $s_\varepsilon = \Omega \cap \partial v_\varepsilon$, $e_\varepsilon = \partial\Omega \cap v_\varepsilon$. The basic idea followed is to study the limit $\varepsilon \to 0$ in the representation formula (2.18) taken for the domain Ω_ε. Since $\boldsymbol{x}$ is exterior to Ω_ε, the latter reads:

$$0 = \int_{\Omega_\varepsilon} b(\boldsymbol{y})G(\boldsymbol{x}, \boldsymbol{y})\,\mathrm{d}V_y + \int_{(\partial\Omega - e_\varepsilon)+s_\varepsilon} \{q(\boldsymbol{y})G(\boldsymbol{x}, \boldsymbol{y}) - u(\boldsymbol{y})H(\boldsymbol{x}, \boldsymbol{y})\}\,\mathrm{d}S_y \quad (2.21)$$

First, one notices that identity (2.7) implies:

$$0 = u(\boldsymbol{x}) \int_{(\partial\Omega - e_\varepsilon)+s_\varepsilon} H(\boldsymbol{x}, \boldsymbol{y})\,\mathrm{d}S_y$$

Taking the difference of the last two equations, one then gets:

$$0 = \int_{\Omega_\varepsilon} b(\boldsymbol{y})G(\boldsymbol{x},\boldsymbol{y})\,\mathrm{d}V_y + \int_{(\partial\Omega - e_\varepsilon)+s_\varepsilon} \{q(\boldsymbol{y})G(\boldsymbol{x},\boldsymbol{y}) - [u(\boldsymbol{y}) - u(\boldsymbol{x})]H(\boldsymbol{x},\boldsymbol{y})\}\,\mathrm{d}S_y \quad (2.22)$$

Assume at this point that u is $C^{0,\alpha}$-continuous at $\boldsymbol{x}$, i.e.:[7]

$$\exists(\alpha, C) > 0, \qquad |u(\boldsymbol{y}) - u(\boldsymbol{x})| \leq C\,|\boldsymbol{y}-\boldsymbol{x}|^{\alpha}$$

Then the normal derivative q is such that

$$\exists(\alpha, C) > 0, \qquad |q(\boldsymbol{y})| \leq C\,|\boldsymbol{y}-\boldsymbol{x}|^{\alpha-1}$$

Note that the above holds even if the unit normal (and thus the flux) is not uniquely determined at $\boldsymbol{x}$, provided $\boldsymbol{y}$ and $\boldsymbol{x}$ are located on the same, regular, component of $\partial\Omega$. Then, the singularity properties (2.14) imply that:

$$\begin{aligned} |[u(\boldsymbol{y}) - u(\boldsymbol{x})]H(\boldsymbol{x},\boldsymbol{y})| &\leq C\,|\boldsymbol{y}-\boldsymbol{x}|^{\alpha-2} \\ |q(\boldsymbol{y})G(\boldsymbol{x},\boldsymbol{y})| &\leq C\,|\boldsymbol{y}-\boldsymbol{x}|^{\alpha-2} \end{aligned} \quad (2.23)$$

At this point, the limiting process $\varepsilon \to 0$ in eq. (2.22) can be considered. The integrals over s_ε, $\partial\Omega - e_\varepsilon$ and Ω_ε are analysed separately.

Integral over s_ε. From (2.23) and since $|\boldsymbol{y}-\boldsymbol{x}| \sim \varepsilon$ and $\mathrm{d}S_y \sim \varepsilon^2\,\mathrm{d}\omega$ (where ω is the solid angle seen from the point $\boldsymbol{x}$), one has:

$$\{[u(\boldsymbol{y}) - u(\boldsymbol{x})]H(\boldsymbol{x},\boldsymbol{y}) - q(\boldsymbol{y})G(\boldsymbol{x},\boldsymbol{y})\}\,\mathrm{d}S_y = O(\varepsilon^{\alpha})\,\mathrm{d}\omega$$

and thus:

$$\lim_{\varepsilon\to 0}\int_{s_\varepsilon} \{[u(\boldsymbol{y}) - u(\boldsymbol{x})]H(\boldsymbol{x},\boldsymbol{y}) - q(\boldsymbol{y})G(\boldsymbol{x},\boldsymbol{y})\}\,\mathrm{d}S_y = 0$$

Integral over $\partial\Omega - e_\varepsilon$. From (2.23), the integrand behaves near $\boldsymbol{x}$ like:

$$[u(\boldsymbol{y}) - u(\boldsymbol{x})]H(\boldsymbol{x},\boldsymbol{y}) - q(\boldsymbol{y})G(\boldsymbol{x},\boldsymbol{y}) \sim |\boldsymbol{y}-\boldsymbol{x}|^{\alpha-2}$$

and the limit when $\varepsilon \to 0$ of the integral over $\partial\Omega - e_\varepsilon$ in (2.22) is simply the corresponding integral over $\partial\Omega$, which is weakly singular (i.e. convergent in the ordinary sense):

$$\begin{aligned} &\lim_{\varepsilon\to 0}\int_{\partial\Omega - e_\varepsilon} \{[u(\boldsymbol{y}) - u(\boldsymbol{x})]H(\boldsymbol{x},\boldsymbol{y}) - q(\boldsymbol{y})G(\boldsymbol{x},\boldsymbol{y})\}\,\mathrm{d}S_y \\ &\qquad = \int_{\partial\Omega} \{[u(\boldsymbol{y}) - u(\boldsymbol{x})]H(\boldsymbol{x},\boldsymbol{y}) - q(\boldsymbol{y})G(\boldsymbol{x},\boldsymbol{y})\}\,\mathrm{d}S_y \end{aligned}$$

[7] This condition is known as the Hölder continuity of exponent α; it is stronger than continuity and weaker than differentiability.

Integral over Ω_ε. In view of the weak singularity $\sim |\boldsymbol{y}-\boldsymbol{x}|^{-1}$ of $G(\boldsymbol{x},\boldsymbol{y})$, no difficulty arises and we get in the limit the corresponding, weakly singular, integral over Ω:

$$\lim_{\varepsilon \to 0} \int_{\Omega_\varepsilon} b(\boldsymbol{y})G(\boldsymbol{x},\boldsymbol{y})\,\mathrm{d}V_y = \int_{\Omega} b(\boldsymbol{y})G(\boldsymbol{x},\boldsymbol{y})\,\mathrm{d}V_y$$

Boundary integral equation

Gathering the previous partial results, the limiting process $\varepsilon \to 0$ in identity (2.22) thus leads to the desired *boundary integral equation*, which relates the values of (u, q) on the boundary surface $\partial\Omega$ only:

$$\int_{\partial\Omega} \{[u(\boldsymbol{y})-u(\boldsymbol{x})]H(\boldsymbol{x},\boldsymbol{y}) - q(\boldsymbol{y})G(\boldsymbol{x},\boldsymbol{y})\}\,\mathrm{d}S_y = \int_{\Omega} b(\boldsymbol{y})G(\boldsymbol{x},\boldsymbol{y})\,\mathrm{d}V_y \qquad (2.24)$$

The above equation, which constitutes the main result presented in this chapter, governs *any* solution, compatible with the assumptions (2.23), to the Poisson equation over Ω. It is formulated in terms of convergent surface integrals, having a weak singularity at $\boldsymbol{y}=\boldsymbol{x}$.

The boundary integral equation (2.24), when taken for $\boldsymbol{x} \notin \partial\Omega$, reduces to the integral representation formula (2.18). In other words, (2.24) is valid whether the source point $\boldsymbol{x}$ is interior, exterior or on the boundary surface $\partial\Omega$ (this is a consequence of eq. (2.7)); it is thus applicable to any $\boldsymbol{x}$ in the domain of definition E of the fundamental solution at hand.

Classical presentation

It consists of taking (2.21) in the limit $\varepsilon \to 0$ directly, without invoking identity (2.7), and leads to the following boundary integral formulation:

$$\kappa(\boldsymbol{x})u(\boldsymbol{x}) + \int_{\partial\Omega} \{u(\boldsymbol{y})H(\boldsymbol{x},\boldsymbol{y}) - q(\boldsymbol{y})G(\boldsymbol{x},\boldsymbol{y})\}\,\mathrm{d}S_y = \int_{\Omega} b(\boldsymbol{y})G(\boldsymbol{x},\boldsymbol{y})\,\mathrm{d}V_y \qquad (2.25)$$

The kernel $H(\boldsymbol{x},\boldsymbol{y})$, although of apparent order $|\boldsymbol{y}-\boldsymbol{x}|^{-2}$, turns out to be integrable (see Appendix 2.C).[8] The *free term* $\kappa(\boldsymbol{x})$ is defined by:

$$\kappa(\boldsymbol{x}) = \lim_{\varepsilon \to 0} \int_{s_\varepsilon} H(\boldsymbol{x},\boldsymbol{y})\,\mathrm{d}S_y \qquad (2.26)$$

[8] This is not true for vector problems, e.g. elasticity.

and is equal to 1/2 if the boundary $\partial\Omega$ is smooth at $\boldsymbol{x}$.

Our chosen way to derive and establish a boundary integral equation in the *regularized* form (2.24) is motivated by consistency with the remainder of this book, because (in anticipation of Chapter 4) it will turn out that in elasticity the non-regularized formulation similar to (2.25) involves integrals that are *not* convergent in the ordinary sense.

Why use singular fundamental solutions?

In view of the technical difficulties associated with the use of singular fundamental solutions, it is legitimate to wonder about the motivations behind this choice. To support this discussion, let us review some possible choices for the geometrical support D of the source b^2 of the auxiliary state (u^2, q^2, b^2) in the reciprocity identity (2.3).

$D \subset \Omega$ is an open subset of $\mathbb{R}^3$. In that case, (2.3) leads to:

$$\int_{\partial\Omega} (q^1 u^2 - q^2 u^1) \, \mathrm{d}S = \int_D b^2 u^1 \, \mathrm{d}V - \int_\Omega b^1 u^2 \, \mathrm{d}V$$

This equation relates not only the values of u^1, q^1 on the boundary $\partial\Omega$ but also involves u^1 in D. As a consequence, neither a boundary integral equation nor a representation formula for u^1 at an interior point $\boldsymbol{x} \in \Omega$ can be obtained.

$D = \emptyset$ (absence of source in the auxiliary state). In this case, (2.3) gives:

$$\int_{\partial\Omega} (q^1 u^2 - q^2 u^1) \, \mathrm{d}S = -\int_\Omega b^1 u^2 \, \mathrm{d}V$$

This equation relates the unknowns u^1, q^1 on the boundary $\partial\Omega$ only, and thus constitutes a boundary integral equation. On the other hand, it does not lead to a representation formula for u^1 at interior points $\boldsymbol{x} \in \Omega$. Moreover, the boundary integral equation thus obtained is of the Fredholm, first-kind type, which is a well-known class of mathematically ill-posed problems that are very sensitive to data perturbations (Tricomi [5], Wing [6]).

$D = \{\boldsymbol{x}\}$, $\boldsymbol{x} \in \mathbb{R}^3$: (unit) point source applied at $\boldsymbol{x}$. We have seen that this choice leads to both integral representations *and* integral equations. Moreover, the latter can be formulated as Fredholm integral equations of the second kind (2.34), which are stable with respect to data perturbations.

2.3 EXTERIOR PROBLEMS AND CONDITIONS AT INFINITY

Consider now a domain $\bar{\Omega} = \mathbb{R}^3 - \Omega$, where Ω is bounded, non-empty open subset of $\mathbb{R}^3$; the unit normal $\bar{\boldsymbol{n}}$ to $\partial\bar{\Omega}$ is taken exterior to $\bar{\Omega}$, and suppose

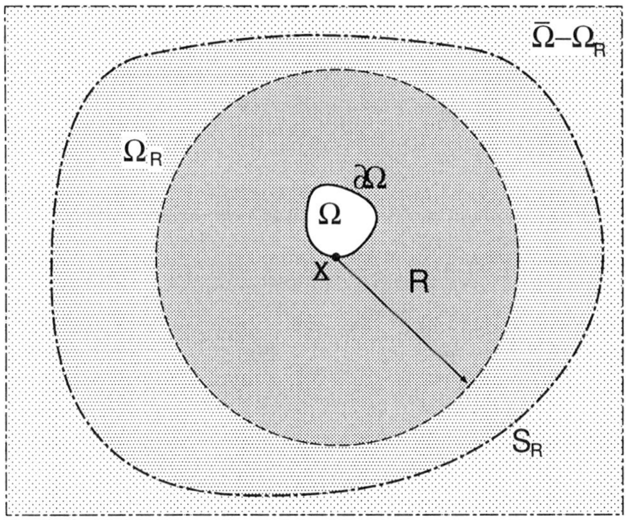

Figure 2.5 Domain Ω_R for the limiting process $R \to \infty$.

(temporarily) $b = 0$. This section aims at showing that the boundary integral equation (2.24) and the integral representation (2.18) remain valid when applied for an unbounded domain, provided u, q satisfy certain conditions at infinity. In order to do so, we consider the limiting process when a bounded domain becomes arbitrarily large. Let $\boldsymbol{x}$ denote any fixed point of $\mathbb{R}^3$ and consider a closed regular surface S_R chosen such that its interior contains both the ball of centre $\boldsymbol{x}$ and (large) radius R and the boundary $\partial\Omega$. The boundary integral equation (2.24) then holds for the *finite* domain $\Omega_R \subset \bar{\Omega}$ bounded by the surfaces $\partial\Omega$ and S_R (Figure 2.5):

$$\int_{\partial\bar{\Omega}} \{[u(\boldsymbol{y}) - u(\boldsymbol{x})]H(\boldsymbol{x},\boldsymbol{y}) - q(\boldsymbol{y})G(\boldsymbol{x},\boldsymbol{y})\}\ \mathrm{d}S_y = \int_{S_R} \{q(\boldsymbol{y})G(\boldsymbol{x},\boldsymbol{y}) - [u(\boldsymbol{y}) - u(\boldsymbol{x})]H(\boldsymbol{x},\boldsymbol{y})\}\ \mathrm{d}S_y$$

where, since the limiting case $R \to \infty$ is considered, the free-space fundamental solution is used. The point $\boldsymbol{x}$ is, by construction, interior to S_R; application of identity (2.7) for S_R thus gives:

$$\int_{\partial\bar{\Omega}} \{[u(\boldsymbol{y}) - u(\boldsymbol{x})]H(\boldsymbol{x},\boldsymbol{y}) - q(\boldsymbol{y})G(\boldsymbol{x},\boldsymbol{y})\}\ \mathrm{d}S_y = -u(\boldsymbol{x}) + \int_{S_R} \{q(\boldsymbol{y})G(\boldsymbol{x},\boldsymbol{y}) - u(\boldsymbol{y})H(\boldsymbol{x},\boldsymbol{y})\}\ \mathrm{d}S_y \quad (2.27)$$

The fundamental solution (2.8) and the differential area element on S_R satisfy:

$$\begin{aligned} G(\boldsymbol{x},\boldsymbol{y}) &= O(1/R) \\ H(\boldsymbol{x},\boldsymbol{y}) &= O(1/R^2) \qquad\qquad (\boldsymbol{y}\in S_R) \\ \mathrm{d}S_y &= O(R^2)\,\mathrm{d}\omega \end{aligned}$$

(ω: algebraic solid angle seen from origin $\boldsymbol{x}$). It follows that the relations

$$u = o(1) \qquad q = o(R^{-1}) \qquad (\boldsymbol{y}\in S_R) \tag{2.28}$$

define sufficient conditions on u, q in order to ensure

$$\lim_{R\to\infty} \int_{S_R} \{u(\boldsymbol{y})H(\boldsymbol{x},\boldsymbol{y}) - q(\boldsymbol{y})G(\boldsymbol{x},\boldsymbol{y})\}\,\mathrm{d}S_y = 0$$

Moreover the above limit does not depend on the shape of S_R.

Boundary integral equation for exterior problems

The limiting process $R\to\infty$ in (2.27) thus results in formulating a *boundary integral equation which governs any solution to the Laplace equation over an unbounded domain* $\bar{\Omega}$ satisfying the conditions at infinity (2.28):

$$u_k(\boldsymbol{x}) + \int_{\partial\Omega} \{[u(\boldsymbol{y}) - u(\boldsymbol{x})]H(\boldsymbol{x},\boldsymbol{y}) - q(\boldsymbol{y})G(\boldsymbol{x},\boldsymbol{y})\}\,\mathrm{d}S_y = 0 \tag{2.29}$$

The above formulation differs from its counterpart (2.24) for bounded domains in the presence of a 'free term' $u(\boldsymbol{x})$ outside the integral sign and the orientation of the unit normal $\boldsymbol{n}$, exterior to $\bar{\Omega}$ and therefore interior to $\partial\Omega$.

Integral representation formulas

Equation (2.29) holds for any choice of the source point $\boldsymbol{x}$. When $\boldsymbol{x}\notin\partial\Omega$, identity (2.7) applied to $\partial\bar{\Omega}$ gives, using the fact that $\boldsymbol{n}$ is *interior* to $\partial\bar{\Omega}$:

$$(1-\kappa) - \int_{\partial\bar{\Omega}} H(\boldsymbol{x},\boldsymbol{y})\,\mathrm{d}S_y = 0 \qquad \text{with} \left\{ \begin{array}{ll} \kappa = 1 & \boldsymbol{x}\in\bar{\Omega} \\ \kappa = 0 & \boldsymbol{x}\notin\bar{\Omega} \end{array} \right.$$

This formula allows us to recast (2.29) into

$$\kappa u(\boldsymbol{x}) = \int_{\partial\bar{\Omega}} \{q(\boldsymbol{y})G(\boldsymbol{x},\boldsymbol{y}) - u(\boldsymbol{y})H(\boldsymbol{x},\boldsymbol{y})\}\,\mathrm{d}S_y$$

The above representation formula is identical to (2.18), except for the difference of unit normal orientation. The same representation formula is then applicable to bounded as well as unbounded domains.

Asymptotic behaviour at infinity of the integral representation formulas

From the properties of the free-space fundamental solution, it is easy to see that conditions (2.28) hold for any potential u and flux q expressed by an integral representation of the form (2.18) and its normal derivative.

Nonzero sources

The limiting process $R \to +\infty$ and formula (2.18) still hold when a distributed source b is present, provided:

$$\lim_{R\to\infty} \left| \int_{\Omega_R} b(\boldsymbol{y})G(\boldsymbol{x},\boldsymbol{y})\,\mathrm{d}V_y \right| < +\infty$$

Otherwise, one must resort to superposition, using some known particular solution u^P to the Poisson equation and putting $u = u^0 + u^P$, where the new unknown u^0 solves the Laplace (i.e. homogeneous) equation. A similar approach must be used when non-vanishing data at infinity is considered, like for potential flow with given stream velocity.

Comments

When the geometrical support of the source b is a *bounded* subset of $\bar{\Omega}$ and without data at infinity, the asymptotic behaviour of (u, q) is identical (up to a multiplicative constant) to that of the fundamental solution (G, H), i.e. R^{-1} and R^{-2} respectively. The conditions at infinity (2.28) are therefore satisfied.

The formulations obtained in this section show that an exterior problem reduces to seeking the solution of an integral equation over a bounded surface. In particular, one is not faced with the necessity of either artificially truncating the domain or using 'infinite elements' (Bettess [1]) that incorporate the known asymptotic behaviour of the unknowns. Integral equation methods thus are very attractive for exterior problems.

2.4 INDIRECT INTEGRAL FORMULATIONS

Consider simultaneously a bounded domain Ω and its unbounded complement $\bar{\Omega} = \mathbb{R}^3 - \Omega$, and let $u, \bar{u}$ denote solutions to the Laplace equation over Ω and $\bar{\Omega}$ respectively (*interior* and *exterior* problems, with $b = \bar{b} = 0$). For any fixed point $\boldsymbol{x}$ interior to Ω, the representation formula (2.18) gives:

$$u(\boldsymbol{x}) = \int_{\partial\Omega} \{q(\boldsymbol{y})G(\boldsymbol{x},\boldsymbol{y}) - u(\boldsymbol{y})H(\boldsymbol{x},\boldsymbol{y})\}\,\mathrm{d}S_y$$

$$0 = \int_{\partial\Omega} \{\bar{q}(\boldsymbol{y})G(\boldsymbol{x},\boldsymbol{y}) - \bar{u}(\boldsymbol{y})\bar{H}(\boldsymbol{x},\boldsymbol{y})\}\,\mathrm{d}S_y$$

(where $\bar{q}$ and $\bar{H}$ refer to the unit normal outward from $\bar{\Omega}$, i.e. inward to Ω). Taking the sum of the above equations and conventionally using in H the unit normal exterior to Ω, one gets:

$$u(\boldsymbol{x}) = \int_{\partial\Omega} \{(q+\bar{q})(\boldsymbol{y})G(\boldsymbol{x},\boldsymbol{y}) - (u-\bar{u})(\boldsymbol{y})H(\boldsymbol{x},\boldsymbol{y})\} \,\mathrm{d}S_y \tag{2.30}$$

Single-layer representation formula. Suppose that the potential u is continuous across $\partial\Omega$: $u = \bar{u}$ over $\partial\Omega$, and introduce the discontinuity ψ of normal derivative :

$$\psi \equiv \bar{q} + q = -(\boldsymbol{\nabla}\bar{u} - \boldsymbol{\nabla}u).\boldsymbol{n}$$

Identity (2.30) thus allows us to express u as a *single-layer potential of density* ψ, as follows:

$$u(\boldsymbol{x}) = \int_{\partial\Omega} \psi(\boldsymbol{y})G(\boldsymbol{x},\boldsymbol{y}) \,\mathrm{d}S_y \tag{2.31}$$

Double-layer representation formula. Suppose that the normal derivative q is continuous across $\partial\Omega$: $q + \bar{q} = 0$ over $\partial\Omega$, and introduce the discontinuity ϕ of potential:

$$\phi \equiv \bar{u} - u$$

Identity (2.30) thus allows us to express u as a *double-layer potential of density* ϕ, as follows:

$$u(\boldsymbol{x}) = \int_{\partial\Omega} \phi(\boldsymbol{y})H(\boldsymbol{x},\boldsymbol{y}) \,\mathrm{d}S_y$$

The above two kinds of representation formulas are termed *indirect* because they use intermediary unknowns (density ψ or ϕ) instead of the primary variables u, q. The previous definitions show that *any* harmonic function in Ω or $\bar{\Omega}$ admits an indirect representation of either type. Conversely, from the properties of the free-space fundamental solution $G(\boldsymbol{x},\boldsymbol{y})$, one easily sees that any single- or double-layer potential is harmonic in Ω and $\bar{\Omega}$ (but not on $\partial\Omega$), i.e.:

$$\Delta_x \int_{\partial\Omega} \psi(\boldsymbol{y})G(\boldsymbol{x},\boldsymbol{y}) \,\mathrm{d}S_y = 0 \qquad \Delta_x \int_{\partial\Omega} \phi(\boldsymbol{y})H(\boldsymbol{x},\boldsymbol{y}) \,\mathrm{d}S_y = 0$$

Formulas (2.31), (2.4) taken for $\boldsymbol{x}$ interior to $\bar{\Omega}$ define indirect representations for the solution $\bar{u}$ of an exterior problem, the definition of ψ, ϕ as well as the orientation of $\boldsymbol{n}$ being the same.[9]

[9] One may also put $\psi = q + \bar{q}$, $\phi = u - \bar{u}$ and use the normal $\boldsymbol{n} = \bar{\boldsymbol{n}}$ exterior to $\bar{\Omega}$, i.e. transpose in a consistent way the conventions used for the interior problem.

Indirect boundary integral equations

Let $\boldsymbol{x}$ be a fixed point of $\partial\Omega$, and consider the integral equations (2.24) for $(u, q, b = 0)$ (interior problem) and (2.29) for $(\bar{u}, \bar{q}, \bar{b} = 0)$ (exterior problem). Taking the sum of those two equations gives:

$$\bar{u}(\boldsymbol{x}) - \int_{\partial\Omega} \psi(\boldsymbol{y})G(\boldsymbol{x}, \boldsymbol{y})\, \mathrm{d}S_y - \int_{\partial\Omega} [\phi(\boldsymbol{y}) - \phi(\boldsymbol{x})]H(\boldsymbol{x}, \boldsymbol{y})\, \mathrm{d}S_y = 0 \qquad (2.32)$$

where $H(\boldsymbol{x}, \boldsymbol{y})$ refers to the unit normal $\bar{\boldsymbol{n}}$ exterior to $\partial\Omega$. Now, consider as an example the interior Dirichlet problem for the Laplace equation:

$$\Delta u = 0 \quad (\boldsymbol{y} \in \Omega) \qquad u = u^D \quad (\boldsymbol{y} \in \partial\Omega)$$

Single-layer potential formulation. Assuming the potential u to be continuous across $\partial\Omega$, one has $u = \bar{u} = u^D$, $\phi = 0$ over $\partial\Omega$. eq. (2.32) then yields

$$\int_{\partial\Omega} G(\boldsymbol{x}, \boldsymbol{y})\psi(\boldsymbol{y})\, \mathrm{d}S_y = u^D(\boldsymbol{x}) \qquad (\boldsymbol{x} \in \partial\Omega)$$

which is a Fredholm integral equation of the first kind (Tricomi [5]) for the unknown density ψ, for the Dirichlet problem. Then, u at interior points of Ω is expressed by the representation (2.31).

Double-layer potential formulation. Assuming the flux q to be continuous across $\partial\Omega$, one has $u = \bar{u} = u^D$, $\psi = 0$ over $\partial\Omega$. eq. (2.32) then yields

$$-\phi(\boldsymbol{x}) + \int_{\partial\Omega} [\phi(\boldsymbol{y}) - \phi(\boldsymbol{x})]H(\boldsymbol{x}, \boldsymbol{y})\, \mathrm{d}S_y = u^D(\boldsymbol{x}) \qquad (\boldsymbol{x} \in \partial\Omega) \qquad (2.33)$$

The following equivalent, non-regularized, integral equation can be obtained starting from (2.25) instead of (2.24):

$$-\frac{1}{2}\phi(\boldsymbol{x}) + \int_{\partial\Omega} \phi(\boldsymbol{y})H(\boldsymbol{x}, \boldsymbol{y})\, \mathrm{d}S_y = u^D(\boldsymbol{x}) \qquad (\boldsymbol{x} \in \partial\Omega) \qquad (2.34)$$

which is a Fredholm integral equation of the second kind for the unknown density ϕ, for the Dirichlet problem. Then, u at interior points of Ω is expressed by the representation (2.4).

Indirect integral equations for the Neumann problem will be discussed in chapter 5.

2.5 TWO-DIMENSIONAL PROBLEMS

Up to now the exposition has concentrated on three-dimensional (3D) problems. Of course, a reciprocity identity is also available in two dimensions, and fundamental solutions of two-dimensional (2D) problems as well. Hence integral representations and equations can be formulated following the same steps as before. Another viewpoint consists of treating 2D problems as special 3D situations which are invariant under all translations along the direction $\boldsymbol{e}_3$, so that the 2D formulation results from an integration of the 3D formulation with respect to y_3. The latter viewpoint is followed here as it gives some insight about the meaning of a two-dimensional fundamental solution.

Accordingly, consider a 3D domain Ω of the form $\Omega = D \times \mathbb{R}$, where D is an open subset of the $(y_1 y_2)$-plane, and suppose that the state (u, q, b) does not depend on y_3. The integral equation (2.24) for the domain Ω, with boundary $\partial D \times \mathbb{R}$, reads:

$$\int_{\partial D} [u(\tilde{y}) - u(\tilde{x})] \int_{\mathbb{R}} H(\tilde{x}, \boldsymbol{y})\, \mathrm{d}y_3\, \mathrm{d}s_y - \int_{\partial D} q(\tilde{y}) \int_{\mathbb{R}} G(\tilde{x}, \boldsymbol{y})\, \mathrm{d}y_3\, \mathrm{d}s_y = \int_D b(\tilde{y}) \int_{\mathbb{R}} G(\tilde{x}, \boldsymbol{y})\, \mathrm{d}y_3\, \mathrm{d}D_y \quad (2.35)$$

where $\tilde{x} = (x_1, x_2)$, $\tilde{y} = (y_1, y_2)$ denote points in the $(y_1 y_2)$-plane. In the above equation, 2D fundamental solutions result from integration of their 3D counterparts along y_3:

$$[G(\tilde{x}, \tilde{y})]_{2D} = \int_{\mathbb{R}} G(\tilde{x}, \boldsymbol{y})\, \mathrm{d}y_3 \qquad [H(\tilde{x}, \tilde{y})]_{2D} = \int_{\mathbb{R}} H(\tilde{x}, \boldsymbol{y})\, \mathrm{d}y_3 \quad (2.36)$$

provided those integrals are convergent. The 3D fundamental solution is here assumed to be invariant under translations along $\boldsymbol{e}_3$[10], and therefore depends on x_3, y_3 through $(y_3 - x_3)$ only; as a consequence, $\{G, H\}(\tilde{x}, \boldsymbol{y}) = \{G, H\}(\boldsymbol{x}, \tilde{y})$. One can then put (still pending verification of the convergence):

$$[G(\tilde{x}, \tilde{y})]_{2D} = \int_{\mathbb{R}} G(\boldsymbol{x}, \tilde{y})\, \mathrm{d}x_3 \qquad [H(\tilde{x}, \tilde{y})]_{2D} = \int_{\mathbb{R}} H(\boldsymbol{x}, \tilde{y})\, \mathrm{d}x_3 \quad (2.37)$$

The three-dimensional interpretation of $[G, H]_{2D}$ is thus the response of E to a superposition of unit point sources located at all points of the line passing through $\tilde{x}$ and of direction $\boldsymbol{e}_3$. This system of sources has an infinite resultant.

Anticipating the possibility of divergent integrals in (2.37), let us first reformulate equation (2.35): since the unknown flux across ∂D must equilibrate the source b, one has:

$$\int_{\partial D} q(\tilde{y})\, \mathrm{d}s_y + \int_D b(\tilde{y})\, \mathrm{d}D_y = 0$$

[10] In other words, the 3D solution refers to a domain E which is invariant under translations along $\boldsymbol{e}_3$; although strictly speaking this assumption is not necessarily true, weakening it has no practical interest or relevance.

Let $\tilde{z} \neq \tilde{x}$ denote a fixed point of D. Upon multiplication of the previous equation by $G(\boldsymbol{x}, \tilde{\boldsymbol{z}})$ followed by integration with respect to y_3, then subtraction of the result from (2.35), one gets:

$$\int_{\partial D} [u(\tilde{y}) - u(\tilde{x})] \int_{\mathbb{R}} H(\boldsymbol{x}, \tilde{y}) \mathrm{d}x_3 \mathrm{d}s_y - \int_{\partial D} q(\tilde{y}) \int_{\mathbb{R}} [G(\boldsymbol{x}, \tilde{y}) - G(\boldsymbol{x}, \tilde{\boldsymbol{z}})] \mathrm{d}x_3 \mathrm{d}s_y$$
$$= \int_D b(\tilde{y}) \int_{\mathbb{R}} [G(\boldsymbol{x}, \tilde{y}) - G(\boldsymbol{x}, \tilde{\boldsymbol{z}})] \, \mathrm{d}x_3 \, \mathrm{d}D_y \quad (2.38)$$

Fundamental solution for the infinite plane

It is obtained from the 3D free-space fundamental solution. First, notice that definition (2.36) for $[G]_{2D}$ leads to a divergent integral:

$$\int_{\mathbb{R}} \frac{1}{r} \, \mathrm{d}x_3 = \int_{\mathbb{R}} \frac{1}{\sqrt{\rho^2 + x_3^2}} \, \mathrm{d}x_3 = +\infty \quad (2.39)$$

(with $\rho^2 = (y_1 - x_1)^2 + (y_2 - x_2)^2$). One must therefore define $[G]_{2D}$ from eq. (2.38), so that:

$$\begin{aligned}
[G(\tilde{x}, \tilde{y})]_{2D} &= \frac{1}{4\pi} \int_{\mathbb{R}} \left(\frac{1}{\sqrt{\rho^2 + x_3^2}} - \frac{1}{\sqrt{\rho'^2 + x_3^2}} \right) \mathrm{d}x_3 \\
&= \frac{1}{4\pi} \left[\ln \left(\frac{x_3 + \sqrt{\rho^2 + x_3^2}}{x_3 + \sqrt{\rho'^2 + x_3^2}} \frac{\rho'}{\rho} \right) \right]_{-\infty}^{+\infty} \\
&= -\frac{1}{2\pi} \log \frac{\rho}{\rho'}
\end{aligned}$$

On the other hand, one has

$$\begin{aligned}
[H(\tilde{x}, \tilde{y})]_{2D} &= -\frac{1}{4\pi} \int_{\mathbb{R}} \frac{(\tilde{y} - \tilde{x}).\boldsymbol{n}}{(\rho^2 + x_3^2)^{3/2}} \, \mathrm{d}x_3 \\
&= -\frac{1}{2\pi} \frac{(\tilde{y} - \tilde{x}).\boldsymbol{n}}{\rho^2}
\end{aligned}$$

Finally, the fundamental solution for the infinite plane is given by:

$$[G(\tilde{x}, \tilde{y})]_{2D} = \frac{1}{2\pi} \log \frac{\rho'}{\rho} \qquad [H(\tilde{x}, \tilde{y})]_{2D} = -\frac{1}{2\pi} \frac{\rho_{,n}}{\rho} \quad (2.40)$$

Trying to derive the fundamental solution to the two-dimensional Poisson equation directly instead of using the y_3-integration method of the 3D solution described here would lead to the same result (2.40). The resulting solution $[G]_{2D}$ is defined up to an additive constant, via the arbitrary choice of $\tilde{z}$ (and hence of ρ'). It is singular like $\log r$ at $\tilde{y} = \tilde{x}$. Its asymptotic behaviour at infinity is surprising at first: $[G]_{2D} \to \infty$ when $\rho \to \infty$. This property, as well as the divergent character of the integral (2.39), are consequences of the fact that $[G]_{2D}$ is the potential created in the 3D free space by a source of *infinite resultant*.

Integral equation

The boundary integral equation satisfied by any solution to the Poisson equation in two dimensions thus assumes the same form as in three-dimensional problems:

$$\int_{\partial D} \{[u(\tilde{y}) - u(\tilde{x})]H(\tilde{x},\tilde{y}) - q(\tilde{y})G(\tilde{x},\tilde{y})\} \, \mathrm{d}s_y = \int_D b(\tilde{y})G(\tilde{x},\tilde{y}) \, \mathrm{d}D_y \quad (2.41)$$

where $(G, H) \equiv ([G]_{2D}, [H]_{2D})$. When $\tilde{x} \in \mathbb{R}^2 \setminus D$, the expected integral representation formula for $u(\tilde{x})$ is obtained:

$$\kappa u(\tilde{x}) = \int_{\partial D} \{q(\tilde{y})G(\tilde{x},\tilde{y}) - u(\tilde{y})H(\tilde{x},\tilde{y})\} \, \mathrm{d}s_y + \int_D b(\tilde{y})G(\tilde{x},\tilde{y}) \, \mathrm{d}D_y \quad (2.42)$$

Exterior problems

The extension to unbounded domains of the form $\bar{D} = \mathbb{R}^2 - D$ (with $b = 0$) proceeds as in the three-dimensional situation. It follows from the asymptotic behaviour at infinity of the 2D fundqmental solution (2.40) that any u satisfying the following sufficient conditions at infinity

$$u(\boldsymbol{y}) = O(R^{-\alpha}), \quad \boldsymbol{\nabla} u(\boldsymbol{y}) = O(R^{-1-\alpha}) \qquad (\boldsymbol{y} \in S_R, R \to +\infty) \qquad (2.43)$$

is governed by the integral equation:

$$u(\tilde{x}) + \int_{\partial D} \{[u(\tilde{y}) - u(\tilde{x})]H(\tilde{x},\tilde{y}) - q(\tilde{y})G(\tilde{x},\tilde{y})\} \, \mathrm{d}s_y = 0$$

Like in three-dimensional situations, when $\tilde{x} \notin \partial D$, the integral representation of $u(\tilde{x})$ assumes the same form whether D or $\mathbb{R}^2 \setminus D$ is bounded, except that the unit normal changes sign.

Appendix 2.A SOME USEFUL FORMULAS

Let $\boldsymbol{x}, \boldsymbol{y}$ denote two points in $\mathbb{R}^n$. The Euclidean norm r of the relative vector $\boldsymbol{r}$ is given by

$$\boldsymbol{r} = \boldsymbol{y} - \boldsymbol{x} = (y_i - x_i)\boldsymbol{e}_i = r_i \boldsymbol{e}_i \qquad r = |\boldsymbol{r}|$$

and its partial derivatives with respect to y-coordinates are given by:

$$r_{,i} = \frac{y_i - x_i}{r} \qquad r_{,ij} = \frac{1}{r}(\delta_{ij} - r_{,i}r_{,j}) \tag{2.A.1}$$

$$r_{,ijk} = \frac{1}{r^2}(3r_{,i}r_{,j}r_{,k} - \delta_{ij}r_{,k} - \delta_{ik}r_{,j} - \delta_{jk}r_{,i}) \tag{2.A.2}$$

Notice that the vector $\boldsymbol{\nabla r} = r_{,i}\boldsymbol{e}_i$ has unit norm: $r_{,i}r_{,i} = 1$. From the above formulas, one gets:

$$\Delta r = \frac{n-1}{r} \quad \Delta\frac{1}{r} = \frac{n-3}{r^3} \qquad (\boldsymbol{y} \neq \boldsymbol{x}) \tag{2.A.3}$$

Appendix 2.B DERIVATION OF THE FUNDAMENTAL SOLUTION OF THE LAPLACIAN

The fundamental solution $G(\boldsymbol{x}, \boldsymbol{y})$ for 3D free space solves:

$$G_{,aa}(\boldsymbol{x}, \boldsymbol{y}) + \delta(\boldsymbol{y} - \boldsymbol{x}) = 0 \qquad (\boldsymbol{y} \in \mathbb{R}^3 - \{\boldsymbol{x}\}) \tag{2.A.4}$$

The three-dimensional Fourier transform, defined by

$$\hat{f}(\boldsymbol{\xi}) = \frac{1}{(2\pi)^3}\int_{\mathbb{R}^3} f(\boldsymbol{y})\mathrm{e}^{-\mathrm{i}\xi_j y_j}\,\mathrm{d}V_y \tag{2.A.5}$$

($\boldsymbol{\xi}$: transformed variable of $\boldsymbol{y}$) is applied to eq. (2.A.4). As a result, the Fourier transform $\hat{G}(\boldsymbol{x}, \boldsymbol{\xi})$ of $G(\boldsymbol{x}, \boldsymbol{y})$ is given by:

$$-\xi^2\hat{G}(\boldsymbol{x}, \boldsymbol{\xi}) + \frac{1}{(2\pi)^3}\mathrm{e}^{-\mathrm{i}\xi\eta_j x_j} = 0$$

where $\boldsymbol{\xi} = |\boldsymbol{\xi}|\,\boldsymbol{\eta} = \xi\boldsymbol{\eta}$ and $|\eta| = 1$. The fundamental solution $G(\boldsymbol{x}, \boldsymbol{y})$ is then given by Fourier inversion:

$$G(\boldsymbol{x}, \boldsymbol{y}) = \frac{1}{(2\pi)^3}\int_0^{+\infty}\int_S \mathrm{e}^{\mathrm{i}\xi\eta_j(y_j - x_j)}\,\mathrm{d}\omega\mathrm{d}\xi \tag{2.A.6}$$

in which $\mathrm{d}\omega$ denotes the solid angle differential measure on the unit sphere S defined by $|\eta| = 1$. Moreover, noticing that the change of variable $\boldsymbol{\eta} \to -\boldsymbol{\eta}$ leaves S globally invariant, one has:

$$\begin{aligned}
&\int_0^{+\infty}\left\{\int_S \mathrm{e}^{\mathrm{i}\xi\eta_j(y_j - x_j)}\,\mathrm{d}\omega\right\}\mathrm{d}\xi \\
&= \frac{1}{2}\int_0^{+\infty}\left\{\int_S \mathrm{e}^{\mathrm{i}\xi\eta_j(y_j - x_j)}\,\mathrm{d}\omega + \int_S \mathrm{e}^{-\mathrm{i}\xi\eta_j(y_j - x_j)}\,\mathrm{d}\omega\right\}\mathrm{d}\xi \\
&= \frac{1}{2}\left\{\int_0^{+\infty}\int_S \mathrm{e}^{\mathrm{i}\xi\eta_j(y_j - x_j)}\,\mathrm{d}\omega\mathrm{d}\xi + \int_S\int_{-\infty}^0 \mathrm{e}^{\mathrm{i}\xi\eta_j(y_j - x_j)}\,\mathrm{d}\omega\mathrm{d}\xi\right\} \\
&= \frac{1}{2}\int_{-\infty}^{+\infty}\int_S \mathrm{e}^{\mathrm{i}\xi\eta_j(y_j - x_j)}\,\mathrm{d}\omega\mathrm{d}\xi
\end{aligned} \tag{2.A.7}$$

Integration with respect to ξ brings in a Dirac delta distribution:

$$\int_{-\infty}^{+\infty} \mathrm{e}^{\mathrm{i}\xi\eta_j(y_j - x_j)}\,\mathrm{d}\xi = 2\pi\delta(\eta_j(y_j - x_j)) \tag{2.A.8}$$

Introduce now a direct orthonormal frame $(\boldsymbol{a}_1, \boldsymbol{a}_2, \boldsymbol{a}_3)$ such that $\boldsymbol{y} - \boldsymbol{x} = r\boldsymbol{a}_3$, so that $\boldsymbol{\eta} = r\sin\varphi(\cos\theta\boldsymbol{a}_1 + \sin\theta\boldsymbol{a}_2) + r\cos\varphi\boldsymbol{a}_3$ and $\mathrm{d}\omega = \sin\varphi\,\mathrm{d}\theta\mathrm{d}\varphi$ (θ, φ: spherical angles with respect to origin $\boldsymbol{x}$ and direction $\boldsymbol{a}_3$), and then $\boldsymbol{\eta}.(\boldsymbol{y} - \boldsymbol{x}) = r\cos\varphi$. As a consequence, eqs (2.A.6), (2.A.7) and (2.A.8) lead to the desired result:

$$\begin{aligned} G(\boldsymbol{x}, \boldsymbol{y}) &= \frac{1}{8\pi^2}\int_0^{2\pi} \mathrm{d}\theta \int_0^{\pi} \delta(r\cos\varphi)\sin\varphi\,\mathrm{d}\varphi \\ &= \frac{1}{8\pi^2}\int_0^{2\pi} \mathrm{d}\theta \int_{-r}^{r} \delta(u)\frac{du}{r} \\ &= \frac{1}{4\pi r} \end{aligned}$$

Fundamental solution for the bi-Laplacian

Since $\Delta r = 2/r$, the previous result immediately implies that

$$G(\boldsymbol{x}, \boldsymbol{y}) = \frac{r}{8\pi}$$

is the free-space fundamental solution of the biharmonic equation

$$\Delta\Delta u = 0.$$

Anisotropic medium

Consider the equation

$$k_{ij}u_{,ij} = 0 \tag{2.A.9}$$

where the (conductivity) tensor $\boldsymbol{k}$ is symmetric and positive definite. Introduce the new coordinates Y_I:

$$\boldsymbol{Y} = \boldsymbol{k}^{-1/2}.\boldsymbol{y}$$

Then, putting $v(\boldsymbol{Y}) = u(\boldsymbol{y})$, eq. (2.A.9) becomes:

$$u_{,i} = v_{,I}k_{Ii}^{-1/2} \qquad u_{,ij} = v_{,IJ}k_{Ii}^{-1/2}k_{Jj}^{-1/2} \qquad k_{ij}u_{,ij} = v_{,II}$$

This manipulation shows that the anisotropic conduction equation (2.A.9) is transformed into the Laplace equation in the coordinates Y_I. It follows that the fundamental solution $G^A(\boldsymbol{x}, \boldsymbol{y})$ of (2.A.9) is obtained by putting $\boldsymbol{Y} = \boldsymbol{k}^{-1/2}.\boldsymbol{y}$ in the isotropic solution $G(\boldsymbol{X}, \boldsymbol{Y})$, giving as a result:

$$G(\boldsymbol{x}, \boldsymbol{y}) = \frac{1}{4\pi\,|\boldsymbol{k}|^{1/2}}\frac{1}{\bar{r}} \qquad (E = \mathbb{R}^3)$$

$$G(\boldsymbol{x}, \boldsymbol{y}) = \frac{1}{2\pi\,|\boldsymbol{k}|^{1/2}}\log\frac{1}{\bar{r}} \qquad (E = \mathbb{R}^2)$$

where:

$$|\boldsymbol{k}| = \det(\boldsymbol{k}) \qquad \bar{r}^2 = (\boldsymbol{y} - \boldsymbol{x})^T \boldsymbol{K}^{-1}(\boldsymbol{y} - \boldsymbol{x})$$

Moreover, this analysis shows that the anisotropic Green's functions (i.e. with Dirichlet boundary conditions) for the half-space and half-plane are still obtained by the method of images.

Appendix 2.C INTEGRABILITY OF $r_{,n}/r^2$

Let S denote a regular surface, and denote by $\boldsymbol{n}$ the unit normal consistent with an *a priori* chosen orientation,[11] and let $\boldsymbol{x} \in S$. Assume that S admits in a neighbourhood of $\boldsymbol{x}$ a parametric description of the form

$$\boldsymbol{r} = \boldsymbol{y} - \boldsymbol{x} = r(\varphi, \theta)\boldsymbol{e}_r \tag{2.A.10}$$

where $\varphi \in [0, \pi], \theta \in [0, 2\pi]$ are the spherical angles associated with origin $\boldsymbol{x}$ and direction $\boldsymbol{e}_3$ and $r(\varphi, \theta)$ is a differentiable function.[12] The relation

$$\tilde{\boldsymbol{n}}\,\mathrm{d}S_y = (\boldsymbol{r}_{,\varphi} \wedge \boldsymbol{r}_{,\theta})\,\mathrm{d}\varphi\mathrm{d}\theta$$

holds for any surface parametrization, $\tilde{\boldsymbol{n}}$ being the unit normal associated with (2.A.10) and thus not necessarily of the same sign as $\boldsymbol{n}$. Since

$$\boldsymbol{r}_{,\varphi} = r_{,\varphi}\boldsymbol{e}_r + r\boldsymbol{e}_\varphi \qquad \boldsymbol{r}_{,\theta} = r_{,\theta}\boldsymbol{e}_r + r\sin\varphi\boldsymbol{e}_\theta$$

one gets

$$\tilde{\boldsymbol{n}}\,\mathrm{d}S_y = (r^2\sin\varphi\boldsymbol{e}_r - rr_{,\varphi}\sin\varphi\boldsymbol{e}_\varphi - rr_{,\theta}\boldsymbol{e}_\theta)\,\mathrm{d}\varphi\mathrm{d}\theta$$

and therefore

$$r_{,\tilde{n}}\frac{\mathrm{d}S_y}{r^2} = \boldsymbol{r}.\tilde{\boldsymbol{n}}\frac{\mathrm{d}S_y}{r^3} = r\boldsymbol{e}_r.\tilde{\boldsymbol{n}}\frac{\mathrm{d}S_y}{r^3} = \sin\varphi\,\mathrm{d}\varphi\mathrm{d}\theta$$

[11] For example, if S is closed, $\boldsymbol{n}$ is the unit normal exterior to S.
[12] Thus, any ray of origin $\boldsymbol{x}$ has at most one intersection with S.

(notice that, by construction, $\tilde{\boldsymbol{n}}.\boldsymbol{e}_r \geq 0$ always holds). It follows that

$$\int_S r_{,n} \frac{\mathrm{d}S_y}{r^2} = \int_S \mathrm{d}\omega = \omega(\boldsymbol{x}, S) \tag{2.A.11}$$

where $\mathrm{d}\omega$ is the differential measure of algebraic solid angle (i.e. $\mathrm{d}\omega$ has the same sign as $\boldsymbol{n}.\boldsymbol{e}_r$) and $\omega(\boldsymbol{x}, S)$ the algebraic solid angle under which S is seen from $\boldsymbol{x}$. In particular, the integral $\int_S r^{-2} r_{,n} \,\mathrm{d}S_y$ is convergent. If S is closed, one has

$$\int_S r_{,\tilde{n}} \frac{\mathrm{d}S_y}{r^2} = \int_S \mathrm{d}\omega = 2\pi \tag{2.A.12}$$

For any arbitrary regular surface S there exists a partition $S = S_1, \ldots, S_N$ such that each S_j has a description of the form (2.A.10). Therefore, equation (2.A.11) holds for any regular surface.

REFERENCES

1. Bettess, P. *Infinite elements*. Penshaw Press (1992).
2. Guiggiani, M., Krishnasamy, G., Rudolphi, T. J., Rizzo, F. J. A general algorithm for the numerical solution of hypersingular boundary integral equations. *ASME J. Appl. Mech.*, **59**, 604–614 (1992).
3. Kupradze, V. D. (ed.). *Three-dimensional problems of the mathematical theory of elasticity and thermoelasticity*. North-Holland (1979).
4. Sladek, J., Sladek, V. Three-dimensional curved crack in an elastic body. *Int. J. Solids Struct.*, **19**, 425–436 (1983).
5. Tricomi, F. G. *Integral equations*. Interscience Publishers, New York (1957).
6. Wing, G. M. *A primer on integral equations of the first kind*. SIAM Publications, Philadelphia, USA (1991).

3

Boundary element method for potential problems

This chapter addresses the boundary element discretization method for the Poisson equation $\Delta u + b = 0$ on a domain Ω of boundary $\partial\Omega$. For ease of presentation, a zero source b is considered; how to incorporate a nonzero b will be shown in Section 3.6. In the absence of sources, the potential u and flux q are governed by the regularized boundary integral equation (2.24), i.e.:

$$\int_{\partial\Omega} \{[u(\boldsymbol{y}) - u(\boldsymbol{x})]H(\boldsymbol{x}, \boldsymbol{y}) - q(\boldsymbol{y})G(\boldsymbol{x}, \boldsymbol{y})\} \, \mathrm{d}S_y = 0 \tag{3.1}$$

where the source point $\boldsymbol{x}$ lies on the boundary $\partial\Omega$. The *collocation method* consists of generating a finite set of equations from (3.1) by enforcing eq. (3.1) at finitely many *collocation points* $\boldsymbol{x}$ of $\partial\Omega$. Moreover the boundary surface $\partial\Omega$ and boundary variables (u, q) are discretized. Of course, the number of collocation points should (at least) match that of boundary unknowns.

The boundary element (BE) discretization approach results essentially from a direct transposition of the interpolation concepts initially developed for finite element methods (FEMs), based on the use of nodes and shape functions. This choice is motivated by the same considerations as in FEMs, i.e. generality and flexibility: the solution technique should be able to deal, in a unified framework, with arbitrary shapes and arbitrary boundary data.

3.1 DISCRETIZATION OF THE BOUNDARY

The starting point of the discretization process consists of introducing a partition of the boundary surface $\partial\Omega$ into N_E non-intersecting *boundary elements*

$E_1, \ldots, E_{N_E}$: $\partial\Omega = \bigcup_{e=1}^{N_E} E_e$. The elements E_e are thus surface or curve patches, depending on whether Ω is a three- or two-dimensional domain. The integral equation (3.1) then assumes the form of a sum of *element integrals*:

$$\sum_{e=1}^{N_E} \int_{E_e} \{[u(\boldsymbol{y}) - u(\boldsymbol{x})]H(\boldsymbol{x}, \boldsymbol{y}) - q(\boldsymbol{y})G(\boldsymbol{x}, \boldsymbol{y})\} \, \mathrm{d}S_y = 0 \tag{3.2}$$

Actual evaluation of those element integrals requires that each element E_e be analytically described. Usually a mapping of each physical element E_e onto a *parent element* Δ_e in a parameter space is introduced:

$$\boldsymbol{\xi} \in \Delta_e \rightarrow \boldsymbol{y}(\boldsymbol{\xi}) \in E_e \qquad (1 \leq e \leq N_E)$$

Δ_e most often assumes a simple shape, i.e. a segment of the ξ-line (if $\partial\Omega$ is a curve) or a triangle or square in the (ξ_1, ξ_2)-plane (if $\partial\Omega$ is a surface).

Element description using nodes and shape functions

Most often one does not attempt to reproduce exactly the boundary $\partial\Omega$, because this would entail increased complexity as well as loss of generality. Instead, the introduction of generic parametric representations (usually formulated in terms of polynomials of the intrinsic coordinate(s) $\boldsymbol{\xi}$), where a given element is defined through a small number of coefficients, provides a systematic and general approach for generating an approximation $\tilde{\partial\Omega} = \bigcup_{e=1}^{N_E} E_e$ of the true boundary $\partial\Omega$.

Of course, exactly the same considerations led in the past to the concepts and techniques now commonly used in finite element methods. The usual concept of boundary element, in a mere transposition of that of a finite element, is based on the use of *nodes* and *shape functions*. The parent element assumes a simple shape: line segment $\xi \in [-1, 1]$ for plane problems, square

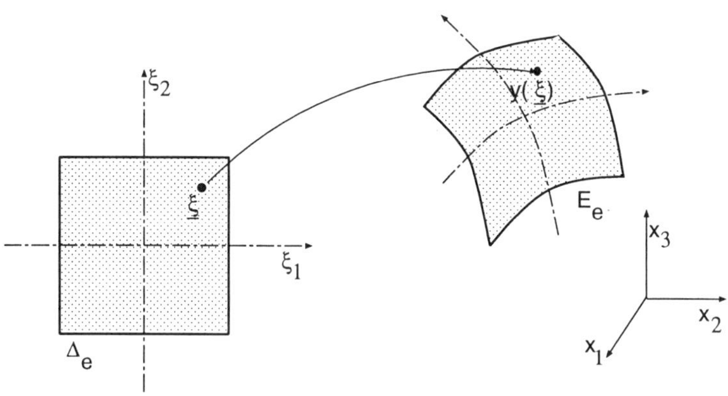

Figure 3.1 Boundary element E_e and parent element Δ_e.

$-1 \leq \xi_1, \xi_2 \leq 1$ or triangle $\xi_1 \geq 0, \xi_2 \geq 0, 1-\xi_1-\xi_2 \geq 0$ for three-dimensional problems. Each physical element E_e of the approximate boundary $\tilde{\partial\Omega}$ is mapped onto its parent element Δ_e using the following formula:

$$\boldsymbol{\xi} \in \Delta_e \rightarrow \boldsymbol{y}(\boldsymbol{\xi}) = \sum_{m=1}^{N_e} N_m(\boldsymbol{\xi}) \boldsymbol{y}^m \qquad \boldsymbol{\xi} \in \Delta_e \tag{3.3}$$

in terms of

- N_e given points $\boldsymbol{y}^1, \ldots, \boldsymbol{y}^{N_e}$ of $\partial\Omega$ called *geometrical nodes*; the index m $(1 \leq m \leq N_e)$ defines the *local numbering* of the nodes on element e.
- N_e *shape functions* $N_1(\boldsymbol{\xi}), \ldots, N_{N_e}(\boldsymbol{\xi})$, usually of polynomial type, and subjected to the following restrictions:

$$N_p(\boldsymbol{\xi}^q) = \delta_{pq} \qquad \sum_{m=1}^{N_e} N_q(\boldsymbol{\xi}) = 1 \quad \forall \boldsymbol{\xi} \in \Delta_e \tag{3.4}$$

  where $\boldsymbol{\xi}^q \in \Delta_e$ is the *antecedent* of the physical node $\boldsymbol{y}^q \in E_e$.

Of course, the specific choice of shape functions and the number N_e of nodes defining the element are related.

This formalism allows us to define curved surface or line patches without difficulty. Besides, the evaluation of element integrals in (3.2) can be performed in a systematic manner using the mapping (3.3): all curvilinear segments (resp. triangular, quadrilateral surface patches) are mapped onto the *same* reference line segment (resp. triangle, square).

Generation of the approximate boundary

The following operations allow us to define a boundary element *mesh* that approximates the true boundary:

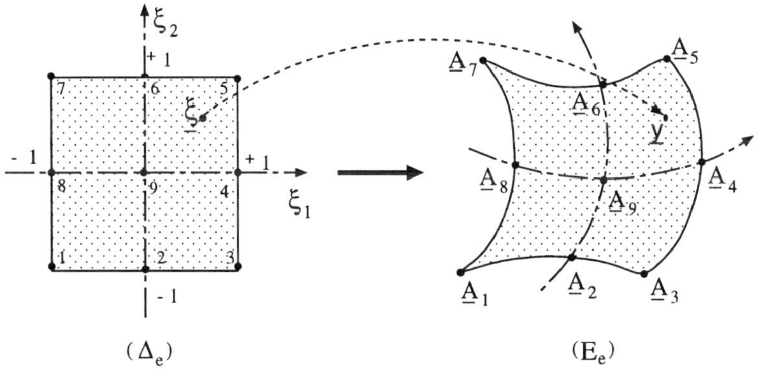

Figure 3.2 Mapping of a nine-noded quadrilateral boundary element.

1. Choose N_N *geometrical nodes* $\boldsymbol{y}^1, \ldots, \boldsymbol{y}^{N_N}$ located on the true boundary $\partial\Omega$. This information is stored in the form of a nodal coordinate table, which gives for each p, $1 \leq p \leq N_N$, the Cartesian coordinates of the node $\boldsymbol{y}^p$ in the global reference frame.

2. Partition $\partial\Omega$ into N_E elements (curved line segments, triangular or quadrilateral surface patches). The *connectivity table* $e, q \rightarrow m(e, q)$ gives the *global number* $m = m(e, q)$ of the node of *local number* q on element E_e. Conversely, the element E_e is defined by the set of N_e nodes $\boldsymbol{y}^{m(e,1)}, \ldots, \boldsymbol{y}^{m(e,N_e)}$

A boundary *mesh* is completely defined by (a) the nodal coordinate and connectivity tables, (b) the specific shape functions used on each element to interpolate the nodes according to (3.3).

Construction of the shape functions

For a boundary element approximation of $\partial\Omega$ to be valid, any two neighbouring elements must connect exactly, i.e. with neither overlapping nor 'holes' between them. The shape functions commonly used in finite element methods fulfil those criteria; some usual shape functions are listed in Appendix B.2. Two adjacent elements must necessarily share all nodes located on their common side: the mesh is *conformal*. The approximation of $\partial\Omega$ by nodes and shape functions thus guarantees that

- The unit normal and tangent plane are unique and continuous on each element;

- The surface $\tilde{\partial\Omega}$ is continuous and non-overlapping at the junction between two elements.

The unit normal and tangent plane are usually discontinuous across element boundaries.

Tangent and unit normal

From the classical differential geometry of curves, the unit tangent $\boldsymbol{\tau}$ and the length of an infinitesimal line element $\mathrm{d}s$ at a point $\boldsymbol{y}(\xi)$ on a curvilinear boundary element are given by:

$$\boldsymbol{\tau} = \frac{1}{J} \sum_{q=1}^{N_e} N_q'(\xi) \boldsymbol{y}^q \qquad \mathrm{d}s = J(\boldsymbol{\xi})\,\mathrm{d}\boldsymbol{\xi} \qquad J = \left| \sum_{q=1}^{N_e} N_q'(\xi) \boldsymbol{y}^q \right| \tag{3.5}$$

The unit normal is thus:

$$\boldsymbol{n} = \boldsymbol{e}_3 \wedge \boldsymbol{\tau} \qquad \text{i.e.} \quad n_j = e_{ij} \tau_i$$

Similarly, from the differential geometry of surfaces, the unit normal $\boldsymbol{n}$ and the area of a differential surface element $\mathrm{d}S$ at a point $\boldsymbol{y}(\boldsymbol{\xi})$ on a surface boundary element are given by:

$$\boldsymbol{n} = \frac{1}{J}(\boldsymbol{a_1} \wedge \boldsymbol{a_2}) \qquad \mathrm{d}S = J(\boldsymbol{\xi})\,\mathrm{d}\boldsymbol{\xi} \tag{3.6}$$

where $J(\boldsymbol{\xi}))$ is the Jacobian of the mapping (3.3), i.e.

$$J(\boldsymbol{\xi}) = |\boldsymbol{a_1} \wedge \boldsymbol{a_2}| = \left\{ |\boldsymbol{a}_1|^2 \, |\boldsymbol{a}_2|^2 - (\boldsymbol{a}_1.\boldsymbol{a}_2)^2 \right\}^{1/2} \tag{3.7}$$

and $(\boldsymbol{a}_1, \boldsymbol{a}_2)$ are the two vectors of the *natural basis* of the tangent plane to E_e at $\boldsymbol{y}(\boldsymbol{\xi})$ associated with the mapping (3.3):

$$\boldsymbol{a}_\alpha(\boldsymbol{\xi}) = \sum_{q=1}^{N_e} N_{q,\alpha}(\boldsymbol{\xi})\boldsymbol{y}^q \qquad (\alpha = 1, 2) \tag{3.8}$$

Obviously, the vectors $(\boldsymbol{a}_1, \boldsymbol{a}_2)$ must be linearly independent for any $\boldsymbol{\xi} \in \Delta_e$ in order to ensure $J(\boldsymbol{\xi}) > 0$ everywhere on the boundary element, so that the mapping (3.3) is one-to-one.

Example: plane triangular boundary elements

The simplest way to approximate a given surface by boundary elements consists of dividing it into (curvilinear) triangles. Then, the vertices $\boldsymbol{y}^1, \ldots, \boldsymbol{y}^{N_N}$ (in global numbering) are the geometrical nodes, and the approximate surface is defined as the set of flat triangles with straight sides and whose vertices are those of the corresponding curvilinear triangle. A given boundary element is thus defined in terms of three nodes (the triangle vertices); its mapping (3.3) onto the reference triangle uses linear shape functions:

$$\boldsymbol{y} = (1 - \xi_1 - \xi_2)\boldsymbol{y}^1 + \xi_1\boldsymbol{y}^2 + \xi_2\boldsymbol{y}^3 \tag{3.9}$$

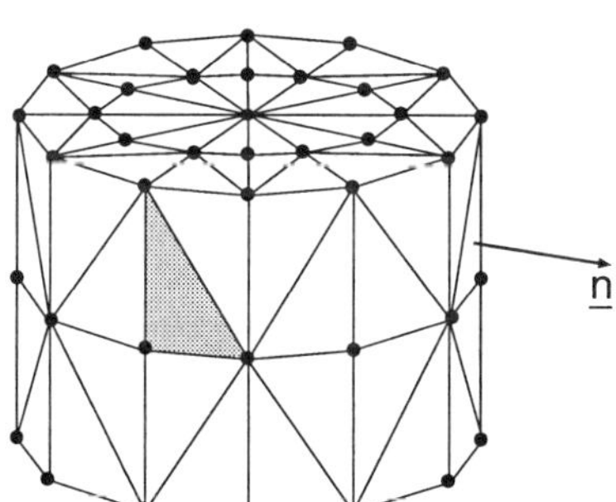

Figure 3.3 Approximation of the boundary of a three-dimensional domain using plane triangular boundary elements.

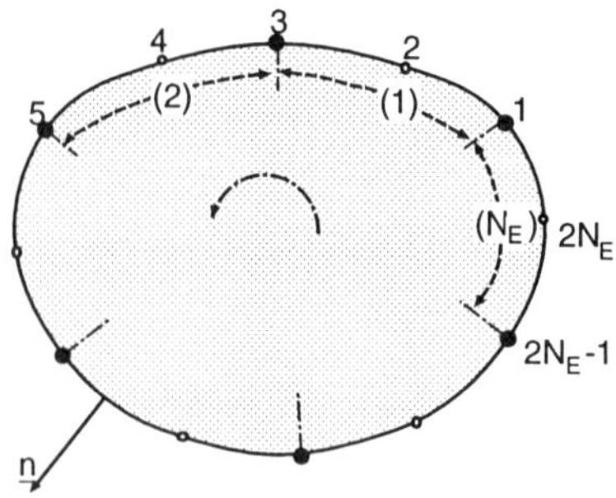

Figure 3.4 Approximation of the boundary of a two-dimensional domain using curved boundary elements.

The corresponding *natural basis* is given by:

$$\boldsymbol{a}_1 = \boldsymbol{y}^1 - \boldsymbol{y}^3 \qquad \boldsymbol{a}_2 = \boldsymbol{y}^2 - \boldsymbol{y}^3$$

The evaluation of J, $\boldsymbol{n}$ and dS using (3.9), (3.6) and (3.7) is easy, noting that $\boldsymbol{a}_1 = \boldsymbol{y}_2 - \boldsymbol{y}_1$ and $\boldsymbol{a}_2 = \boldsymbol{y}_3 - \boldsymbol{y}_1$.

The accuracy of the approximate surface thus defined of course increases as the triangular mesh is refined, i.e. with decreasing element size and increasing numbers of nodes and elements. It is also important to point out that the modes *must* be located at the triangle vertices to ensure overall continuity and avoid element overlapping, when the mapping (3.9) is used for each triangle.

Example: curved segments

A closed curve is divided into N_E curved line segments. For each of them, three nodes are introduced: the two endpoints $\boldsymbol{y}^1, \boldsymbol{y}^3$ and a chosen interior point $\boldsymbol{y}^2$ (in local numbering). The whole curve thus contains $N_N = 2N_E$ distinct nodes. Using the global numbering described in Figure 3.4, the connectivity table reads:

e		1			2			3		...		N_E	
q	1	2	3	1	2	3	1	2	3	...	1	2	3
$m(e,q)$	1	2	3	3	4	5	5	6	7	...	$2N_E - 1$	$2N_E$	1

For each element, the mapping (3.3) is formulated in terms of quadratic shape functions, so that one has (with $-1 \le \xi \le 1$):

$$\boldsymbol{y} = \frac{1}{2}\xi(\xi - 1)\boldsymbol{y}^1 + (1 - \xi^2)\boldsymbol{y}^2 + \frac{1}{2}\xi(\xi + 1)\boldsymbol{y}^3 \tag{3.10}$$

$$\boldsymbol{\tau}\,\mathrm{d}s = [\frac{1}{2}(\boldsymbol{y}^3 - \boldsymbol{y}^1) + \xi(\boldsymbol{y}^1 + \boldsymbol{y}^3 - 2\boldsymbol{y}^2)]\,\mathrm{d}\xi \tag{3.11}$$

3.2 DISCRETIZATION OF THE UNKNOWNS

The variables u, q over the boundary are also represented using interpolation techniques initially developed for the finite element method. Choose a set of N_I *interpolation nodes* $\boldsymbol{z}^1, \ldots, \boldsymbol{z}^{N_I}$ (in global numbering) on $\partial\Omega$. A given boundary element E_e contains $N_I(e)$ such nodes $\boldsymbol{z}^k$ ($1 \leq k \leq N_I(e)$, in local numbering), associated with $N_I(e)$ *interpolations functions* $M_k(\boldsymbol{\xi})$ assumed to verify the requirements (3.4). Then u, q are approximated on E_e by:

$$\tilde{u}(\boldsymbol{y}) = \sum_{k=1}^{N_I(e)} M_k(\boldsymbol{\xi}) u^k \qquad \tilde{q}(\boldsymbol{y}) = \sum_{k=1}^{N_I(e)} M_k(\boldsymbol{\xi}) q^k \qquad (\boldsymbol{\xi} \in \Delta_e) \tag{3.12}$$

where $\boldsymbol{y}$ and $\boldsymbol{\xi}$ are related by the mapping (3.3) and $(u^k = \tilde{u}(\boldsymbol{z}^k), q^k = \tilde{q}(\boldsymbol{z}^k))$ are the *nodal values* of the approximations $\tilde{u}, \tilde{q}$ of (u, q).

Conforming or non-conforming isoparametrical boundary elements

A boundary element interpolation that ensures overall (i.e. on element and across interelement boundaries) continuity of the interpolated variables is termed *conforming*. The nodes associated to conforming interpolations must be selected according to the same rules as the geometrical nodes.

It is *a priori* conceivable to approximate u, continuous over $\partial\Omega$, using a non-continuous (e.g. elementwise constant) interpolation $\tilde{\boldsymbol{u}}$. However, the boundary integral equation (3.1) is mathematically valid only if $u \in C^{0,\alpha}$ ($\alpha > 0$) at $\boldsymbol{x}$, which implies the continuity of u at $\boldsymbol{x}$. Hence, for given interpolation functions $M_k(\boldsymbol{\xi})$, the collocation points $\boldsymbol{x}$ must be chosen consistently. For example, the use of elementwise constant interpolation of a boundary variable precludes the possibility of collocation points located at element edges or vertices. Conforming interpolations, the simplest choice, are thus by far the most commonly used for u; a comparative study (Manolis and Banerjee [11]) strongly supports this option. On the other hand, a non-conforming interpolation for q is acceptable, physically as well as mathematically. Non-conforming interpolations tend to significantly increase the number of unknown degrees of freedom, for a given type (i.e. polynomial degree) of interpolation on each element. Nevertheless, non-conforming interpolations of u are sometimes used for gradient or traction boundary integral equations because the collocation at interelement nodes then raises difficulties related to stronger smoothness requirements on u (see Chapter 5).

In definition (3.12), the same interpolation functions are used for u and q: this choice is usual but not mandatory. Distinct interpolations for u and q are sometimes chosen, for example near edges or corners (see pp. 86-88).

Most often, the geometrical nodes are used as interpolation nodes (i.e. $N_N = N_I$ and $\boldsymbol{y}^i = \boldsymbol{z}^i$, $1 \leq i \leq N_N$), and the shape and interpolation functions are identical: $N_k(\boldsymbol{\xi}) = M_k(\boldsymbol{\xi})$, $(1 \leq k \leq N_N)$. This kind of interpolation, termed *isoparametrical*, is necessarily conforming for both u and q.

3.3 ASSEMBLY AND SOLUTION OF THE DISCRETE PROBLEM

Discretized boundary integral equation

The *collocation* method consists of enforcing exactly the integral equation (3.1) at finitely many *collocation points* $\boldsymbol{x}^c$. Once the mesh is generated, the discretization is performed by inserting the elementwise representations (3.3) for the geometry and (3.12) for the boundary unknowns in:

$$\sum_{e=1}^{N_E} \int_{E_e} \{[\tilde{u}(\boldsymbol{y}) - \tilde{u}(\boldsymbol{x}^c)]H(\boldsymbol{x}^c, \boldsymbol{y}) - \tilde{q}(\boldsymbol{y})G(\boldsymbol{x}^c, \boldsymbol{y})\}\, \mathrm{d}S_y = 0 \qquad (1 \le c \le N_c)$$

The above equation involves element integrals of both singular ($\boldsymbol{x}^c \in E_e$) and nonsingular ($\boldsymbol{x}^c \notin E_e$) nature. For a given point $\boldsymbol{x} \in \partial\Omega$ introduce the index subset $I(\boldsymbol{x})$ such that an integral over element E_e is singular for $e \in I(\boldsymbol{x})$ and nonsingular for $e \in \bar{I}(\boldsymbol{x})$:

$$I(\boldsymbol{x}) = \{e \in [1, N_E], \boldsymbol{x} \in E_e] \qquad \bar{I}(\boldsymbol{x}) = [1, N_E] - I(\boldsymbol{x})$$

Then, the geometrical support of singular element integrals for a given collocation point $\boldsymbol{x}$ is the set $\cup_{e \in I(\boldsymbol{x})} E_e$ of elements.

The discretized boundary integral equation finally has the following form, where the distinction between singular and nonsingular integrations is made:

$$\sum_{e \in \bar{I}(\boldsymbol{x})} \sum_{p=1}^{N_I(e)} \left\{ AR(e,p)u^{m(e,p)} - B(e,p)q^{m(e,p)} \right\} - u(\boldsymbol{x}^c) \sum_{e \in \bar{I}(\boldsymbol{x})} \widehat{AR}(e)$$
$$+ \sum_{e \in I(\boldsymbol{x})} \sum_{p=1}^{N_I(e)} \left\{ AS(e,p)u^{m(e,p)} - B(e,p)q^{m(e,p)} \right\} = 0 \quad (3.13)$$

the element integrals are given by:

$$AS(e,p) = \int_{E_e} [M_p(\boldsymbol{\xi}) - M_p(\boldsymbol{\eta}_e)]H(\boldsymbol{x}^c, \boldsymbol{y})\, \mathrm{d}S_y$$
$$B(e,p) = \int_{E_e} M_p(\boldsymbol{\xi})G(\boldsymbol{x}^c, \boldsymbol{y})\, \mathrm{d}S_y$$
$$AR(e,p) = \int_{E_e} M_p(\boldsymbol{\xi})H(\boldsymbol{x}^c, \boldsymbol{y})\, \mathrm{d}S_y$$
$$\widehat{AR}(e) = \int_{E_e} H(\boldsymbol{x}^c, \boldsymbol{y})\, \mathrm{d}S_y$$

Their numerical evaluation is addressed later in Sections 3.4 (nonsingular integrations) and 3.5 (singular integrations).

For a given element E_e, $e \in I(\boldsymbol{x}^c)$, the antecedent of $\boldsymbol{x}^c$ under the mapping (3.3) is denoted $\boldsymbol{\eta}_e \in \Delta_e$, or simply $\boldsymbol{\eta}$ when no ambiguity arises about the particular boundary element under consideration. In the above expression for the element integral AS, use has been made of eq. (3.12) at $\boldsymbol{y} = \boldsymbol{x}^c$, i.e.:

$$\forall e \in I(\boldsymbol{x}^c) \qquad u(\boldsymbol{x}^c) = \sum_{p=1}^{N_I(e)} M_p(\boldsymbol{\eta}_e) u^{m(e,p)}$$

(where M_p are the interpolation functions associated with element E_e), which in turn implies:

$$\forall e \in I(\boldsymbol{x}) \qquad u(\boldsymbol{y}) - u(\boldsymbol{x}^c) = \sum_{p=1}^{N_I(e)} [M_p(\boldsymbol{\xi}) - M_p(\boldsymbol{\eta}_e)] u^{m(e,p)} \tag{3.14}$$

Assembly of the linear system of equations

The set of N_c scalar equations (3.13) have the following matrix structure:

$$[\boldsymbol{A}]\{\boldsymbol{u}\} + [\boldsymbol{B}]\{\boldsymbol{q}\} = [\boldsymbol{0}] \tag{3.15}$$
$$\{\boldsymbol{u}\} = \{u^m\}_{1 \le m \le N_I} \qquad \{\boldsymbol{q}\} = \{q^m\}_{1 \le m \le N_I}$$

where $[\boldsymbol{A}]$ and $[\boldsymbol{B}]$ are full $N_c \times N_I$ matrices, with no symmetry. This (lack of) property, strongly at variance with finite element methods, stems from the presence of fundamental solutions $G(\boldsymbol{x}, \boldsymbol{y})$, $H(\boldsymbol{x}, \boldsymbol{y})$ in the integral equation: any two portions of the boundary have influence over each other, regardless of the particular interpolation used for the variables (u, q).

Equation (3.15) is linear with respect to the $2N_I$ nodal values (u^m, q^m). Besides, as a consequence of well-posed boundary conditions, the nodal values (u^m, q^m) at a given node $\boldsymbol{z}^m$ are not independent, so that only N_I independent unknowns (i.e. one per node) remain. This means that a minimum of N_I independent equations are needed in (3.13). The simplest, and most commonly used, approach consists in collocating the boundary integral equation at all interpolation nodes: $\boldsymbol{x}^c = \boldsymbol{z}^c$,$(1 \le c \le N_I)$. This generates $N_c = N_I$ equations, i.e as many equations as unknowns.

The matrices $[\boldsymbol{A}]$, $[\boldsymbol{B}]$ are set up through the following *assembly* procedure:

Loop over the N_c collocation points $\boldsymbol{x}^c$:
  Loop over the N_E elements: $1 \le e \le N_E$:
    TEST: $e \in I(\boldsymbol{x}^c)$?
      If YES: singular integrations over E_e, then affectation:
        $A(c, m(e,p)) \leftarrow A(c, m(e,p)) + AS(e,p)$
        $B(c, m(e,p)) \leftarrow B(c, m(e,p)) + B(e,p)$
      If NO: regular integrations over E_e, then affectation:
        $A(c, m(e,p)) \leftarrow A(c, m(e,p)) + AR(e,p)$
        $B(c, m(e,p)) \leftarrow B(c, m(e,p)) + B(e,p)$
        $A(c,c) \leftarrow A(c,c) + \widehat{AR}(e)$
  End of loop over elements
End of loop over collocation points

The total number of element integrals to be evaluated is proportional to $N_E \times N_c$, N_c itself being proportional to N_e.

Incorporation of boundary data. The simplest method consists of substituting the nodal values of u or q that are prescribed by the boundary conditions into $\{\boldsymbol{u}\}, \{\boldsymbol{t}\}$ in eq. (3.15). The columns of this matrix equation are then reordered: those associated to unknown nodal values are kept in the left-hand side, those associated to prescribed nodal values are added (with a minus sign) to the right-hand side. This manipulation leads to an $N_I \times N_I$ square linear system of equations of the form:

$$[\boldsymbol{K}]\{\boldsymbol{v}\} = \{\boldsymbol{f}\} \tag{3.16}$$

where the vector $\{\boldsymbol{v}\}$ gathers the unknown components of $\{\boldsymbol{u}\}, \{\boldsymbol{t}\}$, the matrix $[\boldsymbol{K}]$ contains the columns of $[\boldsymbol{A}], [\boldsymbol{B}]$ associated with those unknown components, the right-hand side $\{\boldsymbol{f}\}$ is equal to minus the sum of the columns of $[\boldsymbol{A}], [\boldsymbol{B}]$ associated with known components of $\{\boldsymbol{u}\}, \{\boldsymbol{t}\}$ multiplied by the value of the relevant component. For example, consider a mixed boundary-value problem ($q = q^D$ over S_q, $u = u^D$ over $S_u = \partial\Omega - S_q$). Assuming a suitable numbering of the nodes, eq. (3.15) becomes:

$$[\boldsymbol{A}_1, \boldsymbol{A}_2]\{\boldsymbol{u}', \boldsymbol{u}^D\}^T + [\boldsymbol{B}_1, \boldsymbol{B}_2]\{\boldsymbol{q}^D, \boldsymbol{q}'\}^T = \boldsymbol{0}$$

and the final system of equations, with unknowns $\{\boldsymbol{u}'\}, \{\boldsymbol{q}'\}$, is:

$$[\boldsymbol{A}_1, \boldsymbol{B}_2]\{\boldsymbol{u}', \boldsymbol{q}'\}^T = -[\boldsymbol{A}_2, \boldsymbol{B}_1]\{\boldsymbol{u}^D, \boldsymbol{q}^D\}^T$$

i.e. equation (3.16) with

$$[\boldsymbol{K}] = [\boldsymbol{A}_1, \boldsymbol{B}_2] \qquad \{\boldsymbol{v}\} = \{\boldsymbol{u}', \boldsymbol{q}'\}^T \qquad \{\boldsymbol{f}\} = -[\boldsymbol{A}_2, \boldsymbol{B}_1]\{\boldsymbol{u}^D, \boldsymbol{q}^D\}^T$$

Properties and solution of the linear system of equations

The matrix $[\boldsymbol{K}]$ in eq. (3.16) is *fully populated* and *nonsymmetric*. This is in sharp contrast with finite element methods, for which the stiffness matrix is symmetric, non-negative and banded (Zienkiewicz [17]).

The system (3.16) is then usually solved numerically using a Gauss elimination, whereby the matrix $[\boldsymbol{K}]$ is transformed into an upper triangular matrix and the solution $\{\boldsymbol{v}\}$ is then obtained through backsubstitution. Iterative methods have also been proposed (Mullen and Rencis [12], Kane et al. [9]) but their use is not widespread. The Gauss elimination technique is well-known, and the reader is referred to Ciarlet [3] or Golub and Van Loan, among others; subroutines are provided in many public-domain software libraries like LINPACK [5] or its successor LAPACK [1].

Comments. The boundary data u^D, q^D can be substituted directly into the element integrals in the discretized boundary integral equation (3.13) in the form of a mathematical function of the integration point $\boldsymbol{y}$, if known. For the case of a nonzero prescribed potential, factoring out a positive power of $|\boldsymbol{y}-\boldsymbol{x}|$ in the mathematical expression of $u^D(\boldsymbol{y})-u^D(\boldsymbol{x})$ is recommended for a proper evaluation of the relevant singular element integrals.

Besides, the explicit setting up of matrices $[\boldsymbol{A}],[\boldsymbol{B}]$ is not mandatory. The final matrix $[\boldsymbol{K}]$ and right-hand side $\{\boldsymbol{f}\}$ can be built directly, using a modification of the assembly procedure in which the results of element integrations are directly affected to $[\boldsymbol{K}],\{\boldsymbol{f}\}$:

TEST: $e \in I(\boldsymbol{x}^c)$?
  If YES: singular integrations over E_e, then affectation:
    If $u^{m(e,q)}$ is known and $q^{m(e,q)}$ is unknown, then:
      $K(n,j(e,q)) \leftarrow K(n,j(e,q)) + B(e,q)$
      $f(n) \leftarrow f(n) + AS(e,q)$
    Else
      $K(n,j(e,q)) \leftarrow K(n,j(e,q)) + AS(e,q)$
      $f(n) \leftarrow f(n) + B(e,q)$
  If NO: regular integrations over E_e, then affectation:
      $(\ldots)$

where $i(e,q)$ and $j(e,q)$ are the sequential numberings of the unknown components of $u^{m(e,q)}$ and $q^{m(e,q)}$, respectively.

Post-processing: evaluation of field variables at interior points

Once the system (3.16) is solved, the nodal values $\{u^m\},\{t^m\}$ $(1 \le m \le N_I)$ are known. Numerical evaluation of the integral representation formulas (2.18), (2.19) is possible and gives approximate values for $u(\boldsymbol{x}), \boldsymbol{\nabla}\boldsymbol{u}(\boldsymbol{x})$ at any given observation point $\boldsymbol{x}$ interior to Ω:

$$u(\boldsymbol{x}) = \sum_{e=1}^{N_E}\sum_{p=1}^{N_I(e)}\left\{q^{m(e,p)}\int_{E_e} M_p(\boldsymbol{\xi})G(\boldsymbol{x},\boldsymbol{y})J(\boldsymbol{\xi})\,\mathrm{d}\boldsymbol{\xi} - u^{m(e,p)}\int_{E_e} M_p(\boldsymbol{\xi})H(\boldsymbol{x},\boldsymbol{y})J(\boldsymbol{\xi})\,\mathrm{d}\boldsymbol{\xi}\right\} \tag{3.17}$$

$$u_{,i}(\boldsymbol{x}) = \sum_{e=1}^{N_E}\sum_{p=1}^{N_I(e)}\left\{q^{m(e,p)}\int_{E_e} M_p(\boldsymbol{\xi})G_{,\bar{\imath}}(\boldsymbol{x},\boldsymbol{y})J(\boldsymbol{\xi})\,\mathrm{d}\boldsymbol{\xi} - u^{m(e,p)}\int_{E_e} M_p(\boldsymbol{\xi})G_{\bar{\imath}j}(\boldsymbol{x},\boldsymbol{y})n_j(\boldsymbol{y})J(\boldsymbol{\xi})\,\mathrm{d}\boldsymbol{\xi}\right\} \tag{3.18}$$

The element integrals in these formulas depend on $\boldsymbol{x}$ through the fundamental solutions. They should therefore be recomputed for each new observation point and cannot be reused. The numerical evaluation of field variables at many observation points may thus represent a fair amount of computer time.

Besides, the presence of $O(|\boldsymbol{y}-\boldsymbol{x}|^{-1})$ and $O(|\boldsymbol{y}-\boldsymbol{x}|^{-2})$ terms in (3.17) and $O(|\boldsymbol{y}-\boldsymbol{x}|^{-2})$, $O(|\boldsymbol{y}-\boldsymbol{x}|^{-3})$ terms in (3.18) induce some numerical difficulties when observation points $\boldsymbol{x}$ close to the boundary $\partial\Omega$ are considered. Although the element integrals are not singular in this case, their numerical evaluation becomes inaccurate for $\boldsymbol{x}$ too close to $\partial\Omega$. This problem, known as *near-singular integration*, will be considered in the next section.

When using free-space fundamental solutions, it is preferable to use the alternative formulation (2.20) of eq. (2.19) where an integration by parts has been performed, so that only $O(1/r)$ and $O(1/r^2)$ terms appear in it:

$$u_{,i}(\boldsymbol{x}) = \sum_{e=1}^{N_E} \sum_{p=1}^{N_I(e)} \left\{ u^{m(e,p)} \int_{E_e} D_{ij} M_p(\boldsymbol{\xi}) G_{,j}(\boldsymbol{x}, \boldsymbol{y}) \, \mathrm{d}S_y \right. \\ \left. - q^{m(e,p)} \int_{E_e} M_p(\boldsymbol{\xi}) G_{,i}(\boldsymbol{x}, \boldsymbol{y}) \, \mathrm{d}S_y \right\} \quad (3.19)$$

3.4 EVALUATION OF NONSINGULAR INTEGRALS

The numerical evaluation of the nonsingular element integrals that appear in the discretized boundary integral equation (3.13) and integral representation formulas (3.17), (3.18) is usually based, like in finite element methods, on Gaussian quadrature formulas.

Principle of Gaussian numerical quadrature

Only a brief account of this very classical subject is given here; the reader is referred to textbooks on numerical integration, e.g. Davis and Rabinowitz [4] or Stroud and Secrest [15].

Line integrals. The approximate value of an integral over the line segment $[-1, 1]$ is given by the formula:

$$\int_{-1}^{1} f(x) \, \mathrm{d}x \approx \sum_{i=1}^{N} w_i f(x_i)$$

where (x_i) and (w_i) respectively denote the abscissae and weights of the *Gauss points*. They are uniquely determined for any given value of N (the x_i are the zeros of the Nth-order Legendre polynomial) and tabulated in the literature. The (x_i, w_i) are listed in Appendix B.3 for $N \leq 7$. The Gaussian rules have the following characteristic property: the N-point formula is exact for all polynomials of degree $\leq 2N - 1$:

$$\forall k, 0 \leq k \leq 2N - 1, \qquad \int_{-1}^{1} x^k \, \mathrm{d}x = \sum_{i=1}^{N} w_i x_i^k \quad (3.20)$$

In this respect, Gaussian quadrature formulas are optimal among quadrature formulas using N points and weights.

The Gauss points are interior to the segment: $\forall(i,N), x_i \in]-1,1[$, their distribution is symmetric with respect to the segment midpoint: $\forall(i,N), x_i = x_{N-i}, w_i = w_{N-i}$; besides, the weights are strictly positive: $\forall(i,N), w_i > 0$.

The numerical evaluation of an integral over $[a,b]$ requires a change of variable:

$$\int_a^b f(x)\,\mathrm{d}x = \frac{b-a}{2}\int_{-1}^{1} f(\frac{b+a}{2} + \frac{b-a}{2}\xi)\,\mathrm{d}\xi$$

Modified Gaussian quadrature formulas are available for integrals involving a *weight* $g(x) \geq 0$ integrable over $[-1,1]$:

$$\int_a^b f(x)g(x)\,\mathrm{d}x \approx \sum_{i=1}^{N} \hat{w}_i f(\hat{x}_i)$$

They are useful for some types of weakly singular integrals, the singularity being incorporated in the weight function $g(x)$. Table B.1 of Appendix B provides the abscissae and weights for one-dimensional integration with logarithmic singularities:

$$[a,b] = [0,1] \qquad g(x) = \ln\frac{1}{x}$$

Two-dimensional integration over the unit square $[-1,1]^2$. It is usually performed by means of the 'product formula':

$$\int_{-1}^{1}\int_{-1}^{1} f(x,y)\,\mathrm{d}x\,\mathrm{d}y \approx \sum_{i=1}^{N}\sum_{j=1}^{N} w_i w_j f(x_i, x_j) \tag{3.21}$$

where the (x_i, w_i) are the points and weights for the line segment $[-1,1]$ used in (3.20).

Two-dimensional integration over the triangle $\mathcal{T}$: $\xi_1 \geq 0, \xi_2 \geq 0, 1-\xi_1-\xi_2 \geq 0$. The change of variables $\xi_1 = 0.25(1+u)(1-v)$, $\xi_2 = 0.5(1+v)$ maps the triangle $\mathcal{T}$ onto the unit square $[-1,1]^2$. The product formula (3.21) is then applicable:

$$\begin{aligned}\int_{\mathcal{T}} f(x,y)\,\mathrm{d}x\,\mathrm{d}y &= \int_{-1}^{1}\int_{-1}^{1} g(u,v)\frac{1}{8}(1-v)\mathrm{d}u\mathrm{d}v \\ &= \frac{1}{8}\sum_{i=1}^{N}\sum_{j=1}^{N} w_i w_j g(x_i, x_j)(1-y_j)\end{aligned}$$

However, this approach breaks the ternary symmetry of the triangle by concentrating integration points near the vertex $x=0, y=1$, which is thereby privileged. Because of this, other Gaussian quadrature formulas, using points

(x_i^T, y_i^T) and weights w_i^T, have been specifically designed for the triangle (Lyness and Jespersen [10]); they respect the ternary symmetry. Numerical evaluations of integrals are then made according to:

$$\int_T f(x, y) \, \mathrm{d}x \, \mathrm{d}y \approx \sum_{i=1}^{N} w_i^T f(x_i^T, y_i^T)$$

The values of (x_i^T, y_i^T, w_i^T) are listed in Appendix B.3 for several values of N.

Choice of the number of Gauss points

The minimum number N of Gauss points needed for a nonsingular element integral depends on whether the boundary is a curve or a surface (i.e. on the dimensionality of the problem at hand).

It also depends, like in finite element methods, on the chosen representation for the geometry (variations of the Jacobian $J(\boldsymbol{\xi})$) and the unknowns (order of the interpolation functions M_k used). In the usual situations (usual polynomial interpolation functions,[1] elements not severely distorted), a moderate number of Gauss points (2×2 or 3×3 over a square) is sufficient.

It finally depends on the relative positions of the observation point $\boldsymbol{x}$ and the element E; in contrast to the previous ones, this consideration is specific to boundary element methods. Indeed, the presence of r^{-1}, r^{-2}, r^{-3} terms ($r = |\boldsymbol{y} - \boldsymbol{x}|$) in the fundamental solutions $G(\boldsymbol{x}, \boldsymbol{y}), H(\boldsymbol{x}, \boldsymbol{y})$ and their derivatives entails severe variations of the integrand over E if $\boldsymbol{x}$ is close to E. On the other hand, when $\boldsymbol{x}$ is far away from E, those terms are roughly constant over E; in this case, the order of the Gauss quadrature formula can be adjusted solely on the basis of the type of interpolation functions and expected variations of the Jacobian. The latter adjustment becomes insufficient when $\boldsymbol{x}$ is close to E. Needless to say, the 'closeness' of $\boldsymbol{x}$ to E should be understood in relative terms, i.e. based on the comparison of a characteristic distance $d(\boldsymbol{x}, E)$ between $\boldsymbol{x}$ and E to a typical length D on E.

It is also not advisable, albeit correct, to compute *all* nonsingular element integrals using N_{max} Gauss points, where N_{max} is adjusted so as to be satisfactory for the closest $\boldsymbol{x}$-E configuration, because this would entail unnecessarily high computation costs. The finer the mesh, the smaller the proportion of pairs $(\boldsymbol{x}, E)$ close to each other, thus the greater the proportion of element integrals for which N_{max} grossly overevaluates N. It is thus better to adjust N to the relative positions of $\boldsymbol{x}$ and E, in order to maintain a satisfactory accuracy in numerical quadrature at the lowest computational cost.

Several approaches have been proposed to address this difficult issue. For example, a theorical upper bound for the quadrature error, which involves N-th order derivatives of the integrand, is used to monitor the adjustment in

[1] In particular those listed in Appendix B.2.

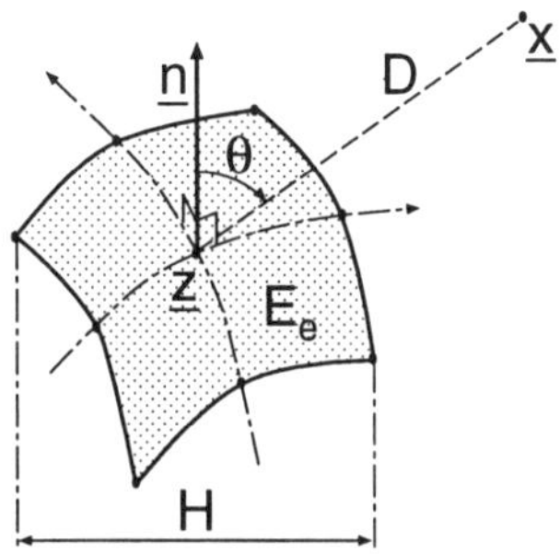

Figure 3.5 Severity indicator: geometrical notations.

[2]. In another approach, Rezayat et al. [14] have undertaken an 'empirical' numerical study in a rather systematic way, on many test cases and using Gaussian formulas of all practically meaningful orders; then a rather simple formula has been fitted with the data thus gathered. As a result, when triangular six-noded or quadrilateral eight-noded boundary elements are used, an 'static severity' indicator SSI is defined by:

$$\text{SSI} = (2.37 + 0.424\cos\theta)D/d \tag{3.22}$$

with D: a characteristic element size (longest diagonal for a quadrilateral, longest side for a triangle), $d = |\boldsymbol{z} - \boldsymbol{x}|$: distance between source point $\boldsymbol{x}$ and element centre $\boldsymbol{z}$) and θ: angle between vectors $\boldsymbol{z} - \boldsymbol{x}$ and $\boldsymbol{n}(\boldsymbol{z})$ chosen such that $\cos\theta > 0$ (Figure 3.5). Recommended values for N depend on SSI according to Table 3.1. This approach is very simple to implement and fairly general.

Table 3.1 Correspondence between static severity indicator SSI and number of Gauss points (from [14]).

SSI	Number of Gauss points per element quadrilatcral ($N_e = 8$)	triangular ($N_e = 6$)
1	2×2	6
2	3×3	7
3	4×4	16
4	5×5	19
5	6×6	28
6	$4 \times (4 \times 4)$	$4 \times 16 = 64$
7	$4 \times (5 \times 5)$	$4 \times 19 = 76$
8	$4 \times (6 \times 6)$	$4 \times 28 = 104$

Table 3.2 Near-singular integration: exact and approximate values (using a 7-point Gauss rule) of I_1, I_2, I_3.

	exact	numerical	$\|\Delta I/I\|$
$I_1/(2\pi h)$	19.02498	19.01439	5.6(-4)
I_2/π	6.00003	5.99396	1.0(-4)
$I_3 h/2\pi$	0.98275	0.95006	3.4(-2)

Near-singular integrals

The accurate evaluation of nonsingular element integrals for $\boldsymbol{x}$ very close to $E(d(\boldsymbol{x}, E) \ll D)$ is difficult due to the high variability of the kernel functions about the orthogonal projection $\boldsymbol{z}$ of $\boldsymbol{x}$ onto E, even if many Gauss points are used; this situation is usually termed *near-singular integration*. As an illustration of this problem, consider the following integrals, similar to those appearing in integral representation formulas like (3.17), (3.18):

$$I_m = \int_0^{2\pi} \mathrm{d}\theta \int_0^R \frac{r\mathrm{d}r}{(h^2 + r^2)^{m/2}} \qquad (m = 1, 2, 3)$$

$h > 0$ symbolizes the distance of $\boldsymbol{x}$ to (its orthogonal projection onto) E; $h/R \ll 1$ is the near-singular situation. The case $m = 3$ corresponds to the strongest near-singularity usually met. Table 3.2 displays the exact and approximate values of the I_m for $R = 20h$; the latter are obtained using a 7-point Gauss quadrature rule, which is thus exact for polynomials of degree ≤ 13.

In particular, the observed 3.4% relative error on I_3 greatly exceeds the usual accuracy requirements for element integrations. Remedies aimed at improving the accuracy of near-singular integrals evaluation have thus been proposed:

Telles transformation. Telles [16] introduces a transformation of the real interval $[-1, 1]$ onto itself which aims at concentrating the Gauss points near $\boldsymbol{x}_0$, the orthogonal projection of $\boldsymbol{x}$ on the element. This transformation is applied to near-singular integration on curves arising in two-dimensional boundary integral formulations or on quadrilateral elements (using a product-type quadrature rule).

For integrals over the line segment $\{-1 \leq \xi_1 \leq 1, \xi_2 = 0)\}$ of the (ξ_1, ξ_2)-plane, let $\boldsymbol{x} = (x_0, d)$, so that the orthogonal projection of $\boldsymbol{x}$ onto the segment is $\boldsymbol{x}_0 = (x_0, 0)$. Then, the integrals

$$\int_{-1}^{1} f(\xi) \frac{\mathrm{d}\xi}{[(\xi - \eta)^2 + d^2]^{n/2}} \qquad (n = 1, 2)$$

(putting $\xi_1 = \xi$ for brevity) are transformed using the change of variable $\xi \in [-1,1] \to \gamma \in [-1,1]$ defined by:

$$\xi \in [-1,1] = \frac{1-J}{1+3R^2}\left\{\gamma^3 - 3R\gamma^2 + \frac{J+3R^2}{1-J}\gamma + 3R\right\} \qquad (\gamma \in [-1,1])$$

where the parameter R is given in terms of x_0 and $J = \mathrm{d}\xi/\mathrm{d}\gamma\,|_{\xi=x_0}$ by:

$$\begin{aligned} R &= \left[-q + \sqrt{q^2+p^3}\right]^{1/3} + \left[-q - \sqrt{q^2+p^3}\right]^{1/3} + \frac{x_0}{1+2J} \\ q &= \frac{1}{2(1+2J)}\left[\frac{1}{2(1+2J)}\left(x_0(3-2J) - \frac{2x_0^3}{1+2J}\right) - x_0\right] \\ p &= \frac{1}{3(1+2J)^2}\left[4J(1-J) + 3(1-x_0^2)\right] \end{aligned}$$

and the value of $J = \mathrm{d}\xi/\mathrm{d}\gamma\,|_{\xi=x_0}$ is evaluated according to the following rule:

$$\begin{aligned} &J = 0.85 + 0.24\ln d && 0.05 \le d \le 1.3 \\ &J = 0.893 + 0.0832\ln d && 1.3 \le d \le 3.618 \\ &J = 1 && 3.618 \le d \end{aligned}$$

Projection and Angular and Radial Transformation (PART) method. This treatment, proposed by Hayami and Matsumoto [7], concerns the evaluation of nearly singular surface integrals of the form:

$$I = \int_S f(\boldsymbol{y}) \frac{\mathrm{d}S_y}{|\boldsymbol{y} - \boldsymbol{x}|^p}$$

(usually $p = 1, 2, 3$ for three-dimensional problems), i.e. when the point $\boldsymbol{x}$ is close to the surface S. It consists of the following steps:

1. Compute the orthogonal projection $\boldsymbol{z}$ of $\boldsymbol{x}$ onto S (using for example a Newton–Raphson technique); the distance $|\boldsymbol{x} - \boldsymbol{z}|$ is denoted d.

2. Project the curved surface S onto the plane tangent to S at $\boldsymbol{z}$; let $\tilde{S}$ denote the projected surface.

3. Introduce polar coordinates (ρ, α) of origin $\boldsymbol{z}$ on $\tilde{S}$. The integral I becomes:

$$I = \int_0^{2\pi} \mathrm{d}\alpha \int_0^{\rho_{max}(\alpha)} f(\rho,\alpha)\frac{1}{r^p} J\rho\,\mathrm{d}\rho$$

4. Apply a radial transformation $R(\rho)$ chosen so as to weaken the near-singularity, which manifests itself chiefly in the ρ-direction.

5. Apply the angular transformation $\alpha \to t$:

$$t(\alpha) = \frac{h_j}{2} \ln \left\{ \frac{1 + \sin(\alpha - \alpha_j)}{1 - \sin(\alpha - \alpha_j)} \right\} \qquad \mathrm{d}\alpha = \frac{\cos(\alpha - \alpha_j)}{h_j} \,\mathrm{d}t = \frac{\mathrm{d}t}{\rho_{max}}$$

6. Evaluate numerically integral I using a Gauss rule in the R, t transformed plane: I has the form

$$I = \int_{t(0)}^{t(2\pi)} \frac{\mathrm{d}t}{\rho_{max}} \int_{R(0)}^{R(\rho_{max}(\alpha))} f(\rho, \alpha) \frac{1}{r^p} J \rho \frac{\mathrm{d}\rho}{\mathrm{d}R} \,\mathrm{d}R$$

Hayami and Matsumoto propose several radial transformations, among which:

- The $\mathrm{Log} - L_2$ transformation: $R(\rho) = \ln \sqrt{\rho^2 + d^2}$
- The $\mathrm{Log} - L_1$ transformation: $R(\rho) = \ln(\rho + d)$

The $\mathrm{Log} - L_1$ transformation is considered superior by Hayami and Matsumoto, who also provides convergence results with respect to the number of Gauss points.

3.5 EVALUATION OF SINGULAR INTEGRALS

The element integrals are singular when $\boldsymbol{x}$ is located on the current element E. Classical boundary integral formulations involve Cauchy principal value integrals; appropriate and general techniques for their numerical evaluation have been formulated only a few years ago, by Guiggiani and Gigante [6]. The approach presented in Chapter 2, based on indirect regularization, leads to integrals that are at most weakly singular. However, for this advantage to be of practical value, an appropriate numerical integration technique must be used since usual Gaussian quadrature rules are applicable only to nonsingular integrals. Let us now describe one such technique, for the two kinds of singular element integrals that appear in the discretized boundary integral equation (3.13):

$$I_1 = \int_{E_e} [N_p(\boldsymbol{\xi}) - N_p(\boldsymbol{\eta})] H(\boldsymbol{x}, \boldsymbol{y}) J(\boldsymbol{\xi}) \,\mathrm{d}\boldsymbol{\xi} \tag{3.23}$$

$$I_2 = \int_{E_e} N_p(\boldsymbol{\xi}) G(\boldsymbol{x}, \boldsymbol{y}) J(\boldsymbol{\xi}) \,\mathrm{d}\boldsymbol{\xi} \tag{3.24}$$

Three-dimensional problems

The integrals are taken over surface elements. The kernel H is singular like $|\boldsymbol{y} - \boldsymbol{x}|^{-2}$. The singular integration technique is based on the introduction of polar coordinates (ρ, α) of origin $\boldsymbol{\eta}$ in the parent element Δ_e (Figure 3.6):

$$(\xi_1, \xi_2) = (\eta_1, \eta_2) + \rho(\cos\alpha, \sin\alpha) \tag{3.25}$$

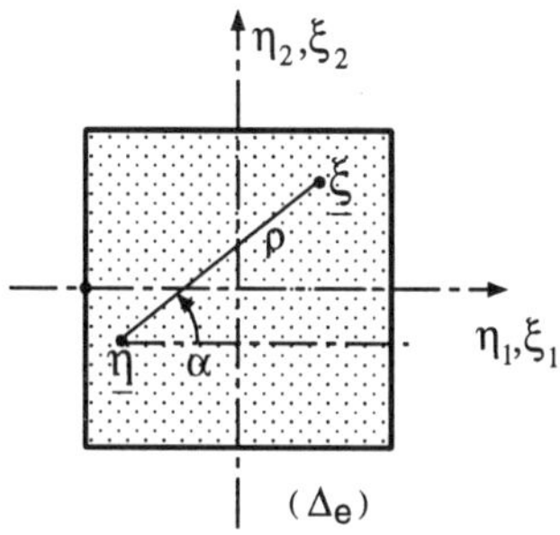

Figure 3.6 Polar coordinates (ρ, α) of origin $\boldsymbol{\eta}$ in the parent element.

Modified shape functions. From the Taylor formula for polynomials in two variables, there exist polynomials $N^I_{p;\alpha}(\boldsymbol{\xi};\boldsymbol{\eta})$ such that:

$$N_p(\boldsymbol{\xi}) - N_p(\boldsymbol{\eta}) = (\xi_\alpha - \eta_\alpha) N^I_{p;\alpha}(\boldsymbol{\xi};\boldsymbol{\eta}) \tag{3.26}$$

Then, substitution of the polar coordinate transformation (3.25) gives:

$$\begin{aligned} N_p(\boldsymbol{\xi}) - N_p(\boldsymbol{\eta}) &= \rho\left[\cos\alpha N^I_{p;1}(\boldsymbol{\xi};\boldsymbol{\eta}) + \sin\alpha N^I_{p;2}(\boldsymbol{\xi};\boldsymbol{\eta})\right] \\ &= \rho \hat{N}_p(\rho, \alpha; \boldsymbol{\eta}) \end{aligned} \tag{3.27}$$

defining the modified shape functions $\hat{N}_p(\rho, \alpha; \boldsymbol{\eta})$. One notices that $\hat{N}_p$ is multiplied by a ρ factor in the process; this is precisely the desired effect of this manipulation. For example, for the simplest case of linear shape functions (3.9) associated with triangular elements, it is easy to show that the corresponding modified shape functions $\hat{N}$ are:

$$\hat{N}_1 = -(\cos\alpha + \sin\alpha) \qquad \hat{N}_2 = \cos\alpha \qquad \hat{N}_3 = \sin\alpha$$

The modified shape functions associated with usual shape functions are listed in Appendix B.2.

Expression of the singular integrals in the (ρ, α) system. Applying the above definitions to the mapping (3.3), it is easy to formulate the distance $|\boldsymbol{y} - \boldsymbol{x}|$ using the modified shape functions:

$$\boldsymbol{y} - \boldsymbol{x} = \sum_{q=1}^{N_e} [N_q(\boldsymbol{\xi}) - N_q(\boldsymbol{\eta})]\boldsymbol{y}^q = \sum_{q=1}^{N_e} \rho \hat{N}_q \boldsymbol{y}^q$$

so that ρ appears as a multiplicative factor in $(\boldsymbol{y} - \boldsymbol{x})$. Using the expansion (3.26) one order further, one has:

$$N^I_{q;\alpha}(\boldsymbol{\xi};\boldsymbol{\eta}) = N_{q,\alpha}(\boldsymbol{\eta}) + \frac{1}{2}(\xi_\beta - \eta_\beta) N^{II}_{q;\alpha\beta}(\boldsymbol{\xi};\boldsymbol{\eta})$$

which in turn implies:

$$\sum_{q=1}^{N_e} \hat{N}_q \boldsymbol{y}^q = \boldsymbol{a}_1(\boldsymbol{\eta})\cos\alpha + \boldsymbol{a}_2(\boldsymbol{\eta})\sin\alpha + \rho O(1) \tag{3.28}$$

On the other hand, the natural basis vectors $\boldsymbol{a}_1, \boldsymbol{a}_2$ must be linearly independent, to avoid element degeneracy. This implies that $\boldsymbol{a}_1(\boldsymbol{\eta})\cos\alpha + \boldsymbol{a}_2(\boldsymbol{\eta})\sin\alpha \neq \boldsymbol{0}$ holds for any value of α. Thus, we have proved that $r = |\boldsymbol{y} - \boldsymbol{x}|$ assumes the form:

$$r = \rho\hat{r}(\rho, \alpha; \boldsymbol{\eta}) = \rho \left| \sum_{q=1}^{N_e} \hat{N}_q \boldsymbol{y}^q \right| \tag{3.29}$$

where the factor $\hat{r}$ is nonzero for $\rho = 0$. The fundamental solutions G, H can then be recast as:

$$H(\boldsymbol{x}, \boldsymbol{y}) = \frac{1}{\rho^2}\hat{H}(\rho, \alpha; \boldsymbol{\eta}) \qquad G(\boldsymbol{x}, \boldsymbol{y}) = \frac{1}{\rho}\hat{G}(\rho, \alpha; \boldsymbol{\eta}) \tag{3.30}$$

where $\hat{G}, \hat{H}$ are nonsingular at $\rho = 0$. Finally, the surface differential area element becomes:

$$\mathrm{d}\boldsymbol{\xi} = \rho\, \mathrm{d}\rho \mathrm{d}\alpha \tag{3.31}$$

As a result of equations (3.27), (3.29), (3.30) and (3.31), the singular integrals (3.23), (3.24) are transformed into nonsingular expressions in the system ρ, α:

$$I_1 = \int_{E_e} \hat{M}_p(\rho, \alpha; \boldsymbol{\eta})\hat{H}(\rho, \alpha; \boldsymbol{\eta})J(\boldsymbol{\xi})\, \mathrm{d}\rho \mathrm{d}\alpha \tag{3.32}$$

$$I_2 = \int_{E_e} M_p(\boldsymbol{\xi})\hat{G}(\rho, \alpha; \boldsymbol{\eta})J(\boldsymbol{\xi})\, \mathrm{d}\rho \mathrm{d}\alpha \tag{3.33}$$

Thus, the regularizing effect of the term $[N_p(\boldsymbol{\xi}) - N_p(\boldsymbol{\eta})]$ in (3.23) is materialized by the factoring out of ρ associated with the introduction of the modified shape functions (3.27).

Evaluation of I_1, I_2 using Gauss points. In order to effectively evaluate of integrals (3.32), (3.33) above, a last change of variables is needed. Indeed, the parent element Δ_e must be subdivided into triangles (four , respectively three, triangles for a square, resp. triangular, parent element) when described using polar coordinates (ρ, α), see Figure 3.7. Each triangle can be characterized by a set of four parameters $\rho_0, \alpha_1, \alpha_2, \alpha_m$ and is defined as the set of points $\boldsymbol{\xi}(\rho, \alpha)$ such that:

$$\alpha_1 \leq \alpha \leq \alpha_2, 0 \leq \rho \leq \rho_0 / \cos(\alpha - \alpha_m)$$

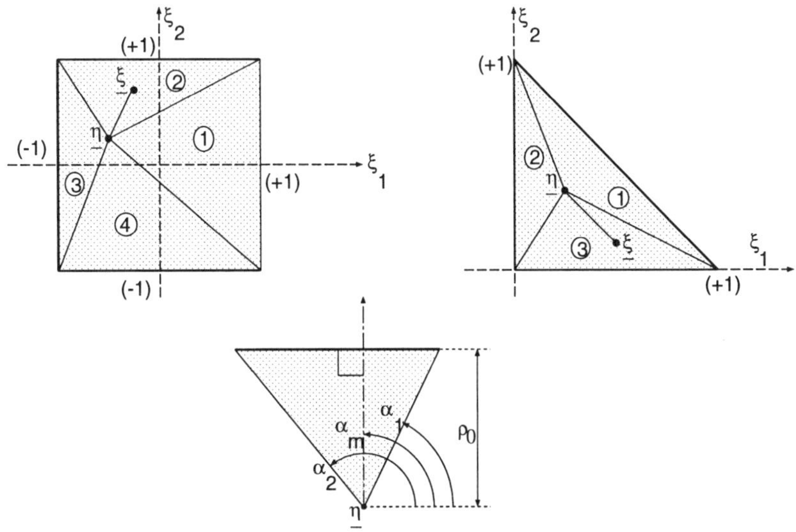

Figure 3.7 Subdivision of the parent element into triangles.

In the general case where $\boldsymbol{\eta}$ is interior to Δ_e, one has $\alpha_m = 0, \pi/2, \pi, 3\pi/2$ (square parent element) or $\alpha_m = \pi/4, \pi, 3\pi/2$ (triangular parent element). For each triangle used to subdivide Δ_e (Figure 3.7), the transformation $(v_1, v_2) \to (\rho, \alpha)$ defined by:

$$\alpha = \frac{\alpha 2 - \alpha 1}{2} v_2 + \frac{\alpha 2 + \alpha 1}{2} \qquad \rho = \frac{1}{2} \frac{\rho_0}{\cos(\alpha - \alpha_m)} (1 + v_1)$$

maps the triangle onto the square $-1 \leq v_1, v_2 \leq 1$. The differential area element becomes:

$$\mathrm{d}\rho \mathrm{d}\alpha = \frac{1}{4}(\alpha_2 - \alpha_1) \frac{\rho_0}{\cos(\alpha - \alpha_m)} dv_1 dv_2$$

The effect of this last transformation is to enable one to compute the singular integrals I_1, I_2 using a Gaussian quadrature rule.

Two-dimensional problems

The element are now curvilinear segments, so that the integration is one-dimensional. The previous technique is applicable and takes a simpler form; it is invoked only for the integral I_1 containing $H(\boldsymbol{x}, \boldsymbol{y})$, which has a $|\boldsymbol{y} - \boldsymbol{x}|^{-1}$ singularity:

$$I_1 = \int_{E_e} [N_p(\boldsymbol{\xi}) - N_p(\boldsymbol{\eta})] H(\boldsymbol{x}, \boldsymbol{y}) \, \mathrm{d}s_y \tag{3.34}$$

The shape functions M_p are (usually) polynomials in the ξ variable. From the Taylor formula, there exist polynomials $\hat{N}_p(\xi;\eta)$ such that:

$$N_p(\xi) - N_p(\eta) = (\xi - \eta)\hat{N}_p(\xi;\eta)$$

Then one has:

$$\boldsymbol{r} = (\boldsymbol{y} - \boldsymbol{x}) = (\xi - \eta)\hat{\boldsymbol{r}} \qquad \text{with} \quad \hat{\boldsymbol{r}} = \sum_{q=1}^{N_e} \hat{M}_q(\xi;\eta)\boldsymbol{y}^q$$

Upon putting:[2]

$$H(\boldsymbol{x},\boldsymbol{y}) = \frac{1}{\xi - \eta}\hat{H}(\xi;\eta)$$

one rewrites the singular integral (3.34) in a completely regular form that can be dealt with using standard Gaussian quadrature rules:

$$I_1 = \int_{-1}^{1} \hat{M}_p(\xi;\eta)\hat{H}(\xi;\eta)J \,\mathrm{d}\xi \tag{3.35}$$

Comments

Let us stress again the fact that the singular integration procedure described above takes full advantage of the regularized form of the boundary integral equation. Thanks to the fact that the final (nonsingular) form can be evaluated using Gaussian quadrature rules, a flexible, robust and accurate integration technique is obtained. This approach works for all usual kinds of boundary elements and interpolation functions; in particular, the treatment applies equally well to curved elements with high-degree interpolations. The same qualities can be achieved within the framework of Cauchy principal value integral formulations, provided the limiting process $\varepsilon \to 0$ is handled carefully; this path is successfully followed by Guiggiani and Gigante [6] (see Section 4.9).

The singular character of the boundary integral equation (3.1) is actually advantageous. Indeed, the singular element integrals in (3.13) contribute coefficients to matrices $[\boldsymbol{A}], [\boldsymbol{B}]$ in eq. (3.15) that are both dominant and concentrated near the diagonal, particularly when the sets of interpolation nodes and collocation points, as well as their numbering, are identical. As a consequence, the final linear system of equations is well-conditioned. Clearly, accuracy in the evaluation of singular integrals is crucial, in that it allows us to take full advantage of the resulting well-conditioned behaviour of the matrix

[2]Note that $\xi - \eta$, not $|\xi - \eta|$, is factored out. For the infinite plane fundamental solution, one has $H = -(\boldsymbol{r}.\boldsymbol{n})/(2\pi r^2)$ and then $\hat{H} = -(\hat{\boldsymbol{r}}.\boldsymbol{n})/(2\pi \hat{r}^2)$.

$[\boldsymbol{K}]$; conversely, inaccuracies would affect most the dominant matrix coefficients and might seriously alter the overall solution procedure. This argument emphasizes the interest of using suitable singular integration techniques.

An alternative transformation of intrinsic coordinates may be used in place of the local polar coordinates (ρ, α). For example, the following mapping is introduced for the triangle 1 (Figure 3.7):

$$\xi_1 - \eta_1 = (1 - \eta_1)\frac{1 + v_1}{2} \qquad \xi_2 - \eta_2 = (v_2 - \eta_2)\frac{1 + v_1}{2} \tag{3.36}$$

where the new variables (v_1, v_2) lie in the unit square: $-1 \leq v_1, v_2 \leq 1$. Thus, two-dimensional Gaussian quadrature can be used at once in the (v_1, v_2)-space. Moreover, it is apparent from this definition that $\boldsymbol{y} - \boldsymbol{x}$ takes the form

$$\boldsymbol{y} - \boldsymbol{x} = \frac{1 + v_1}{2}[\hat{r}(\boldsymbol{v}; \boldsymbol{\eta})]$$

while

$$\mathrm{d}\boldsymbol{\xi} = \frac{1 - \eta_1}{2}\frac{1 + v_1}{2}\,\mathrm{d}v_1\mathrm{d}v_2$$

so that the integrals (3.23), (3.24) are nonsingular in the (v_1, v_2) coordinates.

3.6 EVALUATION OF DOMAIN INTEGRALS

When nonzero distributed sources are present, the boundary integral equation (2.24) contains a domain integral term of the form

$$\int_\Omega b(\boldsymbol{y})G(\boldsymbol{x}, \boldsymbol{y})\,\mathrm{d}V_y \tag{3.37}$$

It is advisable to convert it into boundary integrals whenever possible, in order to perform all computations using only the boundary element mesh. This goal is achievable exactly for particular source distributions b, or at least approximately using the so-called *multiple reciprocity method*, when b is given in the form of a known function. Otherwise, one must resort to the more traditional use of *integration cells* in conjunction with numerical, e.g. Gaussian, quadrature techniques on domain patches.

Multiple reciprocity method

This approach, proposed by Brebbia and Nowak [13], is based on the construction of a sequence $G^{(N)}(\boldsymbol{x}, \boldsymbol{y}), N \geq 0$ of kernel functions such that:

$$G^{(N)}(\boldsymbol{x}, \boldsymbol{y}) = \Delta G^{(N+1)}(\boldsymbol{x}, \boldsymbol{y}) \qquad G^{(0)}(\boldsymbol{x}, \boldsymbol{y}) = G(\boldsymbol{x}, \boldsymbol{y})$$

For the usual free-space solution, i.e. $G = G^{(0)} = 1/(4\pi r)$, one finds:

$$G^{(N)} = \frac{1}{4\pi}\frac{r^{2N-1}}{(2N)!} \qquad (n \geq 0) \tag{3.38}$$

Repeated applications of Green's third identity (2.4) give, for any sufficiently regular function b:

$$\begin{aligned}
\int_\Omega bG \, \mathrm{d}V_y &= \int_\Omega \Delta b G^{(1)} \, \mathrm{d}V_y + \int_{\partial\Omega} \left(G^{(1)}_{,n} b - G^{(1)} b_{,n} \right) \mathrm{d}S_y \\
&= \int_\Omega \Delta\Delta b G^{(2)} \, \mathrm{d}V_y + \int_{\partial\Omega} \left(G^{(1)}_{,n} b - G^{(1)} b_{,n} \right) \mathrm{d}S_y \\
&\quad + \int_{\partial\Omega} \left(G^{(2)}_{,n} \Delta b - G^{(2)} (\Delta b)_{,n} \right) \mathrm{d}S_y \\
&= \ldots \\
&= \int_\Omega \Delta^N b G^{(N)} \, \mathrm{d}V_y + \sum_{p=0}^{N-1} \int_{\partial\Omega} \left(G^{(p+1)}_{,n} \Delta^k b - G^{(p+1)} (\Delta^k b)_{,n} \right) \mathrm{d}S_y
\end{aligned}$$

where the integer N is arbitrary and Δ^p means that the Laplacian Δ is iterated p times. The above formula achieves an approximate conversion of the initial domain integral into boundary integrals provided the residual domain integral

$$\int_\Omega \Delta^N b G^{(N)} \, \mathrm{d}V_y$$

becomes sufficiently small for some adequate value of N.

This method does not, however, solve completely the general issue of domain source integral conversion. The appearance of reciprocals of $(2N+2)!$ in the expression of $G^{(N)}$ suggests that the residual domain integral converges to zero with increasing N, but this is not formally proven. The multiple reciprocity method is well suited to deal with analytically known source distributions b. On the other hand, its application to sources that are known only through numerical values at a finite number of sample points, as in some nonlinear problems, looks illusory: one would have in this case to resort to (iterated) numerical differentiation, which is a mathematically ill-posed procedure, for the evaluation of the iterated Laplacians $\Delta^p b$.

Integration cells

When the source term b is known only numerically,[3] the domain integrals must be computed numerically. This is usually done using a partition of the domain Ω into *integration cells*, i.e. finite-element-type domain patches[4]

[3]For example, resulting from the numerical solution of another problem.
[4]The term 'cell' serves to remind us that those patches do not support unknowns.

that are mapped onto a parent cell (cube, prism, tetrahedron ... for a three-dimensional domain). The sole purpose of those cells is to provide adequate geometrical support for numerical quadrature techniques, usually of Gaussian type, for an arbitrarily shaped domain Ω.

The collocation point $\boldsymbol{x}$ may be located on an integration cell: either the geometrical support of nonzero volume sources is connected with the boundary $\partial\Omega$, or an integral representation formula is considered at an interior point. In most cases the domain integral in the integral equation or representation is weakly singular, so that one can resort to a polar coordinate transformation in the parent cell, similar to the technique developed in Section 3.5 for singular element integrals.

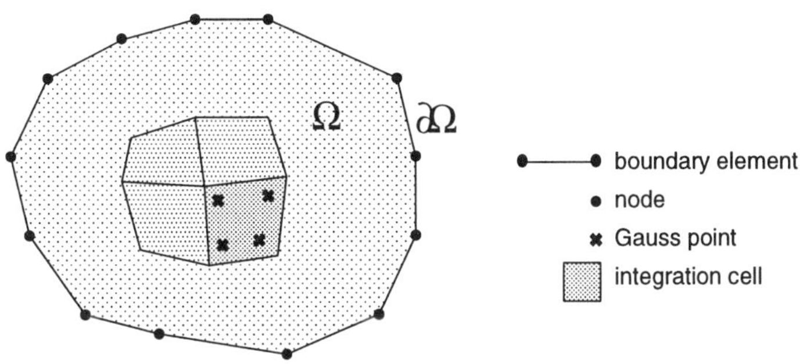

Figure 3.8 Integration cells and boundary elements for a two-dimensional domain.

3.7 USING SUBREGIONS

A partition of the initial domain Ω into disjoint subregions Ω_j, $(1 \leq j \leq N)$ sometimes leads to computational gains in the matrix equation assembly and solution stages. Moreover, its use is almost mandatory when media with piecewise constant physical properties are considered.

The subsequent *multizone analysis* techniques have been investigated in detail by Kane [8]. Let us here develop a simple multizone approach, using $N = 3$ subdomains Ω_1 ,Ω_2, Ω_3, bounded by surfaces (S_1, S_{12}, S_{13}), (S_2, S_{12}, S_{23}) and (S_3, S_{13}, S_{23}) respectively (Figure 3.9). Basically, each subdomain is considered as a separate one, and additional compatibility conditions between the solutions (u^i, q^i) defined on each Ω_i, expressing continuity across the interfaces, must be provided:

$$\begin{aligned} u^1 = u^2 \text{ and } q^1 + q^2 = 0 \qquad & \text{on } S_{12} \\ u^1 = u^3 \text{ and } q^1 + q^3 = 0 \qquad & \text{on } S_{13} \\ u^2 = u^3 \text{ and } q^2 + q^3 = 0 \qquad & \text{on } S_{23} \end{aligned} \tag{3.39}$$

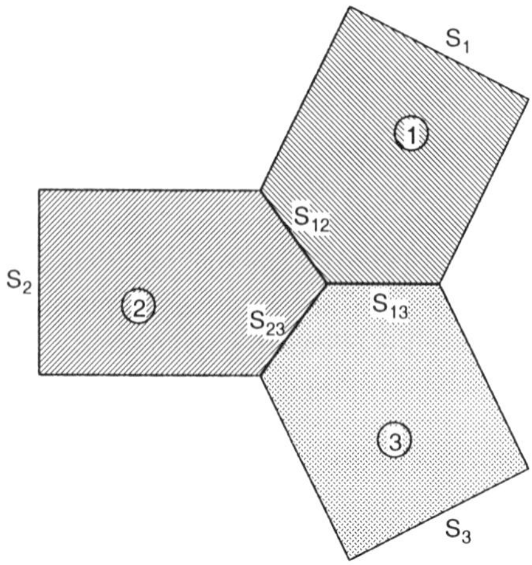

Figure 3.9 Partition of Ω into subregions: geometrical notations ($N = 3$).

Then, the boundary element formulation (3.13) is considered separately for each subregion. Let $\{\boldsymbol{v}_1\}, \{\boldsymbol{v}_{12}\}, \{\boldsymbol{v}_{13}\}$ denote the vectors of unknown degrees of freedom on S_1, S_{12}, S_{13} after introduction of the boundary data. Then, the linear system of equations (3.16) for the subdomain Ω_1 takes the form:

$$\begin{bmatrix} \boldsymbol{K}_s^1 & \boldsymbol{K}_s^{12} & \boldsymbol{K}_s^{13} \\ \boldsymbol{K}_i^1 & \boldsymbol{K}_i^{12} & \boldsymbol{K}_i^{13} \end{bmatrix} \begin{Bmatrix} \boldsymbol{v}_1 \\ \boldsymbol{v}_{12} \\ \boldsymbol{v}_{13} \end{Bmatrix} = \begin{Bmatrix} \boldsymbol{f}_s \\ \boldsymbol{f}_i \end{Bmatrix} \tag{3.40}$$

where the index s (resp. i) indicates that the rows of the global matrix $[\boldsymbol{K}]$ for subdomain Ω_1 corresponding to collocation points located on the surface S_1 (resp. on the interfaces S_{12}, S_{13}). A similar decomposition is made for the subdomains Ω_2, Ω_3.

Those linear systems of equations are coupled through the continuity conditions (3.39). The solution of the complete problem (i.e. for the whole domain Ω) is then obtained through the following steps:

1. *Partial matrix triangulation:* a Gauss elimination is performed in (3.40) to the columns associated with the subvector $\{\boldsymbol{v}_1\}$ of non-interfacial unknowns:

$$\begin{bmatrix} \boldsymbol{T}_s^1 & \boldsymbol{K}_s^{12} & \boldsymbol{K}_s^{13} \\ 0 & \boldsymbol{K}_i^{12} & \boldsymbol{K}_i^{13} \end{bmatrix} \begin{Bmatrix} \boldsymbol{v}_1 \\ \boldsymbol{v}_{12} \\ \boldsymbol{v}_{13} \end{Bmatrix} = \begin{Bmatrix} \boldsymbol{y}_s \\ \boldsymbol{y}_i \end{Bmatrix} \tag{3.41}$$

where the submatrix $[\boldsymbol{T}_s^1]$ is upper-triangular. The purpose of this manoeuvre is to isolate the interfacial unknowns in the equations associated with the submatrices with index i.

2. *Computation of the interfacial unknowns.* Once the latter operations are performed for all subdomains, the following governing linear system of equations for the interfacial unknowns is reached:

$$\begin{bmatrix} \boldsymbol{K}_i^{(1)12} & \boldsymbol{K}_i^{(1)13} & 0 \\ \boldsymbol{K}_i^{(2)12} & 0 & \boldsymbol{K}_i^{(2)23} \\ 0 & \boldsymbol{K}_i^{(3)13} & \boldsymbol{K}_i^{(3)23} \end{bmatrix} \begin{Bmatrix} \boldsymbol{v}_{12} \\ \boldsymbol{v}_{13} \\ \boldsymbol{v}_{23} \end{Bmatrix} = \begin{Bmatrix} \boldsymbol{y}_i^{(1)} \\ \boldsymbol{y}_i^{(2)} \\ \boldsymbol{y}_i^{(3)} \end{Bmatrix} \tag{3.42}$$

It is then solved for $\{\boldsymbol{v}_{12}\}, \{\boldsymbol{v}_{13}\}, \{\boldsymbol{v}_{23}\}$.

3. *Solution of the complete problem.* The solution $(\{\boldsymbol{v}_{12}\}, \{\boldsymbol{v}_{13}\}, \{\boldsymbol{v}_{23}\})$ to the system (3.42) above is finally substituted into the matrix equations (3.40) associated with each subdomain. This gives:

$$[\boldsymbol{T}^1]\{\boldsymbol{v}_1\} = \{\boldsymbol{y}_s\} - [\boldsymbol{K}_s^{12}]\{\boldsymbol{v}_{12}\} - [\boldsymbol{K}_s^{13}]\{\boldsymbol{v}_{13}\}$$

for the subdomaine Ω_1, and so on. The above system of equations can be solved for $\{\boldsymbol{v}_1\}$ using only a backsubstitution, since the matrix $[\boldsymbol{T}^1]$ is upper-triangular. The solution of the initial problem on Ω is complete.

The computational gains provided by this procedure are clearly dependent upon the respective sizes of the exterior S_i and interfacial S_{ij} boundary parts;[5] the latter should ideally be as small as possible compared to the former. In the ideal limiting situation where $|S_{ij}| \ll |S|$, with $|S_1| = |S_2| = |S_3| = |S|/3$ for simplicity, one notices that:

- The number of arithmetical operations needed at the assembly stage decreases from $a|S|^2$ to $3a(|S|/3)^2 = a|S|^2/3$, i.e. is divided by 3 (by N when using N subdomains);

- The number of arithmetical operations needed to solve the linear system (using Gauss elimination) decreases from $b|S|^3/3$ to $b(|S|/3)^3 = |S|^2/27$, i.e. is divided by 9 (by N^2 when using N subdomains).

In practice, for non-negligible interfaces, the relative computational gains are smaller. Substructuring can even lead to *increasing* computational costs if the interfaces are inadequately chosen (in particular when the S_{ij} and S_i have similar sizes); computational gains are not guaranteed. The geometry of the domain Ω at hand must lend itself to substructuring: 'bulky' domain shapes are expected to do poorly in this respect.

[5]More precisely, upon the respective numbers of degrees of freedom supported by the boundary element meshes of the S_i and S_{ij}.

REFERENCES

1. ANDERSON, E., ET AL. LAPACK users' guide, second edition. SIAM, Philadelphia (1995).
2. CETIM, (CENTRE TECHNIQUE DES INDUSTRIES MÉCANIQUES). Application de la méthode des équations intégrales à la mécanique. Internal report, CETIM, Senlis, France (1976).
3. CIARLET, P. G. *Analyse numérique matricielle et optimisation.* Masson, Paris (1982).
4. DAVIS, J. J., RABINOWITZ, P. *Methods of numerical integration.* Academic Press (1975).
5. DONGARRA, J. J., BUNCH, J. R., MOLER, C. B., STEWART, G. B. LINPACK *users' guide.* Society for Industrial and Applied Mathematics (SIAM), Philadelphia, USA (1979).
6. GUIGGIANI, M., GIGANTE, A. A general algorithm for multidimensional Cauchy principal value integrals in the boundary element method. *ASME J. Appl. Mech.*, **57**, 906–915 (1990).
7. HAYAMI, K., MATSUMOTO, H. A numerical quadrature for nearly singular boundary element integrals. *Engng. Anal. with Bound. Elem.*, **13**, 143–154 (1994).
8. KANE, J. H. *Equation solving and multizone BEA*, chapter 9. Prentice-Hall (1994).
9. KANE, J. H., KEYES, D. E., GURU PRASAD, K. Iterative solution techniques in boundary element analysis. *Int. J. Num. Meth. in Eng.*, **31**, 1511–1536 (1991).
10. LYNESS, J. N., JESPERSEN, D. Moderate degree symmetric quadrature rules for the triangle. *J. Inst. Math. Appl.*, **15**, 19–32 (1975).
11. MANOLIS, G. D., BANERJEE, P. K. Conforming versus non-conforming boundary elements in three-dimensional elastostatics. *Int. J. Num. Meth. in Eng.*, **23**, 1885–1904 (1986).
12. MULLEN, R. L., RENCIS, J. J. Iterative methods for solving boundary element equations. *Computers and Structures*, **25**, 713–723 (1987).
13. NOWAK, A. J., BREBBIA, C. A. The multiple-reciprocity method. A new approach for transforming BEM domain integrals to the boundary. *Engng. Anal. with Bound. Elem.*, **6**, 164–168 (1989).
14. REZAYAT, M., SHIPPY, D. J., RIZZO, F. J. On time-harmonic elastic wave analysis by the boundary element method for moderate to high frequencies. *Comp. Meth. in Appl. Mech. Engng.*, **55**, 349–367 (1986).
15. STROUD, A. H., SECREST, D. *Gaussian quadrature formulas.* Prentice-Hall (1966).
16. TELLES, J. C. F. A self-adaptive coordinate transformation for efficient numerical evaluation of general boundary element integrals. *Int. J. Num. Meth. in Eng.*, **24**, 959–973 (1987).
17. ZIENKIEWICZ, O. C. *The finite element method: basic concepts and linear applications.* McGraw-Hill, fourth edn. (1989).

4

Integral equations and representations for elastostatics

This chapter deals with the integral representation formulas and integral equations for equilibrium problems in linear elasticity, assuming infinitesimal perturbations. The relevant basic equations have been reviewed in Section 1.5. The exposition is mainly focused on the three-dimensional situations, the transposition to plane strain being presented in Section 1.6.

Geometry. Let Ω denote a bounded open subset of $\mathbb{R}^3$ whose boundary $\partial\Omega$ is piecewise regular (at least C^2), allowing for the presence of edges or corners,[1] and possibly made of several disjoint closed surfaces, e.g. the case of several internal cavities (the extension to unbounded domains is addressed in Section 4.6). The interior of each connected component of $\partial\Omega$ has nonzero measure, thus precluding cracks (their presence would require specific formulations that are the subject of Chapter 12). The unit normal is conventionally directed towards the exterior of Ω.

4.1 MAXWELL–BETTI RECIPROCITY IDENTITY

The reciprocity identity for linear elasticity stems from the virtual work principle (see e.g. Salençon [13]) whereby for any stress field $\boldsymbol{\sigma}$ in equilibrium with $\boldsymbol{F}$ (i.e. such that $\text{div}\boldsymbol{\sigma} + \boldsymbol{F} = 0$) and any vector ('virtual displacement')

[1] More generally, finitely many points at which finitely many tangent planes meet, e.g. the vertices of a polyhedron, are allowed, but Ω must be on one side of the tangent plane(s).

field $\boldsymbol{u}'$ continuously differentiable over Ω, one has:

$$\int_{\Omega} \boldsymbol{\sigma} : \boldsymbol{\varepsilon}(\boldsymbol{u}')\,\mathrm{d}V - \int_{\partial\Omega} (\boldsymbol{\sigma}.\boldsymbol{n}).\boldsymbol{u}'\,\mathrm{d}S - \int_{\Omega} \rho\boldsymbol{F}.\boldsymbol{u}'\,\mathrm{d}V = 0 \tag{4.1}$$

Then, consider two elastostatic states $(\boldsymbol{u}^1, \boldsymbol{\sigma}^1, \boldsymbol{F}^1)$ and $(\boldsymbol{u}^2, \boldsymbol{\sigma}^2, \boldsymbol{F}^2)$ on Ω. Upon application of the virtual work principle (4.1) to $(\boldsymbol{u}^1, \boldsymbol{\sigma}^1, \boldsymbol{F}^1)$ and the virtual displacement $\boldsymbol{u}^2$, then to $\boldsymbol{u}^2, \boldsymbol{\sigma}^2, \boldsymbol{F}^2)$ and the virtual displacement $\boldsymbol{u}^1$ followed by subtraction of the resulting equalities, one gets:

$$\int_{\Omega} [\boldsymbol{\sigma}^1 : \boldsymbol{\varepsilon}(\boldsymbol{u}^2) - \boldsymbol{\sigma}^2 : \boldsymbol{\varepsilon}(\boldsymbol{u}^1)]\,\mathrm{d}V - \int_{\Omega} [\rho\boldsymbol{F}^1.\boldsymbol{u}^2 - \rho\boldsymbol{F}^2.\boldsymbol{u}^1]\,\mathrm{d}V$$
$$= \int_{\partial\Omega} [\boldsymbol{T}^n(\boldsymbol{u}^1).\boldsymbol{u}^2 - \boldsymbol{T}^n(\boldsymbol{u}^2).\boldsymbol{u}^1]\,\mathrm{d}S$$

Moreover, the symmetry (1.16) of the elastic constitutive law implies:

$$\boldsymbol{\sigma}^1 : \boldsymbol{\varepsilon}(\boldsymbol{u}^2) = \boldsymbol{\sigma}^2 : \boldsymbol{\varepsilon}(\boldsymbol{u}^1)$$

As a result, any pair $(\boldsymbol{u}^1, \boldsymbol{\sigma}^1, \boldsymbol{F}^1)$, $(\boldsymbol{u}^2, \boldsymbol{\sigma}^2, \boldsymbol{F}^2)$ of elastostatic states over Ω satisfy the *Maxwell–Betti reciprocity identity*:

$$\int_{\partial\Omega} [(\boldsymbol{\sigma}^1.\boldsymbol{n}).\boldsymbol{u}^2 - (\boldsymbol{\sigma}^2.\boldsymbol{n}).\boldsymbol{u}^1]\,\mathrm{d}S = \int_{\Omega} [\rho\boldsymbol{F}^2.\boldsymbol{u}^1 - \rho\boldsymbol{F}^1.\boldsymbol{u}^2]\,\mathrm{d}V \tag{4.2}$$

Since both elastostatic states satisfy the Navier equation (1.17), the above identity is in fact the elastic counterpart of Green's third identity (2.4), where the Navier operator $\boldsymbol{\Delta}^\star$ and the traction vector operator $\boldsymbol{T}^n(\boldsymbol{u}) \equiv \boldsymbol{\sigma}(\boldsymbol{u}).\boldsymbol{n}$ replace the Laplacian Δ and the normal derivative $\partial/\partial n$, respectively:

$$\int_{\partial\Omega} [\boldsymbol{T}^n(\boldsymbol{u}^1).\boldsymbol{u}^2 - \boldsymbol{T}^n(\boldsymbol{u}^2).\boldsymbol{u}^1\,\mathrm{d}S] = \int_{\Omega} \left[(\boldsymbol{\Delta}^\star\boldsymbol{u}^1).\boldsymbol{u}^2 - (\boldsymbol{\Delta}^\star\boldsymbol{u}^2).\boldsymbol{u}^1\right]\,\mathrm{d}V \tag{4.3}$$

4.2 FUNDAMENTAL SOLUTIONS OF LINEAR ELASTOSTATICS

Definition

Let E denote an open subset of $\mathbb{R}^3$ such that $\Omega \subseteq E$. Any elastostatic state $\boldsymbol{U}^k, \boldsymbol{\Sigma}^k, \boldsymbol{F}^k)$ associated with a point force of unit magnitude applied at a given fixed point $\boldsymbol{x} \in E$ along the k-direction, i.e.:

$$\rho\boldsymbol{F}^k(\boldsymbol{y}) = \delta(\boldsymbol{y} - \boldsymbol{x})\boldsymbol{e}_k \qquad \boldsymbol{y} \in E \tag{4.4}$$

is called a *fundamental solution* of linear elastostatics. The displacement vector $(\boldsymbol{U}^k(\boldsymbol{x}, \boldsymbol{y})$ and the elastic stress tensor $\boldsymbol{\Sigma}^k(\boldsymbol{x}, \boldsymbol{y})$ are functions of the field

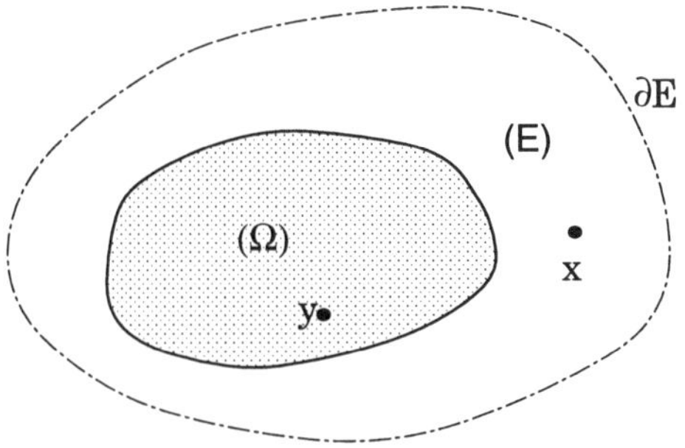

Figure 4.1 The domains Ω and E.

point $\boldsymbol{y}$, the source point $\boldsymbol{x}$ and the direction k of the load (4.4). The components $U_i^k(\boldsymbol{x},\boldsymbol{y}), \Sigma_{ij}^k(\boldsymbol{x},\boldsymbol{y})$ of any fundamental solution satisfy the equilibrium and elastic constitutive equations:

$$\Sigma_{ij,j}^k(\boldsymbol{x},\boldsymbol{y}) + \delta(\boldsymbol{y}-\boldsymbol{x})\delta_{ik} = 0 \qquad \Sigma_{ij}^k(\boldsymbol{x},\boldsymbol{y}) = C_{ijab}U_{a,b}^k(\boldsymbol{x},\boldsymbol{y}) \tag{4.5}$$

where $f_{,\ell}(\boldsymbol{z}^1,\boldsymbol{z}^2)$ conventionally denotes the partial derivative of f with respect to the ℓ-coordinate of its second argument z^2 ($f_{,\ell}(\boldsymbol{x},\boldsymbol{y})$ thus indicates a partial derivative with respect to the ℓ-coordinate of the field point $\boldsymbol{y}$), while $T_i^k(\boldsymbol{x},\boldsymbol{y})$ denotes the traction vector at $\boldsymbol{y}$ associated with the fundamental solution:

$$T_i^k(\boldsymbol{x},\boldsymbol{y}) = \Sigma_{ij}^k(\boldsymbol{x},\boldsymbol{y})n_j(\boldsymbol{y}) = \left(T^n U^k(\boldsymbol{x},\boldsymbol{y})\right)_i \tag{4.6}$$

Global equilibrium. For any fundamental solution of linear elastostatics, any domain $\Omega \subseteq E$ and any fixed point $\boldsymbol{x} \in E, \boldsymbol{x} \notin \partial\Omega$, integrating the equilibrium equation $(4.5)_1$ over Ω and applying the divergence formula lead to the following identity, which merely expresses that the load T_i^k distributed over $\partial\Omega$ equilibrates the unit point force applied at $\boldsymbol{x}$ ($\boldsymbol{n}$ being the unit normal exterior to Ω):

$$\kappa\delta_{ik} + \int_{\partial\Omega} T_i^k(\boldsymbol{x},\boldsymbol{y})\,\mathrm{d}S_y = 0 \qquad \text{with} \left\{ \begin{array}{ll} \kappa = 1 & \boldsymbol{x} \in \Omega \\ \kappa = 0 & \boldsymbol{x} \notin \Omega \end{array} \right. \tag{4.7}$$

Infinite body: the Kelvin solution

Any choice of domain E and boundary conditions on ∂E compatible with the equilibrium condition (4.7) defines a fundamental solution. In the simplest case, known as the Kelvin solution, E is an isotropic infinite body $E = \mathbb{R}^3$ with decay conditions at infinity. The Kelvin displacement, whose derivation is a classical problem of elasticity theory (see Appendix 4.A), is given by:

$$U_i^k(\boldsymbol{x},\boldsymbol{y}) = \frac{1}{16\pi\mu(1-\nu)r}\left[r_{,i}r_{,k} + (3-4\nu)\delta_{ik}\right] \tag{4.8}$$

where $r = |\boldsymbol{r}| = |\boldsymbol{y} - \boldsymbol{x}|$ denotes the Euclidean distance between the source and field points. The Kelvin stress tensor then follows from Hooke's law:

$$\Sigma_{ij}^{k}(\boldsymbol{x}, \boldsymbol{y}) = \frac{-1}{8\pi(1-\nu)r^2}\left[3r_{,i}r_{,k}r_{,j} + (1-2\nu)(\delta_{ik}r_{,j} + \delta_{jk}r_{,i} - \delta_{ij}r_{,k})\right] \quad (4.9)$$

General properties of the Kelvin solution. Being defined in the entire infinite space $\mathbb{R}^3$, the Kelvin solution is invariant under all translations, i.e. depends on $\boldsymbol{x}, \boldsymbol{y}$ only through $\boldsymbol{r} = \boldsymbol{y} - \boldsymbol{x}$. It is also *homogeneous* with respect to $\boldsymbol{y} - \boldsymbol{x}$:

$$\begin{aligned} U_i^k(\lambda\boldsymbol{x}, \lambda\boldsymbol{y}) &= \lambda^{-1}U_i^k(\boldsymbol{x}, \boldsymbol{y}) \\ \Sigma_{ij}^k(\lambda\boldsymbol{x}, \lambda\boldsymbol{y}) &= \lambda^{-2}\Sigma_{ij}^k(\boldsymbol{x}, \boldsymbol{y}) \end{aligned} \quad (4.10)$$

In particular, the Kelvin solution is singular at $\boldsymbol{y} = \boldsymbol{x}$ ($r = 0$):

$$\begin{aligned} U_i^k(\boldsymbol{x}, \boldsymbol{y}) &= O(r^{-1}) \\ \Sigma_{ij}^k(\boldsymbol{x}, \boldsymbol{y}) &= O(r^{-2}) \end{aligned} \quad (4.11)$$

This singular behaviour was predictable: a *concentrated* load of finite magnitude must create infinite displacements and elastic stress at the load point. Moreover, any fundamental solution has infinite strain energy.

From the previous remarks and expressions (4.8), (4.9), other properties are readily established: symmetry with respect to indices:

$$U_i^k(\boldsymbol{x}, \boldsymbol{y}) = U_k^i(\boldsymbol{x}, \boldsymbol{y}) \quad (4.12)$$

and effect of a transposition of $\boldsymbol{x}, \boldsymbol{y}$:

$$\begin{aligned} U_i^k(\boldsymbol{x}, \boldsymbol{y}) &= U_i^k(\boldsymbol{y}, \boldsymbol{x}) & \Sigma_{ij}^k(\boldsymbol{x}, \boldsymbol{y}) &= -\Sigma_{ij}^k(\boldsymbol{y}, \boldsymbol{x}) \\ U_{i,\bar{\ell}}^k(\boldsymbol{x}, \boldsymbol{y}) &= -U_{i,\ell}^k(\boldsymbol{x}, \boldsymbol{y}) & \Sigma_{ij,\bar{\ell}}^k(\boldsymbol{x}, \boldsymbol{y}) &= -\Sigma_{ij,\ell}^k(\boldsymbol{x}, \boldsymbol{y}) \end{aligned} \quad (4.13)$$

where $f_{,\bar{\ell}}(\boldsymbol{z}^1, \boldsymbol{z}^2)$ conventionally denotes the partial derivative of f with respect to the ℓ-coordinate of its first argument z^1 ($f_{,\bar{\ell}}(\boldsymbol{x}, \boldsymbol{y})$ thus indicates a partial derivative with respect to the ℓ-coordinate of the source point $\boldsymbol{x}$). Properties (4.12), (4.13) are consequences of the invariance under translation, i.e. of the fact that $E = \mathbb{R}^3$, and do not hold if $E \neq \mathbb{R}^3$.

Integral identity for the Kelvin solution. As shown in Appendix 4.B, the integral of the traction vector over any (open or closed) surface S that does not contain $\boldsymbol{x}$ is given by the expression:

$$\begin{aligned} &\int_S T_i^k(\boldsymbol{x}, \boldsymbol{y})\, \mathrm{d}S_y \\ &\quad = \frac{e_{ijp}}{8\pi(1-\nu)} \int_{\partial S} \left[r_{,j}r_{,k} - (1-2\nu)\delta_{jk}\right] \tau_p \frac{\mathrm{d}s_y}{r} - \delta_{ik}\frac{\omega(S)}{4\pi} \end{aligned} \quad (4.14)$$

where ω denotes the algebraic solid angle under which S is seen from $\boldsymbol{x}$. If S is closed, formula (4.7) is recovered as a special case of (4.14) with $\omega = 4\pi$ or $\omega = 0$ according to whether $\boldsymbol{x}$ is interior or exterior to S.

The Kelvin traction vector is not integrable. The contour integrals of eq. (4.14) are invariant under scaling of ∂S (the centre $\boldsymbol{x}$ being fixed) but depend on the shape of ∂S. Since the value of any convergent improper integral is by definition independent of the shape of the vanishing exclusion neighbourhood around the singular point, it follows that the Kelvin traction vector is non-integrable on any surface containing $\boldsymbol{x}$. In contrast, the integral of the normal derivative of the fundamental solution to the Laplace equation $G(\boldsymbol{x}, \boldsymbol{y}) = 1/(4\pi r)$ is convergent (the value of this integral is $-\omega(S)/4\pi$). The non-integrability of $T_i^k(\boldsymbol{x}, \boldsymbol{y})$ thus constitutes an important qualitative difference between the fundamental solutions for scalar and elastic problems.

For the special case where ∂S is a circle of centre $\boldsymbol{x}$, it can be shown that the contour integral in eq. (4.14) vanishes (Appendix 4.B) and that $\omega(S) = 2\pi$. Consider then a domain $\Omega \subset \mathbb{R}^3$ of boundary $\partial\Omega$ and let $\boldsymbol{x}$ be a smooth point of $\partial\Omega$. The previous remark is applicable to the surface $S_\varepsilon(\boldsymbol{x}) = \partial\Omega - B(\boldsymbol{x}, \varepsilon)$, where $B(\boldsymbol{x}, \varepsilon)$ is the ball of centre $\boldsymbol{x}$ and radius ε: the edge of S_ε is, in the limit $\varepsilon \to 0$, a circle of centre $\boldsymbol{x}$ and radius ε. The limit when $\varepsilon \to 0$, if it exists, of an integral over the surface $S_\varepsilon(\boldsymbol{x})$ thus defined is the *Cauchy principal value* (CPV) of the integral over $\partial\Omega$:

$$\text{CPV} \int_{\partial\Omega} ()\,\mathrm{d}S \equiv ⨍_{\partial\Omega} ()\,\mathrm{d}S = \lim_{\varepsilon \to 0} \int_{S_\varepsilon(x)} ()\,\mathrm{d}S$$

Here, the integral of the Kelvin traction vector is convergent in the CPV sense:

$$\lim_{\varepsilon \to 0} \int_{S_\varepsilon(x)} T_i^k(\boldsymbol{x}, \boldsymbol{y})\,\mathrm{d}S_y \stackrel{def}{=} ⨍_{\partial\Omega} T_i^k(\boldsymbol{x}, \boldsymbol{y})\,\mathrm{d}S_y = -\frac{1}{2}\delta_{ik}$$

Kelvin solution for anisotropic elasticity. The infinite-space fundamental solution is known for general anisotropic elasticity, in non-closed form:

$$U_i^k(\boldsymbol{x}, \boldsymbol{y}) = \frac{1}{8\pi^2 r} \int_0^{2\pi} N_{ki}(\boldsymbol{\eta})\,\mathrm{d}\theta \tag{4.15}$$

The unit vector $\boldsymbol{\eta}$ appearing in the integral ranges over the unit circle C centred at the origin and lying in the plane orthogonal to the direction $\boldsymbol{y} - \boldsymbol{x}$ (θ: angular parameter along C); $N_{ki}(\boldsymbol{\eta})$ are the coefficients of the matrix $[C_{ijk\ell}\eta_j\eta_\ell]^{-1}$ (see Appendix 4.A). The integral over θ, which results from a Fourier inversion, can be evaluated analytically only for particular types of anisotropy, e.g. transverse anisotropy, eq. (4.A.5).

Other fundamental solutions

Analytical expressions of fundamental solutions of isotropic linear elasticity are also available, albeit not always in closed form, for the following configurations (E, boundary conditions):

Half-space with traction-free surface: $E = \{y_3 < 0\}$, known as the Mindlin solution (unit concentrated load applied at a point interior to the half-space);

its expression is given in Appendix 4.C. The Boussinesq and Cerruti solutions (point load applied on the free surface in the normal and tangential directions, respectively) are contained as special cases of the Mindlin solution.

Bonded half-spaces: $E = \mathbb{R}^3$, but the half-spaces $y_3 < 0$ and $y_3 > 0$ have distinct elastic parameters and are perfectly bonded; its expression (Rongved [12]) is given in Appendix 4.D.

Elastic layer between two traction-free parallel planes: $E = \{(y_1, y_2) \in \mathbb{R}^2, a \leq y_3 \leq b\}$ (Benitez and Rosakis [2]). Unlike the previously mentioned ones, this fundamental solution is known in the form of (rather lengthy) combinations of inverse Hankel transforms.

Green and Neumann tensors: $E = \Omega$, with zero displacement on $\partial\Omega$ (Green's tensor) or traction distribution over $\partial\Omega$) prescribed so as to equilibrate the point $\delta(\boldsymbol{y} - \boldsymbol{x})\boldsymbol{a}$ (Neumann tensor). Their expression (Pham [11]) involves convergent infinite series of iterated integral operators; the Green's tensor for isotropic elasticity is given in Chapter 6.

General properties of the fundamental solutions

Symmetry properties. Let $\boldsymbol{U}$ denote a fundamental elastic solution for some domain E. Consider two different point loads, respectively applied at $\boldsymbol{x}$ along the k-direction and at $\boldsymbol{y}$ along the ℓ-direction. This defines two elastostatic states, to which the Maxwell–Betti identity (4.2) is applicable and yields:

$$\int_{\partial E} \left[T_i^k(\boldsymbol{x}, \boldsymbol{z}) U_i^\ell(\boldsymbol{y}, \boldsymbol{z}) - T_i^\ell(\boldsymbol{y}, \boldsymbol{z}) U_i^k(\boldsymbol{x}, \boldsymbol{z}) \right] \, \mathrm{d}S_z = U_\ell^k(\boldsymbol{x}, \boldsymbol{y}) - U_k^\ell(\boldsymbol{y}, \boldsymbol{x})$$

An interesting consequence of the identity is the following: any fundamental solution defined in terms of homogeneous boundary conditions on ∂E (including decay conditions at infinity when E is unbounded, e.g. for the Kelvin solution) has the symmetry property:

$$U_\ell^k(\boldsymbol{x}, \boldsymbol{y}) = U_k^\ell(\boldsymbol{y}, \boldsymbol{x}) \tag{4.16}$$

whatever the particular domain E considered. This property in turn implies the following relations, similarly to Eqs (2.15), (2.16) for scalar problems:

$$C_{ijk\ell} U^k_{a,\bar{\ell}}(\boldsymbol{x}, \boldsymbol{y}) = \Sigma^a_{ij}(\boldsymbol{y}, \boldsymbol{x}) \tag{4.17}$$

$$C_{ijk\ell} \Sigma^k_{ab,\bar{\ell}}(\boldsymbol{x}, \boldsymbol{y}) = C_{abk\ell} \Sigma^k_{ij,\ell}(\boldsymbol{y}, \boldsymbol{x}) \tag{4.18}$$

The properties (4.16), (4.17) and (4.18) are weaker than those verified by the Kelvin solution, eqs (4.12) and (4.13), but hold true for a wide class of fundamental solutions that includes the Mindlin, Rongved, and Benitez and Rosakis solutions, as well as the Green's tensor of any given domain.

Singularity properties. All fundamental elastostatic solutions are singular at $\boldsymbol{x}$. Moreover, if $\boldsymbol{x}$ is not located on the boundary of E, the singularity of any fundamental solution defined on E and that of the Kelvin solution coincide, i.e. the difference $\boldsymbol{U} - \boldsymbol{U}^{Kelvin}$ is nonsingular at $\boldsymbol{x}$. Indeed, $\boldsymbol{U} - \boldsymbol{U}^{Kelvin}$ appears to be an elastostatic state on E with zero body force and smooth boundary data over ∂E, hence nonsingular. In contrast, if $\boldsymbol{x} \in \partial E$, the difference $\boldsymbol{U} - \boldsymbol{U}^{Kelvin}$ is singular at $\boldsymbol{y} = \boldsymbol{x}$. This property is for example apparent in the expression (4.A.13) of the Mindlin solution.

Other comments

It appears that the fundamental solutions are analytically known in closed form only for very simple domain geometries; their mathematical expressions become very complex as the domain shape departs from the infinite space. In the case of a domain of arbitrary shape, the expression (2.4) of the Green's tensor, although mathematically explicit, is unsuitable for applications because its evaluation by numerical means represents a task more complex and computationally more costly than a numerical solution of the corresponding elasticity problem by usual methods like finite elements.

In a similar line of thought, any fundamental solution is numerically computable as the solution of some elasticity problem (after possibly having removed the singular load through subtraction of the Kelvin solution). However, this task has the same complexity as that of solving the original boundary-value problem which motivated in the first place the introduction of integral equations and representations; hence following this approach the use of integral representations would lose most of its *a priori* practical appeal.

4.3 INTEGRAL REPRESENTATION FORMULAS

Consider a bounded elastic body Ω (the extension to unbounded domains is addressed in Section 4.6). In addition to the unknown elastostatic state $(\boldsymbol{u}, \boldsymbol{\sigma}, \boldsymbol{F})$ (state 1), introduce a fundamental solution, defined by eqs (4.4–4.5), associated to an auxiliary body $E \supseteq \Omega$. Assuming for the time being that the source point $\boldsymbol{x}$ does not lie on the boundary $\partial\Omega$, the Maxwell–Betti reciprocity theorem (4.2) applied to those two states leads to:

$$\int_\Omega \delta_{ik}\delta(\boldsymbol{y}-\boldsymbol{x})u_i(\boldsymbol{y})\,\mathrm{d}V_y = \int_{\partial\Omega} \{t_i(\boldsymbol{y})U_i^k(\boldsymbol{x},\boldsymbol{y}) - u_i(\boldsymbol{y})T_i^k(\boldsymbol{x},\boldsymbol{y})\}\,\mathrm{d}S_y + \int_\Omega \rho F_i(\boldsymbol{y})U_i^k(\boldsymbol{x},\boldsymbol{y})\,\mathrm{d}V_y \qquad (4.19)$$

where $\boldsymbol{t} = \boldsymbol{T}^n(\boldsymbol{u}) = \boldsymbol{\sigma}(\boldsymbol{u}).\boldsymbol{n}$ is the traction vector associated to the displacement $\boldsymbol{u}$ through eqs (1.13–1.14). Moreover, the definition of the Dirac distribution implies that:

$$\int_\Omega \delta_{ik}\delta(\boldsymbol{y}-\boldsymbol{x})u_i(\boldsymbol{y})\,\mathrm{d}V_y = \kappa u_k(\boldsymbol{x}) \qquad \begin{cases} \kappa = 1 & (\boldsymbol{x} \in \Omega) \\ \kappa = 0 & (\boldsymbol{x} \notin \Omega) \end{cases} \qquad (4.20)$$

Integral representation formulas for displacements and stresses

As an immediate consequence of eqs (4.19–4.20), the displacement at any point $\boldsymbol{x} \in E$ *not located on* $\partial\Omega$ is given in terms of the elastic field variables on the boundary by the formula:

$$\kappa u_k(\boldsymbol{x}) = \int_{\partial\Omega} \{t_i(\boldsymbol{y})U_i^k(\boldsymbol{x},\boldsymbol{y}) - u_i(\boldsymbol{y})T_i^k(\boldsymbol{x},\boldsymbol{y})\} \, \mathrm{d}S_y + \int_{\Omega} \rho F_i(\boldsymbol{y})U_i^k(\boldsymbol{x},\boldsymbol{y}) \, \mathrm{d}V_y \quad (4.21)$$

When $\boldsymbol{x} \notin \partial\Omega$, the integrals in eq. (4.21) are (infinitely) differentiable functions of $\boldsymbol{x}$. The integral representation formula for the displacement gradient is thus obtained by a straightforward differentiation of eq. (4.21) with respect to $\boldsymbol{x}$ under the integral sign:

$$\kappa u_{k,\ell}(\boldsymbol{x}) = \int_{\partial\Omega} \left\{t_a(\boldsymbol{y})U^k_{a,\bar{\ell}}(\boldsymbol{x},\boldsymbol{y}) - u_a(\boldsymbol{y})n_b(\boldsymbol{y})\Sigma^k_{ab,\bar{\ell}}(\boldsymbol{x},\boldsymbol{y})\right\} \, \mathrm{d}S_y + \int_{\Omega} \rho F_a(\boldsymbol{y})U^k_{a,\bar{\ell}}(\boldsymbol{x},\boldsymbol{y}) \, \mathrm{d}V_y \quad (4.22)$$

Then, applying the Hooke law (1.14) and making use of the symmetry properties (4.17) and (4.18), one obtains the integral representation of the elastic stress tensor:

$$\kappa \sigma_{ij}(\boldsymbol{x}) = \int_{\partial\Omega} \{t_a(\boldsymbol{y})\Sigma^a_{ij}(\boldsymbol{y},\boldsymbol{x}) - u_a(\boldsymbol{y})n_b(\boldsymbol{y})C_{abk\ell}\Sigma^k_{ij,\ell}(\boldsymbol{y},\boldsymbol{x})\} \, \mathrm{d}S_y + \int_{\Omega} \rho F_a(\boldsymbol{y})\Sigma^a_{ij}(\boldsymbol{y},\boldsymbol{x}) \, \mathrm{d}V_y \quad (4.23)$$

Equations (4.21), (4.22) and (4.23) for $\boldsymbol{x}$ exterior to Ω (i.e. $\kappa = 0$) are sometimes termed *exterior representation formulas.*

Necessity of an integral equation

Equations (4.21) and (4.23) thus give the value at any observation point $\boldsymbol{x}$ interior to Ω of the displacement $\boldsymbol{u}$ and stress $\boldsymbol{\sigma}$ in terms of the body force distribution ρF and the values taken on the boundary $\partial\Omega$ by $(\boldsymbol{u}, \boldsymbol{t})$. For any well-posed elasticity problem, half of the scalar components of $(\boldsymbol{u}, \boldsymbol{t})\,|_{\partial\Omega}$ are known from the boundary conditions, the other half remaining unknown. Equations (4.21), (4.23) show that the solution of a boundary-value problem of elasticity reduces to the determination of those components of $(\boldsymbol{u}, \boldsymbol{t})\,|_{\partial\Omega}$ that are not prescribed by the boundary conditions; they might be said to achieve a 'quasi-inversion' of the Navier equation (1.17).

Alternative integral representation of the stress tensor

For the case where the Kelvin solution is used, the property (4.13) allows us to change derivatives with respect to the observation point into derivatives with respect to the field point. Hence, equation (2.19) becomes:

$$\kappa u_{k,\ell}(\boldsymbol{x}) = \int_{\partial\Omega} \left\{ u_a(\boldsymbol{y}) n_b(\boldsymbol{y}) \Sigma^k_{ab,\ell}(\boldsymbol{x},\boldsymbol{y}) - t_a(\boldsymbol{y}) U^k_{a,\ell}(\boldsymbol{x},\boldsymbol{y}) \right\} \mathrm{d}S_y - \int_{\Omega} \rho F_a(\boldsymbol{y}) U^k_{a,\ell}(\boldsymbol{x},\boldsymbol{y}) \,\mathrm{d}V_y$$

which allows a subsequent integration by parts:

$$\begin{aligned} &\int_{\partial\Omega} u_a(\boldsymbol{y}) n_b(\boldsymbol{y}) \Sigma^k_{ab,\ell}(\boldsymbol{x},\boldsymbol{y}) \,\mathrm{d}S_y \\ &= \int_{\partial\Omega} u_a(\boldsymbol{y}) \left\{ D_{b\ell} \Sigma^k_{ab}(\boldsymbol{x},\boldsymbol{y}) + n_\ell(\boldsymbol{y}) \Sigma^k_{ab,b}(\boldsymbol{x},\boldsymbol{y}) \right\} \mathrm{d}S_y \\ &= \int_{\partial\Omega} D_{\ell b} u_a(\boldsymbol{y}) \Sigma^k_{ab}(\boldsymbol{x},\boldsymbol{y}) \,\mathrm{d}S_y \end{aligned}$$

This derivation, similar to that performed in Chapter 2 to establish the formula (2.20), uses the fact that the Kelvin solution satisfies the equilibrium equation $\Sigma^k_{ab,b}(\boldsymbol{x},\boldsymbol{y}) = 0$ $(\boldsymbol{y} \neq \boldsymbol{x})$. Thus the representation formulas (2.19), (4.23) of the displacement gradient and the stress tensor have the following alternative form, provided the Kelvin solution is used in them:

$$\kappa u_{k,\ell}(\boldsymbol{x}) = \int_{\partial\Omega} \left\{ D_{\ell b} u_a(\boldsymbol{y}) \Sigma^k_{ab}(\boldsymbol{x},\boldsymbol{y}) - t_a(\boldsymbol{y}) U^k_{a,\ell}(\boldsymbol{x},\boldsymbol{y}) \right\} \mathrm{d}S_y - \int_{\Omega} \rho F_a(\boldsymbol{y}) U^k_{a,\ell}(\boldsymbol{x},\boldsymbol{y}) \,\mathrm{d}V_y \tag{4.24}$$

$$\kappa \sigma_{ij}(\boldsymbol{x}) = \int_{\partial\Omega} \left\{ D_{\ell b} u_a(\boldsymbol{y}) C_{ijk\ell} \Sigma^k_{ab}(\boldsymbol{x},\boldsymbol{y}) - t_a(\boldsymbol{y}) \Sigma^a_{ij}(\boldsymbol{x},\boldsymbol{y}) \right\} \mathrm{d}S_y - \int_{\Omega} \rho F_a(\boldsymbol{y}) \Sigma^a_{ij}(\boldsymbol{x},\boldsymbol{y}) \,\mathrm{d}V_y \tag{4.25}$$

The latter formulas present over the former the advantage of involving only kernels behaving like $|\boldsymbol{y}-\boldsymbol{x}|^{-2}$.

4.4 DISPLACEMENT BOUNDARY INTEGRAL EQUATION

A governing *boundary integral equation* for the boundary traces $(\boldsymbol{u},\boldsymbol{t})\,|_{\partial\Omega}$ must now be formulated. However, merely considering eqs (4.21) or (4.23) for observation points $\boldsymbol{x}$ on the boundary $\partial\Omega$ is not satisfactory, because $\boldsymbol{\Sigma}(\boldsymbol{x},\boldsymbol{y})$ is singular like $|\boldsymbol{y}-\boldsymbol{x}|^{-2}$ and the corresponding boundary integrals are divergent when $\boldsymbol{x} \in \partial\Omega$. Besides, eqs (4.21) and (4.23) were established using (4.20), which holds only for $\boldsymbol{x} \notin \partial\Omega$. Hence, to establish a boundary integral equation, one must resort to a *limiting procedure* which takes into account the singular behaviour of the fundamental solution being used.

Rigid-body identity

In preparation of the limiting process, a particular, simple, solution of elastostatics is first introduced like in Chapter 2, namely a rigid-body translation of Ω (which is bounded) with zero body forces:

$$\boldsymbol{u}(\boldsymbol{y}) = \boldsymbol{u}^0(\text{constant}) \qquad \boldsymbol{\sigma}(\boldsymbol{y}) = 0 \qquad \boldsymbol{y} \in \Omega \tag{4.26}$$

Equation (4.21) applied to this simple solution yields the *rigid-body identity*:

$$\kappa u_k^0 + u_i^0 \int_{\partial\Omega} T_i^k(\boldsymbol{x}, \boldsymbol{y}) \, \mathrm{d}S_y = 0 \tag{4.27}$$

which could also be established from eq. (4.7). Equation (4.27) is in particular applicable for the special choice $\boldsymbol{u}^0 = \boldsymbol{u}(\boldsymbol{x})$, i.e. a rigid-body translation having the value of the unknown displacement at $\boldsymbol{x}$, which gives:

$$\kappa u_k(\boldsymbol{x}) + u_i(\boldsymbol{x}) \int_{\partial\Omega} T_i^k(\boldsymbol{x}, \boldsymbol{y}) \, \mathrm{d}S_y = 0 \tag{4.28}$$

Upon subtraction of the above identity from eq. (4.21), one obtains:

$$\int_{\partial\Omega} \left\{ [u_i(\boldsymbol{y}) - u_i(\boldsymbol{x})] T_i^k(\boldsymbol{x}, \boldsymbol{y}) - t_i(\boldsymbol{y}) U_i^k(\boldsymbol{x}, \boldsymbol{y}) \right\} \, \mathrm{d}S_y = \int_{\Omega} \rho F_i(\boldsymbol{y}) U_i^k(\boldsymbol{x}, \boldsymbol{y}) \, \mathrm{d}V_y \tag{4.29}$$

which for the time being holds for any source point $\boldsymbol{x}$ not located on $\partial\Omega$.

Limiting process

Now let the source point $\boldsymbol{x}$ be located on the boundary $\partial\Omega$. Following again the technique and geometrical notations introduced in Chapter 2, the limiting form when $\varepsilon \to 0$ of identity (4.29) for the domain Ω_ε of boundary $(\partial\Omega - e_\varepsilon) + s_\varepsilon$ is sought:

$$\int_{(\partial\Omega - e_\varepsilon) + s_\varepsilon} \left\{ [u_i(\boldsymbol{y}) - u_i(\boldsymbol{x})] T_i^k(\boldsymbol{x}, \boldsymbol{y}) - t_i(\boldsymbol{y}) U_i^k(\boldsymbol{x}, \boldsymbol{y}) \right\} \, \mathrm{d}S_y = \int_{\Omega_\varepsilon} \rho F_i(\boldsymbol{y}) U_i^k(\boldsymbol{x}, \boldsymbol{y}) \, \mathrm{d}V_y \tag{4.30}$$

By construction, $\boldsymbol{x} \notin \Omega_\varepsilon$. Assume that the unknown elastostatic state is such that $\boldsymbol{u}$ has $C^{0,\alpha}$-regularity at $\boldsymbol{x}$, i.e.:

$$\exists(\alpha, C) > 0, \qquad |\boldsymbol{u}(\boldsymbol{y}) - \boldsymbol{u}(\boldsymbol{x})| \leq C \, |\boldsymbol{y} - \boldsymbol{x}|^\alpha$$

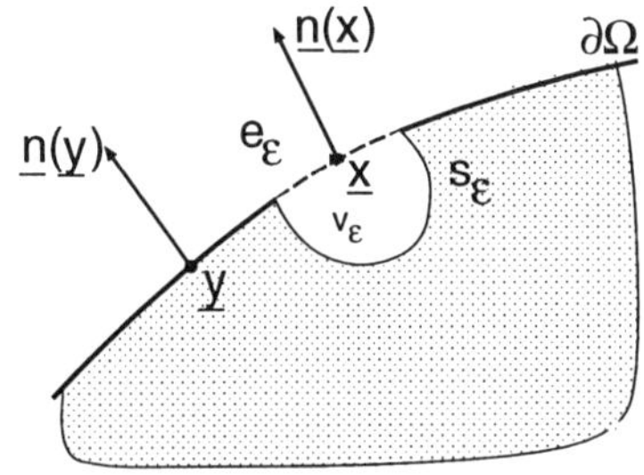

Figure 4.2 Exclusion neighbourhood $v_\varepsilon(x)$ and other notations used for the limiting process $\varepsilon \to 0$ in eq. (4.30).

then, for $\boldsymbol{y}$ sufficiently close to $\boldsymbol{x}$, the stress tensor $\boldsymbol{\sigma}(\boldsymbol{y})$ is such that:

$$|\boldsymbol{\sigma}(\boldsymbol{y})| \le C\,|\boldsymbol{y}-\boldsymbol{x}|^{\alpha-1}$$

Combining the above with the singular behaviour (4.11) of the fundamental solution, one has:

$$\begin{aligned} \left|[u_i(\boldsymbol{y})-u_i(\boldsymbol{x})]T_i^k(\boldsymbol{x},\boldsymbol{y})\right| &\le C\,|\boldsymbol{y}-\boldsymbol{x}|^{\alpha-2} \\ \left|t_i(\boldsymbol{y})U_i^k(\boldsymbol{x},\boldsymbol{y})\right| &\le C\,|\boldsymbol{y}-\boldsymbol{x}|^{\alpha-2} \end{aligned} \tag{4.31}$$

Now the limiting process $\varepsilon \to 0$ in (4.30) can be performed. In order to do so, the integrals over s_ε, $\partial\Omega - e_\varepsilon$ and Ω_ε are considered in turn.

Integral over s_ε. On account of (4.31) and since $|\boldsymbol{y}-\boldsymbol{x}| \sim \varepsilon$, $\mathrm{d}S_y \sim \varepsilon^2\,\mathrm{d}\omega$ (where ω denotes the solid angle seen from $\boldsymbol{x}$), one has:

$$\left\{[u_i(\boldsymbol{y})-u_i(\boldsymbol{x})]T_i^k(\boldsymbol{x},\boldsymbol{y})-t_i(\boldsymbol{y})U_i^k(\boldsymbol{x},\boldsymbol{y})\right\}\,\mathrm{d}S_y = O(\varepsilon^\alpha)\,\mathrm{d}\omega$$

and therefore:

$$\lim_{\varepsilon\to 0}\int_{s_\varepsilon}\left\{[u_i(\boldsymbol{y})-u_i(\boldsymbol{x})]T_i^k(\boldsymbol{x},\boldsymbol{y})-t_i(\boldsymbol{y})U_i^k(\boldsymbol{x},\boldsymbol{y})\right\}\,\mathrm{d}S_y = 0$$

Integral over $\partial\Omega - e_\varepsilon$. On account of (4.31), the integrand behaves in a neighbourhood of $\boldsymbol{x}$ like:

$$\left\{[u_i(\boldsymbol{y})-u_i(\boldsymbol{x})]T_i^k(\boldsymbol{x},\boldsymbol{y})-t_i(\boldsymbol{y})U_i^k(\boldsymbol{x},\boldsymbol{y})\right\} \sim |\boldsymbol{y}-\boldsymbol{x}|^{\alpha-2}$$

so that the limit when $\varepsilon \to 0$ of the integral over $\partial\Omega - e_\varepsilon$ in eq. (4.30) is simply the corresponding improper integral over $\partial\Omega$:

$$\begin{aligned} \lim_{\varepsilon\to 0}\int_{\partial\Omega-e_\varepsilon}&\left\{[u_i(\boldsymbol{y})-u_i(\boldsymbol{x})]T_i^k(\boldsymbol{x},\boldsymbol{y})-t_i(\boldsymbol{y})U_i^k(\boldsymbol{x},\boldsymbol{y})\right\}\,\mathrm{d}S_y \\ &= \int_{\partial\Omega}\left\{[u_i(\boldsymbol{y})-u_i(\boldsymbol{x})]T_i^k(\boldsymbol{x},\boldsymbol{y})-t_i(\boldsymbol{y})U_i^k(\boldsymbol{x},\boldsymbol{y})\right\}\,\mathrm{d}S_y \end{aligned}$$

Integral over Ω_ε. In view of the weak singularity $\sim |\boldsymbol{y}-\boldsymbol{x}|^{-1}$ of the U_i^k, its limit is the corresponding improper domain integral:

$$\lim_{\varepsilon\to 0}\int_{\Omega_\varepsilon} \rho F_i(\boldsymbol{y})U_i^k(\boldsymbol{x},\boldsymbol{y})\,\mathrm{d}V_y = \int_{\Omega} \rho F_i(\boldsymbol{y})U_i^k(\boldsymbol{x},\boldsymbol{y})\,\mathrm{d}V_y$$

Boundary integral equation

The limiting expression for $\varepsilon \to 0$ of identity (4.30), which follows immediately from the above considerations, is the sought-for *displacement boundary integral equation*:

$$\int_{\partial\Omega} \{[u_i(\boldsymbol{y})-u_i(\boldsymbol{x})]T_i^k(\boldsymbol{x},\boldsymbol{y}) - t_i(\boldsymbol{y})U_i^k(\boldsymbol{x},\boldsymbol{y})\}\,\mathrm{d}S_y - \int_{\Omega} \rho F_i(\boldsymbol{y})U_i^k(\boldsymbol{x},\boldsymbol{y})\,\mathrm{d}V_y = 0 \quad (4.32)$$

It relates the values taken by $\boldsymbol{u},\boldsymbol{t}$ on $\partial\Omega$ only, i.e. does not involve the elastic variables at interior points of Ω. Equation (4.32) thus governs *any* elastostatic state on Ω for which assumption (4.31) holds. Moreover, the boundary integral equation (4.32) is formulated in terms of convergent integrals, with weak, integrable, singularities at $\boldsymbol{y}=\boldsymbol{x}$; for this reason it is often said to be *regularized*. Equation (4.32) is the main result of this chapter.

One notices that the boundary integral equation (4.32) is identical to the reformulation (4.29) of the integral representation for the displacement, which was then valid for $\boldsymbol{x} \notin \partial\Omega$. Hence, eq. (4.32) holds whether $\boldsymbol{x}$ is interior, exterior or on the boundary, i.e. for any $\boldsymbol{x}$ lying in the auxiliary domain E for which the fundamental solution is defined.

Somigliana identity

For completeness, recall that a more classical presentation consists of seeking the limit for $\varepsilon \to 0$ in:

$$\int_{(\partial\Omega - e_\varepsilon)+s_\varepsilon} \{u_i(\boldsymbol{y})T_i^k(\boldsymbol{x},\boldsymbol{y}) - t_i(\boldsymbol{y})U_i^k(\boldsymbol{x},\boldsymbol{y})\}\,\mathrm{d}S_y = \int_{\Omega_\varepsilon} \rho F_i(\boldsymbol{y})U_i^k(\boldsymbol{x},\boldsymbol{y})\,\mathrm{d}V_y$$

i.e. without subtracting the rigid-body identity (4.27) and assuming a definite, *spherical* shape (centre $\boldsymbol{x}$, radius ε) for the exclusion neighbourhood $v_\varepsilon(\boldsymbol{x})$. The limiting expression thus obtained is known as the *Somigliana identity*:

$$\kappa_{ki}(\boldsymbol{x})u_i(\boldsymbol{x}) + \mathop{\int\!\!\!\!\!\!-}_{\partial\Omega} \{u_i(\boldsymbol{y})T_i^k(\boldsymbol{x},\boldsymbol{y}) - t_i(\boldsymbol{y})U_i^k(\boldsymbol{x},\boldsymbol{y})\}\,\mathrm{d}S_y - \int_{\Omega} \rho F_i(\boldsymbol{y})U_i^k(\boldsymbol{x},\boldsymbol{y})\,\mathrm{d}V_y = 0 \quad (4.33)$$

The notation $⨍_{\partial\Omega}$ stands for the *Cauchy principal value* (CPV) of a singular integral, i.e. the limit (if it exists):

$$⨍_{\partial\Omega} \{\}\,\mathrm{d}S_y \overset{def}{=} \lim_{\varepsilon\to 0} \int_{\partial\Omega - e_\varepsilon} \{\}\,\mathrm{d}S_y \qquad v_\varepsilon(\boldsymbol{x})\text{: ball of centre } \boldsymbol{x} \text{ and radius } \varepsilon$$

Since this definition of the convergence of integrals relies on a specific choice for the shape of the exclusion neighbourhood, it is weaker than that of ordinary improper integrals (often termed 'weakly singular'), which use exclusion neighbourhoods of arbitrary shape. Besides, the *free term* $\kappa_{ki}(\boldsymbol{x})$ appearing in eq. (4.33), is defined by

$$\kappa_{ki}(\boldsymbol{x}) = \lim_{\varepsilon\to 0} \int_{s_\varepsilon} T_i^k(\boldsymbol{x}, \boldsymbol{y})\,\mathrm{d}S_y$$

It is found to be equal to $1/2\delta_{ki}$ when $\partial\Omega$ is smooth at $\boldsymbol{x}$.

Comments

The boundary integral equation (4.32) is formulated as a limiting case of the exterior representation formula for an exclusion neighbourhood $v_\varepsilon(\boldsymbol{x})$ of vanishing size ε, the source point $\boldsymbol{x}$ being located on the boundary $\partial\Omega$. The shape of $v_\varepsilon(\boldsymbol{x})$ is arbitrary and may vary while $\varepsilon \to 0$.

This presentation departs from the more traditional one, which leads to an integral equation (the Somigliana identity (4.33)) formulated in terms of a specific and restrictive concept of convergence for integrals containing non-integrable singularities. Indeed, the regularized formulation (4.32) is equivalent to the Somigliana identity (4.33) but formulated in terms of weakly singular integrals. This is so essentially because two contributions from non-convergent integrals cancel each other: the sum of the free term and the principal value which appear in the Somigliana identity (4.33) is known, exactly and at once, from the rigid-body identity (4.28). Hence, the boundary integral formulation appears to be actually independent of any particular concept of convergence of singular integrals.

The regularization approach presented here is *indirect*, i.e. it is based on the fact that the fundamental solutions satisfy known integral identities (here: the rigid-body identity). Another approach, proposed by Guiggiani et al. [7]), follows a *direct* course and provides both a correct treatment of the limiting process $\lim_{\varepsilon\to 0}$ and a correct algorithm for the numerical evaluation of the singular element integrals; it is succinctly described in Section 4.8.

Regularity requirements on the unknowns. Another important issue is that $\boldsymbol{u}$ must be $C^{0,\alpha}$-continuous at $\boldsymbol{x}$, because otherwise the integral equation (4.32) is not regularized. This condition arises naturally during the regularization process; it is identical to that ensuring the convergence of the Cauchy principal value integral in the Somigliana identity (4.33), and hence the validity of the

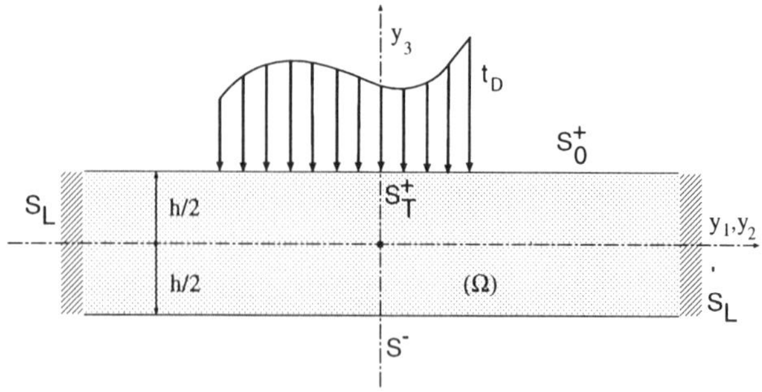

Figure 4.3 Thick elastic plate: geometry and loading.

direct approach for the numerical evaluation of CPV integrals [7]. In this respect, the regularized integral equation (4.32) is also strictly equivalent to the classical Cauchy principal value formulation.

4.5 THE BOUNDARY ELEMENT METHOD IN ELASTOSTATICS

Again, the boundary element method operates in two stages:

1. The *boundary integral equation* is solved, resulting in the knowledge of $(\boldsymbol{u}, \boldsymbol{t})$ everywhere on the boundary $\partial\Omega$.

2. The *integral representation* formulas (4.21), (4.23) then allow an explicit computation of $\boldsymbol{u}(\boldsymbol{x}), \boldsymbol{\varepsilon}(\boldsymbol{x}), \boldsymbol{\sigma}(\boldsymbol{x})$ at any given interior point $\boldsymbol{x}$.

This must be understood as the outline of a *numerical* solution technique, applicable to bodies of arbitrary shape and subjected to arbitrary (well-posed) boundary conditions. Indeed, the integral equations of elastostatics can be solved analytically only exceptionally.

Influence of the choice of fundamental solution

The regularized integral equation (4.32) holds true irrespective of the choice of fundamental solution. However, some choices can be better than others, depending on the shape of the domain Ω and the type of boundary conditions under consideration. To illustrate this, consider two model problems:

Clamped thick elastic plate. A thick plate occupies the region Ω defined by:

$$\Omega = \{\boldsymbol{y} = (y_1, y_2, y_3) \mid (y_1, y_2) \in D, -h/2 \leq y_3 \leq h/2\}$$

where D is an open, connected domain of the y_1, y_2-plane. The thickness h not being small compared to the other plate dimensions, the thick plate is treated as a three-dimensional body. Put $\partial\Omega = S_L \cup S^- \cup S^+$, where S^-, S^+, S_L are the lower, upper and lateral faces of the plate (Figure 4.3). The following boundary conditions are considered, together with vanishing body forces:

- The traction $\boldsymbol{t}^D$ is prescribed over a portion S_T^+ of S^+;
- The surfaces $S_0^+ = S^+ - S_T^+$ and S^- are free of tractions;
- The lateral surface S_L is clamped: $\boldsymbol{u}^D = \boldsymbol{0}$.

The use of fundamental solutions adapted to the geometry and boundary results in a reduction of the actual geometrical support in the integral equation (4.32). Considering in turn several possible choices of fundamental solutions and taking into account the boundary conditions, eq. (4.32) becomes:

1. Kelvin solution:

$$\int_{S^- \cup S_0^+ \cup S_T^+} [u_i(\boldsymbol{y}) - u_i(\boldsymbol{x})] T_i^k(\boldsymbol{x}, \boldsymbol{y}) \, \mathrm{d}S_y - u_i(\boldsymbol{x}) \int_{S_L} T_i^k(\boldsymbol{x}, \boldsymbol{y}) \, \mathrm{d}S_y \\ - \int_{S_L \cup S_T^+} t_i(\boldsymbol{y}) U_i^k(\boldsymbol{x}, \boldsymbol{y}) \, \mathrm{d}S_y = 0$$

2. Mindlin solution (translated so that the free surface is $y_3 = h/2$):

$$\int_{S^-} [u_i(\boldsymbol{y}) - u_i(\boldsymbol{x})] T_i^k(\boldsymbol{x}, \boldsymbol{y}) \, \mathrm{d}S_y - u_i(\boldsymbol{x}) \int_{S_L} T_i^k(\boldsymbol{x}, \boldsymbol{y}) \, \mathrm{d}S_y \\ - \int_{S_L \cup S_T^+} t_i(\boldsymbol{y}) U_i^k(\boldsymbol{x}, \boldsymbol{y}) \, \mathrm{d}S_y = 0$$

3. Solution of Benitez and Rosakis [2]:

$$-u_i(\boldsymbol{x}) \int_{S_L} T_i^k(\boldsymbol{x}, \boldsymbol{y}) \, \mathrm{d}S_y - \int_{S_L \cup S_T^+} t_i(\boldsymbol{y}) U_i^k(\boldsymbol{x}, \boldsymbol{y}) \, \mathrm{d}S_y = 0$$

Soil–structure interaction. In many applications in e.g. geotechnics, the soil in which a construction (foundation, structure, gallery, ...) is embedded is conveniently modelled as an elastic half-space with a free surface. The use (in equilibrium computations) of the Mindlin fundamental solution is thus appropriate. For example, let Ω denote a semi-infinite elastic body defined by the half-space $y_3 \leq 0$ in which a cavity (of boundary S_C and a foundation (of boundary S_F) have been excavated; the remaining free surface is denoted by S_0 (see Figure 4.4). The effect of the structure on the foundation is simulated by a given load applied on S_F. The issue of practical interest is to evaluate the stresses on the cavity surface and in its vicinity induced by the loaded foundation. For this problem:

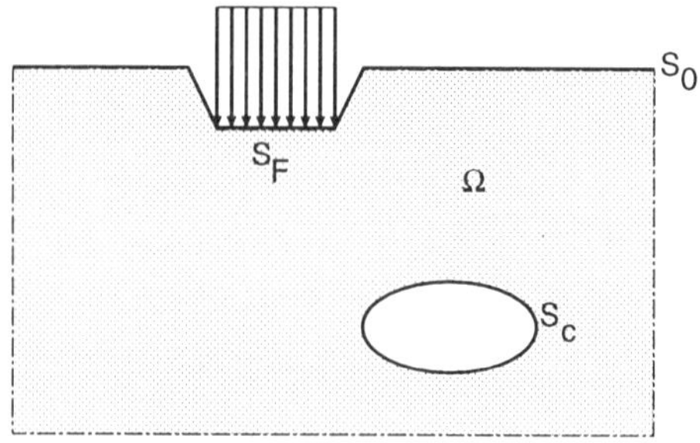

Figure 4.4 Application of the Mindlin solution.

- Using the Kelvin solution, equation (4.32) involves integrals over S_C, S_F *and* S_0. Since the surface S_0 goes to infinity, the discretization of eq. (4.32) implies a choice between (i) truncating S_0 at a distance sufficiently remote from both the foundation and the cavity and (ii) using the knowledge of the asymptotic behaviour of the elastic fields at infinity, for example through infinite boundary elements (Bettess [3]).
- Using the Mindlin solution, integrals over S_0 do not appear in the integral equation; only integrals over the *bounded* surfaces S_C, S_F remain. Furthermore, the causes and effects of practical interest are concentrated in the latter (the solution away from both the foundation and the cavity is usually of little practical value). This approach may bring substantial computational savings and makes the mesh preparation easier.

4.6 INTEGRAL EQUATIONS FOR EXTERIOR PROBLEMS

Consider now an infinite elastic body that occupies the region $\bar{\Omega} = \mathbb{R}^3 - \Omega$, where Ω is a bounded domain. The body force distribution is temporarily assumed to vanish. In this section, the integral equation (4.32) and representation formulas (4.21), (4.23) are shown to remain valid for infinite domains of the form $\bar{\Omega}$ provided the solution $\boldsymbol{u}, \boldsymbol{\sigma}$ satisfy certain conditions at infinity.

The infinite domain $\bar{\Omega}$ is treated as the limiting case of an arbitrarily large bounded domain. Let $\boldsymbol{x} \in \mathbb{R}^3$ be given. Introduce a smooth, closed, surface S_R chosen so as to enclose the ball of centre $\boldsymbol{x}$ and radius R, which in turn should enclose the boundary $\partial\bar{\Omega}$. The integral equation (4.32) is then written for the *bounded* elastic body $\Omega_R \subset \bar{\Omega}$ of external boundary S_R (Figure 4.5):

$$\int_{\partial\bar{\Omega}} \{[u_i(\boldsymbol{y}) - u_i(\boldsymbol{x})]T_i^k(\boldsymbol{x},\boldsymbol{y}) - t_i(\boldsymbol{y})U_i^k(\boldsymbol{x},\boldsymbol{y})\}\, \mathrm{d}S_y$$

$$= \int_{S_R} \{t_i(\boldsymbol{y})U_i^k(\boldsymbol{x},\boldsymbol{y}) - [u_i(\boldsymbol{y}) - u_i(\boldsymbol{x})]T_i^k(\boldsymbol{x},\boldsymbol{y})\}\, \mathrm{d}S_y$$

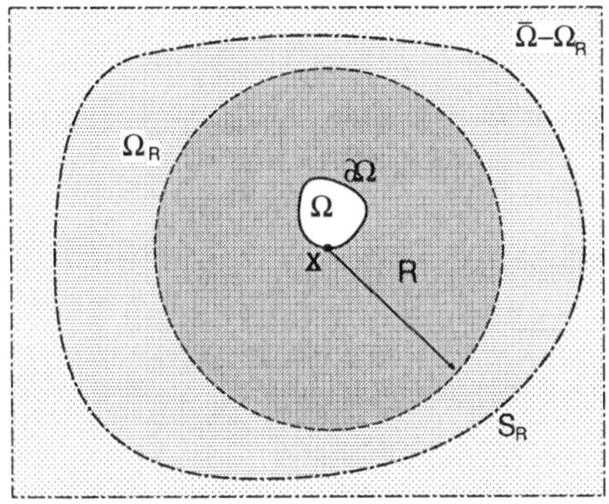

Figure 4.5 Domain Ω_R for the limiting process $r \to \infty$.

Since the limiting situation as Ω_R becomes infinite, i.e. $R \to \infty$, is investigated, the fundamental solution for an infinite elastic space (i.e. Kelvin's) is used. The point $\boldsymbol{x}$ being by construction interior to S_R, the above equation becomes, after application of identity (4.7) with $S = S_R$:

$$\int_{\partial\bar{\Omega}} \{[u_i(\boldsymbol{y}) - u_i(\boldsymbol{x})]T_i^k(\boldsymbol{x},\boldsymbol{y}) - t_i(\boldsymbol{y})U_i^k(\boldsymbol{x},\boldsymbol{y})\} \, \mathrm{d}S_y$$
$$= -u_k(\boldsymbol{x}) + \int_{S_R} \{t_i(\boldsymbol{y})U_i^k(\boldsymbol{x},\boldsymbol{y}) - u_i(\boldsymbol{y})T_i^k(\boldsymbol{x},\boldsymbol{y})\} \, \mathrm{d}S_y \quad (4.34)$$

The Kelvin solution and the differential area element on S_R verify:

$$U_i^k(\boldsymbol{x},\boldsymbol{y}) = O(R^{-1}) \quad T_i^k(\boldsymbol{x},\boldsymbol{y}) = O(R^{-2}) \quad \mathrm{d}S_y = O(R^2)\,\mathrm{d}\omega \qquad (\boldsymbol{y} \in S_R)$$

(ω: solid angle of origin $\boldsymbol{x}$). Hence, under the sufficient conditions on $\boldsymbol{u}, \boldsymbol{\sigma}$:

$$\boldsymbol{u}(\boldsymbol{y}) = o(1) \qquad \boldsymbol{\sigma}(\boldsymbol{y}) = o(R^{-1}) \qquad (\boldsymbol{y} \in S_R) \quad (4.35)$$

one has, independently of any specific assumption about the shape of S_R:

$$\lim_{R\to\infty} \int_{S_R} \{u_i(\boldsymbol{y})T_i^k(\boldsymbol{x},\boldsymbol{y}) - t_i(\boldsymbol{y})U_i^k(\boldsymbol{x},\boldsymbol{y})\} \, \mathrm{d}S_y = 0$$

Integral equation for exterior problems

In the limiting case $R \to \infty$, eq. (4.34) yields the *regularized displacement integral equation for an unbounded body* $\bar{\Omega}$, which governs any elastostatic state over $\bar{\Omega}$ satisfying the conditions at infinity (4.35):

$$u_k(\boldsymbol{x}) + \int_{\partial\bar{\Omega}} \{[u_i(\boldsymbol{y}) - u_i(\boldsymbol{x})]T_i^k(\boldsymbol{x},\boldsymbol{y}) - t_i(\boldsymbol{y})U_i^k(\boldsymbol{x},\boldsymbol{y})\} \, \mathrm{d}S_y = 0 \quad (4.36)$$

This integral equation differs from its counterpart for bounded elastic bodies, eq. (4.32), in two points: the presence of the 'free term' $u_k(\boldsymbol{x})$ outside the integral sign, and the orientation of the unit normal $\boldsymbol{n}$, exterior to the unbounded body $\bar{\Omega}$ and thus interior to $\partial\bar{\Omega}$.

Integral representation formulas

Equation (4.36) is applicable with any choice of the source point $\boldsymbol{x}$. If $\boldsymbol{x} \notin \partial\bar{\Omega}$, identity (4.7) with $S = \partial\bar{\Omega}$ becomes, accounting for the fact that $\boldsymbol{n}$ is here *interior* to $\partial\bar{\Omega}$:

$$(1-\kappa)\delta_{ik} - \int_{\partial\bar{\Omega}} T_i^k(\boldsymbol{x},\boldsymbol{y})\,\mathrm{d}S_y = 0 \qquad \text{with} \begin{cases} \kappa = 1 & \boldsymbol{x} \in \bar{\Omega} \\ \kappa = 0 & \boldsymbol{x} \notin \bar{\Omega} \end{cases} \tag{4.37}$$

The above identity, applied to equation (4.36) with $\boldsymbol{x} \notin \partial\bar{\Omega}$, yields the integral representation formula:

$$\kappa u_k(\boldsymbol{x}) = \int_{\partial\bar{\Omega}} \{t_i(\boldsymbol{y})U_i^k(\boldsymbol{x},\boldsymbol{y}) - u_i(\boldsymbol{y})T_i^k(\boldsymbol{x},\boldsymbol{y})\}\,\mathrm{d}S_y$$

This result is identical to eq. (4.21). The same representation formula for the displacement is thus applicable to bounded as well as unbounded bodies (the unit normal pointing in each case towards the exterior of the domain under consideration). The same is true for the integral representation of the stress tensor, eq. (4.23).

The formulations established in this section show that an exterior problem is solvable by means of an integral equation over a bounded surface. Thus there is no need to introduce fictitious boundaries at finite distances or to resort to infinite elements (Bettess [3]) designed to incorporate the asymptotic behaviour at infinity into finite element formulations. Boundary integral equation formulations are thus very attractive for exterior problems.

Asymptotic behaviour at infinity of elastic integral representations. It is easily seen that the $O(|\boldsymbol{y}-\boldsymbol{x}|^{-1})$ and $O(|\boldsymbol{y}-\boldsymbol{x}|)^{-2}$ behaviour of the Kelvin displacements and stresses imply that any elastostatic state defined by equations (4.21), (4.23) satisfy the conditions at infinity (4.35).

Nonzero body forces

The limiting process $R \to +\infty$ and the formulas (4.21), (4.23) remain valid when nonzero body forces are present subject to the condition:

$$\lim_{R\to\infty} \left| \int_{\Omega_R} \rho F_i(\boldsymbol{y})U_i^k(\boldsymbol{x},\boldsymbol{y})\,\mathrm{d}V_y \right| < +\infty$$

Otherwise, the superposition principle must be invoked, assuming the prior knowledge of a particular solution $\boldsymbol{u}^P$ of the Navier equation (1.17) with body forces. One then puts $\boldsymbol{u} = \boldsymbol{u}^0 + \boldsymbol{u}^P$, the new unknown $\boldsymbol{u}^0$ solving the *homogeneous* Navier equation. A similar procedure should be invoked in the case of loading at infinity (e.g. problems involving infinite elastic media with cracks, see Chapter 13).

As a direct consequence of formulas (4.21,4.23), when finite body forces are distributed over a *bounded* region of Ω and no loading at infinity is present, the fields $\boldsymbol{u}$ and $\boldsymbol{\sigma}$ behave at infinity like Kelvin displacements and stresses created by a concentrated load equal to the resultant of the actual distributed load, i.e. like R^{-1} and R^{-2} respectively, and thus comply with the conditions (4.35). This is a particular instance of the so-called Saint-Venant principle.

4.7 THE BOUNDARY ELEMENT METHOD

The boundary element discretization of equation (4.32) follows the approach developed in Chapter 3 for potential problems and uses essentially the same concepts. The unknown boundary fields (here, the displacement and traction vectors $(\boldsymbol{u}, \boldsymbol{t})$) are interpolated in terms of nodal values $\boldsymbol{u}^k = \tilde{\boldsymbol{u}}(\boldsymbol{z}^k)$ and $\boldsymbol{t}^k = \tilde{\boldsymbol{t}}(\boldsymbol{z}^k)$ at interpolation nodes $\boldsymbol{z}^k$:

$$\tilde{\boldsymbol{u}}(\boldsymbol{y}) = \sum_{k=1}^{N_I(e)} M_k(\boldsymbol{\xi})\boldsymbol{u}^k \qquad \tilde{\boldsymbol{t}}(\boldsymbol{y}) = \sum_{k=1}^{N_I(e)} M_k(\boldsymbol{\xi})\boldsymbol{t}^k \qquad \boldsymbol{\xi} \in \Delta_e \tag{4.38}$$

The nodal unknowns are here *vectors*; this is the only significant difference compared to the treatment of potential problems. The integral equation (4.32) thus takes the following discretized form:

$$\begin{aligned} 0 = &\sum_{e\in I(\boldsymbol{x})} \sum_{p=1}^{N_I(e)} \left\{ AS_i^k(e,p)u_i^{m(e,p)} - B_i^k(e,p)t_i^{m(e,p)} \right\} \\ &+ \sum_{e\in \bar{I}(\boldsymbol{x})} \sum_{p=1}^{N_I(e)} \left\{ AR_i^k(e,p)u_i^{m(e,p)} - B_i^k(e,p)t_i^{m(e,p)} \right\} \\ &- u_i(\boldsymbol{x}) \sum_{e\in \bar{I}(\boldsymbol{x})} \widehat{AR}_i^k(e) \end{aligned} \tag{4.39}$$

in terms of the element matrices:

$$\begin{aligned} AS_i^k(e,p) &= \int_{E_e} [M_p(\boldsymbol{\xi}) - M_p(\boldsymbol{\eta}_e)] T_i^k(\boldsymbol{x},\boldsymbol{y}) J(\boldsymbol{\xi})\, \mathrm{d}\boldsymbol{\xi} \\ B_i^k(e,p) &= \int_{E_e} M_p(\boldsymbol{\xi}) U_i^k(\boldsymbol{x},\boldsymbol{y}) J(\boldsymbol{\xi})\, \mathrm{d}\boldsymbol{\xi} \\ AR_i^k(e,p) &= \int_{E_e} M_p(\boldsymbol{\xi}) T_i^k(\boldsymbol{x},\boldsymbol{y}) J(\boldsymbol{\xi})\, \mathrm{d}\boldsymbol{\xi} \end{aligned}$$

$$\widehat{AR}_i^k(e) = \int_{E_e} T_i^k(\boldsymbol{x}, \boldsymbol{y}) J(\boldsymbol{\xi}) \, d\boldsymbol{\xi}$$

The notation emphasizes the distinction between singular and nonsingular element integration. The linear system of equations (4.39) governs $6N_I$ scalar nodal values $\{u_i^m, t_i^m\}$, $(1 \leq i \leq 3, 1 \leq m \leq N_I)$, which are also constrained by $3N_I$ additional relations expressing the boundary conditions. Using the N_I interpolation nodes $\boldsymbol{z}^k$ as collocation points (the simplest and most common choice), the $3N_I$ scalar equations (4.39) in matrix form are:

$$[\boldsymbol{A}]\{\boldsymbol{u}\} + [\boldsymbol{B}]\{\boldsymbol{t}\} = \boldsymbol{0} \tag{4.40}$$

with

$$\{\boldsymbol{u}\} = \{u_i^m\} \ , \qquad \{\boldsymbol{t}\} = \{t_i^m\}$$

and where $[\boldsymbol{A}]$ and $[\boldsymbol{B}]$ are full $3N_c \times 3N_I$ matrices. Appropriate column swapping between the matrices $[\boldsymbol{A}]$ and $[\boldsymbol{B}]$ allows us to separate knowns (sent in the right-hand side) and the unknowns (kept in the left-hand side). Equation (4.40) thus becomes a square linear system of equations:

$$\boldsymbol{K}\{\boldsymbol{X}\} = \{\boldsymbol{Y}\} \tag{4.41}$$

where $\{\boldsymbol{X}\}$ denotes a 'vector' gathering all the scalar components of $\{\boldsymbol{u}\}, \{\boldsymbol{t}\}$ that remain unknown, the matrix $\boldsymbol{K}$ is made of the columns $[\boldsymbol{A}], [\boldsymbol{B}]$ associated to the degrees of freedom in $\{\boldsymbol{X}\}$ and $\{\boldsymbol{Y}\}$ is the right-hand side equal to (minus) the sum of those columns of $[\boldsymbol{A}], [\boldsymbol{B}]$ associated to known components of $\{\boldsymbol{u}\}, \{\boldsymbol{t}\}$ multiplied by the value taken by these components. The matrix $\boldsymbol{K}$ in eq. (4.41) is full and non-symmetric.

Interpolation of the traction vector

The traction vector $\boldsymbol{t}$ depends linearly upon the unit normal $\boldsymbol{n}$ and hence is expected to undergo jumps across geometrical singularities (edges, corners) or discontinuity lines of boundary conditions.

When $\boldsymbol{t}$ is unknown on one side of the discontinuity line only, one has only to assign traction nodal values at nodes located on such a line to the adjacent surface on which the traction is unknown (i.e. not prescribed by the boundary conditions). The element integrals over adjacent elements on which the traction is given ($\boldsymbol{t} = \boldsymbol{t}^D$) should then be evaluated separately using an analytical expression for $\boldsymbol{t}^D$. Should the boundary data at hand render this treatment impractical, a modification of the interpolation $(4.38)_2$ is advisable in order to account for expected discontinuities of $\boldsymbol{t}$; this can be done using one of several techniques.

Multiple nodes. An interpolation node located on an edge (resp. a corner) supports two (resp. three) vector nodal values $\boldsymbol{t}^{1,2,3}$, each attached to one

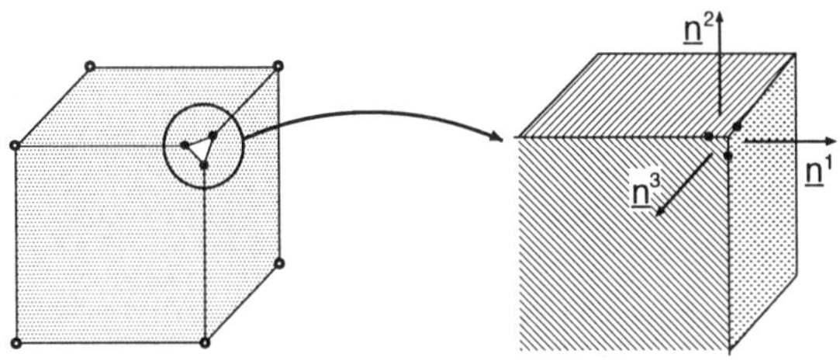

Figure 4.6 Triple node for a corner.

of the smooth surface components adjacent to the geometrical singularity. In this case, assuming the stress *tensor* $\boldsymbol{\sigma}$ is continuous at the edge or corner, the final linear system of equations is supplemented by the relations:

$$\boldsymbol{t}^i.\boldsymbol{n}^j = \boldsymbol{t}^j.\boldsymbol{n}^i$$

Totally non-conformal interpolation. This is the most radical approach to this issue: *all* interpolation nodes for $\boldsymbol{t}$ are chosen interior to the boundary elements, for the whole mesh (Figure 4.7). The interpolation (4.38) is then discontinuous across *any* element edge. For instance, on the square parent element, the interpolation nodes are located on lines $\xi_1 = \xi_1^{\pm}, \xi_2 = \xi_2^{\pm}$ interior to the element $(-1 < \xi_i^- < 0 < \xi_i^+ < 1)$; the special case $\xi_i^{\pm} = \pm 1$ corresponds to the usual conformal interpolation. The interpolation functions $\tilde{M}$ are then defined in terms of 'usual' interpolation functions M by:

$$\tilde{M}_p(\boldsymbol{\xi}) = M_p(\tilde{\boldsymbol{\xi}}) \qquad \text{with} \quad \tilde{\xi}_i = \frac{1-\xi_i}{2}\xi_i^- + \frac{1+\xi_i}{2}\xi_i^+ \tag{4.42}$$

This approach induces a sharp increase in the number of nodal unknowns $\boldsymbol{t}^m$ (compared with the corresponding conformal interpolation), due to the fact that nodes are no longer shared between neighbouring elements, and also

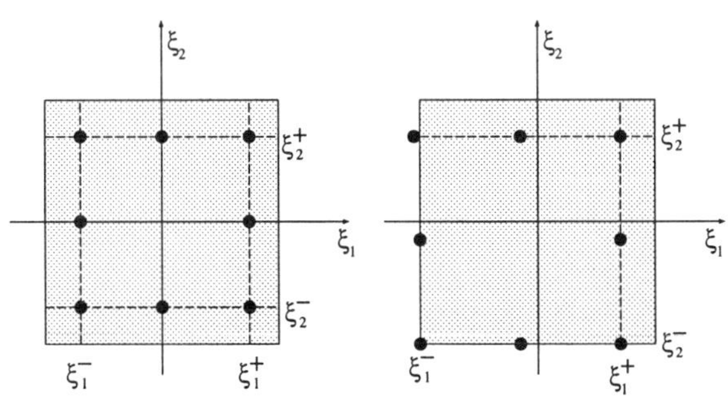

Figure 4.7 Totally (left) or partially (right) non-conformal interpolation.

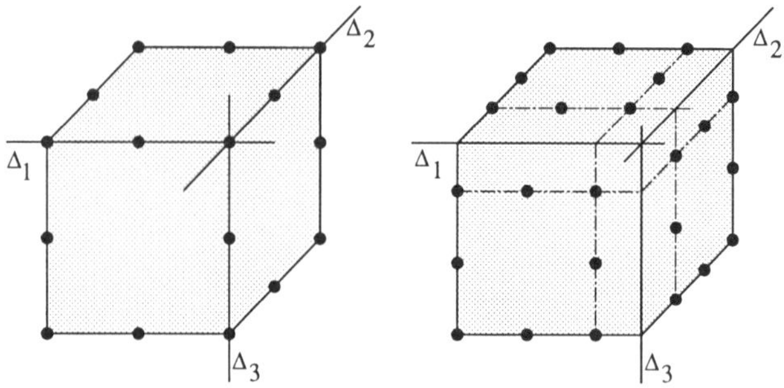

Figure 4.8 Continuous (left) and discontinuous (right) interpolations across edges $\Delta_1, \Delta_2, \Delta_3$.

introduces unwanted discontinuities in the interpolation of $\boldsymbol{t}$; on the other hand its implementation is relatively easy.

Partially non-conformal interpolation. A refinement of the previous approach (Patterson and Sheikh [10]) consists in putting traction interpolation nodes away from edges or lines across which $\boldsymbol{t}$ is known or expected to be discontinuous (e.g. due to change of boundary conditions) while keeping the other interpolation nodes on interelement boundaries, as sketched on Figure 4.8. The modified interpolation functions (4.42) are still used, this time with $\xi_i^{\pm} = \pm 1$ if $\boldsymbol{t}$ is assumed continuous across the line $\xi_i = \pm 1$.

Post-processing: computation of elastic field values at interior points

Once the linear system of equation (4.41) is solved, all nodal values $\boldsymbol{u}^m, \boldsymbol{t}^m$ $(1 \leq m \leq N_I)$ are known. The representation formulas (4.21), (4.23), in discretized form, can then be invoked for explicit evaluation of field variables at (user-specified) interior points $\boldsymbol{x}$. The displacement is thus computed from:

$$u_k(\boldsymbol{x}) = \sum_{e=1}^{N_E} \sum_{p=1}^{N_I(e)} \left\{ t_i^{m(e,p)} \int_{E_e} M_p(\boldsymbol{\xi}) U_i^k(\boldsymbol{x}, \boldsymbol{y}) J(\boldsymbol{\xi}) \, \mathrm{d}\boldsymbol{\xi} - u_i^{m(e,p)} \int_{E_e} M_p(\boldsymbol{\xi}) T_i^k(\boldsymbol{x}, \boldsymbol{y}) J(\boldsymbol{\xi}) \, \mathrm{d}\boldsymbol{\xi} \right\} \tag{4.43}$$

and the displacement gradient (and hence the strain tensor components) from:

$$u_{k,\ell}(\boldsymbol{x}) = \sum_{e=1}^{N_E} \sum_{p=1}^{N_I(e)} \left\{ t_a^{m(e,p)} \int_{E_e} M_p(\boldsymbol{\xi}) U_{a,\bar{\ell}}^k(\boldsymbol{x}, \boldsymbol{y}) J(\boldsymbol{\xi}) \, \mathrm{d}\boldsymbol{\xi} - u_a^{m(e,p)} \int_{E_e} M_p(\boldsymbol{\xi}) n_b(\boldsymbol{y}) \Sigma_{ab,\bar{\ell}}^k(\boldsymbol{x}, \boldsymbol{y}) J(\boldsymbol{\xi}) \, \mathrm{d}\boldsymbol{\xi} \right\} \tag{4.44}$$

the stress tensor components being of course obtained from Hooke's law:

$$\sigma_{ij}(\boldsymbol{x}) = C_{ijk\ell}u_{k,\ell}(\boldsymbol{x}) \tag{4.45}$$

The presence of terms behaving like $|\boldsymbol{y}-\boldsymbol{x}|^{-1}$, $|\boldsymbol{y}-\boldsymbol{x}|^{-2}$ in (4.43) and like $|\boldsymbol{y}-\boldsymbol{x}|^{-2}$, $|\boldsymbol{y}-\boldsymbol{x}|^{-3}$ in (4.44) raises the issue of the accurate evaluation of near-singular integrals (see Section 3.4) for interior points $\boldsymbol{x}$ very close to $\partial\Omega$.

Post-processing: evaluation of stresses on the boundary

It is often necessary to compute the stress tensor $\boldsymbol{\sigma}(\boldsymbol{x})$ at boundary points $\boldsymbol{x} \in \partial\Omega$. However, solving the boundary integral equation (3.13) only gives direct access to the traction vector $\boldsymbol{t} = \boldsymbol{\sigma}.\boldsymbol{n}$, while the representation formula (4.23) is not valid in this form for $\boldsymbol{x} \in \partial\Omega$ due to its hypersingular character. Representation formulas for stresses or displacement gradients at boundary points can be established; this issue is addressed in Chapter 5.

A more classical treatment uses the following: the traction vector $\boldsymbol{t}$, known over $\partial\Omega$, depends linearly through Hooke's law on the displacement gradient $\boldsymbol{\nabla u}$. The latter can be decomposed into a tangential gradient and a normal derivative, i.e.:

$$\boldsymbol{t} = \boldsymbol{n}.\boldsymbol{C} : (\boldsymbol{\nabla}_S \boldsymbol{u} + \boldsymbol{u}_{,n} \otimes \boldsymbol{n}) \tag{4.46}$$

This linear relation can be inverted so as to express the normal derivative $\boldsymbol{u}_{,n}(\boldsymbol{x})$ in terms of $\boldsymbol{t}(\boldsymbol{x})$ and $\boldsymbol{\nabla}_S \boldsymbol{u}(\boldsymbol{x})$; the latter quantities are known once the integral equation (4.32) is solved. Then, the complete displacement gradient $\boldsymbol{\nabla u}(\boldsymbol{x})$ can be evaluated using (4.46), as well as the complete stress tensor $\sigma(\boldsymbol{x})$ using:

$$\boldsymbol{\sigma}(\boldsymbol{x}) = \boldsymbol{C} : (\boldsymbol{\nabla}_S \boldsymbol{u} + \boldsymbol{u}_n \otimes \boldsymbol{n})(\boldsymbol{x})$$

Equation (A.13) allows the practical computation of the tangential gradient $\boldsymbol{\nabla}_S \boldsymbol{u}$ from the displacement interpolation (3.12), resulting in:

$$\boldsymbol{\nabla}_S \boldsymbol{u} = \sum_{k=1}^{N_I(e)} \left[M_{k,\alpha}(\boldsymbol{\eta}) g^{\alpha\beta} \boldsymbol{u}^k \right] \otimes \boldsymbol{a}_\beta(\boldsymbol{\eta})$$

where $g^{\alpha\beta}$ denote the contravariant components of the metric tensor associated with the element parametrization (the covariant components $g_{\alpha\beta}$ are given by $g_{\alpha\beta} = \boldsymbol{a}_\alpha.\boldsymbol{a}_\beta$, and the matrices $[g^{\alpha\beta}]$ and $[g_{\alpha\beta}]$ are inverses of each other), and $\boldsymbol{\eta} = (\eta_1, \eta_2)$ is the variable in the parent element.

This approach assumes that the tangential gradient $\boldsymbol{\nabla}_S \boldsymbol{u}$ is continuous at $\boldsymbol{x}$, and thus is not applicable at points $\boldsymbol{x}$ shared by two or more elements, unless a C^1 interpolation for *both* the displacement and the boundary is used. Therefore, there is no straightforward procedure to compute stresses at interelement nodes (i.e. at most or even all nodes, depending on the particular interpolation functions being used) .

Isotropic elasticity. Explicit formulas are available in this case. Indeed, from eq. (1.18), one has:

$$\begin{aligned} \boldsymbol{t} &= \mu \left[\frac{2\nu}{1-2\nu} (\mathrm{div}\boldsymbol{u})\boldsymbol{n} + \boldsymbol{u}_{,n} + \boldsymbol{n}.\boldsymbol{\nabla}\boldsymbol{u} \right] \\ &= \mu \left[\frac{2\nu}{1-2\nu} (\mathrm{div}_S\boldsymbol{u})\boldsymbol{n} + \boldsymbol{n}.\boldsymbol{\nabla}_S\boldsymbol{u} + \left(\frac{1}{1-2\nu}\boldsymbol{n}\otimes\boldsymbol{n} + \boldsymbol{I} \right).\boldsymbol{u}_n \right] \end{aligned}$$

which inversion leads to:

$$\boldsymbol{u}_{,n} = \frac{1}{\mu} \left[\boldsymbol{I} - \frac{1}{2(1-\nu)} \boldsymbol{n}\otimes\boldsymbol{n} \right] .\boldsymbol{t} - \boldsymbol{n}.\boldsymbol{\nabla}_S\boldsymbol{u} - \frac{\nu}{1-\nu} (\mathrm{div}_S\boldsymbol{u})\boldsymbol{n}$$

Computation of domain integrals

When nonzero body forces are present, a domain integral appears in the right-hand side in the integral equation (4.32):

$$\int_\Omega \rho F_i(\boldsymbol{y}) U_i^k(\boldsymbol{x},\boldsymbol{y}) \, \mathrm{d}V_y \tag{4.47}$$

It is obviously desirable to convert it into boundary integrals, so that the domain boundary supports all computations, like in the absence of body forces. This is possible in several cases, described below. One must otherwise resort to domain integrations, most often using *integration cells*, i.e. finite-element-like subdivisions allowing the use of Gaussian quadrature.

Isotropic elasticity and body forces deriving from a potential. Any fundamental solution $\boldsymbol{U}(\boldsymbol{x},\boldsymbol{y})$ for isotropic elasticity is representable in terms of the *Galerkin tensor*, eq. (1.21) $\boldsymbol{G}(\boldsymbol{x},\boldsymbol{y})$:

$$U_i^k(\boldsymbol{x},\boldsymbol{y}) = 2(1-\nu) G_{i,jj}^k(\boldsymbol{x},\boldsymbol{y}) - G_{j,ji}^k(\boldsymbol{x},\boldsymbol{y}) \tag{4.48}$$

The Galerkin tensor is given by:

$$G_i^k(\boldsymbol{x},\boldsymbol{y}) = F(\boldsymbol{x},\boldsymbol{y})\delta_{ik} = \frac{1}{16\pi(1-\nu)\mu} r\delta_{ik} \tag{4.49}$$

for the three-dimensional Kelvin fundamental solution (see Appendix 4.A) and by equation (4.A.14) for the Mindlin fundamental solution (half-space with a free surface).

Also, assume the existence of a potential Φ and a constant b such that:

$$\rho \boldsymbol{F} = \boldsymbol{\nabla}\Phi \qquad \text{with} \qquad \Delta\Phi = b$$

Then, one can perform an integration by parts, as follows. Noting that:

$$U_{i,i}^k = (1-2\nu) G_{i,ijj}^k$$

one has:

$$\int_{\Omega} U_i^k.\Phi_{,i}\,\mathrm{d}V_y = \int_{\partial\Omega} \Phi U_i^k \boldsymbol{n}_i\,\mathrm{d}S_y - (1-2\nu)\int_{\Omega} \Phi G_{i,ijj}^k\,\mathrm{d}V_y$$
$$= \int_{\partial\Omega} \{\Phi[U_j^k - (1-2\nu)G_{i,ij}^k]n_j + (1-2\nu)\Phi_{,j}n_jG_{i,i}^k\}\,\mathrm{d}S_y$$
$$- (1-2\nu)\int_{\Omega} \Phi_{,jj}G_{i,i}^k\,\mathrm{d}V_y$$

Then:

$$\int_{\Omega} \Phi_{,jj}G_{i,i}^k\,\mathrm{d}V_y = b\int_{\Omega} G_{i,i}^k\,\mathrm{d}V_y = b\int_{\partial\Omega} G_i^k n_i\,\mathrm{d}S_y$$

As a result of these manipulations, the domain integral (4.47) is converted into a sum of boundary integrals:

$$\int_{\Omega} \rho F_i U_i^k\,\mathrm{d}V_y = \int_{\partial\Omega} \{2(1-\nu)\Phi[G_{j,ii}^k - G_{i,ij}^k]n_j + (1-2\nu)\Phi_{,j}n_jG_{i,i}^k\}\,\mathrm{d}S_y$$
$$- (1-2\nu)b\int_{\partial\Omega} G_i^k n_i\,\mathrm{d}S_y$$

This result includes several important situations:

- Gravity forces, with $\Phi = -\rho g y_3$ and $b = 0$.
- Centrifugal inertia, with $\Phi(\boldsymbol{y}) = \frac{1}{2}\rho\omega^2 R^2(\boldsymbol{y})$ and $b = 2\rho\omega^2$ ($\omega/2\pi$: rotation frequency, $R(\boldsymbol{y})$: distance between $\boldsymbol{y}$ and rotation axis).
- Thermal loading induced by a steady-state temperature field T verifying $\Delta T = r$ (usually $r = 0$):

$$\boldsymbol{\sigma} = \boldsymbol{C} : \boldsymbol{\nabla u} - (3\lambda + 2\mu)\alpha T\boldsymbol{I} \quad \text{in } \Omega$$

(α: thermal dilatation coefficient), with $\Phi = (3\lambda + 2\mu)\alpha T$, $b = (3\lambda + 2\mu)\alpha r$. The notation $\boldsymbol{t}$ in the integral equation (4.32) refers to $\boldsymbol{t} = \boldsymbol{T}^n(\boldsymbol{u})$, and thus differs from the traction vector $\boldsymbol{\sigma}.\boldsymbol{n}$ in this case.

Multiple reciprocity method. The approach presented in Section 3.6 is applicable to elasticity problems, provided a sequence $\boldsymbol{U}^{(N)}(\boldsymbol{x},\boldsymbol{y}), N \geq 0$ of fundamental solutions such that:

$$\boldsymbol{U}^{(0)}(\boldsymbol{x},\boldsymbol{y}) = \boldsymbol{U}(\boldsymbol{x},\boldsymbol{y}) \qquad \boldsymbol{U}^{(N)}(\boldsymbol{x},\boldsymbol{y}) = \boldsymbol{\Delta U}^{(N+1)}(\boldsymbol{x},\boldsymbol{y})$$

is available. In particular, considering the three-dimensional Kelvin solution, introduce:

$$G_i^{k(N)}(\boldsymbol{x},\boldsymbol{y}) = \frac{1}{16\pi\mu(1-\nu)}\delta_{ik}\frac{r^{2N+1}}{(2N+2)!}$$

The Galerkin representation (4.48) yields the desired sequence $\boldsymbol{U}^{(N)}$. One then proceeds along the lines of Section 3.6, through repeated applications of the third Green identity (2.4), obtaining:

$$\int_{\Omega} \boldsymbol{F}.\boldsymbol{U}^k \,\mathrm{d}V_y = \int_{\Omega} \boldsymbol{\Delta}^N \boldsymbol{F}.\boldsymbol{U}^{k(N)} \,\mathrm{d}V_y + \sum_{p=0}^{N-1} \int_{\partial\Omega} \left(\boldsymbol{U}^{k(p+1)}_{,n}.\boldsymbol{\Delta}^p \boldsymbol{F} - \boldsymbol{U}^{k(p+1)}.(\boldsymbol{\Delta}^p \boldsymbol{F})_{,n} \right) \mathrm{d}S_y$$

for an arbitrary value of the integer N ($\boldsymbol{\Delta}^p$ denotes the vector Laplace operator iterated p times).

Numerical evaluation using integration cells

One has to resort to them when body forces are known only numerically. Many situations of practical interest fall within this category: thermomechanical coupled problems where the temperature distribution does not satisfy $\Delta T = b$, initial strain or stress distributions associated with incremental treatment of nonlinear constitutive properties, e.g. elastic–plastic problems (Chapter 14).

Substructuring

Applied to elasticity problems, it provides both the ability to deal with heterogeneous, piecewise uniform, elastic properties and reduced computer time and storage requirements for a given problem. Its use is mandatory in the former case, because fundamental solutions are available only for homogeneous media or, at best, geometrically very simple heterogeneities, e.g. parallel stratification.

4.8 THE DIRECT APPROACH FOR SINGULAR ELEMENT INTEGRATION

A numerical evaluation technique of the singular element integrals making use of the indirect regularization idea has been presented in Section 3.5. An alternative approach, termed 'direct', has been developed by Guiggiani and co-workers [6, 5] and is widely used. This section aims at succinctly presenting this direct approach.

The direct approach does not rely on a prior regularization. Instead, the limiting process leading to the statement of the displacement boundary integral equation is revisited. Recall that the limiting form as $\varepsilon \to 0$ of

$$\int_{(\partial\Omega - e_\varepsilon)+s_\varepsilon} \left\{ u_i(\boldsymbol{y}) T_i^k(\boldsymbol{x},\boldsymbol{y}) - t_i(\boldsymbol{y}) U_i^k(\boldsymbol{x},\boldsymbol{y}) \right\} \mathrm{d}S_y = 0 \tag{4.50}$$

is sought (assuming zero body forces for simplicity). At this point, the integrals over s_ε and $\partial\Omega - e_\varepsilon$ are treated separately. First, one has, assuming as usual a $C^{0,\alpha}$ regularity for $\boldsymbol{u}$ at $\boldsymbol{x}$:

$$u_i(\boldsymbol{y})T_i^k(\boldsymbol{x},\boldsymbol{y})\,\mathrm{d}S_y = u_i(\boldsymbol{x})\int_{s_\varepsilon} T_i^k(\boldsymbol{x},\boldsymbol{y})\,\mathrm{d}S_y + O(\varepsilon) \tag{4.51}$$

and from equation (4.14), the limit

$$\kappa_{ik}(\boldsymbol{x}) = \lim_{\varepsilon\to 0}\int_{s_\varepsilon} T_i^k(\boldsymbol{x},\boldsymbol{y})\,\mathrm{d}S_y$$

is known to be bounded and to depend on the shape of s_ε. Next, one has

$$\lim_{\varepsilon\to 0}\int_{\partial\Omega - e_\varepsilon} t_i(\boldsymbol{y})U_i^k(\boldsymbol{x},\boldsymbol{y})\,\mathrm{d}S_y = \int_{\partial\Omega} t_i(\boldsymbol{y})U_i^k(\boldsymbol{x},\boldsymbol{y})\,\mathrm{d}S_y$$

as usual (weakly singular integral, i.e. whose limiting value does not depend on the shape of e_ε). Finally, the limit

$$\lim_{\varepsilon\to 0}\int_{\partial\Omega - e_\varepsilon} u_i(\boldsymbol{y})T_i^k(\boldsymbol{x},\boldsymbol{y})\,\mathrm{d}S_y \tag{4.52}$$

is necessarily bounded (since the last three limits must add up to zero for any ε, eq. (4.50)), and is expected to depend on the shape of e_ε.

At this point, recognizing (as has also been done using indirect regularization) that the limit as $\varepsilon \to 0$ is *globally* (i.e. for the whole identity (4.50)) independent of the shape of the exclusion neighbourhood v_ε, a spherical shape is now used for convenience. In that case, one has $\kappa_{ik}(\boldsymbol{x}) = 1/2$ if $\boldsymbol{x}$ is a smooth boundary point, and is known analytically for other configurations (edges, corners,...). Now, the main idea behind the direct method is to introduce the boundary element discretization into (4.52) *before* considering the limiting process $\varepsilon \to 0$.

Let us assume for simplicity that the collocation point $\boldsymbol{x}$ is interior to some boundary element E. Like in Section 3.5, polar coordinates (ρ, α) are introduced in the parent element Δ_e (Figure 4.9); their range is:

$$-1 \le \xi_1, \xi_2 \le 1 \Leftrightarrow \begin{cases} 0 \le \alpha \le 2\pi \\ 0 \le \rho \le \bar{\rho}(\alpha) \end{cases}$$

where the upper bound function $\bar{\rho}(\alpha)$ is the piecewise regular parametrization (in the (ρ, α) system) of the contour of the element. In practice a subdivision into triangles of E must be introduced (Figure 3.7).

From eqs (3.28) and (3.29), one easily sees that:

$$r = \rho(\boldsymbol{a}_1(\boldsymbol{\eta})\cos\alpha + \boldsymbol{a}_2(\boldsymbol{\eta})\sin\alpha) + \rho^2 O(1) = \rho\hat{r}(\rho,\alpha;\boldsymbol{\eta})$$

Hence, the image in Δ_e of the contour of e_ε is defined by $r = \varepsilon$, i.e.:

$$\rho = A^{-1}(\alpha)\varepsilon(1 + O(\varepsilon)) \qquad \text{with} \qquad A(\alpha) = |\boldsymbol{a}_1(\boldsymbol{\eta})\cos\alpha + \boldsymbol{a}_2(\boldsymbol{\eta})\sin\alpha|$$

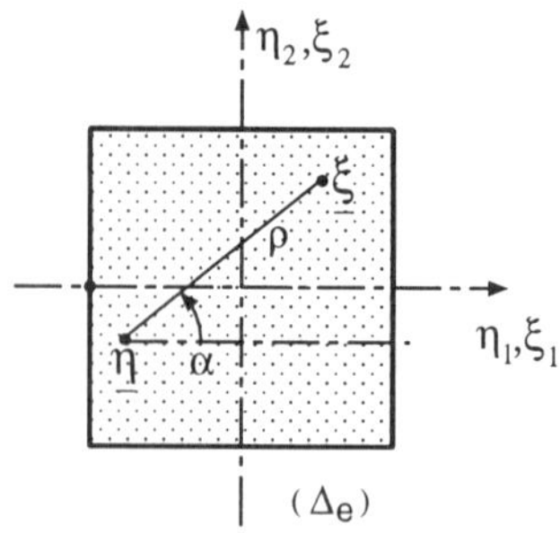

Figure 4.9 Polar coordinates (ρ, α) of origin η in the parent element Δ_e.

Besides, the singular character of $\Sigma_{ij}^k(\boldsymbol{x}, \boldsymbol{y})$ implies that:

$$\Sigma_{ij}^k(\boldsymbol{x}, \boldsymbol{y}) = \frac{1}{\hat{r}^2}\hat{\Sigma}_{ij}^k(\rho, \alpha; \boldsymbol{\eta})$$

Introducing interpolation functions M_p for the displacement, one then has:

$$\begin{aligned}
&\int_{E-e_\varepsilon} M_p(\boldsymbol{y}) T_i^k(\boldsymbol{x}, \boldsymbol{y}) \, \mathrm{d}S_y \\
&= \int_0^{2\pi} \int_{A^{-1}(\alpha)\varepsilon(1+O(\varepsilon))}^{\bar{\rho}(\alpha)} M_p(\boldsymbol{\xi}) \Sigma_{ij}^k(\boldsymbol{x}, \boldsymbol{y}(\boldsymbol{\xi})) n_j(\boldsymbol{y}(\boldsymbol{\xi})) J(\boldsymbol{\xi}) \rho \, \mathrm{d}\rho \mathrm{d}\alpha \\
&= \int_0^{2\pi} \int_{A^{-1}(\alpha)\varepsilon(1+O(\varepsilon))}^{\bar{\rho}(\alpha)} \frac{1}{\rho} \Big\{ M_p(\boldsymbol{\xi}) \hat{\Sigma}_{ij}^k(\rho, \alpha; \boldsymbol{\eta}) n_j(\boldsymbol{y}(\boldsymbol{\xi})) J(\boldsymbol{\xi}) \\
&\qquad - M_p(\boldsymbol{\eta}) \hat{\Sigma}_{ij}^k(0, \alpha; \boldsymbol{\eta}) n_j(\boldsymbol{x}) J(\boldsymbol{\eta}) \Big\} \, \mathrm{d}\rho \mathrm{d}\alpha \\
&\quad + M_p(\boldsymbol{\eta}) n_j(\boldsymbol{x}) J(\boldsymbol{\eta}) \int_0^{2\pi} \hat{\Sigma}_{ij}^k(0, \alpha; \boldsymbol{\eta}) \int_{A^{-1}(\alpha)\varepsilon(1+O(\varepsilon))}^{\bar{\rho}(\alpha)} \frac{1}{\rho} \, \mathrm{d}\rho \mathrm{d}\alpha
\end{aligned}$$

In the last expression, the term in brackets is integrable at $\rho = 0$ (so that the limit for $\varepsilon \to 0$ is the integral over $0 \leq \rho \leq \bar{\rho}(\alpha)$), and the remaining term is treated analytically:

$$\begin{aligned}
&\lim_{\varepsilon \to 0} \int_{E-e_\varepsilon} M_p(\boldsymbol{y}) T_i^k(\boldsymbol{x}, \boldsymbol{y}) \, \mathrm{d}S_y \\
&= \int_0^{2\pi} \int_0^{\bar{\rho}(\alpha)} \frac{1}{\rho} \Big\{ M_p(\boldsymbol{\xi}) \hat{\Sigma}_{ij}^k(\rho, \alpha; \boldsymbol{\eta}) n_j(\boldsymbol{y}(\boldsymbol{\xi})) J(\boldsymbol{\xi}) \\
&\qquad - M_p(\boldsymbol{\eta}) \hat{\Sigma}_{ij}^k(0, \alpha; \boldsymbol{\eta}) n_j(\boldsymbol{x}) J(\boldsymbol{\eta}) \Big\} \, \mathrm{d}\rho \mathrm{d}\alpha \\
&\quad + M_p(\boldsymbol{\eta}) n_j(\boldsymbol{x}) J(\boldsymbol{\eta}) \lim_{\varepsilon \to 0} \int_0^{2\pi} \Big[\ln \bar{\rho}(\alpha) - \ln\Big(A^{-1}(\alpha)\varepsilon\Big) \Big] \hat{\Sigma}_{ij}^k(0, \alpha; \boldsymbol{\eta}) \, \mathrm{d}\alpha
\end{aligned}$$

In fact, the logarithmic term containing ε is not allowed to blow up in the limit $\varepsilon \to 0$ since a bounded overall limit is expected and no other unbounded contribution can possibly arise. Hence, one has necessarily:

$$\int_0^{2\pi} \hat{\Sigma}_{ij}^k(0,\alpha;\boldsymbol{\eta})\,\mathrm{d}\alpha = 0$$

regardless of the particular fundamental solution being used (a direct proof of this last property can be given for the Kelvin solution, among others).

Finally, the strongly singular integral is computed from the formula:

$$\begin{aligned}
&\lim_{\varepsilon\to 0}\int_{E-e_\varepsilon} M_p(\boldsymbol{y})T_i^k(\boldsymbol{x},\boldsymbol{y})\,\mathrm{d}S_y \\
&= \int_0^{2\pi}\int_0^{\bar{\rho}(\alpha)} \frac{1}{\rho}\Big\{M_p(\boldsymbol{\xi})\hat{\Sigma}_{ij}^k(\rho,\alpha;\boldsymbol{\eta})n_j(\boldsymbol{y}(\boldsymbol{\xi}))J(\boldsymbol{\xi}) \\
&\qquad\qquad - M_p(\boldsymbol{\eta})\hat{\Sigma}_{ij}^k(0,\alpha;\boldsymbol{\eta})n_j(\boldsymbol{x})J(\boldsymbol{\eta})\Big\}\,\mathrm{d}\rho\mathrm{d}\alpha \\
&\quad + M_p(\boldsymbol{\eta})n_j(\boldsymbol{x})J(\boldsymbol{\eta})\int_0^{2\pi} \ln\bar{\rho}(\alpha)\hat{\Sigma}_{ij}^k(0,\alpha;\boldsymbol{\eta})\,\mathrm{d}\alpha
\end{aligned} \tag{4.53}$$

4.9 ELASTICITY IN PLANE STRAIN

In the *plane strain* assumption, the elastic body occupies a domain Ω invariant under translation along $\boldsymbol{e}_3$ (say), i.e. $\Omega = D \times \mathbb{R}$, where D is a subset of the plane y_1y_2, while the displacement field has the form:

$$\boldsymbol{u}(\boldsymbol{y}) = u_1(y_1,y_2)\boldsymbol{e}_1 + u_2(y_1,y_2)\boldsymbol{e}_2$$

One can again, as in Section 2.5, integrate the integral equation (4.32) for the *three-dimensional* domain Ω of boundary $\partial D \times \mathbb{R}$ along $\boldsymbol{e}_3$:

$$\begin{aligned}
&\int_{\partial D} [u_\alpha(\tilde{y}) - u_\alpha(\tilde{x})]\int_{\mathbb{R}} T_{\alpha\beta}^\gamma(\boldsymbol{x},\tilde{y})\,\mathrm{d}x_3\,\mathrm{d}s_y \\
&\quad - \int_{\partial D} t_\alpha(\tilde{y})\int_{\mathbb{R}} [U_\alpha^\gamma(\boldsymbol{x},\tilde{y}) - U_\alpha^\gamma(\boldsymbol{x},\tilde{z})]\,\mathrm{d}x_3\,\mathrm{d}s_y \\
&\qquad = \int_D \rho F_\alpha(\tilde{y})\int_{\mathbb{R}} [U_\alpha^\gamma(\boldsymbol{x},\tilde{y}) - U_\alpha^\gamma(\boldsymbol{x},\tilde{z})]\,\mathrm{d}x_3\,\mathrm{d}D_y
\end{aligned} \tag{4.54}$$

where $\boldsymbol{x} \equiv (x_1,x_2,x_3)$, $\tilde{x} = (x_1,x_2),\ldots$ denote points in D; the Greek indices conventionally take the values $1,2$. Use has been made of the following identity, expressing global equilibrium of D:

$$\int_{\partial D} t_\alpha(\boldsymbol{y})\,\mathrm{d}s_y + \int_D \rho F_\alpha(\boldsymbol{y})\,\mathrm{d}D_y = 0$$

The fundamental solution in plane strain is defined as:

$$[U_\alpha^\gamma(\tilde{x},\tilde{y})]_{2D} = \int_{\mathbb{R}} [U_\alpha^\gamma(\boldsymbol{x},\tilde{y}) - U_\alpha^\gamma(\boldsymbol{x},\tilde{\boldsymbol{z}})]\,\mathrm{d}x_3 \tag{4.55}$$

$$[T_\alpha^\gamma(\tilde{x},\tilde{y})]_{2D} = \int_{\mathbb{R}} T_\alpha^\gamma(\boldsymbol{x},\tilde{y})\,\mathrm{d}x_3 \tag{4.56}$$

where $r = |\tilde{y} - \tilde{x}|$ is the Euclidean distance in the $(1,2)$-plane, $\tilde{\boldsymbol{z}} \in D$ is a fixed point $\neq \tilde{x}$, arbitrarily chosen, and $r' = |\tilde{\boldsymbol{z}} - \tilde{x}|$.

Kelvin fundamental solution for plane strain

The explicit form taken by eqs (4.55–4.56) after introduction of the three-dimensional Kelvin solution (4.8–4.9) gives as a result:

$$\left[U_\alpha^\gamma(\tilde{x},\tilde{y})\right]_{2D} = \frac{1}{8\pi\mu(1-\nu)}\left[r_{,\alpha}r_{,\gamma} - (3-4\nu)\delta_{\alpha\gamma}\log\frac{r}{r'}\right] \tag{4.57}$$

$$\left[\Sigma_{\alpha\beta}^\gamma(\tilde{x},\tilde{y})\right]_{2D} = \frac{-1}{4\pi(1-\nu)r}\Big[2r_{,\alpha}r_{,\gamma}r_{,\beta} + (1-2\nu)(\delta_{\alpha\gamma}r_{,\beta} + \delta_{\beta\gamma}r_{,\alpha} - \delta_{\alpha\beta}r_{,\gamma})\Big] \tag{4.58}$$

The solution $[\boldsymbol{U}^\gamma]_{2D}$ so obtained is defined up to an arbitrary translation through the choice of the fixed point $\boldsymbol{z}$ (and hence of ρ'). It has a $\log r$ singular behaviour at $\boldsymbol{y} = \boldsymbol{x}$. The $\log r$ term also means that it becomes infinite for $\rho \to \infty$. This is due to the fact that the two-dimensional fundamental solution is actually the response of a three-dimensional infinite body to an *infinite line* of unit point loads, i.e. a loading of infinite resultant.

The specific plane-strain Kelvin solution (4.57–4.58) was obtained with $\tilde{\boldsymbol{z}}$ chosen such that $z_\gamma = x_\gamma$, the coordinate $z_{3-\gamma}$ being arbitrary, and using the integrals:

$$\int_{\mathbb{R}}\left(\frac{1}{\sqrt{\rho^2 + x_3^2}} - \frac{1}{\sqrt{\rho'^2 + x_3^2}}\right)\mathrm{d}x_3 = -2\log\frac{\rho}{\rho'}$$

$$\int_{\mathbb{R}}\frac{\mathrm{d}x_3}{(\rho^2 + x_3^2)^{n+1/2}} = \frac{2n}{2n-1}\frac{1}{\rho^{2n}} \qquad (n = 1,2)$$

with $\rho^2 = (y_1 - x_1)^2 + (y_2 - x_2)^2$ and $\rho'^2 = (z_{3-\gamma} - x_{3-\gamma})^2$. Seeking directly a fundamental solution to the plane strain elasticity equations yields the same result, i.e. (4.57–4.58).

The Galerkin tensor associated with the plane-strain Kelvin solution is given by:

$$G_\alpha^\gamma(\boldsymbol{x},\boldsymbol{y}) = \frac{-1}{16\pi\mu(1-\nu)}(r^2\ln r)\delta_{\gamma\alpha} \tag{4.59}$$

Integral equation for plane strain

Any plane-strain elastostatic state verifies the following displacement boundary integral equation, obtained directly from eq. (4.54):

$$\int_{\partial D} \{[u_\alpha(\tilde{y}) - u_\alpha(\tilde{x})]T^\gamma_\alpha(\tilde{x},\tilde{y}) - t_\alpha(\tilde{y})U^\gamma_\alpha(\tilde{x},\tilde{y})\}\,\mathrm{d}s_y$$

$$= \int_D \rho F_\alpha(\tilde{y})U^\gamma_\alpha(\tilde{x},\tilde{y})\,\mathrm{d}D_y \qquad (4.60)$$

where the plane-strain fundamental solution $[\boldsymbol{U},\boldsymbol{\Sigma}]_{2D}$ is used. Like its three-dimensional counterpart, it becomes the integral representation for $\boldsymbol{u}(\tilde{x})$ when $\tilde{x} \notin \partial D$. Application of Hooke's law then gives as usual the representation formula for the stress tensor $\boldsymbol{\sigma}(\tilde{x})$.

Exterior problems

Extension to unbounded domains (with $\boldsymbol{F} = 0$) proceeds in the same fashion as in the three-dimensional case. The two-dimensional Kelvin solution behaves at infinity like:

$$\boldsymbol{U}(\tilde{x},\tilde{y}) = O(\ln r) \qquad \boldsymbol{\Sigma}(\tilde{x},\tilde{y}) = O(1/r) \qquad r = |\tilde{y} - \tilde{x}| \to +\infty$$

Using an argument similar to that of Section 4.6, one shows that for any elastostatic state with zero body forces and satisfying the sufficient conditions at infinity:

$$\boldsymbol{u}(\tilde{y}) = O(R^{-\alpha}) \qquad \boldsymbol{\sigma}(\tilde{y}) = O(R^{-1-\alpha}) \qquad \tilde{y} \in S_R, R \to +\infty \quad (4.61)$$

the following integral equation holds:

$$u_\gamma(\tilde{x}) + \int_{\partial D} \{[u_\alpha(\tilde{y}) - u_\alpha(\tilde{x})]T^\gamma_\alpha(\tilde{x},\tilde{y}) - t_\alpha(\tilde{y})U^\gamma_\alpha(\tilde{x},\tilde{y})\}\,\mathrm{d}s_y = 0$$

Like in the three-dimensional case, when $\tilde{x} \notin \partial D$, the integral representation of $\boldsymbol{u}(\tilde{x})$ assumes the same form whether the domain D is bounded or not (the only change being the orientation of the normal to ∂D).

Other fundamental solutions

The method presented above for the Kelvin solution allows us to generate other two-dimensional, plane-strain fundamental solutions from three-dimensional ones; it extends to plane stress and antiplane problems, dynamics, ... It is also applicable for any fundamental three-dimensional solution defined for a domain E invariant under translation along $\boldsymbol{e}_3$ together with the boundary (or infinity) conditions. The two-dimensional fundamental solution is still sought using eq. (4.54); the integrability issue in eq. (4.55) remains present.

For example, the fundamental solution for the half-plane with free boundary (given by eq. (4.A.15), Appendix 4.C) is obtained by integrating along the x_3-axis the three-dimensional Mindlin solution (free surface condition over the plane $y_1 = 0$).

Appendix 4.A THE THREE-DIMENSIONAL KELVIN SOLUTION

Once the Kelvin displacement $U_i^k(\boldsymbol{x}, \boldsymbol{y})$ is known, the Kelvin stresses $\Sigma_{ij}^k(\boldsymbol{x}, \boldsymbol{y})$ follow directly by application of (isotropic) Hooke's law (1.18). The Kelvin displacement solution can be established in several ways:

Limiting process. Consider an infinite body E in which a spherical neighbourhood $B(\boldsymbol{x}, \varepsilon)$ of $\boldsymbol{x}$ (centre $\boldsymbol{x}$, radius ε) has been removed: $E = \mathbb{R}^3 \setminus B(\boldsymbol{x}, \varepsilon)$. Distributed forces of unit resultant $\boldsymbol{e}_k$ are applied on the cavity surface. The displacement and stresses thus generated in E are calculated using the general solution in spherical coordinates (expansion over spherical harmonics). Finally, in the limit $\varepsilon \to 0$ (the resultant load being kept equal to $\boldsymbol{e}_k$), the Kelvin solution (4.8), (4.9) is obtained.

Galerkin representation. In the case of isotropic elasticity, the displacement $\boldsymbol{U}(\boldsymbol{x}, \boldsymbol{y})$ may be sought in the form:

$$U_i^k(\boldsymbol{x}, \boldsymbol{y}) = 2(1-\nu)G_{i,jj}^k(\boldsymbol{x}, \boldsymbol{y}) - G_{j,ji}^k(\boldsymbol{x}, \boldsymbol{y}) \tag{4.A.1}$$

where the components of the *Galerkin tensor* solve:

$$\Delta\Delta G_i^k = -\frac{1}{2\mu(1-\nu)}\delta_{ik}\delta(\boldsymbol{y} - \boldsymbol{x})$$

and thus are proportional to the fundamental solution $r/8\pi$ of the bi-Laplacian operator, i.e.:

$$G_i^k = \frac{1}{16\pi\mu(1-\nu)} r\delta_{ik} \tag{4.A.2}$$

The representation (4.A.1) thus gives eq. (4.8) as a result. This approach does not extend to anisotropic elasticity.

Spatial Fourier transformation. This approach allows us to deal with the general anisotropic case; for this reason it will now be presented extensively.

Anisotropic Kelvin solution: derivation

This development uses the approach and notations of the previous section. For general anisotropic elasticity, the Hooke coefficients $C_{ijk\ell}$ are only constrained by symmetry, eq. (1.16) and strict positiveness, i.e:

$$\forall \boldsymbol{A} \neq \boldsymbol{0} \qquad A_{ij}C_{ijk\ell}A_{k\ell} > 0$$

arising from general thermodynamical requirements. The three-dimensional free-space fundamental solution solves:

$$C_{ijab}U^k_{a,bj}(\boldsymbol{x},\boldsymbol{y}) + \delta(\boldsymbol{y}-\boldsymbol{x})\delta_{ik} = 0 \qquad \boldsymbol{y} \in \mathbb{R}^3 - \{\boldsymbol{x}\} \tag{4.A.3}$$

or, upon application of the spatial Fourier transform:

$$C_{ijab}\xi_b\xi_j\hat{U}^k_a = \frac{1}{(2\pi)^3}\mathrm{e}^{-\mathrm{i}\boldsymbol{\xi}.\boldsymbol{x}}$$

The matrix $\boldsymbol{K}(\boldsymbol{\eta})$ of coefficients $K_{ia}(\boldsymbol{\eta}) = C_{ijab}\eta_b\eta_j$, as well as its inverse $\boldsymbol{N}(\boldsymbol{\eta}) = \boldsymbol{K}^{-1}(\boldsymbol{\eta})$, are symmetrical. One has then, since $C_{ijab}\xi_b\xi_j = \xi^2K_{ia}(\boldsymbol{\eta})$:

$$\hat{U}^k_a(\boldsymbol{x},\boldsymbol{\xi}) = \frac{1}{(2\pi)^3}\mathrm{e}^{-\mathrm{i}\boldsymbol{\xi}.\boldsymbol{x}}\frac{1}{\xi^2}N_{ka}(\boldsymbol{\eta})$$

and the fundamental solution $U^k_a(\boldsymbol{x},\boldsymbol{y})$ is obtained as an inverse Fourier transform:

$$U^k_a(\boldsymbol{x},\boldsymbol{y}) = \frac{1}{(2\pi)^3}\int_0^{+\infty}\int_S \mathrm{e}^{\mathrm{i}\xi\boldsymbol{\eta}.(\boldsymbol{y}-\boldsymbol{x})}N_{ka}(\boldsymbol{\eta})\,\mathrm{d}\omega\mathrm{d}\xi$$

Noting that $\boldsymbol{K}(\boldsymbol{\eta}) = \boldsymbol{K}(-\boldsymbol{\eta})$, $\boldsymbol{N}(\boldsymbol{\eta}) = \boldsymbol{N}(-\boldsymbol{\eta})$ and that S is invariant under the change of variables $\boldsymbol{\eta} \rightarrow -\boldsymbol{\eta}$, one can write:

$$\begin{aligned}
U^k_a(\boldsymbol{x},\boldsymbol{y}) &= \frac{1}{16\pi^3}\int_S\left\{\int_0^{+\infty}\mathrm{e}^{\mathrm{i}\xi\boldsymbol{\eta}.(\boldsymbol{y}-\boldsymbol{x})}N_{ka}(\boldsymbol{\eta})\,\mathrm{d}\omega\right.\\
&\qquad\left.+\int_0^{+\infty}\mathrm{e}^{-\mathrm{i}\xi\boldsymbol{\eta}.(\boldsymbol{y}-\boldsymbol{x})}N_{ka}(-\boldsymbol{\eta})\,\mathrm{d}\omega\right\}\mathrm{d}\xi\\
&= \frac{1}{16\pi^3}\int_S\int_{-\infty}^{+\infty}\mathrm{e}^{\mathrm{i}\xi\boldsymbol{\eta}.(\boldsymbol{y}-\boldsymbol{x})}N_{ka}(\boldsymbol{\eta})\,\mathrm{d}\omega\mathrm{d}\xi\\
&= \frac{1}{8\pi^2}\int_S\delta(\boldsymbol{\eta}.(\boldsymbol{y}-\boldsymbol{x}))N_{ka}(\boldsymbol{\eta})\,\mathrm{d}\omega\\
&= \frac{1}{8\pi^2}\int_0^{2\pi}\mathrm{d}\theta\int_0^{\pi}N_{ka}(r\sin\varphi(\cos\theta\boldsymbol{a}_1+\sin\theta\boldsymbol{a}_2)+r\cos\varphi\boldsymbol{a}_3)\\
&\qquad\qquad\delta(r\cos\varphi)\sin\varphi\,\mathrm{d}\varphi\\
&= \frac{1}{8\pi^2r}\int_0^{2\pi}\mathrm{d}\theta\int_{-r}^{r}\delta(u)N_{ka}(\sqrt{1-u^2}(\cos\theta\boldsymbol{a}_1+\sin\theta\boldsymbol{a}_2)+u\boldsymbol{a}_3)du
\end{aligned}$$

The Kelvin fundamental displacement for general homogeneous anisotropic elasticity is thus given by:

$$U^k_a(\boldsymbol{x},\boldsymbol{y}) = \frac{1}{8\pi^2r}\int_0^{2\pi}N_{ka}(\cos\theta\boldsymbol{a}_1+\sin\theta\boldsymbol{a}_2)\,\mathrm{d}\theta \tag{4.A.4}$$

A closed-form analytical expression for the integral in the above formula cannot be obtained for general anisotropy.

Isotropic elasticity. In this case, one obtains after some calculations:

$$K_{ka}(\boldsymbol{\eta}) = \mu\left[\delta_{ka} + \frac{1}{1-2\nu}\eta_a\eta_k\right] \qquad N_{ka}(\boldsymbol{\eta}) = \frac{1}{\mu}\left[\delta_{ka} - \frac{1}{2(1-\nu)}\eta_a\eta_k\right]$$

and the Kelvin displacement is given by:

$$U_a^k(\boldsymbol{x},\boldsymbol{y}) = \frac{1}{8\pi^2\mu r}\left[2\pi\delta_{ka} - \frac{\pi}{2(1-\nu)}[(\boldsymbol{a}_1.\boldsymbol{e}_a)(\boldsymbol{a}_1.\boldsymbol{e}_k) + (\boldsymbol{a}_2.\boldsymbol{e}_a)(\boldsymbol{a}_2.\boldsymbol{e}_k)\right]$$

Finally, noticing that

$$\boldsymbol{e}_a.\boldsymbol{e}_k = \delta_{ak} = (\boldsymbol{a}_i.\boldsymbol{e}_a)(\boldsymbol{a}_i.\boldsymbol{e}_k)$$

one has, taking into account the definition of $\boldsymbol{a}_3$:

$$(\boldsymbol{a}_1.\boldsymbol{e}_a)(\boldsymbol{a}_1.\boldsymbol{e}_k) + (\boldsymbol{a}_2.\boldsymbol{e}_a)(\boldsymbol{a}_2.\boldsymbol{e}_k)$$
$$= \delta_{ak} - \frac{(y_a - x_a)}{|y-x|}\frac{(y_k - x_k)}{|y-x|} = \delta_{ak} - r_{,a}r_{,k}$$

This identity, substituted into eq. (Appendix 4.A), leads to the expected expression (4.8) for the isotropic Kelvin solution.

Transverse anisotropy. Willis [14] addressed this particular instance of anisotropy, where all $C_{ijk\ell}$ vanish except:

$$\begin{aligned}
&C_{1111} = C_{2222} = \alpha \\
&C_{3333} = \beta \\
&C_{2323} = C_{3131} = \gamma \\
&C_{1133} = C_{2233} = \gamma' - \gamma \\
&C_{1212} = (C_{1111} - C_{1122})/2 = \alpha'
\end{aligned}$$

(i.e. five independent elasticity coefficients $\alpha, \alpha', \gamma, \gamma', \beta$, the symmetry axis being conventionally taken as $\boldsymbol{e}_3$). The Cartesian components of the Kelvin fundamental displacement are then given by:

$$U_i^k(\boldsymbol{x},\boldsymbol{y}) = \frac{C}{\pi r}\frac{(z_3 - z_2)h_{ki}(z_1) + (z_1 - z_3)h_{ki}(z_2) + (z_2 - z_1)h_{ki}(z_3)}{(z_1 - z_2)(z_2 - z_3)(z_3 - z_1)} \tag{4.A.5}$$

where

$$\begin{aligned}
C &= \left\{\beta\gamma^2\sin^6\theta(1-A_1)(1-A_2)(1-A_3)\right\}^{-1} \\
h_{ik}(z) &= D^{-1}N_{ik}((z-1)\cos\theta, -\mathrm{i}(z-1), -(z+1)\sin\theta) \\
D &= (z_1 - z_1^{-1})(z_2 - z_2^{-1})(z_3 - z_3^{-1})
\end{aligned}$$

and

$$
\begin{aligned}
N_{11}(\boldsymbol{\xi}) &= (\alpha'\xi_1^2 + \alpha\xi_2^2 + \gamma\xi_3^2)(\gamma(\xi_1^2 + \xi_2^2) + \beta\xi_3^2) - \gamma'^2\xi_2^2\xi_3^2 \\
N_{12}(\boldsymbol{\xi}) &= \gamma'^2\xi_1\xi_2\xi_3^2 - (\alpha - \alpha')\xi_1\xi_2(\gamma(\xi_1^2 + \xi_2^2) + \beta\xi_3^2) \\
N_{13}(\boldsymbol{\xi}) &= (\alpha - \alpha')\gamma'\xi_1\xi_2^2\xi_3 - \gamma'\xi_1\xi_3(\alpha'\xi_1^2 + \alpha\xi_2^2 + \gamma\xi_3^2) \\
N_{22}(\boldsymbol{\xi}) &= (\alpha\xi_1^2 + \alpha'\xi_2^2 + \gamma\xi_3^2)(\gamma(\xi_1^2 + \xi_2^2) + \beta\xi_3^2) - \gamma'^2\xi_1^2\xi_3^2 \\
N_{23}(\boldsymbol{\xi}) &= (\alpha - \alpha')\gamma'\xi_1^2\xi_2\xi_3 - \gamma'\xi_2\xi_3(\alpha\xi_1^2 + \alpha'\xi_2^2 + \gamma\xi_3^2) \\
N_{33}(\boldsymbol{\xi}) &= (\alpha'\xi_1^2 + \alpha\xi_2^2 + \gamma\xi_3^2)(\alpha\xi_1^2 + \alpha'\xi_2^2 + \gamma\xi_3^2) - (\alpha - \alpha')\xi_1^2\xi_2^2
\end{aligned}
$$

$\theta \in [0, \pi]$ denotes the angle between $\boldsymbol{y} - \boldsymbol{x}$ and $\boldsymbol{e}_3$ and z_i is the root of modulus < 1 of:

$$z^2 + 2B_i z + 1$$

where the coefficient B_i is defined by

$$B_i = \frac{A_i \cos^2\theta + \sin^2\theta + A_i}{(1 - A_i)\sin^2\theta}$$

with

$$A_1 = \frac{\alpha'}{\gamma} \qquad A_2 = \frac{1}{2}\left[\chi + \left(\chi^2 - 4\frac{\alpha}{\beta}\right)^{1/2}\right] \qquad A_3 = \frac{1}{2}\left[\chi - \left(\chi^2 - 4\frac{\alpha}{\beta}\right)^{1/2}\right]$$

and

$$\chi = \frac{\gamma^2 + \alpha\beta - \gamma'^2}{\beta\gamma}$$

Other results are mentioned in Mura [9].

Appendix 4.B INTEGRAL OF THE KELVIN TRACTION VECTOR

The resultant over a given, arbitrary, surface S of the Kelvin traction vector is:

$$
\begin{aligned}
K_i^k(\boldsymbol{x}, S) &= \int_S T_i^k(\boldsymbol{x}, \boldsymbol{y})\,\mathrm{d}S_y \\
&= \frac{-1}{8\pi(1-\nu)} \int_S [(1 - 2\nu)(\delta_{ik} r_{,n} + n_k r_{,i} - n_i r_{,k}) + 3 r_{,i} r_{,k} r_{,n}] \frac{\mathrm{d}S_y}{r^2} \qquad (4.A.6)
\end{aligned}
$$

This quantity can be reformulated using integrations by parts. Indeed, identity (A.8) gives:

$$\int_S (n_k r_i - n_i r_k) \frac{\mathrm{d}S_y}{r^2} = \int_S D_{ik} \frac{1}{r}\,\mathrm{d}S_y = e_{ikp} \int_{\partial j} \tau_p \frac{\mathrm{d}s_y}{r} \qquad (4.A.7)$$

and also:

$$\int_S r_{,i}r_{,k}r_{,n}\frac{\mathrm{d}S_y}{r^2} = \int_S r_{,i}r_{,p}D_{kp}\frac{1}{r}\,\mathrm{d}S_y + \int_S n_k r_{,i}\frac{\mathrm{d}S_y}{r^2}$$
$$= e_{kpj}\int_{\partial S} r_{,i}r_{,p}\tau_j\frac{\mathrm{d}s_y}{r} + \int_S \frac{1}{r}D_{pk}(r_{,i}r_{,p})\,\mathrm{d}S_y + \int_S n_k r_{,i}\frac{\mathrm{d}S_y}{r^2}$$

Taking into account:

$$D_{pk}(r_{,i}r_{,p}) = n_p(r_{,ik}r_{,p} + r_{,i}r_{,pk}) - n_k(r_{,ip}r_{,p} + r_{,i}r_{,pp})$$
$$= \frac{1}{r}\{[\delta_{ik} - r_{,i}r_{,k}]r_{,n} + r_{,i}n_k - r_{,i}r_{,k}r_{,n} - 2n_k r_{,i}\}$$
$$= \frac{1}{r}\{[\delta_{ik} - 2r_{,i}r_{,k}]r_{,n} - n_k r_{,i}\}$$

one gets:

$$3\int_S r_{,i}r_{,k}r_{,n}\frac{\mathrm{d}S_y}{r^2} = e_{kpj}\int_{\partial S} r_{,i}r_{,p}\tau_j\frac{\mathrm{d}s_y}{r} + \delta_{ik}\int_S r_{,n}\frac{\mathrm{d}S_y}{r^2} \tag{4.A.8}$$

Substitution of eqs (4.A.7), (4.A.8) into (4.A.6) then allows $K_i^k(\boldsymbol{x}, S)$ to be reformulated as:

$$K_i^k(\boldsymbol{x}, S) = \frac{e_{ijp}}{8\pi(1-\nu)}\int_{\partial S}[r_{,j}r_{,k} - (1-2\nu)\delta_{jk}]\,\tau_p\frac{\mathrm{d}s_y}{r} - \delta_{ik}\frac{\omega(\boldsymbol{x}, S)}{4\pi} \tag{4.A.9}$$

If the surface S is closed and $\boldsymbol{x} \notin S$, the contour integrals vanish while $\omega(\boldsymbol{x}, S) = 4\pi$ ($\boldsymbol{x}$ interior to S, $\boldsymbol{n}$ being the exterior unit normal) or $\omega(\boldsymbol{x}, S) = 0$ ($\boldsymbol{x}$ exterior to S). Identity (4.7) is thus a special case of eq. (4.A.9).

Non-convergence of K_i^k when $\boldsymbol{x} \in S$

The contour integrals in eq. (4.A.9) are invariant when homotheties of origin $\boldsymbol{x}$ are applied to the contour ∂S. On the other hand, they depend on the *shape* of ∂S. This implies that, for $\boldsymbol{x} \in S$, $K_i^k(\boldsymbol{x}, S)$ cannot be defined as the limiting value of $K_i^k(\boldsymbol{x}, S - s_\varepsilon(\boldsymbol{x}))$ when vanishing neighbourhoods $s_\varepsilon(\boldsymbol{x})$ of arbitrary shape are removed around $\boldsymbol{x}$. The reformulation (4.A.9) thus does not express $K_i^k(\boldsymbol{x}, S)$ as a convergent, weakly singular, integral despite the weakly singular character of the surface integral (solid angle).

Special case: the edge of S is a plane curve

When the curve ∂S lies in a plane and $\boldsymbol{x}$ is located in that plane, a polar coordinate system (r, θ) of origin $\boldsymbol{x}$ can be used:

$$S \ni \boldsymbol{y} = \boldsymbol{x} + r(\theta)\boldsymbol{e}_r$$

and one has besides:

$$\boldsymbol{\tau}\,\mathrm{d}s_y = [r'(\theta)\boldsymbol{e}_r + r(\theta)\boldsymbol{e}_\theta]\,\mathrm{d}\theta$$

$$e_{ijp} r_{,j} \tau_{,p} \frac{\mathrm{d}s_y}{r} = (\boldsymbol{e_r} \wedge \boldsymbol{e_\theta}).\boldsymbol{e}_i\,\mathrm{d}\theta = n_i(\boldsymbol{x})$$

where $\boldsymbol{n}(\boldsymbol{x})$ is the unit normal to the plane of ∂S. Hence:

$$e_{ijp} \int_{\partial S} r_{,j} r_{,k} \tau_p \frac{\mathrm{d}s_y}{r} = n_i(\boldsymbol{x}) \int_0^{2\pi} r_{,k}\,\mathrm{d}\theta = n_i(\boldsymbol{x}) \int_0^{2\pi} (\boldsymbol{e}_r)_k\,\mathrm{d}\theta = 0 \quad (4.A.10)$$

Finally, for this special case, eq. (4.A.9) becomes:

$$K_i^k(\boldsymbol{x}, S) = -\delta_{ik} \frac{\omega(\boldsymbol{x}, S)}{4\pi} - \frac{1-2\nu}{8\pi(1-\nu)} e_{ikp} \int_{\partial S} \tau_p \frac{\mathrm{d}s_y}{r} \quad (4.A.11)$$

The value taken by the remaining contour integral depends on the *shape* of the plane curve ∂S. Indeed:

$$\int_{\partial S} \boldsymbol{\tau} \frac{\mathrm{d}s_y}{r} = \int_0^{2\pi} \left(\frac{r'}{r} \boldsymbol{e}_r + \boldsymbol{e}_\theta \right) \mathrm{d}\theta = \int_0^{2\pi} \frac{r'}{r} \boldsymbol{e}_r\,\mathrm{d}\theta$$

When ∂S is a circle centred at $\boldsymbol{x}$, one has $r'(\theta) = 0$ and the above integral vanishes. This remark allows us to establish the classical formula:

$$\fint_S T_i^k(\boldsymbol{x}, \boldsymbol{y})\,\mathrm{d}S_y = -\frac{\omega(\boldsymbol{x}, S)}{4\pi} \delta_{ik} \quad (4.A.12)$$

where $\fint_S$ denotes an integral taken in the Cauchy principal value sense. Besides, if ∂S is a circle of radius ε and of centre $\boldsymbol{x} + k\varepsilon\boldsymbol{a}$, $0 < k < 1$ ($\boldsymbol{a}$: given unit vector in the circle plane), one finds after some calculations:

$$\int_{\partial S} \boldsymbol{\tau} \frac{\mathrm{d}s_y}{r} = \frac{4}{k} [K(k) - E(k)] (\boldsymbol{n}(\boldsymbol{x}) \wedge \boldsymbol{a})$$

where $K(k), E(k)$ denote respectively the complete elliptic integrals of first and second kinds [1]. This example emphasizes the fact that the contour integrals in eq. (4.A.9) depend on the shape of ∂S; in particulier, $K(k) - E(k) = \pi k^2 + o(k^2)$.

Appendix 4.C HALF-SPACE FUNDAMENTAL SOLUTION

Initially given by Mindlin [8], this fundamental solution is also presented in Mura [9]. Consider the half-space $E = \{(y_1, y_2, y_3) \mid y_3 \geq 0\}$ of boundary $\partial E = \{y_3 = 0\}$. Let $\tilde{x} = (x_1, x_2, -x_3)$ denote the mirror image of the source point $\boldsymbol{x}$ with respect to the plane ∂E and put $\bar{r} = |\boldsymbol{y} - \tilde{x}|$ (Figure 4.10). One then has:

$$r_{,\alpha} = \frac{y_\alpha - x_\alpha}{r} \qquad \bar{r}_{,\alpha} = \frac{y_\alpha - x_\alpha}{\bar{r}} \qquad r_{,3} = \frac{y_3 - x_3}{r} \qquad \bar{r}_{,3} = \frac{y_3 + x_3}{\bar{r}}$$

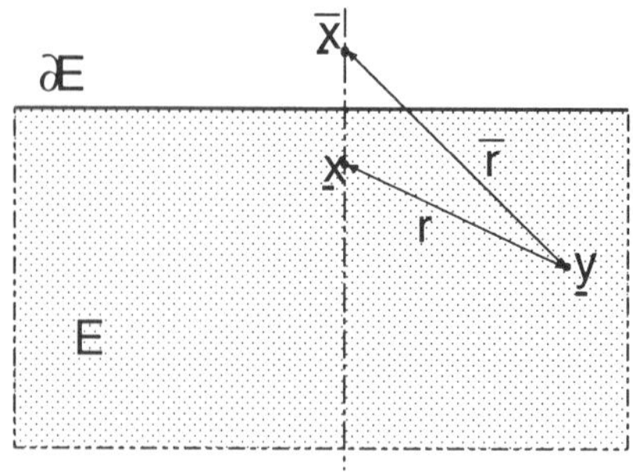

Figure 4.10 Half-space: geometrical notations.

The fundamental solution $\boldsymbol{U}(\boldsymbol{x},\boldsymbol{y})$ solving eq. (4.2) and the free-surface boundary condition

$$T_i^k(\boldsymbol{x},\boldsymbol{y}) = 0 \qquad (y_3 = 0, i = 1,2,3)$$

is given by the following expressions (the indices i,k taking the values 1, 2 only):

$$\begin{aligned}
U_i^k &= \frac{1}{16\pi\mu(1-\nu)}\left\{\frac{\kappa}{r}\delta_{ik} + \frac{1}{\bar{r}}\delta_{ik} + \frac{r_{,i}r_{,k}}{r} + \kappa\frac{\bar{r}_{,i}\bar{r}_{,k}}{\bar{r}} + \frac{2x_3y_3}{\bar{r}^3}\left[\delta_{ik} - 3\bar{r}_{,i}\bar{r}_{,k}\right]\right.\\
&\qquad \left. + \frac{\chi}{\bar{r}(1+\bar{r}_{,3})}\left[\delta_{ik} - \frac{\bar{r}_{,i}\bar{r}_{,k}}{1+\bar{r}_{,3}}\right]\right\} \qquad (4.A.13)\\
U_3^k &= \frac{1}{16\pi\mu(1-\nu)}\left\{\frac{r_{,3}r_{,k}}{r} + \kappa\frac{\bar{r}_{,k}}{\bar{r}}\left[\bar{r}_{,3} - \frac{2x_3}{\bar{r}}\right] - \frac{6x_3y_3}{\bar{r}^3}\bar{r}_{,3}\bar{r}_{,k} + \chi\frac{\bar{r}_{,k}}{\bar{r}(1+\bar{r}_{,3})}\right\}\\
U_i^3 &= \frac{1}{16\pi\mu(1-\nu)}\left\{\frac{r_{,i}r_{,3}}{r} + \kappa\frac{\bar{r}_{,k}}{\bar{r}}\left[\bar{r}_{,3} - \frac{2x_3}{\bar{r}}\right] + \frac{6x_3y_3}{\bar{r}^3}\bar{r}_{,i}\bar{r}_{,3} - \chi\frac{\bar{r}_{,i}}{\bar{r}(1+\bar{r}_{,3})}\right\}\\
U_3^3 &= \frac{1}{16\pi\mu(1-\nu)}\left\{\frac{\kappa}{r} + \frac{1+\chi}{\bar{r}} + \frac{r_{,3}^2}{r} + \kappa\frac{\bar{r}_3\bar{r}_3}{\bar{r}} - \frac{2x_3y_3}{\bar{r}^3}\left[1 - 3\bar{r}_{,3}\bar{r}_{,3}\right]\right\}
\end{aligned}$$

with $\kappa = 3-4\nu$, $\chi = 4(1-\nu)(1-2\nu)$. Some comments are in order:

- For the particular case $x_3 = 0$ (i.e. source point located on the boundary of the half-space), $U_i^3(\boldsymbol{x},\boldsymbol{y})$ as given by eqs (4.A.13) reduces to either the Boussinesq solution ($k=3$: point force normal to the surface) or the Cerruti solution ($k=1,2$: point force tangent to the surface).

- The Mindlin solution (4.A.13) is not equal to the sum of the Kelvin solutions associated with the source points $\boldsymbol{x}$ and $\tilde{\boldsymbol{x}}$: in other words, the method of images used to obtain scalar fundamental half-space solutions does not extend to vector (elasticity) problems.

- The Mindlin solution is invariant under translations along directions in the (y_1, y_2)-plane only, and thus has the following symmetry properties:

$$U_i^k(\boldsymbol{x},\boldsymbol{y}) = U_k^i(\boldsymbol{x},\boldsymbol{y}) \qquad (i,k=1,2)$$
$$U_i^k(\boldsymbol{x},\boldsymbol{y}) = U_i^k(\boldsymbol{y},\boldsymbol{x}) \qquad (i,k=1,2,3 \quad x_3 = y_3)$$

- The fundamental displacement solution (4.A.13) has a $|y-x|^{-1}$ singularity for $\boldsymbol{y} \sim \boldsymbol{x}$; therefore the fundamental stress tensor has a $|y-x|^{-2}$ singularity.
- For any interior source point $\boldsymbol{x} \notin \partial E$, the Kelvin and Mindlin solutions have the same singularity (i.e. their difference is nonsingular).

This solution has been established by Mindlin [8] by seeking its Galerkin tensor (see Section 4.7), which was found to be:

$$\begin{aligned}
G_i^k(\boldsymbol{x},\boldsymbol{y}) &= \frac{1}{16\pi\mu(1-\nu)}\left\{r + (1-\chi)\bar{r} + \chi(y_3+x_3)\ln(\bar{r}+y_3+x_3) - \frac{2x_3^2}{\bar{r}}\right\} \\
G_3^k(\boldsymbol{x},\boldsymbol{y}) &= \frac{(y_k-x_k)}{16\pi\mu(1-\nu)}\left\{2(1-2\nu)\ln(\bar{r}+y_3+x_3) + \frac{2x_3}{\bar{r}}\right\} \qquad (4.A.14) \\
G_3^3(\boldsymbol{x},\boldsymbol{y}) &= \frac{1}{16\pi\mu(1-\nu)}\left\{r + (\kappa-\chi)\bar{r} - \frac{2x_3y_3}{\bar{r}}\right. \\
&\qquad \left. + [\chi(y_3+x_3) - 4(1-2\nu)x_3]\ln(\bar{r}+y_3+x_3)\right\}
\end{aligned}$$

Fundamental solution for the half-plane $E = \{(y_1, y_3), y_3 \geq 0\}$

Its expression can be established by integrating the three-dimensional Mindlin solution (4.A.13) over the line $-\infty < x_2 < +\infty$, following the approach presented in Section 4.9. For plane strain problems, its expression is (see Brebbia, Telles, Wrobel [4]):

$$\begin{aligned}
U_1^1(\boldsymbol{x},\boldsymbol{y}) &= A\left\{-\kappa\ln r - [1+\chi]\ln\bar{r} + r_{,1}^2 + \kappa\bar{r}_{,1}^2 + \frac{2x_3x_3'}{\bar{r}^2}\left[1-2\bar{r}_{,1}^2\right]\right\} \\
U_3^1(\boldsymbol{x},\boldsymbol{y}) &= A\left\{r_{,3}r_{,1} + \kappa\bar{r}_{,1}\left[\bar{r}_{,3} - \frac{2x_3}{\bar{r}}\right] - \frac{4x_3y_3}{\bar{r}^2}\bar{r}_{,3}\bar{r}_{,1} + \chi\theta\right\} \qquad (4.A.15) \\
U_1^3(\boldsymbol{x},\boldsymbol{y}) &= A\left\{r_{,3}r_{,1} + \kappa\bar{r}_{,1}\left[\bar{r}_{,3} - \frac{2x_3}{\bar{r}}\right] + \frac{4x_3x_3'}{\bar{r}^2}\bar{r}_{,3}\bar{r}_{,1} - \chi\theta\right\} \\
U_3^3(\boldsymbol{x},\boldsymbol{y}) &= A\left\{-\kappa\ln r - [1+\chi]\ln\bar{r} + r_{,3}^2 + \kappa\bar{r}_3^2 - \frac{2x_3x_3'}{\bar{r}^2}\left[1-2\bar{r}_{,3}^2\right]\right\}
\end{aligned}$$

with $A = (8\pi\mu(1-\nu))^{-1}$, $\kappa = 3-4\nu$, $\chi = 4(1-\nu)(1-2\nu)$, $r, \bar{r}$ and their derivatives being defined as in the three-dimensional case but with $x_2 = x_2' = y_2 = 0$; moreover $\theta = \tan^{-1}(\bar{r}_{,1}/\bar{r}_{,3})$.

Appendix 4.D FUNDAMENTAL SOLUTION FOR TWO BONDED HALF-SPACES

Consider an infinite elastic body E made of two isotropic elastic half-spaces E_1, E_2 (elasticity constants $\mu_{1,2}, \nu_{1,2}$), perfectly bonded along the separation plane $y_3 = 0$, i.e. $\boldsymbol{u}$ and $\boldsymbol{\sigma}.\boldsymbol{n}$ are continuous across the interface (the domain 1 occupies $y_3 \leq 0$). Assume that a point force is applied at $\boldsymbol{x} \in E_1$. Like in Appendix 4.C, the mirror image $\tilde{\boldsymbol{x}}$ of $\boldsymbol{x}$ with respect to the separation plane $y_3 = 0$ is introduced, and the notations $r, \bar{r}, \dots$ are used again. The fundamental solution $\boldsymbol{U}(\boldsymbol{x}, \boldsymbol{y})$ for this situation, established by Rongved [12], is expressed in terms of Neuber–Papkovich potentials $\boldsymbol{\varphi}, \boldsymbol{\psi}$:

$$U_i^k(\boldsymbol{x}, \boldsymbol{y}) = \frac{1}{2\mu}\left[(\kappa + 1)\Phi_{ik} - (y_j \Phi_{jk} + \psi_k)_{,i}\right]$$

with $\kappa = 3 - 4\nu$. For $\boldsymbol{y} \in E_1$, one has:

$$\Phi_{11} = \Phi_{22} = \frac{1}{2\pi(\kappa_1 + 1)}\left(\frac{1}{r} + \frac{S}{\bar{r}}\right)$$

$$\Phi_{21} = \Phi_{12} = \Phi_{13} = \Phi_{23} = 0$$

$$\Phi_{3k} = \frac{1}{2\pi(\kappa_1 + 1)}\left(\frac{(A\kappa_1 - S)\bar{r}_{,k}}{\bar{r}(1 + \bar{r}_{,3})} - \frac{2Ax_3\bar{r}_{,k}}{\bar{r}^2}\right)$$

$$\Phi_{33} = \frac{1}{2\pi(\kappa_1 + 1)}\left(\frac{1}{r} + \frac{A\kappa_1}{\bar{r}} - \frac{2Ax_3\bar{r}_{,3}}{\bar{r}^2}\right)$$

$$\psi_k = \frac{1}{2\pi(\kappa_1 + 1)}\left(-\frac{(A\kappa_1^2 + B2S\kappa_1)r_{,k}}{2\bar{r}(1 + \bar{r}_{,3})} + \frac{(A\kappa_1 - S)x_3\bar{r}_{,k}}{\bar{r}(1 + \bar{r}_{,3})}\right)$$

$$\psi_3 = \frac{1}{2\pi(\kappa_1 + 1)}\left(-\frac{x_3}{r} - \frac{A\kappa_1 x_3}{\bar{r}} + \frac{1}{2}(A\kappa_1^2 - B)\ln(\bar{r}(1 + \bar{r}_{,3})\right)$$

while for $\boldsymbol{y} \in E_2$, one has:

$$\Phi_{11} = \Phi_{22} = \frac{1}{2\pi(\kappa_1 + 1)}\frac{T}{r}$$

$$\Phi_{21} = \Phi_{12} = \Phi_{13} = \Phi_{23} = 0$$

$$\Phi_{3k} = -\frac{1 - B - T}{2\pi(\kappa_1 + 1)}\frac{r_{,k}}{r(1 - r_{,3})}$$

$$\Phi_{33} = \frac{1}{2\pi(\kappa_1 + 1)}\frac{1 - B}{r}$$

$$\psi_k = \frac{1}{2\pi(\kappa_1 + 1)}\left(-\frac{((A - 1)\kappa_1 + (2T + B - 1)\kappa_2)r_{,k}}{2r(1 - r_{,3})} + \frac{(1 - A - T)x_3 r_{,k}}{r(1 - r_{,3})}\right)$$

$$\psi_3 = \frac{1}{2\pi(\kappa_1 + 1)}\left(\frac{(A - 1)x_3}{r} + \frac{1}{2}[(1 - A)\kappa_1 + (1 - B)\kappa_2]\ln(r(1 - r_{,3})\right)$$

with $k = 1, 2$ and using the notations:

$$A = \frac{1-\Gamma}{1+\kappa_1\Gamma} \qquad B = \frac{\kappa_2 - \kappa_1\Gamma}{\kappa_2 + \Gamma} \qquad S = \frac{1-\Gamma}{1+\Gamma}$$

$$T = \frac{2(\kappa_1 + 1)\Gamma}{(\kappa_2 + 1)(1+\Gamma)} \qquad \Gamma = \frac{\mu_2}{\mu_1}$$

REFERENCES

1. ABRAMOWITZ, M., STEGUN, I. *Handbook of Mathematical Functions.* Dover Publications, New York (1972).
2. BENITEZ, F. G., ROSAKIS, A. J. Three-dimensional elastostatics of a layer and a layered medium. *J. Elast.*, **18**, 3–50 (1987).
3. BETTESS, P. *Infinite elements.* Penshaw Press (1992).
4. BREBBIA, C. A., TELLES, J. C. F., WROBEL, L. C. *Boundary element techniques.* Springer-Verlag (1984).
5. GUIGGIANI, M. Formulation and numerical treatment of boundary integral equations with hypersingular kernels. In V. Sladek (ed.), *Singular integrals in boundary element methods.* Computational Mechanics Publications, Southampton (1998).
6. GUIGGIANI, M., GIGANTE, A. A general algorithm for multidimensional Cauchy principal value integrals in the boundary element method. *ASME J. Appl. Mech.*, **57**, 906–915 (1990).
7. GUIGGIANI, M., KRISHNASAMY, G., RUDOLPHI, T. J., RIZZO, F. J. A general algorithm for the numerical solution of hypersingular boundary integral equations. *ASME J. Appl. Mech.*, **59**, 604–614 (1992).
8. MINDLIN, R.D. Force at a point in the interior of a semi-infinite solid. *Physics*, **7**, 195–201 (1936).
9. MURA, T. *Micromechanics of defects in solids.* Martinus Nijhoff (1982).
10. PATTERSON, C., SHEIKH, M. A. Interelement continuity in the boundary element method. In C.A. Brebbia (ed.), *Topics in boundary element research.*, vol. 1. Springer-Verlag (1984).
11. PHAM, T. L. *Tenseurs de Green et de Neumann.* Ph.D. thesis, (1966).
12. RONGVED, L. Force interior to one of two joined semi-infinite solids. In *Proc. 2nd. Midwestern Conf. Solid Mechanics.*, pp. 1–13 (1955).
13. SALENÇON, J. *Mécanique des milieux continus (vols. 1,2,3).* Presses de l'Ecole Polytechnique, Palaiseau, France (1992).
14. WILLIS, J. R. The elastic interaction energy of dislocation loops in anisotropic media. *Quart. J. Mech. Appl. Math.*, **17**, 157–174 (1965).

5

Representation of gradients and stresses on the boundary

5.1 MOTIVATION

It has been shown in the previous chapters that the boundary integral equation (2.24) in potential theory or (4.32) in elasticity, associated with the appropriate interior representation formulas, together provide a boundary-only approach to the numerical solution of the relevant boundary-value problems.

However, neither the integral equations nor the representation formulas obtained up to now allow the direct computation of derivative quantities (a generic term for potential or displacement gradients, stresses ...) when the source point $\boldsymbol{x}$ is located on the domain boundary. The interior integral representation formulas for derivative quantities indeed have a non-integrable singularity for $\boldsymbol{x} \in \partial\Omega$, which prevents their direct use. Only the traction vector or the flux, i.e. the normal projection of the corresponding derivative quantity, are readily obtainable.

Nevertheless, there are situations where the availability of either an integral equation or a representation formula expressing the gradient of the primary variable (potential, displacement ...) at boundary points is desirable. The purpose of this chapter is to establish and discuss these formulations. To motivate them, some instances where they are either useful or necessary are first reviewed.

Evaluation of tangential stresses on the boundary

Derivative quantities can be evaluated on the boundary by combining q, resp. $\boldsymbol{t}$, with tangential derivatives of u, resp. $\boldsymbol{u}$ (computed from the approximate

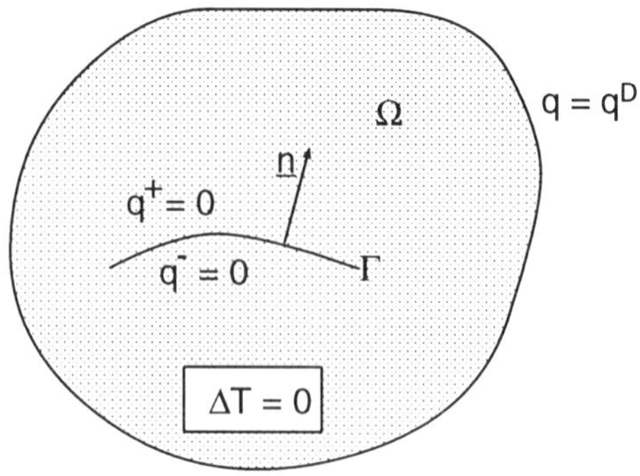

Figure 5.1 Conductive medium and insulating screen.

boundary solution), see Sections 4.7. However, recent investigations (Guiggiani [4]) have shown that this technique, although simple to implement, may be inaccurate whereas evaluations based on gradient representation formulas on the boundary give more satisfactory results. One must, however, use great care in implementing the latter.

Screens, cracks

Consider a thermally conductive body Ω, in which an insulating screen is embedded (Figure 5.1). Such a screen is mathematically realized by an open surface Γ such that the heat flux vanishes on both sides:

$$q^{\pm} = 0 \qquad \text{on } \Gamma^{\pm}$$

whereas a temperature jump $\phi = u^{+} - u^{-}$ is allowed across Γ. Assuming to fix our ideas a Neumann boundary condition over the external boundary: $q = q^{D}$ on $\partial\Omega$, the integral equation (2.24) becomes:

$$\int_{\partial\Omega} \left\{ [u(\boldsymbol{y}) - u(\boldsymbol{x})] H(\boldsymbol{x},\boldsymbol{y}) - q^{D}(\boldsymbol{y}) G(\boldsymbol{x},\boldsymbol{y}) \right\} \mathrm{d}S_y + \int_{\Gamma} \phi(\boldsymbol{y}) H(\boldsymbol{x},\boldsymbol{y}) \, \mathrm{d}S_y = 0$$

in which the unit normal $\boldsymbol{n}$ to Γ is conventionally oriented from Γ^{-} to Γ^{+}.

Denote by u_0 the temperature field for the same boundary data q^{D} but without the screen Γ. The field u_0 is then continuous across the geometrical surface Γ and solves the integral equation:

$$\int_{\partial\Omega} \left\{ [u_0(\boldsymbol{y}) - u_0(\boldsymbol{x})] H(\boldsymbol{x},\boldsymbol{y}) - q^{D}(\boldsymbol{y}) G(\boldsymbol{x},\boldsymbol{y}) \right\} \mathrm{d}S_y + \int_{\Gamma} \phi(\boldsymbol{y}) H(\boldsymbol{x},\boldsymbol{y}) \, \mathrm{d}S_y = 0$$

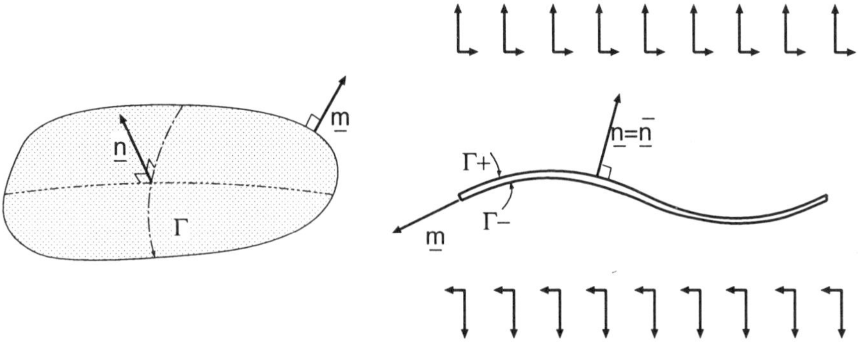

Figure 5.2 Symmetrically loaded crack embedded in an infinite body.

so that the difference $\Delta u = u - u_0$ solves (for $\boldsymbol{x}$ not located on Γ):

$$\int_{\partial\Omega} [\Delta u(\boldsymbol{y}) - \Delta u(\boldsymbol{x})] H(\boldsymbol{x}, \boldsymbol{y}) \, \mathrm{d}S_y + \int_{\Gamma} [\Delta u^+ - \Delta u^-](\boldsymbol{y}) H(\boldsymbol{x}, \boldsymbol{y}) \, \mathrm{d}S_y = 0$$

One readily notices that the above equation has the form $\mathcal{L}\Delta u = 0$, where $\mathcal{L}$ is a linear operator. Such a homogeneous linear equation has either infinitely many solutions (this is the case here) or no solution. The integral equation (2.24) thus *degenerates* in the presence of discontinuity surfaces for u and does not provide a means to solve the problem of heat conduction with a screen. The same equations, and hence the same conclusions, apply for e.g. potential flow around an impenetrable, infinitely thin, screen Γ.

Boundary integral equation formulations for the equilibrium of an elastic body Ω containing a crack Γ with faces $\Gamma^{\pm}$ raise similar issues. For example, considering for simplicity a crack Γ embedded in an infinite body and loaded on the crack faces by tractions $\pm \boldsymbol{t}^{D}$, the representation formula (4.21) for the displacement $\boldsymbol{u}$ at a given point $\boldsymbol{x}$ located away from the crack faces gives:

$$u_k(\boldsymbol{x}) = \int_{\Gamma} \Big\{ \phi_i(\boldsymbol{y}) T_i^k(\boldsymbol{x}, \boldsymbol{y}) + \psi_i(\boldsymbol{y}) U_i^k(\boldsymbol{x}, \boldsymbol{y}) \Big\} \, \mathrm{d}S_y$$

with $\boldsymbol{\phi} = \boldsymbol{u}^+ - \boldsymbol{u}^-$ et $\boldsymbol{\psi} = \boldsymbol{t}^+ + \boldsymbol{t}^-$. A similar, regularized, formula is available when $\boldsymbol{x} \in \Gamma$:

$$\frac{1}{2}[u_k^+ + u_k^-](\boldsymbol{x}) = \phi_i(\boldsymbol{x}) K_i^k(\boldsymbol{x}, \Gamma) + \int_{\Gamma} \Big\{ [\phi_i(\boldsymbol{y}) - \phi_i(\boldsymbol{x})] T_i^k(\boldsymbol{x}, \boldsymbol{y}) - \psi_i(\boldsymbol{y}) U_i^k(\boldsymbol{x}, \boldsymbol{y}) \Big\} \, \mathrm{d}S_y$$

The above integral equation is unaffected when the loading applied to the crack faces is $\boldsymbol{t}^{\pm} \pm \boldsymbol{t}^0$ instead of $\boldsymbol{t}^{\pm}$; for the important special case of *symmetrical loading* (Figure 5.2), one has $\boldsymbol{\psi} = \boldsymbol{0}$. The above integral equation thus degenerates in the same way as its counterpart for screens.

The logical answer is then to use boundary integral equation formulations based on the integral representation for stresses, or *traction boundary integral equations*, as established in this chapter. Their applications to the very important area of fracture mechanics problems is addressed in Chapter 13.

Indirect integral equations

Indirect boundary integral formulations for Neumann problems are obtained using integral representations of the gradient of single- or double-layer potentials (see Section 2.4) on the boundary.

Error indicators

Consider, as an example, the Neumann interior problem (prescribed flux $q = q^D$ on $\partial\Omega$) for the Laplace equation. One wishes to be able to evaluate *a posteriori*, by means of a distributed error indicator, the *local* accuracy of a boundary element approximate solution $u\,|_{\partial\Omega}$ to the integral equation (2.24). One among several ways of devising such an error indicator, following Ingber and Mitra [9], consists of comparing the prescribed $q^D(\boldsymbol{x})$ boundary data to the value $q(\boldsymbol{x})$ computed *a posteriori* using the knowledge of a numerical solution u; $q^D(\boldsymbol{x}) = q(\boldsymbol{x})$ ideally, i.e. if the solution u is exact. An error functional J, additive with respect to the boundary elements, may be defined by e.g.:

$$J = \int_{\partial\Omega} |q - q^D|^2 \, \mathrm{d}S = \sum_e \int_{E_e} |q - q^D|^2 \, \mathrm{d}S = \sum_e J_e$$

The element contributions J_e are then the local error indicators, each associated with one element. Their practical evaluation needs a representation formula for q on the boundary, given u (and q^D) on the boundary. Error estimation techniques using hypersingular residuals have also been developed by Guiggiani [5], Paulino *et al.* [14], among others.

5.2 DEFINITION USING A LIMITING PROCESS

Consider a field variable u governed over some domain Ω by the Laplace equation $\Delta u = 0$. The integral representation of $u_{,i}(\boldsymbol{x})$ ($\boldsymbol{x} \notin \partial\Omega$) is obtained by a differentiation of eq. (2.18) with respect to x_i, i.e.:

$$\kappa u_{,i}(\boldsymbol{x}) = \int_{\partial\Omega} \left\{ q(\boldsymbol{y}) G_{,\bar{\imath}}(\boldsymbol{x}, \boldsymbol{y}) - u(\boldsymbol{y}) G_{,\bar{\imath}j}(\boldsymbol{x}, \boldsymbol{y}) n_j(\boldsymbol{y}) \right\} \mathrm{d}S_y \qquad (5.1)$$

where G is any fundamental solution compatible with Ω (the notations $()_{,i}$ and $()_{,\bar{\imath}}$ conventionally indicate partial differentiation with respect to components of the first and second argument, respectively, of functions $f(\boldsymbol{x}, \boldsymbol{y})$). The kernels $G_{,\bar{\imath}}(\boldsymbol{x}, \boldsymbol{y})$ and $G_{,\bar{\imath}j}(\boldsymbol{x}, \boldsymbol{y})$ have respectively $|\boldsymbol{y} - \boldsymbol{x}|^{-2}$ and $|\boldsymbol{y} - \boldsymbol{x}|^{-3}$

singularities. The latter, of particular importance here, is usually referred to as *hypersingular* in the literature. Hence, to extend the validity of eq. (5.1) to observation points $\boldsymbol{x} \in \partial\Omega$, a regularization approach, of a more advanced nature than that previously developed in connection with e.g. the integral equation (2.24) due to the presence of the hypersingular kernel, is called for.

The limiting process to follow uses the same notations as in Section 4.4: exclusion neighbourhood $v_\varepsilon(\boldsymbol{x}) = v_\varepsilon$ of diameter $\leq \varepsilon$, 'punctured' domain $\Omega_\varepsilon = \Omega - v_\varepsilon$ of boundary $\partial\Omega_\varepsilon = (\partial\Omega - e_\varepsilon) + s_\varepsilon$ (with $s_\varepsilon = \Omega \cap \partial v_\varepsilon$, $e_\varepsilon = \partial\Omega \cap v_\varepsilon$). The integral representation for the gradient of u on the boundary is sought as the limiting form when $\varepsilon \to 0$ of eq. (5.1) for the domain Ω_ε, i.e.:

$$0 = \int_{(\partial\Omega - e_\varepsilon)+s_\varepsilon} \left\{ q(\boldsymbol{y}) G_{,\bar{\imath}}(\boldsymbol{x},\boldsymbol{y}) - u(\boldsymbol{y}) G_{,\bar{\imath}j}(\boldsymbol{x},\boldsymbol{y}) n_j(\boldsymbol{y}) \right\} \mathrm{d}S_y \tag{5.2}$$

Two approaches are presented: second-order regularization (Section 5.3) or integration by parts followed by a first-order regularization (Section 5.4).

5.3 SECOND-ORDER REGULARIZATION

This approach is based on a reformulation of eq. (5.2) using an auxiliary identity. Indeed, any potential $u^\star$ of the form:

$$u^\star = u_0 + a_i(y_i - x_i) \qquad (u_0, \boldsymbol{a})\text{: arbitrary constants} \tag{5.3}$$

is a particular solution of $\Delta u^\star = 0$, and thus verifies eq. (5.2), i.e.:

$$\int_{(\partial\Omega - e_\varepsilon)+s_\varepsilon} a_a n_a(\boldsymbol{y}) G_{,\bar{\imath}}(\boldsymbol{x},\boldsymbol{y}) \,\mathrm{d}S_y \\ - \int_{(\partial\Omega - e_\varepsilon)+s_\varepsilon} [u_0 + a_a(y_a - x_a)] G_{,\bar{\imath}j}(\boldsymbol{x},\boldsymbol{y}) n_j(\boldsymbol{y}) \,\mathrm{d}S_y = 0 \tag{5.4}$$

Subtraction of the above identity, considered for the specific values $\{u_0, \boldsymbol{a}\} = \{u(\boldsymbol{x}), \boldsymbol{\nabla} u(\boldsymbol{x})\}$, from eq. (5.2) leads to the equality:

$$\int_{(\partial\Omega - e_\varepsilon)+s_\varepsilon} [q(\boldsymbol{y}) - u_{,a}(\boldsymbol{x}) n_a(\boldsymbol{y})] G_{,\bar{\imath}}(\boldsymbol{x},\boldsymbol{y}) \,\mathrm{d}S_y \\ - \int_{(\partial\Omega - e_\varepsilon)+s_\varepsilon} [u(\boldsymbol{y}) - u(\boldsymbol{x}) - u_{,a}(\boldsymbol{x})(y_a - x_a)] G_{,\bar{\imath}j}(\boldsymbol{x},\boldsymbol{y}) n_j(\boldsymbol{y}) \,\mathrm{d}S_y = 0 \tag{5.5}$$

in which the limiting process $\varepsilon \to 0$ is now investigated. Assume that the solution u has a $C^{1,\alpha}$ smoothness at $\boldsymbol{x}$, i.e.:

$$\exists (C > 0, \alpha > 0) \quad \begin{cases} |u(\boldsymbol{y}) - u(\boldsymbol{x}) - u_{,a}(\boldsymbol{x})(y_a - x_a)| \leq C\,|\boldsymbol{y} - \boldsymbol{x}|^{1+\alpha} \\ |q(\boldsymbol{y}) - u_{,a}(\boldsymbol{x}) n_a(\boldsymbol{y})| \leq C\,|\boldsymbol{y} - \boldsymbol{x}|^{1+\alpha} \end{cases} \tag{5.6}$$

The limiting behaviours of the integrals over s_ε and $\partial\Omega - e_\varepsilon$ of eq. (5.5) are now examined separately.

Integral over s_ε. From the fact that $|\boldsymbol{y}-\boldsymbol{x}| \sim \varepsilon$ and $\mathrm{d}S_y \sim \varepsilon^2\,\mathrm{d}\omega$ (ω: solid angle of origin $\boldsymbol{x}$) and using the assumption (5.6), the integrals over s_ε in eq. (5.5) vanish when $\varepsilon \to 0$:

$$\int_{s_\varepsilon} [q(\boldsymbol{y}) - u_{,a}(\boldsymbol{x})n_a(\boldsymbol{y})]G_{,\bar{\imath}}(\boldsymbol{x},\boldsymbol{y})\,\mathrm{d}S_y = O(\varepsilon^\alpha)$$

$$\int_{s_\varepsilon} [u(\boldsymbol{y}) - u(\boldsymbol{x}) - u_{,a}(\boldsymbol{x})(y_a - x_a)]G_{,\bar{\imath}j}(\boldsymbol{x},\boldsymbol{y})n_j(\boldsymbol{y})\,\mathrm{d}S_y = O(\varepsilon^\alpha)$$

Integrals over $\partial\Omega - e_\varepsilon$. The two integrands of eq. (5.5) behave like $|\boldsymbol{y}-\boldsymbol{x}|^{\alpha-2}$ for $\boldsymbol{y}$ approaching $\boldsymbol{x}$, by virtue of the smoothness assumption, eq. (5.6). Therefore, they are integrable over $\partial\Omega$: both integrals over $\partial\Omega - e_\varepsilon$ in eq. (5.5) are convergent when $\varepsilon \to 0$:

$$\begin{aligned}
&\lim_{\varepsilon\to 0}\int_{\partial\Omega - e_\varepsilon} [q(\boldsymbol{y}) - u_{,a}(\boldsymbol{x})n_a(\boldsymbol{y})]G_{,\bar{\imath}}(\boldsymbol{x},\boldsymbol{y})\,\mathrm{d}S_y \\
&\qquad = \int_{\partial\Omega} [q(\boldsymbol{y}) - u_{,a}(\boldsymbol{x})n_a(\boldsymbol{y})]G_{,\bar{\imath}}(\boldsymbol{x},\boldsymbol{y})\,\mathrm{d}S_y \\
&\lim_{\varepsilon\to 0}\int_{\partial\Omega - e_\varepsilon} [u(\boldsymbol{y}) - u(\boldsymbol{x}) - u_{,a}(\boldsymbol{x})(y_a - x_a)]G_{,\bar{\imath}j}(\boldsymbol{x},\boldsymbol{y})n_j(\boldsymbol{y})\,\mathrm{d}S_y \\
&\qquad = \int_{\partial\Omega} [u(\boldsymbol{y}) - u(\boldsymbol{x}) - u_{,a}(\boldsymbol{x})(y_a - x_a)]G_{,\bar{\imath}j}(\boldsymbol{x},\boldsymbol{y})n_j(\boldsymbol{y})\,\mathrm{d}S_y
\end{aligned}$$

Integral representation for the gradient

As a result of the previous analysis, eq. (5.5) becomes, in the limit $\varepsilon \to 0$:

$$\begin{aligned}
&\int_{\partial\Omega} [q(\boldsymbol{y}) - u_{,a}(\boldsymbol{x})n_a(\boldsymbol{y})]G_{,\bar{\imath}}(\boldsymbol{x},\boldsymbol{y})\,\mathrm{d}S_y \\
&\quad - \int_{\partial\Omega} [u(\boldsymbol{y}) - u(\boldsymbol{x}) - u_{,a}(\boldsymbol{x})(y_a - x_a)]G_{,\bar{\imath}j}(\boldsymbol{x},\boldsymbol{y})n_j(\boldsymbol{y})\,\mathrm{d}S_y = 0 \qquad (5.7)
\end{aligned}$$

and involves weakly singular, convergent, integrals. It is however preferable to reformulate the above equation by rearranging terms as follows:

$$\begin{aligned}
&u(\boldsymbol{y}) - u(\boldsymbol{x}) - u_{,a}(\boldsymbol{x})(y_a - x_a) \\
&\qquad = [u(\boldsymbol{y}) - u(\boldsymbol{x}) - D_a u(\boldsymbol{x})(y_a - x_a)] - q(\boldsymbol{x})n_a(\boldsymbol{x})(y_a - x_a) \\
&q(\boldsymbol{y}) - u_{,a}(\boldsymbol{x})n_a(\boldsymbol{y}) \\
&\qquad = [q(\boldsymbol{y}) - q(\boldsymbol{x})] - u_{,a}[n_a(\boldsymbol{y}) - n_a(\boldsymbol{x})]
\end{aligned}$$

where $D_a u = u_{,a} - n_a q$ is the tangential part of the partial derivative $u_{,a}$ (Appendix A). Equation (5.7) thus becomes:

$$\begin{aligned}
A_{ia}(\boldsymbol{x})u_{,a}(\boldsymbol{x}) &= \int_{\partial\Omega} [q(\boldsymbol{y}) - q(\boldsymbol{x})]G_{,\bar{\imath}}(\boldsymbol{x},\boldsymbol{y})\,\mathrm{d}S_y \\
&\quad - \int_{\partial\Omega} [u(\boldsymbol{y}) - u(\boldsymbol{x}) - D_a u(\boldsymbol{x})(y_a - x_a)]G_{,\bar{\imath}j}(\boldsymbol{x},\boldsymbol{y})n_j(\boldsymbol{y})\,\mathrm{d}S_y \qquad (5.8)
\end{aligned}$$

where $A_{ia}(\boldsymbol{x})$, the free term, is given by:

$$A_{ia}(\boldsymbol{x}) = \int_{\partial\Omega} [n_a(\boldsymbol{y}) - n_a(\boldsymbol{x})] G_{,\bar{i}}(\boldsymbol{x},\boldsymbol{y}) \, \mathrm{d}S_y \\ - n_a(\boldsymbol{x}) n_b(\boldsymbol{x}) \int_{\partial\Omega} (y_b - x_b) G_{,\bar{i}j}(\boldsymbol{x},\boldsymbol{y}) n_j(\boldsymbol{y}) \, \mathrm{d}S_y \quad (5.9)$$

and depends only on $\boldsymbol{x}$ and on the shape of Ω at $\boldsymbol{x}$. Equations (5.8)–(5.9) thus define an integral representation formula for the gradient $\boldsymbol{\nabla} u(\boldsymbol{x})$ at a point $\boldsymbol{x} \in \partial\Omega$ in terms of the quantities:

$$u(\boldsymbol{y}) - u(\boldsymbol{x}) - \boldsymbol{\nabla}_S u(\boldsymbol{x}).(\boldsymbol{y} - \boldsymbol{x}) \qquad q(\boldsymbol{y}) - q(\boldsymbol{x})$$

which are completely determined by the knowledge of (u, q) on $\partial\Omega$ (only the tangential part $\boldsymbol{\nabla}_S u$ of $\boldsymbol{\nabla} u$ appears under the integral sign). They are convergent if u is $C^{1,\alpha}$-regular at $\boldsymbol{x}$ (sufficient condition). Moreover, the integrals appearing in the free term A_{ia}, eq. (5.9), are convergent if $\partial\Omega$ is suffiently smooth at $\boldsymbol{x}$, more precisely if the unit normal is $C^{0,\alpha}$-regular at $\boldsymbol{x}$, i.e.:

$$\exists (C, \alpha) \qquad |\boldsymbol{n}(\boldsymbol{y}) - \boldsymbol{n}(\boldsymbol{x})| \leq |\boldsymbol{y} - \boldsymbol{x}|^{\alpha} \quad (5.10)$$

Finally, the effective computation of $\boldsymbol{\nabla} u(\boldsymbol{x})$ using formula (5.8) implies solving a 3×3 linear system of equations for the scalar unknowns $u_{,a}$, which in turn requires the invertibility of the matrix $A_{ia}(\boldsymbol{x})$.

The result (5.8–5.9), valid irrespective of the specific fundamental solution $G(\boldsymbol{x},\boldsymbol{y})$ being used, allows us to evaluate the complete gradient $\boldsymbol{\nabla} u(\boldsymbol{x})$ at $\boldsymbol{x} \in \partial\Omega$ knowing (u, q) over $\partial\Omega$ (usually from a boundary element solution of the integral equation (2.24).

5.4 INTEGRATION BY PARTS AND FIRST-ORDER REGULARIZATION

Assume that the infinite-space fundamental solution $G = 1/4\pi r$ is used in the integral identity (5.2). G then verifies the property (2.9):

$$G_{,\bar{i}}(\boldsymbol{x},\boldsymbol{y}) = -G_{,i}(\boldsymbol{x},\boldsymbol{y})$$

(interchangeability of derivatives of G with respect to $\boldsymbol{x}$ and $\boldsymbol{y}$), which allows us to reformulate (5.2) as:

$$0 = \int_{(\partial\Omega - e_\varepsilon) + s_\varepsilon} \Big\{ q(\boldsymbol{y}) G_{,i}(\boldsymbol{x},\boldsymbol{y}) - u(\boldsymbol{y}) G_{,ij}(\boldsymbol{x},\boldsymbol{y}) n_j(\boldsymbol{y}) \Big\} \, \mathrm{d}S_y \quad (5.11)$$

whereupon the following integration by parts, valid for any *closed* surface S such that $\boldsymbol{x} \notin S$, is applicable. One has:

$$\int_S u(\boldsymbol{y}) G_{,ij}(\boldsymbol{x},\boldsymbol{y}) n_j(\boldsymbol{y}) \, \mathrm{d}S_y = \int_S u(\boldsymbol{y}) [D_{ji} G_{,j}(\boldsymbol{x},\boldsymbol{y}) + G_{,jj}(\boldsymbol{x},\boldsymbol{y}) n_i(\boldsymbol{y})] \, \mathrm{d}S_y$$

introducing the tangential differential operator $D_{ij}() = n_i()_{,j} - n_j()_{,i}$, defined in Appendix A. Noting that G solves the field equation $G_{,jj}(\boldsymbol{x}, \boldsymbol{y}) = 0$ ($\boldsymbol{y} \neq \boldsymbol{x}$) and applying identity (A.8) to the remaining term, one obtains:

$$\int_S u(\boldsymbol{y}) D_{ji} G_{,j}(\boldsymbol{x}, \boldsymbol{y})\, \mathrm{d}S_y = \int_S D_{ij} u(\boldsymbol{y}) G_{,j}(\boldsymbol{x}, \boldsymbol{y})\, \mathrm{d}S_y$$

This integration by parts scheme has been used in e.g. Kupradze [11], Sladek and Sladek [16]. Here it allows us to recast the integral identity (5.2) into:

$$0 = \int_{(\partial\Omega - e_\varepsilon)+s_\varepsilon} \left\{ q(\boldsymbol{y}) G_{,i}(\boldsymbol{x}, \boldsymbol{y}) - D_{ij} u(\boldsymbol{y}) G_{,j}(\boldsymbol{x}, \boldsymbol{y}) \right\} \mathrm{d}S_y \tag{5.12}$$

Now the limiting form of this relation when $\varepsilon \to 0$ is sought.

First-order regularization

The simple (affine) solution $u^\star$ defined by eq. (5.3) satisfies identity (5.12). Now the specific choice $\{u_0, \boldsymbol{a}\} = \{u(\boldsymbol{x}), \boldsymbol{\nabla} u(\boldsymbol{x})\}$ is made, and one has:

$$\begin{aligned} D_{ij} u^\star(\boldsymbol{y}) &= n_i(\boldsymbol{y}) u_{,j}(\boldsymbol{x}) - n_j(\boldsymbol{y}) u_{,i}(\boldsymbol{x}) \\ &= D_{ij} u(\boldsymbol{x}) + [n_i(\boldsymbol{y}) - n_i(\boldsymbol{x})] u_{,j}(\boldsymbol{x}) - [n_j(\boldsymbol{y}) - n_j(\boldsymbol{x})] u_{,i}(\boldsymbol{x}) \end{aligned}$$

$$\begin{aligned} q^\star(\boldsymbol{y}) &= n_j(\boldsymbol{y}) u_{,j}(\boldsymbol{x}) \\ &= q(\boldsymbol{x}) + u_{,j}(\boldsymbol{x})[n_j(\boldsymbol{y}) - n_j(\boldsymbol{x})] \end{aligned}$$

Substituting the simple solution $(u^\star, q^\star)$ into (5.12) and using the above relations leads to the identity:

$$\begin{aligned} 0 = &\int_{\partial\Omega - e_\varepsilon} \left\{ q(\boldsymbol{x}) G_{,i}(\boldsymbol{x}, \boldsymbol{y}) - D_{ij} u(\boldsymbol{x}) G_{,j}(\boldsymbol{x}, \boldsymbol{y}) n_j(\boldsymbol{y}) \right\} \mathrm{d}S_y \\ &+ \int_{\partial\Omega - e_\varepsilon} u_{,j}(\boldsymbol{x})[n_j(\boldsymbol{y}) - n_j(\boldsymbol{x})] G_{,i}(\boldsymbol{x}, \boldsymbol{y})\, \mathrm{d}S_y \\ &- \int_{\partial\Omega - e_\varepsilon} \left\{ [n_i(\boldsymbol{y}) - n_i(\boldsymbol{x})] u_{,j}(\boldsymbol{x}) - [n_j(\boldsymbol{y}) - n_j(\boldsymbol{x})] u_{,i}(\boldsymbol{x}) \right\} G_{,j}(\boldsymbol{x}, \boldsymbol{y})\, \mathrm{d}S_y \\ &+ \int_{s_\varepsilon} \left\{ n_j(\boldsymbol{y}) u_{,j}(\boldsymbol{x}) G_{,i}(\boldsymbol{x}, \boldsymbol{y}) - [n_i(\boldsymbol{y}) u_{,j}(\boldsymbol{x}) - n_j(\boldsymbol{y}) u_{,i}(\boldsymbol{x})] G_{,j}(\boldsymbol{x}, \boldsymbol{y}) \right\} \mathrm{d}S_y \end{aligned}$$

which is then subtracted from eq. (5.12). After some suitable term grouping, one isolates the free term:

$$\begin{aligned} \tilde{A}_{ij}(\boldsymbol{x}, \varepsilon) = &\int_{\partial\Omega - e_\varepsilon} [n_i(\boldsymbol{y}) - n_i(\boldsymbol{x})] G_{,j}(\boldsymbol{x}, \boldsymbol{y})\, \mathrm{d}S_y \\ &- \delta_{ij} \int_{\partial\Omega - e_\varepsilon} [n_a(\boldsymbol{y}) - n_a(\boldsymbol{x})] G_{,a}(\boldsymbol{x}, \boldsymbol{y})\, \mathrm{d}S_y \\ &- \int_{\partial\Omega - e_\varepsilon} [n_j(\boldsymbol{y}) - n_j(\boldsymbol{x})] G_{,i}(\boldsymbol{x}, \boldsymbol{y})\, \mathrm{d}S_y \end{aligned} \tag{5.13}$$

and obtains the identity:

$$\tilde{A}_{ia}(\boldsymbol{x},\varepsilon)u_{,a}(\boldsymbol{x})$$

$$= \int_{\partial\Omega - e_\varepsilon} \Big\{ \Big[D_{ij}u(\boldsymbol{y}) - D_{ij}u(\boldsymbol{x}) \Big] G_{,j}(\boldsymbol{x},\boldsymbol{y}) - [q(\boldsymbol{y}) - q(\boldsymbol{x})] G_{,i}(\boldsymbol{x},\boldsymbol{y}) \Big\} \, \mathrm{d}S_y$$
$$+ \int_{s_\varepsilon} \Big\{ n_i(\boldsymbol{y})[u_{,j}(\boldsymbol{y}) - u_{,j}(\boldsymbol{x})] - n_j(\boldsymbol{y})[u_i(\boldsymbol{y}) - u_{,i}(\boldsymbol{x})] \Big\} G_{,j}(\boldsymbol{x},\boldsymbol{y}) \, \mathrm{d}S_y$$
$$- \int_{s_\varepsilon} n_j(\boldsymbol{y})[u_{,j}(\boldsymbol{y}) - u_{,j}(\boldsymbol{x})] \, \mathrm{d}S_y \tag{5.14}$$

Assuming that u has $C^{1,\alpha}$ smoothness at $\boldsymbol{x}$, property (5.6) is applicable and readily leads to the conclusion that, in the limit $\varepsilon \to 0$, the integral over $\partial\Omega - e_\varepsilon$ in (5.14) yields a weakly singular (i.e. convergent) integral over $\partial\Omega$ while all integrals over s_ε vanish. Moreover, again assuming that $\boldsymbol{x}$ is a smooth point of $\partial\Omega$ satisfying (5.10), the coefficients $\tilde{A}_{ij}(\boldsymbol{x},\varepsilon)$ are expressed in terms of convergent integrals and their limiting values $\tilde{A}_{ij}(\boldsymbol{x})$ for $\varepsilon \to 0$ are:

$$\tilde{A}_{ij}(\boldsymbol{x}) = \int_{\partial\Omega} \Big\{ [n_i(\boldsymbol{y}) - n_i(\boldsymbol{x})] G_{,j}(\boldsymbol{x},\boldsymbol{y}) - [n_j(\boldsymbol{y}) - n_j(\boldsymbol{x})] G_{,i}(\boldsymbol{x},\boldsymbol{y}) \Big\} \, \mathrm{d}S_y$$
$$- \delta_{ij} \int_{\partial\Omega} [n_a(\boldsymbol{y}) - n_a(\boldsymbol{x})] G_{,a}(\boldsymbol{x},\boldsymbol{y}) \, \mathrm{d}S_y \tag{5.15}$$

In fact, it is shown in Appendix 5.B that the free term $\tilde{A}_{ij}(\boldsymbol{x})$ is given by:

$$\tilde{A}_{ij}(\boldsymbol{x}) = \frac{1}{2}\delta_{ij} + \frac{1}{4\pi} \Big[n_i(\boldsymbol{x}) I_j(\boldsymbol{x},\partial\Omega) - n_j I_i(\boldsymbol{x}) - \delta_{ij} n_a(\boldsymbol{x}) I_a(\boldsymbol{x}) \Big] \tag{5.16}$$

where (Appendix 5.A, eq. (5.A.7)):

$$I_a(\boldsymbol{x},S) = \int_S \left(\frac{1}{r} r_{,n} - K \right) n_a \frac{\mathrm{d}S_y}{r} - \int_{\partial S} \nu_a \frac{\mathrm{d}s_y}{r} \tag{5.17}$$

Integral representation of the potential gradient

As a result of the foregoing analysis, the regularized integral representation of $\boldsymbol{\nabla} u(\boldsymbol{x})$ at any (smooth) boundary point $\boldsymbol{x}$ satisfying (5.10) reads:

$$\frac{1}{2}u_{,i}(\boldsymbol{x}) = \int_{\partial\Omega} \Big[D_{ij}u(\boldsymbol{y}) - D_{ij}u(\boldsymbol{x}) \Big] G_{,j}(\boldsymbol{x},\boldsymbol{y}) \, \mathrm{d}S_y$$
$$- \int_{\partial\Omega} [q(\boldsymbol{y}) - q(\boldsymbol{x})] G_{,i}(\boldsymbol{x},\boldsymbol{y}) \, \mathrm{d}S_y$$
$$+ D_{ij}u(\boldsymbol{x}) I_j(\boldsymbol{x},\partial\Omega) - q(\boldsymbol{x}) I_i(\boldsymbol{x},\partial\Omega) \tag{5.18}$$

The fundamental solution is $G(\boldsymbol{x},\boldsymbol{y}) = 1/4\pi r$. All integrals are weakly singular. The result (5.18) holds under the sufficient smoothness assumptions $u \in C^{1,\alpha}$ et $\boldsymbol{n} \in C^{0,\alpha}$ at $\boldsymbol{x}$.

Comments. Equation (5.18) expresses the gradient $\nabla u(\boldsymbol{x})$ at $\boldsymbol{x} \in \partial\Omega$ in terms of the values taken on $\partial\Omega$ by (u, q). Since all integrals are weakly singular, the numerical evaluation of this formula is feasible.

5.5 EXTENSION TO ELASTOSTATICS

The foregoing approach applies in similar fashion to elastostatics, resulting in regularized integral representation formulas for the displacement gradient or the elastic stress tensor at a smooth boundary point $\boldsymbol{x}$. It has been shown in Section 4.3 that the displacement gradient admits, for $\boldsymbol{x} \notin \partial\Omega$, the integral representation formula:

$$\kappa u_{k,\ell}(\boldsymbol{x}) = \int_{\partial\Omega} \left\{ t_a(\boldsymbol{y}) U^k_{a,\bar{\ell}}(\boldsymbol{x}, \boldsymbol{y}) - u_a(\boldsymbol{y}) n_b(\boldsymbol{y}) \Sigma^k_{ab,\bar{\ell}}(\boldsymbol{x}, \boldsymbol{y}) \right\} \mathrm{d}S_y$$

Thus, using again the limiting process approach based on the introduction of an exclusion neighbourhood $v_\varepsilon(\boldsymbol{x})$, the integral representation of the displacement gradient is sought as the limiting form for $\varepsilon \to 0$ of the previous identity with domain Ω_ε, i.e.:

$$0 = \int_{(\partial\Omega - e_\varepsilon) + s_\varepsilon} \left\{ t_a(\boldsymbol{y}) U^k_{a,\bar{\ell}}(\boldsymbol{x}, \boldsymbol{y}) - u_a(\boldsymbol{y}) n_b(\boldsymbol{y}) \Sigma^k_{ab,\bar{\ell}}(\boldsymbol{x}, \boldsymbol{y}) \right\} \mathrm{d}S_y \qquad (5.19)$$

Both the second-order regularization and the integration by parts followed by first-order regularization approaches are again considered.

Second-order regularization

Consider particular elastostatic states $\boldsymbol{u}_\star$ of the form:

$$u^\star_a(\boldsymbol{y}) = u^0_a + a_{ac}(y_c - x_c) \qquad t_i(\boldsymbol{y}) = C_{ijab} n_j(\boldsymbol{y}) a_{ab} \qquad (5.20)$$

(u^0_i, a_{ij}: arbitrary constants), i.e. superpositions of rigid-body translations and uniform strain. The state $\boldsymbol{u}_\star$ satisfies the integral identity (5.19), and one then has:

$$\int_{(\partial\Omega - e_\varepsilon) + s_\varepsilon} C_{ijab} n_j(\boldsymbol{y}) a_{ab} U^k_{i,\bar{\ell}}(\boldsymbol{x}, \boldsymbol{y}) \, \mathrm{d}S_y$$
$$- \int_{(\partial\Omega - e_\varepsilon) + s_\varepsilon} [u^0_a + a_{ac}(y_c - x_c)] n_j(\boldsymbol{y}) \Sigma^k_{aj,\bar{\ell}}(\boldsymbol{x}, \boldsymbol{y}) \, \mathrm{d}S_y = 0 \qquad (5.21)$$

Taking the above equation with $\{\boldsymbol{u}^0, \boldsymbol{a}\} = \{\boldsymbol{u}(\boldsymbol{x}), \nabla \boldsymbol{u}(\boldsymbol{x})\}$ (i.e. setting the constants in $\boldsymbol{u}_\star$ equal to the unknown displacement and displacement gradient at the collocation point $\boldsymbol{x}$) and subtracting the result from (5.19) gives:

$$\int_{(\partial\Omega-e_\varepsilon)+s_\varepsilon} \Big[u_a(\boldsymbol{y}) - u_a(\boldsymbol{x}) - u_{a,c}(\boldsymbol{x})(y_c - x_c)\Big] n_b(\boldsymbol{y}) \Sigma^k_{ab,\bar{\ell}}(\boldsymbol{x},\boldsymbol{y})\,\mathrm{d}S_y \\ - \int_{(\partial\Omega-e_\varepsilon)+s_\varepsilon} [t_i(\boldsymbol{y}) - \sigma_{ij}(\boldsymbol{x}) n_j(\boldsymbol{y})] U^k_{i,\bar{\ell}}(\boldsymbol{x},\boldsymbol{y})\,\mathrm{d}S_y = 0$$

Assuming $\boldsymbol{u}$ to have $C^{1,\alpha}$ smoothness at $\boldsymbol{x}$, the limiting process $\varepsilon \to 0$ in the above identity can be performed in the fashion of Section 5.4, resulting in the identity:

$$\int_{\partial\Omega} \Big[u_a(\boldsymbol{y}) - u_a(\boldsymbol{x}) - u_{a,c}(\boldsymbol{x})(y_c - x_c)\Big] n_b(\boldsymbol{y}) \Sigma^k_{ab,\bar{\ell}}(\boldsymbol{x},\boldsymbol{y})\,\mathrm{d}S_y \\ - \int_{\partial\Omega} [t_i(\boldsymbol{y}) - \sigma_{ij}(\boldsymbol{x}) n_j(\boldsymbol{y})] U^k_{i,\bar{\ell}}(\boldsymbol{x},\boldsymbol{y})\,\mathrm{d}S_y = 0 \quad (5.22)$$

which involves weakly singular integrals only. Then the following relations are introduced in order to rearrange terms:

$$u_a(\boldsymbol{y}) - u_a(\boldsymbol{x}) - u_{a,c}(\boldsymbol{x})(y_c - x_c) \\ = [u_a(\boldsymbol{y}) - u_a(\boldsymbol{x}) - D_c u_a(\boldsymbol{x})(y_c - x_c)] - u_{a,n}(\boldsymbol{x})(y_c - x_c) n_c(\boldsymbol{x})$$

$$t_i(\boldsymbol{y}) - \sigma_{ij}(\boldsymbol{x}) n_j(\boldsymbol{y}) \\ = [t_i(\boldsymbol{y}) - t_i(\boldsymbol{x})] - \sigma_{ij}(\boldsymbol{x})[n_j(\boldsymbol{y}) - n_j(\boldsymbol{x})]$$

This results in the following regularized integral representation formula for the displacement gradient at a boundary point $\boldsymbol{x}$:

$$\int_{\partial\Omega} [u_a(\boldsymbol{y}) - u_a(\boldsymbol{x}) - D_c u_a(\boldsymbol{x})(y_c - x_c)] n_b(\boldsymbol{y}) \Sigma^k_{ab,\bar{\ell}}(\boldsymbol{x},\boldsymbol{y})\,\mathrm{d}S_y \\ - \int_{\partial\Omega} [t_a(\boldsymbol{y}) - t_a(\boldsymbol{x})] U^k_{a,\bar{\ell}}(\boldsymbol{x},\boldsymbol{y})\,\mathrm{d}S_y = A_{abk\ell}(\boldsymbol{x}) u_{a,b}(\boldsymbol{x}) \quad (5.23)$$

where the free term $A_{abk\ell}(\boldsymbol{x})$ is given by:

$$A_{abk\ell}(\boldsymbol{x}) = \int_{\partial\Omega} n_b(\boldsymbol{x}) n_c(\boldsymbol{x})(y_c - x_c) n_j(\boldsymbol{y}) \Sigma^k_{aj,\bar{\ell}}(\boldsymbol{x},\boldsymbol{y})\,\mathrm{d}S_y \\ - C_{ijab} \int_{\partial\Omega} [n_j(\boldsymbol{y}) - n_j(\boldsymbol{x})] U^k_{i,\ell}(\boldsymbol{x},\boldsymbol{y})\,\mathrm{d}S_y \quad (5.24)$$

The terms $\boldsymbol{u}(\boldsymbol{y}) - \boldsymbol{u}(\boldsymbol{x}) - \boldsymbol{\nabla}_S \boldsymbol{u}(\boldsymbol{x}).(\boldsymbol{y} - \boldsymbol{x})$ and $\boldsymbol{t}(\boldsymbol{y}) - \boldsymbol{t}(\boldsymbol{x})$ depend only on the values taken by $(\boldsymbol{u}, \boldsymbol{t})$ on the boundary (the knowledge of $\boldsymbol{u}$ on $\partial\Omega$ implies that of its surface gradient $\boldsymbol{\nabla}_S \boldsymbol{u}$).

The result (5.23) relies on a $C^{1,\alpha}$ smoothness assumption for $\boldsymbol{u}$ at $\boldsymbol{x}$ and also hinges on the invertibility of the fourth-order tensor $\boldsymbol{A}(\boldsymbol{x})$ (5.24). Moreover, the integrals appearing in the free term $A_{abk\ell}$ are weakly singular provided the smoothness assumption (5.10) for $\partial\Omega$ is met.

The representation formula (5.23) can be used with any fundamental solution (provided it is defined on a domain $E \supseteq \Omega$). It links the Cartesian displacement gradient $\boldsymbol{\nabla u}(\boldsymbol{x})$ at $\boldsymbol{x} \in \partial\Omega$ to the boundary values of the (tangential gradient of the) displacement $\boldsymbol{u}$ and the traction vector $\boldsymbol{t}$, in the form of a system of nine linear equations ($k, \ell = 1, 2, 3$) on the nine components $u_{a,b}(\boldsymbol{x})$.

Integration by parts and first-order regularization

When using the Kelvin (infinite-space) fundamental solution for isotropic or anisotropic media, the integrated-by-parts variant (4.24) of the integral representation formula for the displacement gradient can be used as a starting point. One seeks in this case the limiting form when $\varepsilon \to 0$ of:

$$0 = \int_{(\partial\Omega - e_\varepsilon)+s_\varepsilon} \Big\{ D_{\ell b}u_a(\boldsymbol{y})\Sigma^k_{ab}(\boldsymbol{x},\boldsymbol{y}) - t_a(\boldsymbol{y})U^k_{a,\ell}(\boldsymbol{x},\boldsymbol{y}) \Big\} \,\mathrm{d}S_y \tag{5.25}$$

This equation is in particular true for the simple solution (5.20). Taking as before $\{\boldsymbol{u}^0, \boldsymbol{a}\} = \{\boldsymbol{u}(\boldsymbol{x}), \boldsymbol{\nabla u}(\boldsymbol{x})\}$, one obtains the identity:

$$\begin{aligned}
0 = &\int_{\partial\Omega - e_\varepsilon} \Big\{ [n_\ell(\boldsymbol{y}) - n_\ell(\boldsymbol{x})]u_{a,b}(\boldsymbol{x}) - [n_b(\boldsymbol{y}) - n_b(\boldsymbol{x})]u_{a,\ell}(\boldsymbol{x}) \Big\} \Sigma^k_{ab}(\boldsymbol{x},\boldsymbol{y}) \,\mathrm{d}S_y \\
&- \int_{\partial\Omega - e_\varepsilon} \sigma_{ab}(\boldsymbol{x})[n_b(\boldsymbol{y}) - n_b(\boldsymbol{x})]U^k_{a,\ell}(\boldsymbol{x},\boldsymbol{y}) \,\mathrm{d}S_y \\
&+ \int_{\partial\Omega - e_\varepsilon} \Big\{ D_{\ell b}u_a(\boldsymbol{x})\Sigma^k_{ab}(\boldsymbol{x},\boldsymbol{y}) - t_a(\boldsymbol{x})U^k_{a,\ell}(\boldsymbol{x},\boldsymbol{y}) \Big\} \,\mathrm{d}S_y \\
&+ \int_{s_\varepsilon} \Big\{ n_\ell(\boldsymbol{y})u_{a,b}(\boldsymbol{x}) - n_b(\boldsymbol{y})u_{a,\ell}(\boldsymbol{x}) \Big\} \Sigma^k_{ab}(\boldsymbol{x},\boldsymbol{y}) \,\mathrm{d}S_y \\
&- \int_{s_\varepsilon} \sigma_{ab}(\boldsymbol{x})n_b(\boldsymbol{y})U^k_{a,\ell}(\boldsymbol{x},\boldsymbol{y}) \,\mathrm{d}S_y
\end{aligned}$$

Its subtraction from (5.25) then yields the equation:

$$\begin{aligned}
\tilde{A}_{abk\ell}(\boldsymbol{x},\varepsilon)u_{a,b}(\boldsymbol{x}) = &\int_{\partial\Omega - e_\varepsilon} \Big[D_{\ell b}u_a(\boldsymbol{y}) - D_{\ell b}u_a(\boldsymbol{x}) \Big] \Sigma^k_{ab}(\boldsymbol{x},\boldsymbol{y}) \,\mathrm{d}S_y \\
&- \int_{\partial\Omega - e_\varepsilon} [t_a(\boldsymbol{y}) - t_a(\boldsymbol{x})]U^k_{a,\ell}(\boldsymbol{x},\boldsymbol{y}) \,\mathrm{d}S_y \\
&+ \int_{s_\varepsilon} n_\ell(\boldsymbol{y})[u_{a,b}(\boldsymbol{y}) - u_{a,b}(\boldsymbol{x})]\Sigma^k_{ab}(\boldsymbol{x},\boldsymbol{y}) \,\mathrm{d}S_y \\
&- \int_{s_\varepsilon} n_b(\boldsymbol{y})[u_{a,\ell}(\boldsymbol{y}) - u_{a,\ell}(\boldsymbol{x})]\Sigma^k_{ab}(\boldsymbol{x},\boldsymbol{y}) \,\mathrm{d}S_y \\
&- \int_{s_\varepsilon} [t_a(\boldsymbol{y}) - C_{abcd}n_b(\boldsymbol{y})u_{c,d}(\boldsymbol{x})]U^k_{a,\ell}(\boldsymbol{x},\boldsymbol{y}) \,\mathrm{d}S_y
\end{aligned} \tag{5.26}$$

Assuming that $\boldsymbol{u}$ and $\boldsymbol{n}$ have respectively $C^{1,\alpha}$ and $C^{0,\alpha}$ smoothness at $\boldsymbol{x}$, the limiting process $\varepsilon \to 0$ is performed in the same way as in Section 5.4: the integrals over $(\partial\Omega - e_\varepsilon)$ become weakly singular integrals over $\partial\Omega$ while the integrals over s_ε vanish. The free term:

$$\begin{aligned}\tilde{A}_{abk\ell}(\boldsymbol{x},\varepsilon) = &\int_{\partial\Omega - e_\varepsilon} [n_\ell(\boldsymbol{y}) - n_\ell(\boldsymbol{x})]\Sigma^k_{ab}(\boldsymbol{x},\boldsymbol{y})\,\mathrm{d}S_y \\ &- \int_{\partial\Omega - e_\varepsilon} [n_j(\boldsymbol{y}) - n_j(\boldsymbol{x})]\delta_{b\ell}\Sigma^k_{aj}(\boldsymbol{x},\boldsymbol{y})\,\mathrm{d}S_y \\ &- C_{ijab}\int_{\partial\Omega - e_\varepsilon} [n_j(\boldsymbol{y}) - n_j(\boldsymbol{x})]U^k_{i,\ell}(\boldsymbol{x},\boldsymbol{y})\,\mathrm{d}S_y \end{aligned} \tag{5.27}$$

becomes, in the limit $\varepsilon \to 0$, the corresponding weakly singular integral over $\partial\Omega$. It is in fact shown in Appendix 5.C that:

$$\begin{aligned}&\lim_{\varepsilon\to 0} \tilde{A}_{abk\ell}(\boldsymbol{x},\varepsilon)u_{a,b}(\boldsymbol{x}) \\ &\qquad = \frac{1}{2}u_{k,\ell}(\boldsymbol{x}) - D_{\ell j}u_i(\boldsymbol{x})A^k_{ij}(\boldsymbol{x},\partial\Omega) + t_i(\boldsymbol{x})B^k_{i\ell}(\boldsymbol{x},\partial\Omega)\end{aligned} \tag{5.28}$$

where $A^k_{ij}(\boldsymbol{x},\partial\Omega)$ and $B^k_{i\ell}(\boldsymbol{x},\partial\Omega)$ are expressed in terms of weakly singular integrals (eqs (5.A.1) and (5.A.2)) by:

$$A^k_{ij}(\boldsymbol{x},S) = \frac{-1}{8\pi(1-\nu)}[(1-2\nu)(\delta_{ik}I_j + \delta_{sk}I_i - \delta_{ij}I_k) + 3J_{ijk}] \tag{5.29}$$

$$B^k_{i\ell}(\boldsymbol{x},S) = \frac{-1}{16\pi\mu(1-\nu)}[(3-4\nu)\delta_{ik}I_\ell + 3J_{ik\ell} - \delta_{\ell k}I_i - \delta_{i\ell}I_k] \tag{5.30}$$

where:

$$I_a(\boldsymbol{x},S) = \int_S \left(\frac{1}{r}r_{,n} - K\right)n_a\frac{\mathrm{d}S_y}{r} - \int_{\partial S}\nu_a\frac{\mathrm{d}s_y}{r} \tag{5.31}$$

$$\begin{aligned}3J_{abc}(\boldsymbol{x},S) = &\,\delta_{ac}I_b(\boldsymbol{x},S) + \delta_{bc}I_a(\boldsymbol{x},S) \\ &+ \int_S (r_{,b}r_{,p}D_{pa}n_c - D_a(n_b n_c) - D_b(n_a n_c))\frac{\mathrm{d}S_y}{r} \\ &+ \int_S (2n_a n_b - r_{,a}r_{,b})n_{,c}K\frac{\mathrm{d}S_y}{r} + \delta_{ab}\int_S n_c r_{,n}\frac{\mathrm{d}S_y}{r^2} \\ &- 2\int_S n_a n_c n_b r_{,n}\frac{\mathrm{d}S_y}{r^2} + \int_{\partial S}(\nu_a n_c n_b + \nu_b n_c n_a - r_{,a}r_{,b}\nu_c)\frac{\mathrm{d}s_y}{r} \\ &+ \int_{\partial S}(\nu_p n_a - \nu_a n_p)r_{,p}r_{,b}n_c\frac{\mathrm{d}s_y}{r}\end{aligned} \tag{5.32}$$

As a result of this analysis, the following, explicit, regularized integral representation formula for the displacement gradient at a boundary point is established:

$$\begin{aligned}\frac{1}{2}u_{k,\ell}(\boldsymbol{x}) = &\int_{\partial\Omega} [D_{\ell b}u_a(\boldsymbol{y}) - D_{\ell b}u_a(\boldsymbol{x})]\Sigma^k_{ab}(\boldsymbol{x},\boldsymbol{y})\,\mathrm{d}S_y \\ &- \int_{\partial\Omega} [t_a(\boldsymbol{y}) - t_a(\boldsymbol{x})]U^k_{a,\ell}(\boldsymbol{x},\boldsymbol{y})\,\mathrm{d}S_y \\ &+ D_{\ell b}u_a(\boldsymbol{x})A^k_{ab}(\boldsymbol{x},\partial\Omega) - t_a(\boldsymbol{x})B^k_{a\ell}(\boldsymbol{x},\partial\Omega)\end{aligned} \tag{5.33}$$

in which all integrals are weakly singular. The residual terms A^k_{ab}, $B^k_{a\ell}$ are given by eqs (5.29), (5.30). A subsequent application of Hooke's law $\boldsymbol{\sigma}(\boldsymbol{x}) = \boldsymbol{C} : \boldsymbol{\nabla u}(\boldsymbol{x})$ then yields the regularized integral representation formula for the stress tensor at a boundary point, in explicit form:

$$\begin{aligned}\frac{1}{2}\sigma_{ij}(\boldsymbol{x}) = C_{ijk\ell}&\int_{\partial\Omega} [D_{\ell b}u_a(\boldsymbol{y}) - D_{\ell b}u_a(\boldsymbol{x})]\Sigma^k_{ab}(\boldsymbol{x},\boldsymbol{y})\,\mathrm{d}S_y \\ &- \int_{\partial\Omega} [t_a(\boldsymbol{y}) - t_a(\boldsymbol{x})]\Sigma^i_{ij}(\boldsymbol{x},\boldsymbol{y})\,\mathrm{d}S_y \\ &+ D_{\ell b}u_a(\boldsymbol{x})C_{ijk\ell}A^k_{ab}(\boldsymbol{x},\partial\Omega) - t_a(\boldsymbol{x})A^a_{ij}(\boldsymbol{x},\partial\Omega)\end{aligned} \tag{5.34}$$

This expression uses the symmetry property (4.12) of the isotropic Kelvin solution, through the relations:

$$C_{ijk\ell}U^k_{a,\ell}(\boldsymbol{x},\boldsymbol{y}) = \Sigma^a_{ij}(\boldsymbol{x},\boldsymbol{y}) \;\Rightarrow\; C_{ijk\ell}B^k_{a\ell}(\boldsymbol{x},\partial\Omega) = A^a_{ij}(\boldsymbol{x},\partial\Omega)$$

5.6 INDIRECT BOUNDARY INTEGRAL EQUATIONS

The concept of indirect integral representation has been introduced in Section 2.4 for potential problems and observation points $\boldsymbol{x}$ away from the boundary $\partial\Omega$. In this section, indirect integral representation formulas for observation points $\boldsymbol{x}$ on the boundary, i.e. in the singular case, are established as by-products of regularized integral equations.

Scalar problems

Consider a bounded domain Ω and its bounded complement $\bar{\Omega} = \mathbb{R}^3 - \Omega$. Let u and $\bar{u}$ be solutions to the Laplace equation in Ω and $\bar{\Omega}$, respectively (*interior* and *exterior* problems, with $b = \bar{b} = 0$). Applying the integral equation (2.24) to both the interior and exterior problems and adding the resulting relations gives:

$$\bar{u}(\boldsymbol{x}) = \int_{\partial\Omega} \Big\{\psi(\boldsymbol{y})G(\boldsymbol{x},\boldsymbol{y}) + [\phi(\boldsymbol{y}) - \phi(\boldsymbol{x})]H(\boldsymbol{x},\boldsymbol{y})\Big\}\,\mathrm{d}S_y$$

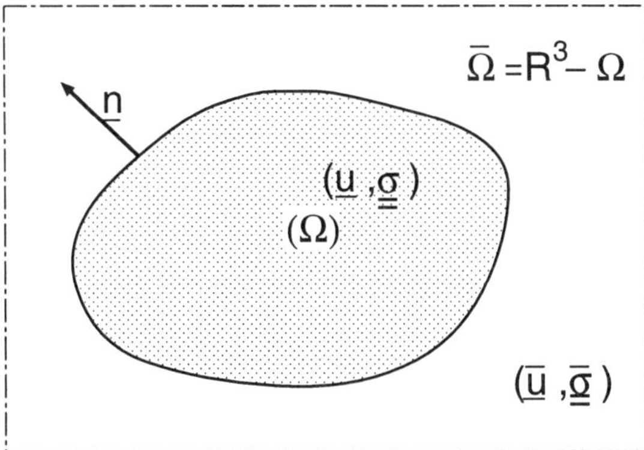

Figure 5.3 Interior and exterior problems: geometrical notations.

The same operation performed on the integral equation (5.18) gives:

$$\frac{1}{2}(\bar{u}_{,i}+u_{,i})(\boldsymbol{x}) = -D_{ij}\phi(\boldsymbol{x})I_j(\boldsymbol{x},\partial\Omega) - \psi(\boldsymbol{x})I_i(\boldsymbol{x},\partial\Omega) - \int_{\partial\Omega}\Big[D_{ij}\phi(\boldsymbol{y}) - D_{ij}\phi(\boldsymbol{x})\Big]G_{,j}(\boldsymbol{x},\boldsymbol{y})\,\mathrm{d}S_y - \int_{\partial\Omega}[\psi(\boldsymbol{y})-\psi(\boldsymbol{x})]G_{,i}(\boldsymbol{x},\boldsymbol{y})\,\mathrm{d}S_y$$

using the notations of Section 2.4, i.e. putting $\psi = \bar{q}+q$ (flux discontinuity across $\partial\Omega$) and $\phi = \bar{u}-u$ (potential discontinuity across $\partial\Omega$); $\boldsymbol{n}$ is the unit normal to $\partial\Omega$ exterior to Ω.

The indirect integral representation at an observation point $\boldsymbol{x}\in\partial\Omega$ of a solution u to the Laplace equation, expressed in terms of the single-layer and double-layer densities ψ and ϕ follow at once:

$$\bar{u}(\boldsymbol{x}) = \int_{\partial\Omega}\psi(\boldsymbol{y})G(\boldsymbol{x},\boldsymbol{y})\,\mathrm{d}S_y \;=\; u(\boldsymbol{x}) \tag{5.35}$$

$$\bar{u}(\boldsymbol{x}) = \int_{\partial\Omega}[\phi(\boldsymbol{y})-\phi(\boldsymbol{x})]H(\boldsymbol{x},\boldsymbol{y})\,\mathrm{d}S_y = u(\boldsymbol{x})+\phi(\boldsymbol{x}) \tag{5.36}$$

The potential gradient $\boldsymbol{\nabla}u$ is then found to satisfy:

$$\frac{1}{2}(u_{,i}+\bar{u}_{,i})(\boldsymbol{x})$$

$$= -\psi(\boldsymbol{x})I_i(\boldsymbol{x},\partial\Omega) - \int_{\partial\Omega}[\psi(\boldsymbol{y})-\psi(\boldsymbol{x})]G_{,i}(\boldsymbol{x},\boldsymbol{y})\,\mathrm{d}S_y \tag{5.37}$$

$$= D_{ji}\phi(\boldsymbol{x})I_j(\boldsymbol{x},\partial\Omega) - \int_{\partial\Omega}\Big[D_{ij}\phi(\boldsymbol{y}) - D_{ij}\phi(\boldsymbol{x})\Big]G_{,j}(\boldsymbol{x},\boldsymbol{y})\,\mathrm{d}S_y \tag{5.38}$$

The indirect integral representations on the boundary allow to formulate regularized indirect integral equations for the unknown density ψ or ϕ.

Interior Dirichlet problem. The potential u is prescribed on the boundary: $u = u^D$ on $\partial\Omega$. As a direct consequence of eqs (5.35) and (5.36), an indirect integral equation for this boundary-value problem can be formulated from either a single-layer representation of unknown density ψ:

$$u^D(\boldsymbol{x}) = \int_{\partial\Omega} \psi(\boldsymbol{y}) G(\boldsymbol{x}, \boldsymbol{y}) \, \mathrm{d}S_y \tag{5.39}$$

or a double-layer representation of unknown density ϕ:

$$u^D(\boldsymbol{x}) = \int_{\partial\Omega} [\phi(\boldsymbol{y}) - \phi(\boldsymbol{x})] G_{,n}(\boldsymbol{x}, \boldsymbol{y}) \, \mathrm{d}S_y - \phi(\boldsymbol{x}) \tag{5.40}$$

Interior Neumann problem. The flux q is prescribed on the boundary: $q = q^D$ on $\partial\Omega$. As a direct consequence of eqs (5.37) and (5.38), an indirect integral equation for this boundary-value problem can be formulated from either a single-layer representation of unknown density ψ:

$$q^D(\boldsymbol{x}) = \psi(\boldsymbol{x}) \Big(\frac{1}{2} - n_j(\boldsymbol{x}) I_j(\boldsymbol{x}, \partial\Omega) \Big) - n_j(\boldsymbol{x}) \int_{\partial\Omega} [\psi(\boldsymbol{y}) - \psi(\boldsymbol{x})] G_{,j}(\boldsymbol{x}, \boldsymbol{y}) \, \mathrm{d}S_y \tag{5.41}$$

or a double-layer representation of unknown density ϕ:

$$q^D(\boldsymbol{x}) = -D_j \phi(\boldsymbol{x}) I_j(\boldsymbol{x}, \partial\Omega) - n_i(\boldsymbol{x}) \int_{\partial\Omega} \Big[D_{ij}\phi(\boldsymbol{y}) - D_{ij}\phi(\boldsymbol{x}) \Big] G_{,j}(\boldsymbol{x}, \boldsymbol{y}) \, \mathrm{d}S_y \tag{5.42}$$

Elastostatics

Consider two elastostatic states $(\boldsymbol{u}, \boldsymbol{\sigma})$ and $(\bar{\boldsymbol{u}}, \bar{\boldsymbol{\sigma}})$ without body forces, defined respectively for a bounded body Ω and its unbounded complement $\bar{\Omega} = \mathbb{R}^3 - \Omega$. Isotropic elasticity is assumed for simplicity.

Let $\boldsymbol{x}$ denote a point on the boundary $\partial\Omega$. Applying the displacement integral equation (4.32) to both the interior and exterior problems and adding the resulting relations leads to:

$$\bar{u}_k(\boldsymbol{x}) = \int_{\partial\Omega} \Big\{ \psi_i(\boldsymbol{y}) U_i^k(\boldsymbol{x}, \boldsymbol{y}) + [\phi_i(\boldsymbol{y}) - \phi_i(\boldsymbol{x})] T_i^k(\boldsymbol{x}, \boldsymbol{y}) \Big\} \, \mathrm{d}S_y$$

The same operation performed on the gradient integral equation (5.33) gives:

$$\begin{aligned} \frac{1}{2}[u_{k,\ell}(\boldsymbol{x}) + \bar{u}_{k,\ell}(\boldsymbol{x})] = & -D_{\ell b}\phi_a(\boldsymbol{x}) A_{ab}^k(\boldsymbol{x}, \partial\Omega) - \psi_a(\boldsymbol{x}) B_{a\ell}^k(\boldsymbol{x}, \partial\Omega) \\ & - \int_{\partial\Omega} [D_{\ell b}\phi_a(\boldsymbol{y}) - D_{\ell b}\phi_a(\boldsymbol{x})] \Sigma_{ab}^k(\boldsymbol{x}, \boldsymbol{y}) \, \mathrm{d}S_y \\ & - \int_{\partial\Omega} [\psi_a(\boldsymbol{y}) - \psi_a(\boldsymbol{x})] U_{a,\ell}^k(\boldsymbol{x}, \boldsymbol{y}) \, \mathrm{d}S_y \end{aligned}$$

having put $\boldsymbol{\psi} = \bar{\boldsymbol{t}} + \boldsymbol{t}$ (traction vector discontinuity across $\partial\Omega$) and $\boldsymbol{\phi} = \bar{\boldsymbol{u}} - \boldsymbol{u}$ (displacement discontinuity across $\partial\Omega$); $\boldsymbol{n}$ is the unit normal to $\partial\Omega$ exterior to Ω.

The indirect integral representation at an observation point $\boldsymbol{x} \in \partial\Omega$ of the elastostatic state, expressed in terms of the single-layer and double-layer densities $\boldsymbol{\psi}$ and $\boldsymbol{\phi}$, follows at once. The displacement $\boldsymbol{u}(\boldsymbol{x})$ is given by:

$$\bar{u}_k(\boldsymbol{x}) = \int_{\partial\Omega} \psi_i(\boldsymbol{y}) U_i^k(\boldsymbol{x}, \boldsymbol{y}) \, \mathrm{d}S_y = u_k(\boldsymbol{x}) \tag{5.43}$$

$$\bar{u}_k(\boldsymbol{x}) = \int_{\partial\Omega} [\phi_i(\boldsymbol{y}) - \phi_i(\boldsymbol{x})] T_i^k(\boldsymbol{x}, \boldsymbol{y}) \, \mathrm{d}S_y = u_k(\boldsymbol{x}) + \phi_k(\boldsymbol{x}) \tag{5.44}$$

while the deplacement gradient is such that:

$$\frac{1}{2}(u_{k,\ell} + \bar{u}_{k,\ell})(\boldsymbol{x})$$

$$= -\psi_a(\boldsymbol{x}) B_{a\ell}^k(\boldsymbol{x}, \partial\Omega) - \int_{\partial\Omega} [\psi_a(\boldsymbol{y}) - \psi_a(\boldsymbol{x})] U_{a,\ell}^k(\boldsymbol{x}, \boldsymbol{y}) \, \mathrm{d}S_y \tag{5.45}$$

$$= -D_{\ell b}\phi_a(\boldsymbol{x}) A_{ab}^k(\boldsymbol{x}, \partial\Omega) - \int_{\partial\Omega} [D_{\ell b}\phi_a(\boldsymbol{y}) - D_{\ell b}\phi_a(\boldsymbol{x})] \Sigma_{ab}^k(\boldsymbol{x}, \boldsymbol{y}) \, \mathrm{d}S_y \tag{5.46}$$

The indirect integral representations on the boundary allow to formulate regularized indirect integral equations for the unknown density $\boldsymbol{\psi}$ or $\boldsymbol{\phi}$.

Simplifications occur when the above indirect representation formulas are considered for interior points $\boldsymbol{x}$, the regularization being no longer necessary.

Interior Dirichlet problem. The displacement is prescribed on the boundary: $\boldsymbol{u} = \boldsymbol{u}^D$ on $\partial\Omega$. As a direct consequence of eqs (5.43) and (5.44), an indirect integral equation for this boundary-value problem can be formulated from either a single-layer representation of unknown density $\boldsymbol{\psi}$:

$$u_k^D(\boldsymbol{x}) = \int_{\partial\Omega} \psi_i(\boldsymbol{y}) U_i^k(\boldsymbol{x}, \boldsymbol{y}) \, \mathrm{d}S_y \tag{5.47}$$

or a double-layer representation of unknown density $\boldsymbol{\phi}$:

$$u_k^D(\boldsymbol{x}) = \int_{\partial\Omega} [\phi_i(\boldsymbol{y}) - \phi_i(\boldsymbol{x})] T_i^k(\boldsymbol{x}, \boldsymbol{y}) \, \mathrm{d}S_y - \phi_k(\boldsymbol{x}) \tag{5.48}$$

Interior Neumann problem. The traction is prescribed on the boundary: $\boldsymbol{t} = \boldsymbol{t}^D$ on $\partial\Omega$. As a direct consequence of eqs (5.45) and (5.46), an indirect integral equation for this boundary-value problem can be formulated from either a single-layer representation of unknown density $\boldsymbol{\psi}$:

$$t_i^D(\boldsymbol{x}) = \frac{1}{2}\psi_i(\boldsymbol{x}) - \psi_a(\boldsymbol{x}) n_j(\boldsymbol{x}) A_{ij}^a(\boldsymbol{x}, \partial\Omega) - C_{ijk\ell} n_j(\boldsymbol{x}) \int_{\partial\Omega} [\psi_a(\boldsymbol{y}) - \psi_a(\boldsymbol{x})] U_{a,\ell}^k(\boldsymbol{x}, \boldsymbol{y}) \, \mathrm{d}S_y \tag{5.49}$$

or a double-layer representation of unknown density ϕ:

$$t_i^D(\boldsymbol{x}) = -C_{ijk\ell}n_j(\boldsymbol{x})D_{\ell b}\phi_a(\boldsymbol{x})A_{ab}^k(\boldsymbol{x},\partial\Omega) - C_{ijk\ell}n_j(\boldsymbol{x})\int_{\partial\Omega}[D_{\ell b}\phi_a(\boldsymbol{y}) - D_{\ell b}\phi_a(\boldsymbol{x})]\Sigma_{ab}^k(\boldsymbol{x},\boldsymbol{y})\,\mathrm{d}S_y \quad (5.50)$$

Indirect integral equations for the exterior Dirichlet and Neumann problems are established in a similar way.

Link to the classical potential theory

The regularization approach applied to the formulation of indirect integral equations provides an alternative way of finding the classical jump properties of scalar and elastic potentials and their derivatives (see Kupradze et al. [11] for a detailed treatment of elastic potentials).

5.7 COMMENTS

Definition of the integral equations

The various integral equations presented in this book are limiting forms of exterior integral representation formulas as the diameter ε of an exclusion neighbourhood $v_\varepsilon(\boldsymbol{x})$ becomes arbitrarily small. The shape of $v_\varepsilon(\boldsymbol{x})$ is arbitrary and may change as $\varepsilon \to 0$. The collocation point $\boldsymbol{x}$ is maintained on the boundary $\partial\Omega$ during the limiting process. The limiting process $\varepsilon \to 0$ can be performed by either using an indirect regularization, i.e. taking advantage of integral identities associated with simple solutions (as in this book) or following the direct approach (Guiggiani et al. [7], Guiggiani [6]).

Traction boundary integral equations can also be defined as the limiting form of integral representation formulas for the traction vector at interior points as they approach the boundary. This viewpoint is followed e.g. by Krishnasamy et al. [10] or in [1]. The resulting formulations involve Hadamard finite part (FP) integrals when no regularization is used, Cauchy principal value (CPV) integrals after partial regularization by means of an integration by parts (Sladek et Sladek [16]), or are completely regularized [10]. In some cases, the 'limit-to-the-boundary' approach is performed analytically using symbolic algebra.

The non- or partially regularized approaches lead to integral formulations that are apparently dependent on a specific, and restrictive, definition for the convergence of integrals involving non-integrable (in the usual sense) singularities. The integral equations (2.24), (4.32), (5.18), (5.33) resulting from the presentation adopted here are equivalent to the former ones but rely on the convergence of singular integrals in the usual sense only (i.e. weakly singular integrals) and are formulated in terms of integrable singularities only (this is also true for the other works on regularization quoted earlier). This is due to the fact that the limiting process $\varepsilon \to 0$ is considered for exclusion neighbourhoods v_ε of arbitrary shape whereas convergence in the CPV or FP senses relies specifically on spherical neighbourhoods v_ε. This equivalence between CPV or FP formulations and regularized (weakly singular) ones is essentially due to the fact that the various contributions of strongly singular or hypersingular terms cancel each other: for example, the sum of the free term and the CPV integral in the Somigliana identity (4.33) is known, exactly and without calculations, from the rigid-body identity (4.28). Thus, the integral formulations do not actually depend on a specific definition of the convergence of a singular integral and appear less arbitrary in this respect. The strongly singular (CPV) or hypersingular (FP) character of, respectively, the displacement and traction integral equations is apparent rather than real. From a more practical viewpoint, numerical integration is usually applicable only to convergent integrals (although some numerical quadrature formulas exist for CPV integrals over line segments), and the potentially non-integrable contributions must thus be dealt with separately, by means of an indirect regularization, a direct approach or analytic limit-to-the-boundary procedures.

Smoothness assumptions

Another important issue is that the primary unknown (potential or displacement) and the unit normal are required to possess $C_{1,\alpha}$ and $C_{0,\alpha}$ smoothness, respectively, at the collocation point $\boldsymbol{x} \in \partial\Omega$. These requirements ensure that the singular integrals in the gradient representation formula on the boundary, eq. (5.18) or (5.33), are convergent, i.e. that their regularization is effective.

In fact, exactly the same smoothness requirements appear for finite-part integrals to be convergent in non-regularized formulations. For instance, the Lyapunov–Tauber theorem in potential theory for scalar problems (Gunther [8]) or elasticity (Kupradze et al. [11]), states that the gradient of the double-layer potential:

$$\int_{\partial\Omega} \psi(\boldsymbol{y}) H(\boldsymbol{x},\boldsymbol{y})\,\mathrm{d}S_y \qquad \text{or} \qquad \int_{\partial\Omega} \psi_i(\boldsymbol{y}) T_i^k(\boldsymbol{x},\boldsymbol{y})\,\mathrm{d}S_y$$

is defined for $\boldsymbol{x} \in \partial\Omega$ only if the density ψ or $\boldsymbol{\psi}$ has $C^{1,\alpha}$ smoothness at $\boldsymbol{y} = \boldsymbol{x}$. This requirement also appears in the direct approach (see Section 5.9).

On the other hand, the original identity (5.2) or (5.19) holds true for *any* $\varepsilon > 0$ without having to make any specific smoothness assumption on the densities. This suggests that a limiting form could be established even without the $C^{1,\alpha}$ smoothness assumption on the potential or displacement. In that case, the derivation should follow different lines (for example, the $C^{1,\alpha}$ assumption is necessary to eliminate the integrals over s_ε in the limit $\varepsilon \to 0$). This smoothness issue has been the subject of some debate in the recently published literature, see Martin and Rizzo [12], Cruse and Richardson [3], Martin et al. [13].

5.8 BOUNDARY ELEMENT DISCRETIZATION

To fix our ideas, the discretization of the integral equation (5.18) is considered, assuming that $u \in C^{1,\alpha}$ and $\boldsymbol{n} \in C^{0,\alpha}$ at $\boldsymbol{y} = \boldsymbol{x}$; the latter conditions in turn imply that $q \in C^{0,\alpha}$ at $\boldsymbol{y} = \boldsymbol{x}$. It is easy to define a $C^{0,\alpha}$ interpolation of the form (3.12) for q (all usual interpolations meet this condition, provided they are conformal if $\boldsymbol{x}$ lies on an element edge). On the other hand, defining conformal $C^{1,\alpha}$ interpolations is relatively easy for plane problems (boundary elements are curvilinear segments), e.g. using Hermite shape functions or Overhauser elements [2] but much more difficult for three-dimensional situations (boundary elements are surface patches). Besides, failing to enforce the $C^{1,\alpha}$ smoothness at $\boldsymbol{x}$ of the interpolation of u may lead to inaccurate numerical evaluation (Krishnasamy et al. [10]).

When an interpolation of the usual form (3.12) is used for u, no collocation point $\boldsymbol{x}$ is thus allowed to lie on interelement edges: any usual interpolation of u is $C^{1,\alpha}$ at any point interior to an element. This restriction forces us to choose between:

- Keeping conformal interpolations , but collocating the integral equation at element interior points only. This may lead to overdetermined linear systems of equations, to be solved in the least square sense.
- Using non-conformal interpolations for u, defined using element interior nodes only (Section 4.7). One may then use the interpolation nodes for u as collocation points while retaining the $C^{1,\alpha}$ smoothness at these points. This approach leads *a priori* to square linear systems of equations.
- Using simultaneously a $C^{0,\alpha}$ interpolation for u and another $C^{0,\alpha}$ interpolation for the tangential derivatives $D_{ij}u$.

Expression of the operator D_{ij}

For a surface boundary element, the operator D_{ij} is expressed in terms of the parameter $\boldsymbol{\xi}$ on the parent element by eq. (A.17):

$$D_{ij} f \, \mathrm{d}S_y = e_{aij} \Big[f_{,1}(\boldsymbol{a}_2.\boldsymbol{e}_a) - f_{,2}(\boldsymbol{a}_1.\boldsymbol{e}_a) \Big]$$

which, when applied to the interpolation (3.12) of u, leads to:

$$J(\boldsymbol{\xi}) D_{ij} u_i(\boldsymbol{y}(\boldsymbol{\xi})) = e_{aij} \sum_{q=1}^{N_e} \Big[(\boldsymbol{a}_2.\boldsymbol{e}_a) M_{q,1}(\boldsymbol{\xi}) - (\boldsymbol{a}_1.\boldsymbol{e}_a) M_{q,2}(\boldsymbol{\xi}) \Big] u^q \qquad (5.51)$$

Numerical evaluation of integrals

The nonsingular integrations are performed using the standard techniques (Gaussian quadrature, see Section 3.4). In addition, the integral equation

(5.18) involves singular integrals of the form:

$$I = \int_E [q(\boldsymbol{y}) - q(\boldsymbol{x})] G(\boldsymbol{x}, \boldsymbol{y}) \, \mathrm{d}S_y \tag{5.52}$$

$$J = \int_E \Big[D_{ij} u(\boldsymbol{y}) - D_{ij} u(\boldsymbol{x}) \Big] H(\boldsymbol{x}, \boldsymbol{y}) \, \mathrm{d}S_y \tag{5.53}$$

with $\boldsymbol{x} \in E$. Let $\boldsymbol{\eta}$ denote the point in the parent element associated to $\boldsymbol{x}$ through the mapping of E.

The singular integral I can be evaluated by means of the procedure developed in Section 3.5, introducing a polar coordinate system on the reference element by putting $(\xi_1, \xi_2) = (\eta_1 + \rho\cos\alpha, \eta_2 + \rho\sin\alpha)$ and using the modified shape functions $\hat{M}(\rho, \alpha : \boldsymbol{\eta})$ in the interpolation of q.

The singular integral J is of a new type. If u is modelled using standard shape functions, J is defined only if $\boldsymbol{x}$ is interior to the element. In this case, the polar coordinate system on the reference element is again introduced, and one has to reformulate

$$J(\boldsymbol{\xi}) \Big[D_{ij} u_a(\boldsymbol{y}) - D_{ij} u_a(\boldsymbol{x}) \Big]$$

in terms of ρ, α and so ρ appears as a factor. In order to do so, one must recall that the Taylor formula for polynomials implies that, for any polynomial shape function $N(\boldsymbol{\xi})$, there exist functions $N^I_\alpha(\boldsymbol{\xi}; \boldsymbol{\eta})$, polynomial in $\boldsymbol{\xi}$, such that:

$$N(\boldsymbol{\xi}) - N(\boldsymbol{\eta}) = (\xi_\alpha - \eta_\alpha) N^I_\alpha(\boldsymbol{\xi}; \boldsymbol{\eta}) = \rho \hat{N}(\rho, \alpha; \boldsymbol{\eta}) \tag{5.54}$$

This fact can be applied to the following expressions, polynomial in $\boldsymbol{\xi}$:

$$\boldsymbol{a}_\alpha(\boldsymbol{\xi}) - \boldsymbol{a}_\alpha(\boldsymbol{\eta}) = (\xi_\beta - \eta_\beta)(N_{k,\alpha})^I_\beta(\boldsymbol{\xi}; \boldsymbol{\eta}) \boldsymbol{y}^k = \rho \hat{\boldsymbol{a}}_\alpha(\rho, \alpha; \boldsymbol{\eta})$$

$$M_{q,\alpha}(\boldsymbol{\xi}) - M_{q,\alpha}(\boldsymbol{\eta}) = (\xi_\beta - \eta_\beta)(M_{q,\alpha})^I_\beta(\boldsymbol{\xi}; \boldsymbol{\eta}) = \rho \hat{M}_{q,\alpha}(\rho, \alpha; \boldsymbol{\eta})$$

Similarly, the Jacobian J of the element mapping is given by:

$$g = J^2 = g_{11} g_{22} - g_{12}^2 \qquad g_{\alpha\beta} = \boldsymbol{a}_\alpha . \boldsymbol{a}_\beta$$

and one then obtains:

$$g(\boldsymbol{\xi}) - g(\boldsymbol{\eta}) = \rho \Big[g_{11}(\boldsymbol{\xi}) \hat{g}_{22} + \hat{g}_{11} g_{22}(\boldsymbol{\eta}) - \hat{g}_{12} [g_{12}(\boldsymbol{\xi}) + g_{12}(\boldsymbol{\eta})] \Big]$$

$$\hat{g}_{\beta\gamma} = \boldsymbol{a}_\beta(\boldsymbol{\xi}) \hat{\boldsymbol{a}}_\gamma + \hat{\boldsymbol{a}}_\beta \hat{\boldsymbol{a}}_\gamma(\boldsymbol{\eta})$$

Some algebraic manipulations allow us to show that:

$$M_{q,1}(\boldsymbol{\xi}) \boldsymbol{a}_2(\boldsymbol{\xi}) - \frac{J(\boldsymbol{\xi})}{J(\boldsymbol{\eta})} M_{q,1}(\boldsymbol{\eta}) \boldsymbol{a}_2(\boldsymbol{\eta})$$

$$= \rho \left\{ M_{q,1}(\boldsymbol{\xi}) \hat{\boldsymbol{a}}_2 + \hat{M}_{q,1} \boldsymbol{a}_2(\boldsymbol{\eta}) - \frac{\hat{g}}{J(\boldsymbol{\eta})[J(\boldsymbol{\eta}) + J(\boldsymbol{\xi})]} M_{q,1}(\boldsymbol{\eta}) \boldsymbol{a}_2(\boldsymbol{\eta}) \right\}$$

$$M_{q,2}(\boldsymbol{\xi}) \boldsymbol{a}_1(\boldsymbol{\xi}) - \frac{J(\boldsymbol{\xi})}{J(\boldsymbol{\eta})} M_{q,2}(\boldsymbol{\eta}) \boldsymbol{a}_1(\boldsymbol{\eta})$$

$$= \rho \left\{ M_{q,2}(\boldsymbol{\xi}) \hat{\boldsymbol{a}}_1 + \hat{M}_{q,2} \boldsymbol{a}_1(\boldsymbol{\eta}) - \frac{\hat{g}}{J(\boldsymbol{\eta})[J(\boldsymbol{\eta}) + J(\boldsymbol{\xi})]} M_{q,2}(\boldsymbol{\eta}) \boldsymbol{a}_1(\boldsymbol{\eta}) \right\}$$

Gathering the previous results, one can identify expressions for $\widehat{D_{ij}M_q}(\rho,\alpha;\boldsymbol{\eta})$ so that:

$$J[D_{ij}u(\boldsymbol{y}) - D_{ij}u(\boldsymbol{x})] = \sum_{q=1}^{N_e} \rho\widehat{D_{ij}M_q}(\rho,\alpha;\boldsymbol{\eta})u^q$$

Hence the singular integral (5.53) takes the nonsingular form:

$$J = \sum_{q=1}^{N_e} u \int_E \widehat{D_{ij}M_q}(\rho,\alpha;\boldsymbol{\eta})\hat{\Sigma}_{is}^k(\rho,\alpha;\boldsymbol{\eta})\,\mathrm{d}\rho\mathrm{d}\alpha \tag{5.55}$$

Collocation at geometrical nodes

Another technique, proposed by Cruse, Polch and Huang [15], consists in introducing a conformal $C^{0,\alpha}$ interpolation for the derivatives $D_{ij}u$ while retaining a conformal, $C^{0,\alpha}$ isoparametrical interpolation for u. Thus, two interpolations of $D_{ij}u$ are considered simultaneously: one obtained by differentiating (3.12), associated to nodal values of u and given by (5.51); the other directly associated to nodal values of $(D_{ij}u)^p$ and interpolation functions $\tilde{M}(\boldsymbol{\xi})$, i.e.:

$$D_{ij}u(\boldsymbol{y}) = \sum_{p=1}^{N_I(e)} M_p(\boldsymbol{\xi})u^p \qquad \text{and} \qquad D_{ij}u(\boldsymbol{y}) = \sum_{p=1}^{N_I(e)} \tilde{M}_p(\boldsymbol{\xi})(D_{ij}u)^p$$

The nodal values $(D_{ij}u)^p$ are then linked to the primal unknowns u^p, by minimizing the least-squares discrepancy $\mathcal{J}$ between the two interpolations:

$$\mathcal{J} = \frac{1}{2}\sum_{e=1}^{N_E}\sum_{p=1}^{N_e}\int_{E_e}\left\{u^p[D_{ij}M_p(\boldsymbol{\xi})] - (D_{ij}u_i)^p\tilde{M}_p(\boldsymbol{\xi})\right\}^2 J(\boldsymbol{\xi})\,\mathrm{d}\boldsymbol{\xi} \tag{5.56}$$

Taking the gradient of $\mathcal{J}$ with respect to the $(D_{ij}u)^p$, one is readily led to solve for $(D_{ij}u)^p$ the linear system of equations:

$$[MM]_{pk}\{(D_{ij}u)^k\} = [M\tilde{M}_{ij}]_{pk}\{u^k\} \tag{5.57}$$

with:

$$[MM]_{pk} = \int_{\Gamma_{pk}} M_p(\boldsymbol{\xi})M_k(\boldsymbol{\xi})J(\boldsymbol{\xi})\,\mathrm{d}\boldsymbol{\xi}$$
$$\left[M\tilde{M}_{ij}\right]_{pk} = \int_{\Gamma_{pk}} M_p(\boldsymbol{\xi})D_{ij}\tilde{M}_k(\boldsymbol{\xi})J(\boldsymbol{\xi})\,\mathrm{d}\boldsymbol{\xi}$$

and where Γ_{pk} denotes the intersection of the geometrical supports of the interpolation functions M_p and M_k. This procedure thus establishes a linear relationship between the nodal values $(D_{ij}u)^p$ and u^p. The matrices $[MM]$

and $[M\tilde{M}]$ have size $N_N \times N_N$. Also, $[MM]$ is symmetric and banded, thus setting up and solving (using the Choleski factorization) the linear system (5.57) requires only a moderate computational effort. This technique enables one to use the geometrical or displacement nodes (a high proportion of which are located on element edges) as collocation points, which is simpler, usually leads to square boundary element matrices, and generates less unknowns.

Elasticity

The previous considerations are valid for elasticity problems as well, the only difference lying in the vector character of the nodal unknowns. They are in particular relevant for the boundary element numerical treatment of cracked elastic solids with cracks (linear fracture mechanics), see Chapter 13.

5.9 THE DIRECT APPROACH

The direct approach for the evaluation of singular element integrals has been presented in Section 4.9 for strongly singular kernels. Its extension to hypersingular integral formulations (Guiggiani et al. [7], Guiggiani [4]) is the subject of this section. Scalar potential problems are considered for simplicity.

The limiting process leading to the statement of the gradient boundary integral equation is based on the identity (5.2), recalled for convenience:

$$0 = \int_{(\partial\Omega - e_\varepsilon)+s_\varepsilon} \{q(\boldsymbol{y})G_{,\bar{\ell}}(\boldsymbol{x},\boldsymbol{y}) - u(\boldsymbol{y})G_{,\bar{\ell}j}(\boldsymbol{x},\boldsymbol{y})n_j(\boldsymbol{y})\} \, \mathrm{d}S_y \tag{5.58}$$

Again, u and $\boldsymbol{n}$ are assumed to possess $C^{1,\alpha}$ and $C^{0,\alpha}$ regularity at $\boldsymbol{x}$, respectively (thus, q has also a $C^{0,\alpha}$ regularity at $\boldsymbol{x}$). Like in Section 4.8, the boundary element discretization is introduced into (5.58) *before* actually considering the limiting process $\varepsilon \to 0$. Denoting by E the single boundary element containing the collocation point $\boldsymbol{x}$, eq. (5.58) becomes:

$$\begin{aligned}
0 = &\int_{\partial\Omega - E} \{q(\boldsymbol{y})G_{,\bar{\ell}}(\boldsymbol{x},\boldsymbol{y}) - u(\boldsymbol{y})G_{,\bar{\ell}j}(\boldsymbol{x},\boldsymbol{y})n_j(\boldsymbol{y})\} \, \mathrm{d}S_y \\
&+ \int_{E-e_\varepsilon} q(\boldsymbol{y})G_{,\bar{\ell}}(\boldsymbol{x},\boldsymbol{y}) \, \mathrm{d}S_y - \int_{E-e_\varepsilon} [u(\boldsymbol{y}) - u(\boldsymbol{x})]G_{,\bar{\ell}j}(\boldsymbol{x},\boldsymbol{y})n_j(\boldsymbol{y}) \, \mathrm{d}S_y \\
&- b_i(\boldsymbol{x},\varepsilon)u(\boldsymbol{x}) + \kappa_{im}(\boldsymbol{x},\varepsilon)u_{,m}(\boldsymbol{x}) + O(\varepsilon)
\end{aligned} \tag{5.59}$$

where $b_i(\boldsymbol{x},\varepsilon)$ and $\kappa_{im}(\boldsymbol{x},\varepsilon)$ are defined by:

$$b_i(\boldsymbol{x},\varepsilon) = \int_{(E-e_\varepsilon)+s_\varepsilon} G_{,\bar{\ell}j}(\boldsymbol{x},\boldsymbol{y})n_j(\boldsymbol{y}) \, \mathrm{d}S_y \tag{5.60}$$

$$\kappa_{im}(\boldsymbol{x},\varepsilon) = \int_{s_\varepsilon} \left\{n_m(\boldsymbol{y})G_{,\bar{\ell}}(\boldsymbol{x},\boldsymbol{y}) - (y_m - x_m)G_{,\bar{\ell}j}(\boldsymbol{x},\boldsymbol{y})n_j(\boldsymbol{y})\right\} \mathrm{d}S_y \tag{5.61}$$

The $O(\varepsilon)$ term in eq. (5.59) gathers those integrals over s_ε which vanish in the limit $\varepsilon \to 0$ by virtue of the assumed smoothness of u and $\boldsymbol{n}$.

From the singularity properties of the derivatives of G, $\kappa_{im}(\boldsymbol{x}, \varepsilon)$ remains finite in the limit $\varepsilon \to 0$, the limiting value $\kappa_{im}(\boldsymbol{x})$ being expected to depend on the shape of s_ε. Moreover, we let $\tilde{E}$ denote a smooth arbitrary surface such that $\tilde{E} + (E - e_\varepsilon) + s_\varepsilon$ is a closed surface and the distance of $\boldsymbol{x}$ from $\tilde{E}$ is finite. Then, since $G_{,\bar{\ell}jj} = 0$, one has:

$$0 = \int_{\tilde{D}} G_{,\bar{\ell}jj}(\boldsymbol{x}, \boldsymbol{y})\, \mathrm{d}V_y = \int_{\tilde{E}} G_{,\bar{\ell}j}(\boldsymbol{x}, \boldsymbol{y}) n_j(\boldsymbol{y})\, \mathrm{d}S_y + b_i(\boldsymbol{x}, \varepsilon)$$

(where $\tilde{D}$ is the domain enclosed by $\tilde{E} + (E - e_\varepsilon) + s_\varepsilon$). Since the integral over $\tilde{E}$ is finite and does not depend on ε, nor on the shape of s_ε, $b_i(\boldsymbol{x}, \varepsilon)$ has a finite value $b_i(\boldsymbol{x})$ which is independent of s_ε.

The limiting process $\varepsilon \to 0$ for the integrals over $(E - e_\varepsilon)$ in eq. (5.59) is now performed by introducing polar coordinates (ρ, α) in the parent element Δ_e, in the same fashion (and using the same notations) as in Section 4.8. At this point, recognizing (as has also been done using indirect regularization and in Section 4.8 for the direct approach to strongly singular integrals) that the limiting process $\varepsilon \to 0$ is *globally* independent of the shape of v_ε, a spherical shape is used for convenience.

The singular character of $G_{,\bar{\ell}}(\boldsymbol{x}, \boldsymbol{y})$ and $G_{,\bar{\ell}j}(\boldsymbol{x}, \boldsymbol{y})$ implies that:

$$G_{,\bar{\ell}}(\boldsymbol{x}, \boldsymbol{y}) = \frac{1}{\hat{r}^2}\hat{\mathcal{G}}_i(\rho, \alpha; \boldsymbol{\eta}) \qquad G_{,\bar{\ell}j}(\boldsymbol{x}, \boldsymbol{y}) = \frac{1}{\hat{r}^3}\hat{\mathcal{G}}_{ij}(\rho, \alpha; \boldsymbol{\eta})$$

Introduce (for simplicity) the same interpolation functions M_p for u and q. First, following the same line of reasoning as in Section 4.8, the integral involving the strongly singular kernel $G_{,\bar{\ell}}(\boldsymbol{x}, \boldsymbol{y})$ becomes:

$$\begin{aligned}
&\int_{E-e_\varepsilon} M_p(\boldsymbol{y}) G_{,\bar{\ell}}(\boldsymbol{x}, \boldsymbol{y})\, \mathrm{d}S_y \\
&= \int_0^{2\pi} \int_{A^{-1}(\alpha)\varepsilon(1+O(\varepsilon))}^{\bar{\rho}(\alpha)} M_p(\boldsymbol{\xi}) G_{,\bar{\ell}}(\boldsymbol{x}, \boldsymbol{y}(\boldsymbol{\xi})) J(\boldsymbol{\xi}) \rho\, \mathrm{d}\rho \mathrm{d}\alpha \\
&= \int_0^{2\pi} \int_{A^{-1}(\alpha)\varepsilon(1+O(\varepsilon))}^{\bar{\rho}(\alpha)} \frac{1}{\rho}\Big\{ M_p(\boldsymbol{\xi}) \hat{\mathcal{G}}_i(\rho, \alpha; \boldsymbol{\eta}) J(\boldsymbol{\xi}) \\
&\qquad - M_p(\boldsymbol{\eta}) \hat{\mathcal{G}}_i(0, \alpha; \boldsymbol{\eta}) J(\boldsymbol{\eta}) \Big\}\, \mathrm{d}\rho \mathrm{d}\alpha \\
&\quad + M_p(\boldsymbol{\eta}) J(\boldsymbol{\eta}) \int_0^{2\pi} \ln \bar{\rho}(\alpha) \hat{\mathcal{G}}_i(0, \alpha; \boldsymbol{\eta})\, \mathrm{d}\alpha
\end{aligned} \tag{5.62}$$

Next, noting that

$$M_p(\boldsymbol{\xi}) - M_p(\boldsymbol{\eta}) = \rho \hat{M}_p(\rho, \alpha; \boldsymbol{\eta}) = \rho[\hat{M}_p(0, \alpha; \boldsymbol{\eta}) + O(\rho)]$$

(where the $\hat{M}_p(\rho, \alpha; \boldsymbol{\eta})$ are the 'modified shape functions' introduced in Section 3.5), the integral involving the hypersingular kernel $G_{,\bar{\ell}j}(\boldsymbol{x}, \boldsymbol{y})$ becomes:

$$\begin{aligned}
&\int_{E-e_\varepsilon} [M_p(\boldsymbol{y}) - M_p(\boldsymbol{x})] G_{,\bar{\ell}j}(\boldsymbol{x}, \boldsymbol{y}) n_j(\boldsymbol{y})\, \mathrm{d}S_y \\
&= \int_0^{2\pi} \int_{A^{-1}(\alpha)\varepsilon(1+O(\varepsilon))}^{\bar{\rho}(\alpha)} \frac{1}{\rho^2} \Big\{ [M_p(\boldsymbol{\xi}) - M_p(\boldsymbol{\eta})] \hat{\mathcal{G}}_{ij}(\rho, \alpha; \boldsymbol{\eta}) n_j(\boldsymbol{y}(\boldsymbol{\xi})) J(\boldsymbol{\xi}) \\
&\qquad - \hat{M}_p(0, \alpha; \boldsymbol{\eta}) \hat{\mathcal{G}}_{ij}(0, \alpha; \boldsymbol{\eta}) n_j(\boldsymbol{y}(\boldsymbol{\eta})) J(\boldsymbol{\eta}) \Big\}\, \mathrm{d}\rho \mathrm{d}\alpha \\
&\quad + \hat{M}_p(0, \alpha; \boldsymbol{\eta}) n_j(\boldsymbol{y}(\boldsymbol{\eta})) J(\boldsymbol{\eta}) \int_0^{2\pi} \ln \bar{\rho}(\alpha) \hat{\mathcal{G}}_{ij}(0, \alpha; \boldsymbol{\eta})\, \mathrm{d}\alpha
\end{aligned} \tag{5.63}$$

Equations (5.62) and (5.63) are the limiting values when $\varepsilon \to 0$ of the respective integrals over $(E - e_\varepsilon)$ in eq. (5.59). To establish the limiting form of eq. (5.59), one only has to evaluate $b_i(\boldsymbol{x})$ and $\kappa_{im}(\boldsymbol{x})$. For general fundamental solutions, this can be done numerically. When eqs (5.60) and (5.61) are defined in terms of the infinite-space fundamental solution, simple expressions for $b_i(\boldsymbol{x})$ and $\kappa_{im}(\boldsymbol{x})$ are established as follows.

This presentation, while following the same general ideas, slightly departs from that of [7, 4]. In [7, 4], the term $[M_p(\boldsymbol{\xi}) - M_p(\boldsymbol{\eta})]$ is broken and the expansion with respect to powers of ρ in the hypersingular integral is carried out one step further.

Infinite-space fundamental solution

In the most common situations when the infinite-space fundamental solution $G = 1/4\pi r$ is used in the integral identity (5.11), one has

$$G_{,\bar{\ell}j}(\boldsymbol{x}, \boldsymbol{y}) = -G_{,\ell j}(\boldsymbol{x}, \boldsymbol{y})$$

and the expressions of $b_i(\boldsymbol{x})$ and $\kappa_{im}(\boldsymbol{x}, \varepsilon)$ can be obtained using analytical manipulations. First, a now familiar integration by parts pattern yields:

$$\int_{E-e_\varepsilon+s_\varepsilon} G_{,\bar{\ell}j}(\boldsymbol{x}, \boldsymbol{y}) n_j(\boldsymbol{y})\, \mathrm{d}S_y = e_{\ell jm} \int_{\partial E} \tau_m(\boldsymbol{y}) G_{,j}(\boldsymbol{y}, \boldsymbol{x})\, \mathrm{d}s_y = b_i(\boldsymbol{x}) \tag{5.64}$$

Then, one has:

$$\begin{aligned}
&n_m(\boldsymbol{y}) G_{,\bar{\ell}}(\boldsymbol{x}, \boldsymbol{y}) - (y_m - x_m) G_{,\bar{\ell}j}(\boldsymbol{x}, \boldsymbol{y}) n_j(\boldsymbol{y}) \\
&\quad = n_m(\boldsymbol{y}) G_{,\bar{\ell}}(\boldsymbol{x}, \boldsymbol{y}) + D_{j\ell}\Big((y_m - x_m) G_{,j}(\boldsymbol{x}, \boldsymbol{y}) \Big) - G_{,j}(\boldsymbol{x}, \boldsymbol{y}) D_{j\ell}(y_m - x_m) \\
&\quad = n_m(\boldsymbol{y}) G_{,\bar{\ell}}(\boldsymbol{x}, \boldsymbol{y}) + D_{j\ell}\Big((y_m - x_m) G_{,j}(\boldsymbol{x}, \boldsymbol{y}) \Big) \\
&\qquad\qquad + G_{,j}(\boldsymbol{x}, \boldsymbol{y}) (\delta_{jm} n_\ell(\boldsymbol{y}) - \delta_{\ell m} n_j(\boldsymbol{y})) \\
&\quad = D_{\ell m} G(\boldsymbol{y}, \boldsymbol{x}) + D_{j\ell}\Big((y_m - x_m) G_{,j}(\boldsymbol{x}, \boldsymbol{y}) \Big) - H(\boldsymbol{y}, \boldsymbol{x}) \delta_{\ell m}
\end{aligned}$$

so that, upon integration over s_ε and using the integration by parts identity (A.8), one finds:

$$\kappa_{im}(\boldsymbol{x},\varepsilon) = \int_{e_\varepsilon} \{e_{\ell mk}G(\boldsymbol{y},\boldsymbol{x}) + e_{j\ell k}(y_m - x_m)G_{,j}(\boldsymbol{x},\boldsymbol{y})\}\,\tau_k(\boldsymbol{y})\,\mathrm{d}s_y - \frac{1}{2}\delta_{im} + O(\varepsilon) \quad (5.65)$$

using eq. (2.26) together with the fact that the surface $\partial\Omega$ is assumed smooth at $\boldsymbol{x}$. Also, due to the assumed spherical shape of s_ε, a direct calculation shows that the contour integrals over e_ε vanish in the limit $\varepsilon \to 0$, so that:

$$\lim_{\varepsilon\to 0} \kappa_{im}(\boldsymbol{x},\varepsilon) = \kappa_{im}(\boldsymbol{x}) = -\frac{1}{2}\delta_{im} \quad (5.66)$$

Elastostatics

In that case, the direct approach proceeds from identity (5.19), and the limiting process $\varepsilon \to 0$ is performed in:

$$\begin{aligned} 0 = &\int_{\partial\Omega - E} \left\{t_i(\boldsymbol{y})U^k_{i,\bar{\ell}}(\boldsymbol{x},\boldsymbol{y}) - u_i(\boldsymbol{y})\Sigma^k_{ij,\bar{\ell}}(\boldsymbol{x},\boldsymbol{y})n_j(\boldsymbol{y})\right\}\,\mathrm{d}S_y \\ &+ \int_{E-e_\varepsilon} t_i(\boldsymbol{y})U^k_{i,\bar{\ell}}(\boldsymbol{x},\boldsymbol{y})\,\mathrm{d}S_y - \int_{E-e_\varepsilon} [u_i(\boldsymbol{y}) - u_i(\boldsymbol{x})]\Sigma^k_{ij,\bar{\ell}}(\boldsymbol{x},\boldsymbol{y})n_j(\boldsymbol{y})\,\mathrm{d}S_y \\ &- b_{ik\ell}(\boldsymbol{x},\varepsilon)u_i(\boldsymbol{x}) + \kappa_{ikm\ell}(\boldsymbol{x},\varepsilon)u_{i,m}(\boldsymbol{x}) + O(\varepsilon) \end{aligned} \quad (5.67)$$

where $b_{ik\ell}(\boldsymbol{x},\varepsilon)$ and $\kappa_{ikm\ell}(\boldsymbol{x},\varepsilon)$ are defined by:

$$b_{ik\ell}(\boldsymbol{x},\varepsilon) = \int_{(E-e_\varepsilon)+s_\varepsilon} \Sigma^k_{ij,\bar{\ell}}(\boldsymbol{x},\boldsymbol{y})n_j(\boldsymbol{y})\,\mathrm{d}S_y \quad (5.68)$$

$$\kappa_{ikm\ell}(\boldsymbol{x},\varepsilon) = \int_{s_\varepsilon} \left\{C_{abim}n_b(\boldsymbol{y})U^k_{a,\bar{\ell}}(\boldsymbol{x},\boldsymbol{y}) - (y_m - x_m)\Sigma^k_{ij,\bar{\ell}}(\boldsymbol{x},\boldsymbol{y})n_j(\boldsymbol{y})\right\}\,\mathrm{d}S_y \quad (5.69)$$

When the Kelvin elastostatic fundamental solution is used, one finds:

$$b_{ik\ell}(\boldsymbol{x}) = e_{\ell jm}\int_{\partial E} \Sigma^k_{ij}(\boldsymbol{x},\boldsymbol{y})\tau_m\,\mathrm{d}s_y \quad (5.70)$$

$$\kappa_{ikm\ell}(\boldsymbol{x},\varepsilon) = -\frac{1}{2}\delta ik\delta m\ell \quad (5.71)$$

The limiting process $\varepsilon \to 0$ for the integrals over $(E - e_\varepsilon)$ in eq. (5.67) follows similar lines and yields formulas having the same structure as eqs (5.62) and (5.63); this is left to the reader.

Appendix 5.A INTEGRAL IDENTITIES FOR THE KELVIN SOLUTION

The auxiliary integrals involving the Kelvin fundamental solution:

$$A^k_{ij}(\boldsymbol{x}, S) = \int_S \Sigma^k_{ij}(\boldsymbol{x}, \boldsymbol{y}) \, \mathrm{d}S_y$$
$$B^k_{i\ell}(\boldsymbol{x}, S) = \int_S U^k_{i,\ell}(\boldsymbol{x}, \boldsymbol{y}) \, \mathrm{d}S_y$$

have been introduced in Section 5.5. From eqs (4.8) and (4.9), they have the form:

$$A^k_{ij}(\boldsymbol{x}, S) = -\frac{1}{8\pi(1-\nu)} \Big[(1-2\nu)(\delta_{ik} I_j + \delta_{jk} I_i - \delta_{ij} I_k) + 3J_{ijk} \Big] \tag{5.A.1}$$
$$B^k_{i\ell}(\boldsymbol{x}, S) = -\frac{1}{16\pi\mu(1-\nu)} \Big[(3-4\nu)\delta_{ik} I_\ell + 3J_{ik\ell} - \delta_{\ell k} I_i - \delta_{i\ell} I_k \Big] \tag{5.A.2}$$

in terms of auxiliary integrals I_a, J_{abc}:

$$I_a(\boldsymbol{x}, S) = \int_S r_{,a} \frac{\mathrm{d}S_y}{r^2} \tag{5.A.3}$$
$$J_{abc}(\boldsymbol{x}, S) = \int_S r_{,a} r_{,b} r_{,c} \frac{\mathrm{d}S_y}{r^2} \tag{5.A.4}$$

Similarly, for the fundamental solution $G = 1/(4\pi r)$ of the Laplace equation:

$$\int_S G_{,a}(\boldsymbol{x}, \boldsymbol{y}) \, \mathrm{d}S_y = -\frac{1}{4\pi} I_a(\boldsymbol{x}, S) \tag{5.A.5}$$

The integrals I_a, J_{abc} exhibit a singularity which is not integrable on a surface, and thus are not *a priori* defined for $\boldsymbol{x} \in S$. On the other hand, assuming $\boldsymbol{x} \notin S$, they can be reformulated by means of an integration by parts process into contour integrals on ∂S plus *convergent* surface integrals.

Reformulation of integral I_a

From the definition of the partial tangential derivative D_a, one has:

$$\frac{r_{,a}}{r^2} = -(\frac{1}{r})_{,a} = -D_a \frac{1}{r} + n_a \frac{1}{r^2} r \tag{5.A.6}$$

Identity (A.10) is then applied to $f = 1/r$, resulting in:

$$I_a(\boldsymbol{x}, S) = \int_S \left(\frac{1}{r} r_{,n} - K \right) n_a \frac{\mathrm{d}S_y}{r} - \int_{\partial S} \nu_a \frac{\mathrm{d}s_y}{r} \tag{5.A.7}$$

with $r_{,n} = n_p r_{,p}$, $K = D_r n_r$.

Reformulation of integral J_{abc}

From eq. (2.A.2), one has:

$$3J_{abc}(\boldsymbol{x}, S) = \delta_{ab}I_c(\boldsymbol{x}, S) + \delta_{ac}I_b(\boldsymbol{x}, S) + \delta_{bc}I_a(\boldsymbol{x}, S) + \int_S r_{,abc}\,\mathrm{d}S_y \quad (5.A.8)$$

Next, upon noticing that $r_{,abc} = D_c r_{,ab} + n_c n_p r_{,abp}$, identity (A.10) with $f = r_{,ab}$ gives:

$$\begin{aligned}
\int_S r_{,abc}\frac{\mathrm{d}S_y}{r^2} &= \int_S K n_c r_{,ab}\,\mathrm{d}S_y + \int_{\partial S} r_{,ab}\nu_c\,\mathrm{d}s_y + \int_S n_c n_p r_{,abp}\,\mathrm{d}S_y \\
&= \delta_{ab}\int_S r_{,n}\frac{\mathrm{d}S_y}{r^2} - \delta_{ab}I_c(\boldsymbol{x}, S) \\
&\quad - \int_S K n_c r_{,a} r_{,b}\frac{\mathrm{d}S_y}{r} - \int_{\partial S} r_{,a} r_{,b}\nu_c\frac{\mathrm{d}s_y}{r} \\
&\quad + \int_S n_c\left(D_{pa} r_{,bp} + n_a r_{,bpp}\right)\mathrm{d}S_y + \int_{\partial S} r_{,ab}\nu_c\,\mathrm{d}s_y \qquad (5.A.9)
\end{aligned}$$

Taking into account that:

$$r_{,ba} = \frac{\delta_{ba} - r_{,b}r_{,a}}{r} \qquad\qquad r_{,bpp} = (\frac{2}{r})_{,b}$$

one then obtains:

$$\begin{aligned}
&\int_S n_c\Big[D_{pa} r_{,bp} + n_a r_{,bpp}\Big]\,\mathrm{d}S_y \\
&= \int_S n_c\Big[D_{ba}\frac{1}{r} + n_a(\frac{2}{r})_{,b} - D_{pa}(r_{,b}r_{,p})\Big]\,\mathrm{d}S_y \\
&= \int_S n_c\Big[n_a(\frac{1}{r})_{,b} + n_b(\frac{1}{r})_{,a} - D_{pa}(\frac{r_{,b}r_{,p}}{r})\Big]\,\mathrm{d}S_y
\end{aligned}$$

or equivalently, using formula (A.10) with $f = n_a n_c(1/r)$ and $f = n_b n_c(1/r)$ and formula (A.8) for the integrand $D_{pa}(r_{,b}r_{,p})$:

$$\begin{aligned}
&\int_S n_c\left(D_{pa} r_{,bp} + n_a r_{,bpp}\right)\mathrm{d}S_y \\
&= \int_S 2n_c n_a n_b(K - \frac{1}{r}r_{,n})\frac{\mathrm{d}S_y}{r} - \int_S [D_b(n_c n_a) + D_a(n_c n_b)]\frac{\mathrm{d}s_y}{r^2} \\
&\quad + \int_S r_{,b}r_{,p}D_{pa}n_c\frac{\mathrm{d}S_y}{r} + \int_{\partial S}[\nu_b n_c n_a + \nu_a n_c n_b]\frac{\mathrm{d}s_y}{r} \\
&\quad - \int_{\partial S} n_c r_{,b}r_{,p}[n_p\nu_a - n_a\nu_p]\frac{\mathrm{d}s_y}{r}
\end{aligned}$$

where use has been made of:

$$e_{paq}\tau_q = n_p\nu_a - n_a\nu_p$$

Finally, substituting this relation into eq. (5.A.9) and the resulting equality into eq. (5.A.8) produces after some manipulation the result:

$$\begin{aligned}3J_{abc}(\boldsymbol{x},S) &= \delta_{ac}I_b(\boldsymbol{x},S)+\delta_{bc}I_a(\boldsymbol{x},S)\\&+\int_S\Big[r_{,b}r_{,p}D_{pa}n_c-D_a(n_bn_c)-D_b(n_an_c)\Big]\frac{\mathrm{d}S_y}{r}\\&+\delta_{ab}\int_S n_cr_{,n}\frac{\mathrm{d}S_y}{r^2}+\int_S(2n_an_b-r_{,a}r_{,b})n_{,c}K\frac{\mathrm{d}S_y}{r}\\&-2\int_S n_an_cn_br_{,n}\frac{\mathrm{d}S_y}{r^2}+\int_{\partial S}(\nu_an_cn_b+\nu_bn_cn_a-r_{,a}r_{,b}\nu_c)\frac{\mathrm{d}s_y}{r}\\&+\int_{\partial S}(\nu_pn_a-\nu_an_p)r_{,p}r_{,b}n_c\frac{\mathrm{d}s_y}{r}\end{aligned}\tag{5.A.10}$$

Comments

The surface integrals in reformulations (5.A.7) and (5.A.10) are convergent for $\boldsymbol{x}\in S, \boldsymbol{x}\notin\partial S$ (with integrands behaving like $O(r^{-1})$ and $O(r^{-2}r_{,n})$).

If S is closed, the contour integrals in eqs (5.A.7), (5.A.10) vanish. If S is piecewise smooth, eqs (5.A.7), (5.A.10) should be applied separately on each smooth component of S and the resulting expressions added.

The symmetry in a,b,c of $J_{abc}(\boldsymbol{x},S)$ defined by (5.A.4) implies that the value of J_{abc} given by eq. (5.A.10) is invariant under permutations of a,b,c, although this is not apparent. On the other hand, this symmetry is not expected for the surface or contour integrals considered separately.

Particular case of a planar surface

If S is planar (lying e.g. in the (y_1,y_2)-plane), one has for any $\boldsymbol{y}\in S$:

$$K=0 \qquad n_a=\delta_{a3} \qquad D_an_b=0 \qquad \nu_3=0 \qquad D_{3\beta}f=f_{,\beta} \qquad D_{\alpha\beta}f=0$$

and equations (5.A.7), (5.A.10) reduce to:

$$I_a(\boldsymbol{x},S)=\delta_{a3}\int_S r_{,3}\frac{\mathrm{d}S_y}{r^2}+(\delta_{a3}-1)\int_{\partial S}\nu_a\frac{\mathrm{d}s_y}{r}\tag{5.A.11}$$

$$3J_{\alpha\beta\gamma}(\boldsymbol{x},S)=\delta_{\alpha\beta}I_\gamma(\boldsymbol{x},S)+\delta_{\beta\gamma}I_\alpha(\boldsymbol{x},S)-\int_{\partial S}r_{,\alpha}r_{,\beta}\nu_{,\gamma}\frac{\mathrm{d}s_y}{r}\tag{5.A.12}$$

$$3J_{\alpha\beta3}(\boldsymbol{x},S)=\delta_{\alpha\beta}I_\gamma(\boldsymbol{x},S)-\int_{\partial S}\nu_{,\alpha}r_{,\beta}r_{,3}\frac{\mathrm{d}s_y}{r}\tag{5.A.13}$$

$$J_{\alpha33}(\boldsymbol{x},S)=-\int_{\partial S}\nu_{,\alpha}r_{,3}^2\frac{\mathrm{d}s_y}{r}\tag{5.A.14}$$

$$J_{333}(\boldsymbol{x},S)=I_3(\boldsymbol{x},S)\tag{5.A.15}$$

If in addition $\boldsymbol{x} \in S$, one has $r_{,3} = 0$ as well, and the only nonzero integrals in the above expressions are $I_\alpha(\boldsymbol{x}, S)$ and $J_{\alpha\beta\gamma}(\boldsymbol{x}, S$.

Appendix 5.B PROOF OF EXPRESSION (5.16) OF THE FREE TERM

The objective is to reformulate the limit when $\varepsilon \to 0$ of the free term:

$$\begin{aligned}\tilde{A}_{ia}(\boldsymbol{x}, \varepsilon) = &\int_{\partial\Omega - e_\varepsilon} [n_i(\boldsymbol{y}) - n_i(\boldsymbol{x})] G_{,a}(\boldsymbol{x}, \boldsymbol{y}) \, \mathrm{d}S_y \\ &- \delta_{ia} \int_{\partial\Omega - e_\varepsilon} [n_j(\boldsymbol{y}) - n_j(\boldsymbol{x})] G_{,j}(\boldsymbol{x}, \boldsymbol{y}) \, \mathrm{d}S_y \\ &- \int_{\partial\Omega - e_\varepsilon} [n_a(\boldsymbol{y}) - n_a(\boldsymbol{x})] G_{,i}(\boldsymbol{x}, \boldsymbol{y}) \, \mathrm{d}S_y\end{aligned}$$

introduced in Section 5.4. First, the various terms are rearranged in order to obtain:

$$\begin{aligned}\tilde{A}_{ia}(\boldsymbol{x}, \varepsilon) = &\int_{\partial\Omega - e_\varepsilon} \Big\{ D_{ia} G(\boldsymbol{x}, \boldsymbol{y}) - \delta_{ia} H(\boldsymbol{x}, \boldsymbol{y}) \Big\} \, \mathrm{d}S_y \\ &+ \frac{1}{4\pi} \int_{\partial\Omega - e_\varepsilon} \Big\{ n_i(\boldsymbol{x}) r_{,a} - \delta_{ia} n_j(\boldsymbol{x}) r_{,j} - n_a(\boldsymbol{x}) r_{,i} \Big\} \frac{\mathrm{d}S_y}{r^2}\end{aligned}$$

Integrating by parts using eq. (A.8) and using formula (5.A.7), one shows from the above equality that $\tilde{A}_{ia}(\boldsymbol{x}, \varepsilon)$ is expressible as a sum of surface integrals which are convergent as $\varepsilon \to 0$ and contour integrals over the edge e_ε of $(\partial\Omega - e_\varepsilon)$; the limiting process $\varepsilon \to 0$ in the contour integrals remains to be investigated. Taking eq. (2.A.12) into account, one obtains:

$$\begin{aligned}\tilde{A}_{ia}(\boldsymbol{x}, \varepsilon) = &\frac{1}{2} \delta_{ia} \\ &+ \frac{1}{4\pi} \Big[n_i(\boldsymbol{x}) I_a(\partial\Omega, \boldsymbol{x}) - \delta_{ia} n_j(\boldsymbol{x}) I_j(\boldsymbol{x}, \partial\Omega) - n_a(\boldsymbol{x}) I_i(\boldsymbol{x}, \partial\Omega) \Big] \\ &+ \frac{1}{4\pi} \int_{e_\varepsilon} \Big\{ e_{iac} \tau_c - n_i(\boldsymbol{x}) \nu_a + \delta_{ia} n_j(\boldsymbol{x}) \nu_j + n_a(\boldsymbol{x}) \nu_i \Big\} \frac{\mathrm{d}s_y}{r}\end{aligned} \tag{5.A.16}$$

Next, since $\boldsymbol{\tau} = \boldsymbol{n} \wedge \boldsymbol{\nu}$, one has:

$$e_{iac} \tau_c = e_{iac} e_{bcd} n_d \nu_b = n_i \nu_a - n_a \nu_i$$

so that the contour integral over e_ε in eq. (5.A.16) becomes:

$$\frac{1}{4\pi} \int_{s_\varepsilon} \Big\{ \delta_{ia} n_j(\boldsymbol{x}) \nu_j + [n_a(\boldsymbol{x}) - n_a \nu_i] - [n_i(\boldsymbol{x}) - n_i] \nu_a \Big\} \frac{\mathrm{d}s_y}{r}$$

Moreover, assuming that $\partial\Omega$ is smooth at $\boldsymbol{x}$, e_ε becomes in the limit $\varepsilon \to 0$ a planar curve lying in the tangent plane to $\partial\Omega$ at $\boldsymbol{x}$, and one has:

$$\boldsymbol{n}(\boldsymbol{y}) - \boldsymbol{n}(\boldsymbol{x}) = O(\varepsilon) \qquad \boldsymbol{n}(\boldsymbol{x}).\boldsymbol{\nu}(\boldsymbol{y}) = O(\varepsilon) \qquad \frac{\mathrm{d}s_y}{r} = O(1)\,\mathrm{d}\theta \tag{5.A.17}$$

where θ denotes the polar angle in the tangent plane to $\partial\Omega$ and with origin $\boldsymbol{x}$. The contour integral thus vanishes in the limit $\varepsilon \to 0$.

Appendix 5.C PROOF OF EXPRESSION (5.28) FOR THE ELASTIC FREE TERM

The limiting value when $\varepsilon \to 0$ of the free term $\tilde{A}_{abk\ell}(\boldsymbol{x}, \varepsilon)$ introduced in equation (5.26) is sought:

$$\begin{aligned}\tilde{A}_{abk\ell}(\boldsymbol{x}, \varepsilon) = & -\delta_{b\ell} \int_{\partial\Omega - e_\varepsilon} [n_j(\boldsymbol{y}) - n_j(\boldsymbol{x})] \Sigma^k_{aj}(\boldsymbol{x}, \boldsymbol{y})\,\mathrm{d}S_y \\ & + \int_{\partial\Omega - e_\varepsilon} [n_\ell(\boldsymbol{y}) - n_\ell(\boldsymbol{x})] \Sigma^k_{ab}(\boldsymbol{x}, \boldsymbol{y})\,\mathrm{d}S_y \\ & - C_{ijab} \int_{\partial\Omega - e_\varepsilon} [n_j(\boldsymbol{y}) - n_j(\boldsymbol{x})] U^k_{i,\ell}(\boldsymbol{x}, \boldsymbol{y})\,\mathrm{d}S_y \end{aligned} \tag{5.A.18}$$

Upon term rearrangement and using the symmetry $C_{ijab} = C_{abij}$ of the Hooke tensor, the above formula is recast into:

$$\begin{aligned}\tilde{A}_{abk\ell}(\boldsymbol{x}, \varepsilon) = & -\delta_{b\ell} \int_{\partial\Omega - e_\varepsilon} T^k_a(\boldsymbol{x}, \boldsymbol{y})\,\mathrm{d}S_y \\ & - n_\ell(\boldsymbol{x}) C_{ijab} \int_{\partial\Omega - e_\varepsilon} U^k_{i,j}(\boldsymbol{x}, \boldsymbol{y})\,\mathrm{d}S_y \\ & + n_j(\boldsymbol{x}) \delta_{b\ell} \int_{\partial\Omega - e_\varepsilon} \Sigma^k_{aj}(\boldsymbol{x}, \boldsymbol{y})\,\mathrm{d}S_y \\ & + n_j(\boldsymbol{x}) C_{ijab} \int_{\partial\Omega - e_\varepsilon} U^k_{i,\ell}(\boldsymbol{x}, \boldsymbol{y})\,\mathrm{d}S_y \\ & + C_{ijab} \int_{\partial\Omega - e_\varepsilon} \left[n_\ell(\boldsymbol{y}) U^k_{i,j}(\boldsymbol{x}, \boldsymbol{y}) - n_j(\boldsymbol{y}) U^k_{i,\ell}(\boldsymbol{x}, \boldsymbol{y}) \right] \mathrm{d}S_y \end{aligned} \tag{5.A.19}$$

Upon integration by parts by means of identity (A.8), the last integral in the above equality becomes a contour integral:

$$\int_{\partial\Omega - e_\varepsilon} \left[n_\ell(\boldsymbol{y}) U^k_{i,j}(\boldsymbol{x}, \boldsymbol{y}) - n_j(\boldsymbol{y}) U^k_{i,\ell}(\boldsymbol{x}, \boldsymbol{y}) \right] \mathrm{d}S_y = \int_{e_\varepsilon} U^k_i(\boldsymbol{x}, \boldsymbol{y}) e_{\ell jm} \tau_m(\boldsymbol{y})\,\mathrm{d}s_y \tag{5.A.20}$$

This result, together with equations (5.A.1–5.A.2) and (5.A.7–5.A.10), shows that $\tilde{A}_{abk\ell}(\boldsymbol{x},\varepsilon)$ is expressible as a sum of surface integrals which are convergent in the limit $\varepsilon \to 0$ and of contour integrals over ∂S and over e_ε; the limiting process $\varepsilon \to 0$ in the latter contour integrals remains to be investigated.

The contour integrals in the limit $\varepsilon \to 0$

Introducing the symbolic notation $\partial\tilde{A}_{abk\ell}(\boldsymbol{x},\varepsilon)$ for the sum of all contour integrals over e_ε in eq. (5.A.19), that equation becomes, taking (5.A.20), (5.A.7) and (5.A.10) into account:

$$\begin{aligned}\partial\tilde{A}_{abk\ell}(\boldsymbol{x},\varepsilon) = &-\delta_{b\ell}\Big[\partial K_i^k(\boldsymbol{x},\varepsilon) - n_j(\boldsymbol{x})\partial A_{aj}^k(\boldsymbol{x},\varepsilon)\Big] \\ &- C_{ijab}\Big[n_\ell(\boldsymbol{x})\partial B_{ij}^k(\boldsymbol{x},\varepsilon) - n_j(\boldsymbol{x})\partial B_{ir}^k(\boldsymbol{x},\varepsilon)\Big] \\ &+ C_{ijab}\int_{e_\varepsilon} U_{i,\ell}^k(\boldsymbol{x},\boldsymbol{y})e_{\ell jm}\tau_m(\boldsymbol{y})\,\mathrm{d}s_y \qquad (5.A.21)\end{aligned}$$

where $\partial A_{ij}^k(\boldsymbol{x},\varepsilon)$, $\partial B_{ij}^k(\boldsymbol{x},\varepsilon)$ and $\partial K_i^k(\boldsymbol{x},\varepsilon)$ symbolically denote the sum of all contour integrals over e_ε in the expressions of $A_{ij}^k(\boldsymbol{x},\partial\Omega - e_\varepsilon)$, $B_{ij}^k(\boldsymbol{x},\partial\Omega - e_\varepsilon)$ and $K_i^k(\boldsymbol{x},\partial\Omega - e_\varepsilon)$ respectively. From eqs (5.A.2), (5.A.7) and (5.A.10), one obtains:

$$\begin{aligned}&n_\ell(\boldsymbol{x})\partial B_{ij}^k(\boldsymbol{x},\varepsilon) - n_j(\boldsymbol{x})\partial B_{i\ell}^k(\boldsymbol{x},\varepsilon) \\ &= D\int_{e_\varepsilon}[n_j(\boldsymbol{x})n_\ell - n_\ell(\boldsymbol{x})n_j][\nu_i n_k + \nu_k n_i + (\nu_p n_i - \nu_i n_p)r_{,p}r_{,k}]\frac{\mathrm{d}s_y}{r} \\ &\quad + D\int_{e_\varepsilon}[(3-4\nu)\delta_{ik} + r_{,i}r_{,k}][n_\ell(\boldsymbol{x})\nu_j - n_j(\boldsymbol{x})\nu_\ell]\frac{\mathrm{d}s_y}{r} \qquad (5.A.22)\end{aligned}$$

where $D = 1/(16\pi\mu(1-\nu))$. Using relations (5.A.17) and $n_\ell(\boldsymbol{x})n_j(\boldsymbol{y}) - n_j(\boldsymbol{x})n_\ell(\boldsymbol{y}) = O(\varepsilon)$, one shows that:

$$\int_{e_\varepsilon}[n_\ell(\boldsymbol{x})n_j - n_j(\boldsymbol{x})n_\ell][\nu_i n_k + \nu_k n_i + (\nu_p n_i - \nu_i n_p)r_{,p}r_{,k}]\frac{\mathrm{d}s_y}{r} = O(\varepsilon)$$

Besides, the Kelvin fundamental displacement, eq. (4.8), appears in the right-hand side of eq. (5.A.22), so that:

$$\begin{aligned}&n_\ell(\boldsymbol{x})\partial B_{ij}^k(\boldsymbol{x},\varepsilon) - n_j(\boldsymbol{x})\partial B_{i\ell}^k(\boldsymbol{x},\varepsilon) \\ &= \int_{e_\varepsilon} U_i^k(\boldsymbol{x},\boldsymbol{y})[n_\ell(\boldsymbol{x})\nu_j - n_j(\boldsymbol{x})\nu_\ell]\frac{\mathrm{d}s_y}{r} + O(\varepsilon) \\ &= e_{\ell jm}\int_{e_\varepsilon}\tau_m U_i^k(\boldsymbol{x},\boldsymbol{y})\frac{\mathrm{d}s_y}{r} + O(\varepsilon) \qquad (5.A.23)\end{aligned}$$

having noticed that:

$$n_\ell(\boldsymbol{x})\nu_j - n_j(\boldsymbol{x})\nu_\ell = n_\ell\nu_j - n_j\nu_\ell + O(\varepsilon) = e_{\ell jm}\tau_m + O(\varepsilon)$$

Besides, from eqs (5.A.1), (5.A.7) and (5.A.10), one also has:

$$\begin{aligned} n_j(\boldsymbol{x})\partial A^k_{aj}(\boldsymbol{x},\varepsilon) &\\ = -\frac{1-2\nu}{8\pi(1-\nu)}\Bigg\{ &\int_{e_\varepsilon} [n_a(\boldsymbol{x})\nu_k - n_k(\boldsymbol{x})\nu_a - \delta_{ak}n_j(\boldsymbol{x})\nu_j]\frac{\mathrm{d}s_y}{r} \\ &+ \int_{e_\varepsilon} [n_k\nu_a(n_j(\boldsymbol{x})n_j - 1) + (n_a n_k - \delta_{ak})n_j(\boldsymbol{x})\nu_j]\frac{\mathrm{d}s_y}{r} \\ &+ \int_{e_\varepsilon} n_j(\boldsymbol{x})r_{,j}[(n_a\nu_p - n_p\nu_a)n_k r_{,p} - r_{,a}\nu_k]\frac{\mathrm{d}s_y}{r}\Bigg\} \end{aligned}$$

Upon noting that $\boldsymbol{n}(\boldsymbol{x}).\boldsymbol{\nu}(\boldsymbol{y}) = O(\varepsilon)$, $\boldsymbol{n}(\boldsymbol{x}).\boldsymbol{n}(\boldsymbol{y}) - 1 = O(\varepsilon)$ and $\boldsymbol{n}(\boldsymbol{x})\boldsymbol{\nabla}r = O(\varepsilon)$, one then obtains:

$$\begin{aligned} n_j(\boldsymbol{x})\partial A^k_{aj}(\boldsymbol{x},\varepsilon) + O(\varepsilon) &= -\frac{1-2\nu}{8\pi(1-\nu)}e_{akm}\int_{e_\varepsilon}\tau_m\frac{\mathrm{d}s_y}{r} \\ &= \partial K^k_i(\boldsymbol{x},\varepsilon) + O(\varepsilon) \end{aligned} \tag{5.A.24}$$

where ${}^k_i(\boldsymbol{x},\varepsilon)$ is given by eq. (4.A.11) and having used the fact that e_ε becomes a planar curve in the limit $\varepsilon \to 0$. Finally, substitution of the partial results (5.A.23) and (5.A.24) into eq. (5.A.21) yields:

$$\partial\tilde{A}_{abk\ell}(\boldsymbol{x},\varepsilon) = O(\varepsilon) \qquad \text{i.e.} \quad \lim_{\varepsilon\to 0}\partial\tilde{A}_{abk\ell}(\boldsymbol{x},\varepsilon) = 0 \tag{5.A.25}$$

In other words, the overall contribution of the contour integrals over e_ε arising in eq. (5.A.19) vanishes in the limit $\varepsilon \to 0$. This fact allows us to state that:

$$\begin{aligned} \tilde{A}_{abk\ell}(\boldsymbol{x}) = \frac{1}{2}\delta_{ak}\delta_{b\ell} - n_\ell(\boldsymbol{x})A^k_{ab}(\boldsymbol{x},\partial\Omega) + n_j(\boldsymbol{x})\delta_{b\ell}A^k_{aj}(\boldsymbol{x},\partial\Omega) \\ + C_{ijab}n_j(\boldsymbol{x})B^k_{i\ell}(\boldsymbol{x},\partial\Omega) \end{aligned}$$

and the inner product with $u_{a,b}(\boldsymbol{x})$ yields expression (5.28):

$$\tilde{A}_{abk\ell}(\boldsymbol{x})u_{a,b}(\boldsymbol{x}) = \frac{1}{2}u_{k,\ell}(\boldsymbol{x}) - D_{\ell j}u_i(\boldsymbol{x})A^k_{ij}(\boldsymbol{x},\partial\Omega) + t_i(\boldsymbol{x})B^k_{i\ell}(\boldsymbol{x},\partial\Omega) \tag{5.A.26}$$

REFERENCES

1. BONNET, M. Regularized boundary integral equations for three-dimensional bounded or unbounded elastic bodies containing curved cracks of arbitrary

shape under dynamic loading. In N. G. Zamani, C. A. Brebbia (eds), *Boundary element techniques: applications in engineering*, pp. 171–189. Computational Mechanics Publications, Southampton (1989).

2. Camp, C. V., Gipson, G. S. Overhauser elements in boundary element analysis. *Mathl. Comput. Modelling*, **15**, 59–69 (1991).

3. Cruse, T. A., Richardson, J. D. Nonsingular Somigliana stress identities in elasticity. *Int. J. Num. Meth. in Eng.*, **39**, 3273–3304 (1996).

4. Guiggiani, M. Hypersingular formulation for boundary stress evaluation. *Engng. Anal. with Bound. Elem.*, **14**, 169–179 (1994).

5. Guiggiani, M. Sensitivity analysis for boundary element error estimation and mesh refinement. *Int. J. Num. Meth. in Eng.*, **39**, 2907–2920 (1996).

6. Guiggiani, M. Formulation and numerical treatment of boundary integral equations with hypersingular kernels. In V. Sladek (ed.), *Singular integrals in boundary element methods*. Computational Mechanics Publications, Southampton (1998).

7. Guiggiani, M., Krishnasamy, G., Rudolphi, T. J., Rizzo, F. J. A general algorithm for the numerical solution of hypersingular boundary integral equations. *ASME J. Appl. Mech.*, **59**, 604–614 (1992).

8. Gunther, N. M. *La théorie du potentiel et ses applications aux problèmes fondamentaux de la physique mathématique*. Gauthier-Villars, Paris (1934).

9. Ingber, M. S., Mitra, A. K. Grid redistribution based on measurable error indicators for the direct boundary element method. *Engng. Anal. with Bound. Elem.*, **9**, 13–19 (1992).

10. Krishnasamy, G., Rizzo, F. J., Rudolphi, T. J. Hypersingular boundary integral equations: their occurrence, interpretation, regularization and computation. In P. K. Banerjee, S. Kobayashi (eds), *Developments in boundary element methods*, vol. 7: Advanced Dynamic Analysis, chapter 7. Elsevier (1992).

11. Kupradze, V. D. (ed.). *Three-dimensional problems of the mathematical theory of elasticity and thermoelasticity*. North-Holland (1979).

12. Martin, P. A., Rizzo, F. J. Hypersingular integrals: how smooth must the density be? *Int. J. Num. Meth. in Eng.*, **39**, 687–704 (1996).

13. Martin, P. A., Rizzo, F. J., Cruse, T. A. Smoothness relaxation strategies for singular and hypersingular integral equations. *Int. J. Num. Meth. in Eng.*, **42**, 885–906 (1998).

14. Paulino, G. H., Gray, L. J., Zarikian, V. Hypersingular residuals – a new approach for error estimation in the boundary element method. *Int. J. Num. Meth. in Eng.*, **39**, 2005–2029 (1996).

15. Polch, E. Z., Cruse, T. A., Huang, C. J. Traction BIE solutions for flat cracks. *Comp. Mech.*, **2**, 253–267 (1987).

16. Sladek, J., Sladek, V. Three-dimensional curved crack in an elastic body. *Int. J. Solids Struct.*, **19**, 425–436 (1983).

6

Some classical general results

6.1 FUNDAMENTAL PROBLEMS OF POTENTIAL THEORY

Indirect boundary integral equations associated with the Dirichlet and Neumann problems for the Laplace equation are obtained from double- and single-layer potential representations of the unknown, respectively (Sections 2.4 and 5.6); they read:

$$-\frac{1}{2}\psi(\boldsymbol{x}) + \int_{\partial\Omega} \psi(\boldsymbol{y}) n_j(\boldsymbol{y}) G_{,j}(\boldsymbol{x}, \boldsymbol{y}) \, \mathrm{d}S_y = u^D(\boldsymbol{x})$$
$$\frac{1}{2}\phi(\boldsymbol{x}) - n_j(\boldsymbol{x}) \int_{\partial\Omega} \phi(\boldsymbol{y}) G_{,j}(\boldsymbol{x}, \boldsymbol{y}) \, \mathrm{d}S_y = q^D(\boldsymbol{x})$$

where the normal derivatives $n_j(\boldsymbol{y})G_{,j}(\boldsymbol{x}, \boldsymbol{y})$ and $n_j(\boldsymbol{x})G_{,j}(\boldsymbol{x}, \boldsymbol{y})$ are integrable provided the boundary $\partial\Omega$ is smooth at $\boldsymbol{x}$. Both are weakly singular Fredholm integral equations of the second kind (Tricomi [4]), and as such verify the *Fredholm alternative.* According to it, for an integral equation of the form:

$$\phi(\boldsymbol{x}) - \int_{\partial\Omega} \phi(\boldsymbol{y}) K(\boldsymbol{x}, \boldsymbol{y}) \, \mathrm{d}S_y = f(\boldsymbol{x}) \tag{6.1}$$

with ϕ unknown and f given, one of the following clauses is true:

1. eq. (6.1) has a unique solution ϕ for any f, in which case the *adjoint equation*:

$$\phi(\boldsymbol{x}) - \int_{\partial\Omega} \phi(\boldsymbol{y}) K(\boldsymbol{y}, \boldsymbol{x}) \, \mathrm{d}S_y = f(\boldsymbol{x}) \tag{6.2}$$

also has a unique solution ϕ for any f;

2. The homogeneous (i.e. $f = 0$) integral equation corresponding to (6.1) has at least one non-trivial solution $\phi \neq 0$. A necessary and sufficient condition for the existence of a solution of equation (6.1) for a given f is that:

$$\int_{\partial\Omega} f(\boldsymbol{y})\bar{u}(\boldsymbol{y})\,\mathrm{d}S_y = 0$$

where $\bar{u}$ is any solution of the adjoint equation (6.2).

This theorem allows us to study the existence and uniqueness of solution for the Dirichlet and Neumann problems.

In contrast, using single- and double-layer potentials for the Dirichlet and Neumann problems, respectively, led to Fredholm integral equations of the first kind, i.e of the form:

$$\int_{\partial\Omega} \phi(\boldsymbol{y})K(\boldsymbol{x},\boldsymbol{y})\,\mathrm{d}S_y = f(\boldsymbol{x})$$

for which no general theorem similar to the Fredholm alternative is known. However, even those first-kind integral equations that are uniquely solvable are known to be ill-conditioned, and hence should preferably be avoided in applications.

6.2 SINGULAR BOUNDARY INTEGRAL EQUATIONS OF ELASTICITY THEORY

Indirect integral equation formulations have been used to prove existence of solution for the Dirichlet and Neumann elastic boundary-value problems. Kupradze ([1], [2]) formulated strongly singular indirect integral equations obtained from double-layer (for the Dirichlet problem) or single-layer (for the Neumann problem) representations for $\boldsymbol{u}$, as follows:

$$u_k^D(\boldsymbol{x}) = -\frac{1}{2}\psi_k(\boldsymbol{x}) + \fint_{\partial\Omega} \psi_i(\boldsymbol{y})n_j(\boldsymbol{y})\Sigma_{ij}^k(\boldsymbol{x},\boldsymbol{y})\,\mathrm{d}S_y$$

$$t_l^D(\boldsymbol{x}) = -\frac{1}{2}\phi_k(\boldsymbol{x}) + n_p(\boldsymbol{x})\fint_{\partial\Omega} \phi_i(\boldsymbol{y})C_{lpkr}U_{i,r}^k(\boldsymbol{x},\boldsymbol{y})\,\mathrm{d}S_y$$

where the notation $\fint_{\partial\Omega}$ stands for the Cauchy principal value of a singular integral (see Section 4.4). Both integral equations are singular integral equations of the second kind, i.e. of the form:

$$(\frac{1}{2}\boldsymbol{I} + \boldsymbol{K})\varphi = \boldsymbol{f} \tag{6.3}$$

(φ: unknown, $\boldsymbol{f}$: data, $\boldsymbol{K}$: integral operator). To prove the existence of a solution for these integral equations, one possible approach is to show that the Fredholm alternative still holds true when strongly singular integral equations are considered. To this end, Kupradze [1],[2] constructs a certain singular integral operator, of kernel $\boldsymbol{H}(\boldsymbol{x},\boldsymbol{y})$, such that in the integral equation:

$$(\frac{1}{2}\boldsymbol{I}-\boldsymbol{H})(\frac{1}{2}\boldsymbol{I}+\boldsymbol{K})\varphi=(\frac{1}{2}\boldsymbol{I}-\boldsymbol{K})\boldsymbol{f} \tag{6.4}$$

the integral operator $\frac{1}{2}\boldsymbol{K}-\frac{1}{2}\boldsymbol{H}-\boldsymbol{H}\boldsymbol{K}$ is nonsingular; $\boldsymbol{H}$ is thus called the *regularizing integral operator* for the singular integral equation (6.3). Following this approach, any given singular equation (6.3) can be transformed into an equivalent but nonsingular second-kind integral equation for which the Fredholm alternative applies. Existence of solution for the Dirichlet and Neumann problems of elasticity theory is thereby established.

No account of the detailed construction of $\boldsymbol{H}$ is given here, because this complex operation, although very interesting from a fundamental viewpoint, is of dubious practical value: the appearance of *iterated* integral kernels $\boldsymbol{H}.\boldsymbol{K}$:

$$((\boldsymbol{H}.\boldsymbol{K}).\varphi)(\boldsymbol{x})\equiv\int_{\partial\Omega}\int_{\partial\Omega}\boldsymbol{H}(\boldsymbol{x},\boldsymbol{z}).\boldsymbol{K}(\boldsymbol{z},\boldsymbol{y}).\varphi(\boldsymbol{y})\,\mathrm{d}S_y\,\mathrm{d}S_z$$

in the integral equation (6.4) indicates that its numerical implementation is expected to entail high numerical integration costs.

6.3 GREEN TENSORS FOR BOUNDED DOMAINS

Pham [3] studied the fundamental solution $\boldsymbol{G}(\boldsymbol{x},\boldsymbol{y})$ for an arbitrary elastic domain E such that

$$\boldsymbol{G}(\boldsymbol{x},\boldsymbol{y})=\boldsymbol{0}\quad(\forall\boldsymbol{y}\in\partial\Omega)$$

which is called the *Green's tensor* for E. Introducing the integral operator $\mathcal{K}$:

$$\varphi\rightarrow\mathcal{K}\varphi=\int_{\partial E}B_i^k(\boldsymbol{x},\boldsymbol{y},\boldsymbol{n}(\boldsymbol{y}))\phi_i\,\mathrm{d}S_y$$

where $B_i^k(\boldsymbol{x},\boldsymbol{y},\boldsymbol{n}(\boldsymbol{y}))$ is the *Kupradze–Bashelishvili* [1] tensor:

$$B_i^k(\boldsymbol{x},\boldsymbol{y},\boldsymbol{n}(\boldsymbol{y}))=-\frac{1}{2\pi r^2}\left[\frac{2-4\nu}{3-4\nu}\delta_{ik}+\frac{3}{3-4\nu}r_{,i}r_{,k}\right]r_{,j}n_j(\boldsymbol{y}) \tag{6.5}$$

Pham obtains the Green's tensor as an infinite series of iterated applications of the operator $\mathcal{K}$:

$$\boldsymbol{G}^k(\boldsymbol{x},\boldsymbol{y})=\boldsymbol{U}_k(\boldsymbol{x},\boldsymbol{y})+\sum_{n=0}^{+\infty}\frac{1}{2^{n+1}}\mathcal{K}(\mathcal{I}+\mathcal{K})^n\boldsymbol{U}^k(\boldsymbol{x},\boldsymbol{y})$$

and establishes the absolutely convergent character of this series.

6.4 FUNDAMENTAL PROBLEMS OF ELASTICITY THEORY

Pham [3] established the exact solutions for the fundamental boundary-value problems of linear elasticity, for smooth domains Ω and sufficiently regular boundary data: traction $\boldsymbol{t}^D$ (Neumann problem) or displacement $\boldsymbol{u}^D$ (Dirichlet problem).

Neumann problem

Its solution is sought as a single-layer potential:

$$u_k(\boldsymbol{x}) = \int_{\partial\Omega} \psi_i(\boldsymbol{y}) U_i^k(\boldsymbol{x},\boldsymbol{y})\,\mathrm{d}S_y$$

which leads to the following indirect singular integral equation:

$$\frac{1}{2}\psi_i(\boldsymbol{x}) - n_j(\boldsymbol{x}) \fint_{\partial\Omega} \psi_a(\boldsymbol{y}) \Sigma_{ij}^a(\boldsymbol{x},\boldsymbol{y})\,\mathrm{d}S_y = t_i^D(\boldsymbol{x}) \tag{6.6}$$

It is shown in [3] that the solution to (6.6) is given by the following infinite series:

$$\psi_k(\boldsymbol{y}) = \sum_{n=0}^{+\infty} g_k^n(\boldsymbol{y}) \quad \text{with} \quad \begin{cases} g_k^0(\boldsymbol{y}) = \dfrac{1}{2} t_k^D(\boldsymbol{y}) \\ g_k^{n+1}(\boldsymbol{y}) = -n_j(\boldsymbol{y}) \fint_{\partial\Omega} g_a^n(\boldsymbol{z}) \Sigma_{ij}^a(\boldsymbol{y},\boldsymbol{z})\,\mathrm{d}S_z \end{cases}$$

which is uniformly and absolutely convergent in the space $C(\partial\Omega)$ of vector functions continuous over $\partial\Omega$, provided that the data $\boldsymbol{t}^D$ is self-equilibrated:

$$\int_{\partial\Omega} \boldsymbol{t}^D\,\mathrm{d}S = \boldsymbol{0}$$

Dirichlet problem

Its solution is sought as a potential of a particular type:

$$u_k(\boldsymbol{x}) = \int_{\partial\Omega} \phi_i(\boldsymbol{y}) B_i^k(\boldsymbol{x},\boldsymbol{y},\boldsymbol{n}(\boldsymbol{y}))\,\mathrm{d}S_y$$

where $B_i^k(\boldsymbol{x},\boldsymbol{y},\boldsymbol{n}(\boldsymbol{y}))$ is the Kupradze–Bashelishvili tensor, eq. (6.5). The Dirichlet problem thus admits the following indirect integral formulation:

$$-\phi_k(\boldsymbol{x}) + \fint_{\partial\Omega} \phi_i(\boldsymbol{y}) B_i^k(\boldsymbol{x},\boldsymbol{y},\boldsymbol{n}(\boldsymbol{y}))\,\mathrm{d}S_y = u_k^D(\boldsymbol{x}) \tag{6.7}$$

It is shown in [3] that the solution to (6.7) is given by the following infinite series:

$$\phi_k(\boldsymbol{y}) = \sum_{n=0}^{+\infty} h_k^n(\boldsymbol{y})$$

with

$$\begin{cases} h_k^0(\boldsymbol{y}) = -\dfrac{1}{2}u_k^D(\boldsymbol{y}) \\ h_k^{n+1}(\boldsymbol{y}) = \dfrac{1}{2}\left\{h_k^n(\boldsymbol{y}) + \displaystyle\fint_{\partial\Omega} h_i^n(\boldsymbol{z})B_i^k(\boldsymbol{y},\boldsymbol{z},\boldsymbol{n}(\boldsymbol{z}))\,\mathrm{d}S_z\right\} \end{cases}$$

which is also convergent in the space $C(\partial\Omega)$.

REFERENCES

1. KUPRADZE, V. D. *Dynamical problems in elasticity*, vol. 3 of *Progress in solids mechanics*. North-Holland (1963).
2. KUPRADZE, V. D. (ed.). *Three-dimensional problems of the mathematical theory of elasticity and thermoelasticity*. North-Holland (1979).
3. PHAM, T. L. *Tenseurs de Green et de Neumann*. Ph.D. thesis, (1966).
4. TRICOMI, F. G. *Integral equations*. Interscience Publishers, New York (1957).

Part II

Wave and evolution problems

7

Scalar and elastic wave problems in the time domain

This chapter deals with integral equation formulations for transient scalar or elastic wave problems formulated in the time domain. The boundary integral formulations established in Chapters 2 and 4 for equilibrium problems play a key role in obtaining regularized boundary integral equations for dynamical problems.

7.1 WAVE EQUATION AND FUNDAMENTAL SOLUTIONS

In wave propagation problems, the unknown field variable $u = u(\boldsymbol{y}, t)$ (e.g. acoustic pressure) solves the scalar wave equation:

$$\Delta u - \frac{1}{c^2}\ddot{u} + F = 0 \qquad \forall (\boldsymbol{y}, t) \in \Omega \times [0, T] \tag{7.1}$$

where c denotes the wave velocity in the (linear, isotropic, homogeneous) physical medium under consideration and F is a known body source (any such pair u, F is termed a *dynamical state* in the sequel). The statement of a well-posed wave problem must include, in addition to the local equation (7.1), initial conditions (conventionally taken at $t = 0$):

$$u(\boldsymbol{y}, 0) = u_0(\boldsymbol{y}) \qquad \dot{u}(\boldsymbol{y}, t) = v_0(\boldsymbol{y})$$

as well as boundary conditions (either u or $q = \boldsymbol{\nabla} u.\boldsymbol{n}$, or possibly some relation between u and q, must be prescribed at any given point of the boundary $\partial\Omega$); in addition, conditions at infinity must be prescribed when dealing with an unbounded domain Ω. The frequent special case $u_0(\boldsymbol{y}) = v_0(\boldsymbol{y}) = 0$ is termed *initial rest* or *quiescent past*.

Reciprocity identity

The Green's formula (2.4) applied to a pair of distinct dynamical states (u^1, F^1), (u^2, F^2) yields:

$$\int_{\partial\Omega} (u^2 q^1 - u^1 q^2)\,\mathrm{d}S = \int_{\Omega} \left\{ u^2 \left(\frac{1}{c^2}\ddot{u}^1 - F^1 \right) - u^1 \left(\frac{1}{c^2}\ddot{u}^2 - F^2 \right) \right\} \mathrm{d}V$$

It holds even if the dynamical states 1 and 2 are considered at different time instants $t_1 = \tau$ and $t_2 = t - \tau$, with $t \geq \tau \geq 0$. The resulting equation, upon integration over $0 \leq \tau \leq t$, leads to the following reciprocity identity, valid for any pair of dynamical states and any $t \geq 0$:

$$\begin{aligned}
&\int_{\partial\Omega} \left[(q^1 \star u^2)(\boldsymbol{y}, t) - (q^2 \star u^1)(\boldsymbol{y}, t) \right] \mathrm{d}S_y \\
&= \int_{\Omega} \left[(F^2 \star u^1)(\boldsymbol{y}, t) - (F^1 \star u^2)(\boldsymbol{y}, t) \right] \mathrm{d}V_y \\
&\quad + \int_{\Omega} \frac{1}{c^2} \left[v_0^2(\boldsymbol{y}) u^1(\boldsymbol{y}, t) - v_0^1(\boldsymbol{y}) u^2(\boldsymbol{y}, t) \right] \mathrm{d}V_y \\
&\quad + \int_{\Omega} \frac{1}{c^2} \left[u_0^2(\boldsymbol{y}) \dot{u}^1(\boldsymbol{y}, t) - u_0^1(\boldsymbol{y}) \dot{u}^2(\boldsymbol{y}, t) \right] \mathrm{d}V_y
\end{aligned} \tag{7.2}$$

where the symbol $\star$ denotes the *time convolution product* of two functions of time, defined as:

$$(a \star b)(t) = \int_0^t a(t - \tau) b(\tau)\,\mathrm{d}\tau = \int_0^t a(\tau) b(t - \tau)\,\mathrm{d}\tau \qquad (t \geq 0) \tag{7.3}$$

The initial conditions appear in eq. (7.2) as a consequence of an integration by parts over the time variable:

$$\begin{aligned}
(u^2 \star \ddot{u}^1)(\cdot, t) &= \left[u^2(\cdot, \tau) \dot{u}^1(\cdot, t - \tau) \right]_0^t - (\dot{u}^2 \star \dot{u}^1)(\cdot, t) \\
&= u^2(\cdot, t) v_0^1(\cdot) - u^1(\cdot, t) v_0^2(\cdot) - (\dot{u}^2 \star \dot{u}^1)(\cdot, t)
\end{aligned}$$

and of the equality $\dot{u}^2 \star \dot{u}^1 = \dot{u}^1 \star \dot{u}^2$.

Fundamental solutions

Consider a domain $E \subseteq \mathbb{R}^3$ and an arbitrary, fixed, point $\boldsymbol{x} \in E$. A fundamental solution, denoted by $G[\boldsymbol{x}, t, \boldsymbol{y}|f]$, is any dynamical state in E corresponding to a point source applied at $\boldsymbol{x}$ with a time-modulated intensity $f(t)$ and quiescent past. Any such $G[\boldsymbol{x}, t, \boldsymbol{y}|f]$ thus solves:

$$\Delta G - \frac{1}{c^2}\ddot{G} + \delta(\boldsymbol{y} - \boldsymbol{x}) f(t) = 0 \tag{7.4}$$

The source function $f(t)$ is assumed to be twice continuously differentiable and identically vanishing for $t < 0$ (which implies $f(0) = \dot{f}(0) = 0$).

The fundamental solution for the infinite space ($E = \mathbb{R}^3$) is given by:

$$G[\boldsymbol{x}, t, \boldsymbol{y}|f] = \frac{f(t - r/c)}{4\pi r} \tag{7.5}$$

$$G_{,j}[\boldsymbol{x}, t, \boldsymbol{y}|f] = \frac{-r_{,j}}{4\pi r^2}\left[f(t - r/c) + \frac{r}{c}\dot{f}(t - r/c)\right] \tag{7.6}$$

It depends on t through the *retarded time* $t - r/c$: this is due to causality, i.e. to the fact that information travels at finite speed c through the medium. Also, the following symmetry property holds:

$$G[\boldsymbol{x}, t, \boldsymbol{y}|f] = G[\boldsymbol{y}, t, \boldsymbol{x}|f] \tag{7.7}$$

The r^{-1} singular behaviour of G is apparent. With the static source $f(t) = 1$, $G[\boldsymbol{x}, t, \boldsymbol{y}|f]$ is equal for any time instant $t \geq r/c$ to the fundamental solution $G(\boldsymbol{x}, \boldsymbol{y}) = 1/(4\pi r)$ of the Laplace equation. To put this in another way, if $f(t) = 1 + \eta(t)$ with $\eta(t) \to 0$ when $t \to +\infty$, then the dynamic fundamental solution is the sum of the static fundamental solution and a decaying transient term:

$$G[\boldsymbol{x}, t, \boldsymbol{y}|f] = G(\boldsymbol{x}, \boldsymbol{y}) + G[\boldsymbol{x}, t, \boldsymbol{y}|\eta] \qquad \lim_{t \to +\infty} |G[\boldsymbol{x}, t, \boldsymbol{y}|\eta]| = 0$$

A second-order Taylor expansion in r about $r = 0$ of $f(t - r/c)$ leads to:

$$G[\boldsymbol{x}, t, \boldsymbol{y}|f] = f(t)G(\boldsymbol{x}, \boldsymbol{y}) - \frac{\dot{f}(t)}{4\pi c} + \frac{\ddot{f}(t)}{8\pi c^2}r + o(r) \tag{7.8}$$

$$G_{,j}[\boldsymbol{x}, t, \boldsymbol{y}|f] = f(t)G_{,j}(\boldsymbol{x}, \boldsymbol{y}) + \frac{r_{,j}}{8\pi c^2}\ddot{f}(t) + o(1) \tag{7.9}$$

In other words, the dynamic and static fundamental solutions have the same singular behaviour.

The notation $G[\cdot|f]$ (Eringen and Suhubi [2]) describes G as an *operator* over the input source f, which will prove convenient.

Other fundamental solutions

Like in equilibrium problems, an infinite number of fundamental solutions can be defined, each one being characterized by the specification of E, f and boundary conditions over ∂E. For example, let E denote a half-space bounded by the plane ∂E. Using again the method of images (see Section 2.1), let $\bar{\boldsymbol{x}}$ be symmetrical to $\boldsymbol{x}$ with respect to ∂E. The fundamental solutions G^D and G^N associated respectively with homogeneous Dirichlet and Neumann boundary conditions on ∂E are given by:

$$G^D[\boldsymbol{x}, t, \boldsymbol{y}|f] = G[\boldsymbol{x}, t, \boldsymbol{y}|f] - G[\bar{\boldsymbol{x}}, t, \boldsymbol{y}|f] \qquad (G^D\ |_{\partial E} = 0)$$

$$G^N[\boldsymbol{x}, t, \boldsymbol{y}|f] = G[\boldsymbol{x}, t, \boldsymbol{y}|f] + G[\bar{\boldsymbol{x}}, t, \boldsymbol{y}|f] \qquad (H^N\ |_{\partial E} = 0)$$

Impulsive fundamental solutions. Consider the special case of an *impulsive* source applied at point $\boldsymbol{y} = \boldsymbol{x}$ and time instant $t = 0$:

$$f(t) = \delta(t)$$

The impulsive fundamental solution $G(\boldsymbol{x}, t, \boldsymbol{y})$, defined by

$$G(\boldsymbol{x}, t, \boldsymbol{y}) = G[\boldsymbol{x}, t, \boldsymbol{y}|\delta]$$

is at initial rest in $E \setminus \{\boldsymbol{x}\}$: $G(\boldsymbol{x}, 0, \boldsymbol{y}) = \dot{G}(\boldsymbol{x}, 0, \boldsymbol{y}) = 0$ for $\boldsymbol{y} \neq \boldsymbol{x}$. Noting that

$$f(t) = \int_0^t f(t-\tau)\delta(\tau)\,\mathrm{d}\tau = (f \star \delta)(t)$$

where $\star$ is the convolution (7.3), the following relationship holds:

$$G[\boldsymbol{x}, t, \boldsymbol{y}|f] = G(\boldsymbol{x}, t, \boldsymbol{y}) \star f \tag{7.10}$$

7.2 INTEGRAL REPRESENTATION FORMULAS

Consider a regular bounded domain $\Omega \subset \mathbb{R}^3$ and let $\boldsymbol{x}$ denote an arbitrarily chosen fixed point *not located on* $\partial\Omega$. In the reciprocity identity (7.2), choose as the two distinct states the unknown and fundamental dynamical states, respectively. Recalling that the fundamental solution is at initial rest, this leads to

$$\begin{aligned} u(\boldsymbol{x}, t) &+ \int_{\partial\Omega} \{H(\boldsymbol{x}, t, \boldsymbol{y}) \star u(\boldsymbol{y}, t) - G(\boldsymbol{x}, t, \boldsymbol{y}) \star q(\boldsymbol{y}, t)\}\,\mathrm{d}S_y \\ &= \int_\Omega G(\boldsymbol{x}, t, \boldsymbol{y}) \star F(\boldsymbol{y}, t)\,\mathrm{d}V_y \\ &\quad + \frac{1}{c^2}\int_\Omega \left\{v_0(\boldsymbol{y})G(\boldsymbol{x}, t, \boldsymbol{y}) + u_0(\boldsymbol{y})\dot{G}(\boldsymbol{x}, t, \boldsymbol{y})\right\}\mathrm{d}V_y \end{aligned}$$

where $H = G_{,j}n_j$ denotes the normal derivative of the fundamental solution. The above equation, upon rearrangement according to eq. (7.10), leads to the following integral representation formula for u:

$$\begin{aligned} \kappa u(\boldsymbol{x}, t) = &\int_{\partial\Omega} \{G[\boldsymbol{x}, t, \boldsymbol{y}|q(\boldsymbol{y}, t)] - H[\boldsymbol{x}, t, \boldsymbol{y}|u(\boldsymbol{y}, t)]\}\,\mathrm{d}S_y \\ &+ \int_\Omega G[\boldsymbol{x}, t, \boldsymbol{y}|F(\boldsymbol{y}, t)]\,\mathrm{d}V_y \\ &+ \frac{1}{c^2}\int_\Omega \left\{v_0(\boldsymbol{y})G(\boldsymbol{x}, t, \boldsymbol{y}) + u_0(\boldsymbol{y})\dot{G}(\boldsymbol{x}, t, \boldsymbol{y})\right\}\mathrm{d}V_y \end{aligned} \tag{7.11}$$

where $\kappa = 1$ ($\boldsymbol{x} \in \Omega$), $\kappa = 0$ ($\boldsymbol{x} \notin \Omega$). The wave function $u(\boldsymbol{x}, t)$ is thus expressed in terms of the variables on the boundary, the body source distribution and the initial data. Since the integrals in this formula are indefinitely differentiable with respect to the observation point coordinates as long as $\boldsymbol{x}$ does not lie on $\partial\Omega$, the integral representation formula for the gradient of u follows at once:

$$\begin{aligned}\kappa u_{,i}(\boldsymbol{x},t) = &\int_{\partial\Omega} \{G_{,\bar{i}}[\boldsymbol{x},t,\boldsymbol{y}|q(\boldsymbol{y},t)] - G_{,\bar{i}j}[\boldsymbol{x},t,\boldsymbol{y}|u(\boldsymbol{y},t)]n_j(\boldsymbol{y})\}\,\mathrm{d}S_y \\ &+ \int_{\Omega} G_{,\bar{i}}[\boldsymbol{x},t,\boldsymbol{y}|F(\boldsymbol{y},t)]\,\mathrm{d}V_y \\ &+ \frac{1}{c^2}\int_{\Omega}\left\{v_0(\boldsymbol{y})G_{,\bar{i}}(\boldsymbol{x},t,\boldsymbol{y}) + u_0(\boldsymbol{y})\dot{G}_{,\bar{i}}(\boldsymbol{x},t,\boldsymbol{y})\right\}\mathrm{d}V_y \qquad (7.12)\end{aligned}$$

where $\kappa = 1$ ($\boldsymbol{x} \in \Omega$), $\kappa = 0$ ($\boldsymbol{x} \notin \Omega$). Formulas (7.11), (7.12) are applicable at any time instant $t \geq 0$ and observation point $\boldsymbol{x} \notin \partial\Omega$. Any fundamental solution compatible with the domain Ω may be used. In practice, one most often resorts to the free-space fundamental solution (7.5). The latter depends on the observation and field points only through $\boldsymbol{y} - \boldsymbol{x}$ (invariance under translations), so that:

$$G_{,\bar{i}}[\boldsymbol{x},t,\boldsymbol{y}|f] = -G_{,i}[\boldsymbol{x},t,\boldsymbol{y}|f]$$

The integral representation of the wave function gradient thus takes the form:

$$\begin{aligned}\kappa u_{,i}(\boldsymbol{x},t) = &\int_{\partial\Omega} \{G_{,ij}[\boldsymbol{x},t,\boldsymbol{y}|u(\boldsymbol{y},t)]n_j(\boldsymbol{y}) - G_{,i}[\boldsymbol{x},t,\boldsymbol{y}|q(\boldsymbol{y},t)]\}\,\mathrm{d}S_y \\ &- \int_{\Omega} G_{,i}[\boldsymbol{x},t,\boldsymbol{y}|F(\boldsymbol{y},t)]\,\mathrm{d}V_y \\ &- \frac{1}{c^2}\int_{\Omega}\left\{v_0(\boldsymbol{y})G_{,i}(\boldsymbol{x},t,\boldsymbol{y}) - u_0(\boldsymbol{y})\dot{G}_{,i}(\boldsymbol{x},t,\boldsymbol{y})\right\}\mathrm{d}V_y\end{aligned}$$

Upon integration by parts using identity (A.8), one then obtains:

$$\begin{aligned}&\int_{\partial\Omega} G_{,ij}[\boldsymbol{x},t,\boldsymbol{y}|u(\boldsymbol{y},t)]n_j(\boldsymbol{y})\,\mathrm{d}S_y \\ &= \int_{\partial\Omega} D_{ji}G_{,j}[\boldsymbol{x},t,\boldsymbol{y}|u(\boldsymbol{y},t)]\,\mathrm{d}S_y + \int_{\partial\Omega} n_i(\boldsymbol{y})G_{,jj}[\boldsymbol{x},t,\boldsymbol{y}|u_k(\boldsymbol{y},t)]\,\mathrm{d}S_y \\ &= \int_{\partial\Omega}\left\{G_{,j}[\boldsymbol{x},t,\boldsymbol{y}|D_{ij}u(\boldsymbol{y},t)] + \frac{1}{c^2}n_j(\boldsymbol{y})G[\boldsymbol{x},t,\boldsymbol{y}|\ddot{u}(\boldsymbol{y},t)]\right\}\mathrm{d}S_y\end{aligned}$$

taking into account eq. (7.4). The integral representation of the wave function gradient thus admits the following alternative form:

$$\begin{aligned}\kappa u_{,i}(\boldsymbol{x},t) = &\int_{\partial\Omega}\left\{G_{,j}[\boldsymbol{x},t,\boldsymbol{y}|D_{ij}u(\boldsymbol{y},t)] + \frac{1}{c^2}G[\boldsymbol{x},t,\boldsymbol{y}|\ddot{u}(\boldsymbol{y},t)]n_j(\boldsymbol{y})\right\}\mathrm{d}S_y \\ &- \int_{\partial\Omega} G_{,i}[\boldsymbol{x},t,\boldsymbol{y}|q(\boldsymbol{y},t)]\,\mathrm{d}S_y - \int_{\Omega} G_{,i}[\boldsymbol{x},t,\boldsymbol{y}|F(\boldsymbol{y},t)]\,\mathrm{d}V_y \\ &- \frac{1}{c^2}\int_{\Omega}\left\{v_0(\boldsymbol{y})G_{,i}(\boldsymbol{x},t,\boldsymbol{y}) - u_0(\boldsymbol{y})\dot{G}_{,i}(\boldsymbol{x},t,\boldsymbol{y})\right\}\mathrm{d}V_y \qquad (7.13)\end{aligned}$$

which is the dynamical analog to eq. (2.20) for statics.

7.3 BOUNDARY INTEGRAL EQUATIONS

Due to the singular behaviour of the dynamical fundamental solutions at $\boldsymbol{y} = \boldsymbol{x}$ when $\boldsymbol{x} \in \partial\Omega$, integral equations must be established through a limiting process, following the same approach as in statics (Section 2.2): a domain $\Omega_\varepsilon = \Omega - v_\varepsilon$ is created by removing a neighbourhood $v_\varepsilon(\boldsymbol{x}) = v_\varepsilon$ of $\boldsymbol{x}$ of diameter $\leq \varepsilon$, and one puts $\partial\Omega_\varepsilon = (\partial\Omega - e_\varepsilon) + s_\varepsilon$ (with $s_\varepsilon = \Omega \cap \partial v_\varepsilon$) and $e_\varepsilon = \partial\Omega \cap v_\varepsilon$. Equations (7.8–7.9) play a key role: the singular behaviour of the dynamic fundamental solution is transferred into its static counterpart, so that the regularization techniques previously devised for static problems (Chapter 2) are applicable to dynamical problems as well.

For simplicity of exposition, assume vanishing body sources and initial rest (i.e. $F = 0, u_0 = v_0 = 0$). A boundary integral equation is sought as the limiting form when $\varepsilon \to 0$ of identity (7.11) applied to the domain Ω_ε:

$$\int_{(\partial\Omega - e_\varepsilon)+s_\varepsilon} \{H[\boldsymbol{x}, t, \boldsymbol{y}|u(\boldsymbol{y}, t)] - G[\boldsymbol{x}, t, \boldsymbol{y}|q(\boldsymbol{y}, t)]\} \, \mathrm{d}S_y = 0 \tag{7.14}$$

Using the *static* identity (2.7) and treating $u(\boldsymbol{x}, t)$ as a scalar constant, one has:

$$u(\boldsymbol{x}, t) \int_{(\partial\Omega - e_\varepsilon)+s_\varepsilon} H(\boldsymbol{x}, \boldsymbol{y}) \, \mathrm{d}S_y = 0$$

so that eq. (7.14) can be reformulated as:

$$\int_{(\partial\Omega - e_\varepsilon)+s_\varepsilon} \{H[\boldsymbol{x}, t, \boldsymbol{y}|u(\boldsymbol{y}, t)] - u(\boldsymbol{x}, t) H(\boldsymbol{x}, \boldsymbol{y})\} \, \mathrm{d}S_y$$
$$- \int_{(\partial\Omega - e_\varepsilon)+s_\varepsilon} G[\boldsymbol{x}, t, \boldsymbol{y}|q(\boldsymbol{y}, t)] \, \mathrm{d}S_y = 0 \tag{7.15}$$

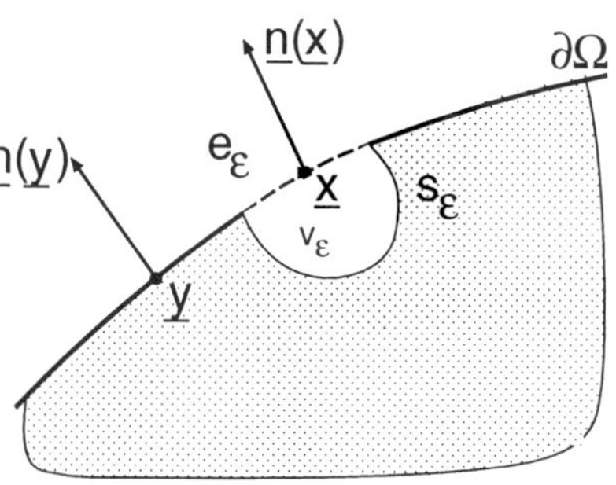

Figure 7.1 Exclusion neighbourhood $v_\varepsilon(x)$ and other geometrical notation used for the limiting process $\varepsilon \to 0$.

Moreover, one notices that:

$$\begin{aligned} &H[\boldsymbol{x},t,\boldsymbol{y}|u(\boldsymbol{y},t)] - u(\boldsymbol{x},t)H(\boldsymbol{x},\boldsymbol{y}) \\ &\qquad = \{H[\boldsymbol{x},t,\boldsymbol{y}|u(\boldsymbol{y},t)] - u(\boldsymbol{y},t)H(\boldsymbol{x},\boldsymbol{y})\} + [u(\boldsymbol{y},t) - u(\boldsymbol{x},t)]H(\boldsymbol{x},\boldsymbol{y}) \end{aligned}$$

Since the difference between corresponding dynamic and static fundamental solutions is nonsingular (Section 7.1), the above equality shows that the regularization issue in dynamics reduces to that in statics, which is solved. In fact, the quantity $[u(\boldsymbol{y},t) - u(\boldsymbol{x},t)]H(\boldsymbol{x},\boldsymbol{y})$ was shown in Chapter 2 to be weakly singular provided u is $C^{0,\alpha}$-continuous at $\boldsymbol{x}$. The latter assumption thus ensures that the integrals in eq. (7.15) are convergent as well, so that the limiting process $\varepsilon \to 0$ leads to weakly singular integrals. As a result of this discussion, any solution to the wave equation (7.1) with vanishing body source distribution and at initial rest solves the following regularized boundary integral equation:

$$\begin{aligned} &\int_{\partial\Omega} \{H[\boldsymbol{x},t,\boldsymbol{y}|u(\boldsymbol{y},t)] - u(\boldsymbol{x},t)H(\boldsymbol{x},\boldsymbol{y})\}\ \mathrm{d}S_y \\ &\qquad\qquad - \int_{\partial\Omega} G[\boldsymbol{x},t,\boldsymbol{y}|q(\boldsymbol{y},t)]\,\mathrm{d}S_y = 0 \end{aligned} \tag{7.16}$$

Regularized integral representation of the wave function gradient

Here, the free-space fundamental solution (7.5) is specifically invoked, so as to resort to the integration by parts, then first-order regularization approach (see Section 5.4). The integral representation formula for $u_{,i}(\boldsymbol{x},t)$ is accordingly sought as the limiting form when $\varepsilon \to 0$ of the identity:

$$\begin{aligned} &\int_{(\partial\Omega-e_\varepsilon)+s_\varepsilon} \left\{G_{,j}[\boldsymbol{x},t,\boldsymbol{y}|D_{ij}u(\boldsymbol{y},t)] + \frac{1}{c^2}G[\boldsymbol{x},t,\boldsymbol{y}|\ddot{u}(\boldsymbol{y},t)]n_i(\boldsymbol{y})\right\}\ \mathrm{d}S_y \\ &\qquad\qquad - \int_{(\partial\Omega-e_\varepsilon)+s_\varepsilon} G_{,i}[\boldsymbol{x},t,\boldsymbol{y}|q(\boldsymbol{y},t)]\,\mathrm{d}S_y = 0 \end{aligned} \tag{7.17}$$

which is rearranged in order to isolate the static fundamental solution:

$$\begin{aligned} &\int_{(\partial\Omega-e_\varepsilon)+s_\varepsilon} \{G_{,j}[\boldsymbol{x},t,\boldsymbol{y}|D_{ij}u(\boldsymbol{y},t)] - D_{ij}u(\boldsymbol{y},t)G_{,j}(\boldsymbol{x},\boldsymbol{y})\}\ \mathrm{d}S_y \\ &\quad + \frac{1}{c^2}\int_{(\partial\Omega-e_\varepsilon)+s_\varepsilon} G[\boldsymbol{x},t,\boldsymbol{y}|\ddot{u}(\boldsymbol{y},t)]n_i(\boldsymbol{y})\ \mathrm{d}S_y \\ &\quad - \int_{(\partial\Omega-e_\varepsilon)+s_\varepsilon} \{G_{,i}[\boldsymbol{x},t,\boldsymbol{y}|q(\boldsymbol{y},t)] - q(\boldsymbol{y},t)G_{,i}(\boldsymbol{x},\boldsymbol{y})\}\ \mathrm{d}S_y \\ &\quad + \int_{(\partial\Omega-e_\varepsilon)+s_\varepsilon} \{D_{ij}u(\boldsymbol{y},t)G_{,j}(\boldsymbol{x},\boldsymbol{y}) - q(\boldsymbol{y},t)U^k_{,i}(\boldsymbol{x},\boldsymbol{y})\}\ \mathrm{d}S_y = 0 \end{aligned}$$

The first three integrals in the above equation, in the limit $\varepsilon \to 0$, are weakly singular, convergent integrals over $\partial\Omega$: the first and third ones as a consequence of the nonsingular character of the difference between dynamic and static fundamental solutions, eq. (7.9); the second one because the singularity of G is integrable. Finally, invoking the regularization result (5.33) for the static fundamental solution, one has:

$$\begin{aligned}
\lim_{\varepsilon\to 0}\int_{(\partial\Omega-e_\varepsilon)+s_\varepsilon} &\{D_{ij}u(\boldsymbol{y},t)G_{,j}(\boldsymbol{x},\boldsymbol{y}) - q(\boldsymbol{y},t)G(\boldsymbol{x},\boldsymbol{y})\}\,\mathrm{d}S_y \\
&= -\frac{1}{2}u_{,i}(\boldsymbol{x},t) + D_{ij}u(\boldsymbol{x},t)I_j(\boldsymbol{x},\partial\Omega) - q(\boldsymbol{x},t)I_i(\boldsymbol{x},\partial\Omega) \\
&\quad + \int_{\partial\Omega}[D_{ij}u(\boldsymbol{y},t) - D_{ij}u(\boldsymbol{x},t)]G_{,j}(\boldsymbol{x},\boldsymbol{y})\,\mathrm{d}S_y \\
&\quad - \int_{\partial\Omega}[q(\boldsymbol{y},t) - q(\boldsymbol{x},t)]G_{,i}(\boldsymbol{x},\boldsymbol{y})\,\mathrm{d}S_y
\end{aligned}$$

where $I_i(\boldsymbol{x},\partial\Omega)$ is given by eq. (5.31). This analysis leads to the following statement: at any point $\boldsymbol{x}$ of $\partial\Omega$ at which u and the unit normal $\boldsymbol{n}$ are respectively $C^{1,\alpha}$- and $C^{0,\alpha}$-continuous, the wave function gradient is given by:

$$\begin{aligned}
\frac{1}{2}u_{,i}(\boldsymbol{x},t) = {} & D_{ij}u(\boldsymbol{x},t)I_j(\boldsymbol{x},\partial\Omega) - q(\boldsymbol{x},t)I_i(\boldsymbol{x},\partial\Omega) \\
& + \int_{\partial\Omega}[G_{,j}[\boldsymbol{x},t,\boldsymbol{y}|D_{ij}u(\boldsymbol{y},t)] - D_{ij}u(\boldsymbol{x},t)G_{,j}(\boldsymbol{x},\boldsymbol{y})]\,\mathrm{d}S_y \\
& - \int_{\partial\Omega}[G_{,i}[\boldsymbol{x},t,\boldsymbol{y}|q(\boldsymbol{y},t)] - q(\boldsymbol{x},t)G_{,i}(\boldsymbol{x},\boldsymbol{y})]\,\mathrm{d}S_y \\
& + \frac{1}{c^2}\int_{\partial\Omega}G[\boldsymbol{x},t,\boldsymbol{y}|\ddot{u}(\boldsymbol{y},t)]n_i(\boldsymbol{y})\,\mathrm{d}S_y
\end{aligned} \tag{7.18}$$

Comments

The nonsingular character of the difference between dynamic and static fundamental solutions, eqs (7.8)–(7.9), plays an instrumental role in establishing regularized integral equations for wave problems. This is true for elastodynamics as well (see Section 7.6 below).

The integral equation (7.16) holds true independently of the particular fundamental solution being used, whereas the gradient representation on the boundary, as given by eq. (7.64), involves specifically the free-space fundamental solution. All integrals in eqs (7.16), (7.18) are at most weakly singular, subject to the same regularity requirements on (u,q) and $\partial\Omega$ as in statics.

Two-dimensional problems

Following the same line of thought as in Section 2.5, one readily finds that the integral equations for the two-dimensional wave equation have the same form as in three dimensions. For example, the integral equation (7.16) transposes itself to two-dimensional problems by putting $\boldsymbol{x} = (x_1, x_2), \boldsymbol{y} = (y_1, y_2)$, the integrals being taken over the *curve* $\partial\Omega$ bounding the domain $\Omega \subset \mathbb{R}^2$; the two-dimensional free-space fundamental solution is given by:

$$G[\tilde{x}, t, \tilde{y}|f] = \frac{1}{2\pi} \int_r^{ct} \frac{f(t - \eta/c)}{\sqrt{\eta^2 - r^2}} \, \mathrm{d}\eta \qquad (r = |\tilde{y} - \tilde{x}|) \tag{7.19}$$

One notices that it involves all retarded times $t - r/v$ for $r/t \leq v \leq c$. This is in contrast to the three-dimensional free-space fundamental solution (7.5), in which only the retarded time $t - r/c$ corresponding to the physical velocity c appears. This 'lagging' effect reflects the fact that the two-dimensional solution results from an integration of its three-dimensional counterpart over $-\infty \leq x_3 \leq +\infty$, i.e. is actually the wave function generated in the three-dimensional free space by an *infinite line* of point sources.

Nonzero sources or initial conditions

The validity of the regularization method is not affected by nonzero sources or initial conditions. The right-hand side of the integral equation (7.16) should simply be replaced by:

$$\int_\Omega G[\boldsymbol{x}, t, \boldsymbol{y}|F(\boldsymbol{y}, t)] \, \mathrm{d}V_y + \frac{1}{c^2} \int_\Omega \left\{ v_0(\boldsymbol{y}) G(\boldsymbol{x}, t, \boldsymbol{y}) + u_0(\boldsymbol{y}) \dot{G}(\boldsymbol{x}, t, \boldsymbol{y}) \right\} \mathrm{d}V_y$$

7.4 EXTERIOR PROBLEMS AND CONDITIONS AT INFINITY

Radiation condition

To extend the validity of the boundary integral equation (7.16) to unbounded domains $\bar{\Omega}$ of the form $\bar{\Omega} = \mathbb{R}^3 - \Omega$ (Ω bounded) will cause certain restrictions on the admissible behaviour at infinity of the solution u. In order to show this, consider a finite domain Ω_R bounded from within by $\partial\Omega$ and from without by the sphere S_R of centre $\boldsymbol{x}$ and (large) radius R. The boundary integral equation (7.16) holds true for any finite value of R and reads:

$$\begin{aligned} u(\boldsymbol{x}, t) + \int_{\partial\bar{\Omega}} & [H[\boldsymbol{x}, t, \boldsymbol{y}|u(\boldsymbol{y}, t)] - u(\boldsymbol{x}, t) H(\boldsymbol{x}, \boldsymbol{y})] \, \mathrm{d}S_y \\ & + \int_{S_R} \{H[\boldsymbol{x}, t, \boldsymbol{y}|u(\boldsymbol{y}, t)] - G[\boldsymbol{x}, t, \boldsymbol{y}|q(\boldsymbol{y}, t)]\} \, \mathrm{d}S_y = 0 \end{aligned} \tag{7.20}$$

where identity (2.7) for the static fundamental solution has been used, in the form:

$$\int_{S_R} H(\boldsymbol{x}, \boldsymbol{y}) \, \mathrm{d}S_y = -1$$

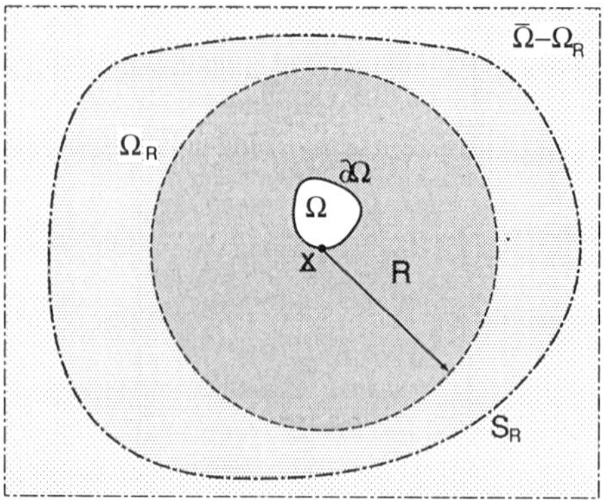

Figure 7.2 Domain Ω_R for the limiting process $R \to \infty$.

The normal $\boldsymbol{n}$ is directed outwards from Ω_R, hence in particular inwards to Ω.

The limiting form when $R \to +\infty$ of eq. (7.20) is sought (Figure 7.2). In particular, conditions on u, q are sought so as to satisfy:

$$\lim_{R\to\infty} I_k = 0$$

with

$$I_k = \int_{S_R} \{H[\boldsymbol{x}, t, \boldsymbol{y}|u(\boldsymbol{y}, t)] - G[\boldsymbol{x}, t, \boldsymbol{y}|q(\boldsymbol{y}, t)]\} \,\mathrm{d}S_y \tag{7.21}$$

One notices that eq. (7.6) allows us to write:

$$H[\boldsymbol{x}, t, \boldsymbol{y}|f] = -\frac{1}{c}G[\boldsymbol{x}, t, \boldsymbol{y}|\dot{f}] - \frac{1}{r}G[\boldsymbol{x}, t, \boldsymbol{y}|f] \qquad (\boldsymbol{y} \in S_R)$$

since $r_{,i}n_{,i} = 1$ over S_R. Then, eq. (7.21) becomes:

$$I_k = -\int_{S_R} \left\{G\left[\boldsymbol{x}, t, \boldsymbol{y} \mid \left(q + \frac{1}{c}\dot{u}\right)(\boldsymbol{y}, t)\right] + \frac{1}{R}G[\boldsymbol{x}, t, \boldsymbol{y}|u(\boldsymbol{y}, t)]\right\} R^2 \,\mathrm{d}\omega$$

using $\mathrm{d}S_y = R^2\,\mathrm{d}\omega$, where ω denotes the differential solid angle of origin $\boldsymbol{x}$. Taking into account the fact that $G = O(1/R)$ at infinity, a set of *sufficient* conditions at infinity for I_k to vanish in the limit $R \to \infty$ is:

$$u = o(1) \qquad \text{(decay)} \tag{7.22}$$

$$q + \frac{1}{c}\dot{u} = o(R^{-1}) \qquad \text{(radiation)} \tag{7.23}$$

These conditions ensure the uniqueness of an exterior problem (Eringen and Suhubi [2]). The second one, known as the Sommerfeld radiation condition, says that the wave is outgoing, i.e. the absence of source at infinity.

Integral equation

As a result of the foregoing analysis, for any solution u to the (homogeneous) wave equation (7.1) on the unbounded domain $\bar{\Omega}$ satisfying the conditions at infinity (7.23), the following boundary integral equation holds:

$$u_k(\boldsymbol{x},t) + \int_{\partial\bar{\Omega}} [H[\boldsymbol{x},t,\boldsymbol{y}|u(\boldsymbol{y},t)] - u(\boldsymbol{x},t)H(\boldsymbol{x},\boldsymbol{y})]\, \mathrm{d}S_y \\ - \int_{\partial\bar{\Omega}} G[\boldsymbol{x},t,\boldsymbol{y}|q(\boldsymbol{y},t)]\, \mathrm{d}S_y = 0 \quad (7.24)$$

The only differences between this equation and eq. (7.62) for bounded domains are the appearance of a free term $u_k(\boldsymbol{x},t)$ and the reversed orientation of the unit normal $\boldsymbol{n}$, here exterior to $\bar{\Omega}$ and hence interior to Ω.

The limiting process $R \to \infty$ and the conditions at infinity (7.23) still hold when nonzero body sources and initial data (F, u_0, v_0) are considered provided the integral

$$\int_{\bar{\Omega}} \left\{ G[\boldsymbol{x},t,\boldsymbol{y}|F(\boldsymbol{y},t)] + \frac{1}{c^2}\left[v_0(\boldsymbol{y})G(\boldsymbol{x},t,\boldsymbol{y}) + u_0(\boldsymbol{y})\dot{G}(\boldsymbol{x},t,\boldsymbol{y})\right]\right\} \mathrm{d}V_y \quad (7.25)$$

is convergent. The right-hand side of the integral equation (7.24) then becomes nonzero and equal to the above integral.

Integral representations

In view of identity (2.7), eq. (7.24) for $\boldsymbol{x} \notin \partial\Omega$ is none other than the integral representation formula (7.11). The same representation formula thus applies to bounded as well as unbounded domains, provided the unit normal is consistently oriented towards the exterior of the body under consideration. Any field u expressed in $\bar{\Omega}$ by (7.24) satisfies the conditions at infinity (7.23).

Nonzero sources

The representation formula (7.11) does not allow us to express any dynamical state (u, F) over an unbounded domain. In particular, a *plane wave*:

$$u^I(\boldsymbol{y},t) = Ag(t - \boldsymbol{y}.\boldsymbol{p}/c) \quad (7.26)$$

($\boldsymbol{p}$ denoting a given unit vector that defines the propagation direction) does not satisfy the conditions at infinity (7.23). One may then use linear superposition. For instance, consider the scattering of an incident plane wave (7.26) by a rigid bounded obstacle occupying the region Ω. The total field u can then be split: $u = u^I + u^S$. The scattered field $\boldsymbol{u}^S$ satisfies $q^S = -q^I$ on $\partial\Omega$ as well as the conditions at infinity (7.23); it thus admits a representation of the form (7.11):

$$\kappa u^S(\boldsymbol{x},t) = -\int_{\partial\Omega} \{G[\boldsymbol{x},t,\boldsymbol{y}|q^I(\boldsymbol{y},t)] + H[\boldsymbol{x},t,\boldsymbol{y}|u^S(\boldsymbol{y},t)]\} \, \mathrm{d}S_y \tag{7.27}$$

where $\kappa = 1$ ($\boldsymbol{x} \in \bar{\Omega}$), $\kappa = 0$ ($\boldsymbol{x} \notin \bar{\Omega}$). Similarly, one has to resort to superposition when the distributions of body sources and initial data render the integral (7.25) divergent.

Integral formulation for the total field

Consider an exterior scattering problem where the incident field u^I is known in $\mathbb{R}^3$, satisfies the homogeneous wave equation: $c^2\Delta u^I - \ddot{u}^I = 0$ and violates the conditions at infinity (7.22), (7.23); zero body sources and initial rest are assumed. Invoking the superposition principle, the scattered field u^S admits the integral representation (7.11).

In addition, since u^I is defined in the bounded domain Ω, the representation formula for u^I *with respect to* Ω can be invoked. It reads:

$$(1-\kappa)u^I(\boldsymbol{x},t) = \int_{\partial\Omega} \{G[\boldsymbol{x},t,\boldsymbol{y}|q^I(\boldsymbol{y},t)] - H[\boldsymbol{x},t,\boldsymbol{y}|u^I(\boldsymbol{y},t)]\} \, \mathrm{d}S_y \tag{7.28}$$

where the unit normal (implicit in q^I and H) is now exterior to Ω. Then, subtracting (7.28) from (7.28), one finds the following integral representation of the *total* field u:

$$\kappa u(\boldsymbol{x},t) - u^I(\boldsymbol{x},t) = -\int_{\partial\Omega} H[\boldsymbol{x},t,\boldsymbol{y}|u(\boldsymbol{y},t)] \, \mathrm{d}S_y \tag{7.29}$$

A boundary integral equation for the total field can be obtained following the same approach. Note that the total field integral formulation has the advantage of reducing the numerical integration burden.

7.5 THE BOUNDARY ELEMENT METHOD IN DYNAMICS

The boundary element method in dynamics is of course based on the concepts presented in Chapter 3 regarding the spatial interpolation of the boundary and the variable it supports. One major difference of course is the fact that now the field variables are also time-dependent. This dictates the necessity of defining interpolations in time and collocating the boundary integral equation at various time instants: wave problems in the time domain are solved, as with other discretization approaches, by means of *time-marching schemes*. For simplicity of exposition, this section addresses only wave problems without body sources and at initial rest, the extension to more general data being straightforward. Also, the interpolation in space is assumed isoparametric.

Interpolation of boundary field variables

Assume a boundary element representation of $\partial\Omega$, in terms of N_N nodes $\boldsymbol{y}^1, \ldots, \boldsymbol{y}^{N_N}$ and N_E elements $E_1, \ldots, E_{N_E}$. The variables (u, q), as well as $\partial\Omega$ itself, are interpolated according to (3.12), i.e. using shape functions and (time-dependent) nodal values:

$$u(\boldsymbol{y}, t) = \sum_{k=1}^{N_i} M_k(\boldsymbol{\xi}) u^k(t) \qquad q(\boldsymbol{y}, t) = \sum_{k=1}^{N_i} M_k(\boldsymbol{\xi}) q^k(t) \qquad (\boldsymbol{\xi} \in \Delta_e) \quad (7.30)$$

The time interval of interest $[0, T]$ is subdivided into N time subintervals:

$$I_\beta = [t_{\beta-1}, t_\beta] \qquad (1 \leq \beta \leq N)$$

For simplicity, the time step $\Delta t = t_\beta - t_{\beta-1}$ is assumed constant, so that $t_\beta = \beta\Delta t$. Interpolation functions $\phi^\beta(t)$ are introduced, so that:

$$u^k(t) = \sum_{\beta=1}^{N} \phi^\beta(t) u^k_\beta \qquad q^k(t) = \sum_{\beta=1}^{N} \phi^\beta(t) q^k_\beta$$

The numerical solution of a wave problem then reduces, as is usually the case, to the determination of the unknown nodal values $u^k_\beta = u(\boldsymbol{y}^k, t_\beta)$ and $q^k_\beta = q(\boldsymbol{y}^k, t_\beta)$. For instance, a continuous and piecewise linear interpolation in time is achieved in terms of a single time interpolation function $S(u)$, $u \in [-1, 1]$:

$$\phi^\beta(t) = S(\frac{t - t_\beta}{\Delta t}) = S(\frac{t}{\Delta t} - \beta) \qquad S(u) = \begin{cases} 1 + u & -1 \leq u \leq 0 \\ 1 - u & 0 \leq u \leq 1 \end{cases} \quad (7.31)$$

Discretized form of the integral equation

Using for definiteness the free-space fundamental solution (7.5), the boundary integral equation (7.16) takes, after discretization of geometry and boundary variables and using notations similar to those of eq. (3.13) of Chapter 3, the following form:

$$\begin{aligned} 0 = & \sum_{e \in I(\boldsymbol{x})} \sum_{p=1}^{N_I(e)} \left\{ [AS_1(e,p,b) + AS_2(e,p,b)] u_b^{m(e,p)} - B(e,p,b) q^{m(e,p)_b} \right\} \\ & + \sum_{e \in \bar{I}(\boldsymbol{x})} \sum_{p=1}^{N_I(e)} \left\{ AR(e,p,b) u_b^{m(e,p)} - B(e,p,b) q_b^{m(e,p)} \right\} \\ & + u(\boldsymbol{x}) \left\{ 1 - \sum_{e \in \bar{I}(\boldsymbol{x})} \widehat{AR}(e) \right\} \\ & + \sum_{\beta=0}^{b-1} \sum_{e \in \bar{I}(\boldsymbol{x})} \sum_{p=1}^{N_I(e)} \left\{ AR(e,p,\beta) u_\beta^{m(e,p)} - B(e,p,\beta) q_\beta^{m(e,p)} \right\} \end{aligned} \quad (7.32)$$

where the element integrals are given by:

$$
\begin{aligned}
AS_1(e,p,\beta) &= \int_{E_e} \left[[S(b-\beta-\frac{r}{c\Delta t})-1] + \frac{r}{c\Delta t}S'(b-\beta-\frac{r}{c\Delta t}) \right] r_{,n}\frac{\mathrm{d}S_y}{4\pi r^2} \\
AS_2(e,p,\beta) &= -\int_{E_e} [M_p(\boldsymbol{\xi}) - M_p(\boldsymbol{\eta}_e)] r_{,n}\frac{\mathrm{d}S_y}{4\pi r^2} \\
AR(e,p,\beta) &= -\int_{E_e} \left[S(b-\beta-\frac{r}{c\Delta t}) + \frac{r}{c\Delta t}S'(b-\beta-\frac{r}{c\Delta t}) \right] M_p(\boldsymbol{\xi}) r_{,n}\frac{\mathrm{d}S_y}{4\pi r^2} \\
B(e,p,\beta) &= \int_{E_e} S(b-\beta-\frac{r}{c\Delta t}) M_p(\boldsymbol{\xi})\frac{\mathrm{d}S_y}{4\pi r} \\
\widehat{AR}(e) &= -\int_{E_e} r_{,n}\frac{\mathrm{d}S_y}{4\pi r^2}
\end{aligned}
$$

The collocation time instant t was chosen as an integer multiple $b\Delta t$ of the time step. Some integrals become (weakly) singular for $r \sim 0$. In this case, $(t - r/c) \sim t$, which explains that singular element integrals occur only as cofactors of nodal values at the collocation time instant, i.e. $\beta = b$; in other words, all element integrals involving a time interpolation function ϕ_β with $\beta < b$ are nonsingular.

Equation (7.32) also highlights the fact that the element integrals depend on β, b only through $b - \beta$. This is a direct consequence of a property shared by all fundamental solutions of the wave equation, namely invariance under translation in time:

$$G[\boldsymbol{x}, t, \boldsymbol{y}|f(t)] = G[\boldsymbol{x}, t + t', \boldsymbol{y}|t(t + t')]$$

together with the use of a constant time step Δt. Notice also that, when using piecewise linear interpolation in time, the interpolation function S, eq. (7.31) is such that:

$$S(b-\beta-\frac{r}{c\Delta t}) - 1 + \frac{r}{c\Delta t}S'(b-\beta-\frac{r}{c\Delta t}) = 0$$

so that the actual evaluation of the singular integral $AS_1(e,p)$ is avoided.

Structure of the time-marching scheme

The discretized integral equation (7.32), collocated for example[1] at the nodes $\boldsymbol{x} = \boldsymbol{y}^1, \ldots \boldsymbol{y}^{N_N}$ and the instant $t = b\Delta t$, takes the following matrix form:

$$[\boldsymbol{A}^0]\{\boldsymbol{U}_b\} + [\boldsymbol{B}^0]\{\boldsymbol{Q}_b\} + \sum_{\beta=1}^{b-1}\left([\boldsymbol{A}^{b-\beta}]\{\boldsymbol{U}_\beta\} + [\boldsymbol{B}^{b-\beta}]\{\boldsymbol{Q}_\beta\}\right) = 0 \qquad (7.33)$$

[1] Other choices may of course be used for the collocation points.

where $\{\boldsymbol{U}_\beta\}$ and $\{\boldsymbol{Q}_\beta\}$ are 'vectors' gathering the nodal values u_β^k and q_β^k at a fixed time step β. As a consequence of the property of invariance under translation in time, the above equation has the form of a *discrete convolution*. The integral equation collocated at $t = t_b = b\Delta t$ involves *a priori* the nodal values at all previous time steps $\beta \leq b$. Equation (7.33) emphasizes the *time-marching* character of the solution strategy: the solution at $t = t_b$ is computed using the accumulated knowledge of the solution at all previous time instants, this process being started with the initial data at $t_0 = 0$.

Half of the variables at $t = t_b$ are true unknowns, the other half being known from the boundary conditions. One must then do column-shifting (see Chapter 3) in eq. (7.33) so that only the matrix columns associated to unknown degrees of freedom remain in the left-hand side. Equation (7.33) thus assumes the form:

$$[\boldsymbol{K}]\{\boldsymbol{X}_b\} = \{\boldsymbol{F}_b\} - \sum_{\beta=1}^{b-1} \left([\boldsymbol{A}^{b-\beta}]\{\boldsymbol{U}_\beta\} + [\boldsymbol{B}^{b-\beta}]\{\boldsymbol{T}_\beta\} \right) \tag{7.34}$$

where $\{\boldsymbol{X}_b\}$ denotes the 'vector' of unknown degrees of freedom at $t = t_b$ and $\{\boldsymbol{F}_b\}$ gathers the contributions from prescribed boundary values at $t = t_b$. The right-hand side of eq. (7.34) emphasizes the fact that the unknowns at $t = t_b$ depend not only on the boundary data at $t = t_b$ but also on the solution history (i.e. u_β^k, q_β^k, $\beta < b$); this is of course due to the presence of the retarded time $t - r/c$ in the fundamental solution (7.5). The 'resolvent matrix' $[\boldsymbol{K}]$, obtained through column-shifting in $[\boldsymbol{A}^0]$, $[\boldsymbol{B}^0]$, has to be assembled and factored only once. The matrix $[\boldsymbol{K}]$ is non-symmetric but, in contrast to static problems, is sparse because nonzero contributions occur only for $r \leq c\Delta t$: nonzero coefficients in $[\boldsymbol{K}]$ occur only for nodes close to the collocation point.[2]

Also, since (i) the free-space fundamental solution depends on time only through the retarded time $t - r/c$ and (ii) r cannot take values larger than $\mathrm{diam}(\partial\Omega) = \max_{x,y\in\partial\Omega} |\boldsymbol{y} - \boldsymbol{x}|)$, there is a critical integer value $N(\partial\Omega, \Delta t)$:

$$N(\partial\Omega, \Delta t) = 1 + (\text{integer part of}) \frac{1}{c\Delta t} \mathrm{diam}(\partial\Omega)$$

such that

$$[\boldsymbol{A}^{b-\beta}] = [\boldsymbol{B}^{b-\beta}] = [0] \qquad (b - \beta > N(\partial\Omega, \Delta t))$$

As a consequence, the summation over β in (7.34) starts at $\beta = \max(0, b - N(\partial\Omega, \Delta t))$. This comment is applicable only to three-dimensional problems; due to the 'lagging effect' previously discussed, the complete solution history must be taken into account when eq. (7.34) is applied to two-dimensional problems.

[2]The 'closeness' depends of course on the time step Δt; typically $c\Delta t$ is a fraction of a typical element diameter.

Numerical solution process

This consists of the following steps:

1. Assemble the matrices $([\boldsymbol{A}^0], [\boldsymbol{B}^0]), \ldots, ([\boldsymbol{A}^{N(\partial\Omega,\Delta t)}], [\boldsymbol{B}^{N(\partial\Omega,\Delta t)}])$.
2. Incorporate the boundary data; build the 'resolvent matrix' $[\boldsymbol{K}]$ by selecting the appropriate columns in $[\boldsymbol{A}^0], [\boldsymbol{B}^0]$; factor $[\boldsymbol{K}]$.
3. Solve (7.34) for $\{\boldsymbol{X}_b\}$ at each time step $b = 0, 1, \ldots, N$.
4. Post-processing (if needed in the problem at hand): compute u or $\boldsymbol{\nabla} u$ at user-prescribed interior points with the integral representations (7.11), (7.13).

One notes that all matrices created in step (1) must be stored because they are repeatedly used in step (3) to contribute to the discrete convolution; they are multiplied by a different vector at each time step b. A closer scrutiny of eq. (7.32) reveals that each matrix has many null coefficients. Indeed, for a given collocation point $\boldsymbol{x} \in \partial\Omega$, the portion of $\partial\Omega$ which contributes through the element integrals to nonzero coefficients of $[\boldsymbol{A}^b]$, $[\boldsymbol{B}^b]$ is:

$$\{\boldsymbol{y} \in \partial\Omega \mid (b-1)\Delta t \leq r/c \leq (b+1)\Delta t\}$$

Careful programming thus allows us to reduce the overall storage requirement to significantly less than the amount required at first glance by the $2 \times N(\partial\Omega, \Delta t)$ matrices.

The fundamental solution possesses causality properties (witnessed by the presence of the retarded time), but using products of space and time functions in the interpolation scheme (7.30) destroys, strictly speaking, the causality at the discretization level: two nodes belonging to the same boundary element 'communicate' instantaneously. The time-marching scheme can nevertheless be considered approximately causal if (a) the 'grid size' h (i.e. for example the largest boundary element diameter) is reasonably small compared to the characteristic size of $\partial\Omega$ and (b) the time step Δt is less than h/c.

Some unsolved issues in time-domain BEM discretization

The foregoing presentation summarizes the main features of the standard approach to the time-domain boundary element method. Most recent published works in this area follow the above general path. The implementation of the time-domain boundary element method involves some technical issues which in our opinion are still not well understood. They are listed below, with some suggested remedies (untested as yet).

Stability and convergence of the time-marching scheme. This very important issue is seldom addressed in the literature (the recent paper [6] is the only one we know of that deals directly with this matter). Indeed, the discretized problem depends on the time step in a very intricate way due to the presence of retarded time in the various fundamental solutions. Some authors consider that the time-stepping scheme is of implicit, and thus unconditionally stable, nature due to this dependency. There is however no rigorous argumentation to support this viewpoint. Indeed, various occurrences of instability in time are documented in recently published works, see [6].

Computation of element integrals. Here also, the occurrence of retarded time in the integral equations creates implementational difficulties that do not appear in static or frequency-domain formulations. The element integrals of (7.32) have the following general form:

$$\int_E K(\boldsymbol{x},\boldsymbol{y})M_p(\boldsymbol{\xi})S(b-\beta-\frac{r}{c\Delta t})J(\boldsymbol{\xi})\,\mathrm{d}\boldsymbol{\xi} \tag{7.35}$$

where the support of the interpolation function $S(\eta)$ is the interval $[-1,1]$; moreover $S(\eta)$ has discontinuous first-order derivatives at $\eta=\pm1$ and $\eta=0$.

Denote by $S^\beta(\boldsymbol{x})$ the spatial region bounded by two spheres centred at $\boldsymbol{x}$ and of radii $(\beta-1)c\Delta t$ and $(\beta+1)c\Delta t$. The geometrical support of the integral (7.35) is in fact the portion $E^\beta(\boldsymbol{x})$ of the boundary element E that intersects the region $S^\beta(\boldsymbol{x})$, i.e. $E^\beta(\boldsymbol{x})=E\cap S^\beta(\boldsymbol{x})$. The integrand in eq. (7.35) is non-differentiable across the edge of $E^\beta(\boldsymbol{x})$ for all usual time interpolation functions ϕ^β. The numerical computation of this integral using Gauss points on the entire element E is thus likely to yield significantly inaccurate values. Consider the following simple example: integrate numerically over $[-1,1]$, using Gauss points, the function ψ_a: $\psi_a(z)=0$ if $z\leq a$ and $\psi_a(z)=z-a$ if $z\geq a$, for various values of a such that $-1<a<1$ and observe that the relative quadrature error is rather important and fluctuates with a. This witnesses the fact that traditional Gaussian quadrature over the entire boundary element is far less reliable when used for time-domain boundary element methods. Besides, the quadrature error using the n-point Gaussian rule is proportional to the $2n$-th derivative of the integrand at some point of the interval; here the integrand is not differentiable over the entire interval.

In order to use Gaussian quadrature satisfactorily, one should ideally determine analytically the subelements $E^\beta(\boldsymbol{x})$ and then integrate numerically using Gauss points over each $E^\beta(\boldsymbol{x})$, thus restoring the usual high accuracy of numerical quadrature. For two-dimensional situations involving simple discretizations (rectilinear elements), the geometrical problem of determining the intersections $E^\beta(\boldsymbol{x})$ seems solvable, whereas this looks intractable for three-dimensional situations.

Another possibility is to resort to 'brute force' (e.g. use large numbers of Gauss points). This entails computational integration costs much greater than for statics or frequency-domain dynamics. Not much attention is devoted in the recent literature to this issue, despite its apparent practical importance. Karabalis [3] uses an algorithm whereby an element is subdivided into many small subelements and Gaussian quadrature is applied separately over each subelement.

Alternative approaches for interpolation in time. The classical time-marching scheme incurs difficulties in numerical integration due to the fact that the time interpolation functions have a bounded support and are only piecewise differentiable. Such problems are expected to disappear, or at least become of much less influence, upon introduction of regular (i.e. several times differentiable) interpolation functions in time, of support $\mathbb{R}^+$. Also, this might facilitate the convergence and stability analysis. Besides, the time-marching has a somewhat 'brute force' character: it yields the details of the evolution process but does not allow us to pre-define partial or global information that then could be computed in a cheaper way using well-chosen interpolation functions in time. Two possibilities seem worthy of investigation:

Spectral interpolation. Following this approach, the temporal variation of (u, q) is polynomial of arbitrarily high degree n in t over the entire time interval of interest $[0, T]$. Spectral methods are common in computational fluid mechanics (Metivet [5]); the numerical solution is known to converge for $n \to \infty$ at a rate which increases with the smoothness of the exact solution.

Interpolation using wavelets. The wavelets (Daubechies [1]) are functions used for the analysis of signals. Most real signals have, for practical purposes, a finite support in both time and frequency domains (for instance, a musical note has a finite duration and a well-defined pitch). On the other hand, the Fourier transform of an impulse has an infinite frequency content, while a time-harmonic signal has a perfectly localized Fourier transform. Hence the idea of introducing other signal analysis functions that are (for practical purposes) localized in both time and frequency domains. The Gauss bell curve fulfils these criteria but cannot be used for other reasons. In wavelet analysis, a function ψ (the wavelet), smooth and localized in both time and frequency, is chosen. A given signal is then analysed through translation and dilatation of ψ. In particular, ψ is required to be of zero mean, allowing to 'see' jumps in the analysed signal. Some wavelets, together with their translated and dilated images, define Hilbertian bases of real-valued functions. This in turn implies the possibility of decomposition/synthesis of real-valued functions into wavelets, as well as of uncoupling dynamical problems into as many independent subproblems as the number of terms retained in the wavelet expansion of the unknown field.

7.6 GOVERNING EQUATIONS OF LINEAR ELASTODYNAMICS

The purpose of this section and the next ones is to present the extension to (vector) time-domain elastodynamic problems of the concepts developed in Sections 7.1 to 7.4 for scalar wave problems. The presentation is limited to isotropic elastodynamics. It is first necessary to recall the basic equations; a detailed exposition of linear elastodynamics is available in e.g. Eringen and Suhubi [2]. For a given force distribution $\boldsymbol{F}(\boldsymbol{y},t)$ over the body Ω, an *elastodynamic state* is any triplet $(\boldsymbol{u},\boldsymbol{\sigma},\boldsymbol{F})$ satisfying the linear elastic constitutive equation and the fundamental equation of motion:

$$\begin{cases} \boldsymbol{\sigma} = \lambda(\text{div}\boldsymbol{u})\mathbf{1} + \mu(\boldsymbol{\nabla}\boldsymbol{u} + \boldsymbol{\nabla}^T\boldsymbol{u}) \\ \text{div}\boldsymbol{\sigma} + \rho(\boldsymbol{F} - \ddot{\boldsymbol{u}}) = 0 \end{cases} \qquad \forall(\boldsymbol{y},t) \in \Omega \times [0,T] \tag{7.36}$$

To define a well-posed elastodynamic problem, one has to provide boundary conditions on $\partial\Omega$: for example, given load $\boldsymbol{t}^D$ over a portion S_T of $\partial\Omega$:

$$\boldsymbol{T}^n(\boldsymbol{u})(\boldsymbol{y},t) = \boldsymbol{t}^D(\boldsymbol{y},t) \qquad \forall(\boldsymbol{y},t) \in S_T \times [0,T]$$

and prescribed displacement over the complementary portion $S_u = \partial\Omega - S_T$:

$$\boldsymbol{u}(\boldsymbol{y},t) = \boldsymbol{u}^D(\boldsymbol{y},t) \qquad \forall(\boldsymbol{y},t) \in S_u \times [0,T]$$

as well as initial conditions:

$$\boldsymbol{u}(\boldsymbol{y},0) = \boldsymbol{u}_0(\boldsymbol{y}) \qquad \dot{\boldsymbol{u}}(\boldsymbol{y},t) = \boldsymbol{v}_0(\boldsymbol{y})$$

Moreover, conditions at infinity must be enforced as well if Ω is unbounded (see Section 7.10).

Lamé decomposition

Recall that any (sufficiently regular) vector field $\boldsymbol{u}$ can be decomposed into irrotational and solenoidal parts [2]:

$$\boldsymbol{u} = \boldsymbol{\nabla}\phi + \mathbf{rot}\boldsymbol{\psi} \tag{7.37}$$

where the scalar and vector potentials $\phi, \boldsymbol{\psi}$ are given by:

$$\begin{aligned} \phi(\boldsymbol{y},t) &= -\frac{1}{4\pi}\int_\Omega \frac{1}{|\boldsymbol{z}-\boldsymbol{y}|}\text{div}\boldsymbol{u}(\boldsymbol{z},t)\,\mathrm{d}V_z \\ \boldsymbol{\psi}(\boldsymbol{y},t) &= \frac{1}{4\pi}\int_\Omega \frac{1}{|\boldsymbol{z}-\boldsymbol{y}|}\mathbf{rot}\boldsymbol{u}(\boldsymbol{z},t)\,\mathrm{d}V_z \end{aligned} \tag{7.38}$$

If $\boldsymbol{u}$ is an elastodynamic displacement field, the *Lamé potentials* ϕ (scalar) and $\boldsymbol{\psi}$ (vector) respectively solve the scalar and vector wave equations:

$$\begin{aligned} c_L^2\Delta\phi - \ddot{\phi} + f_L &= 0 \\ c_T^2\boldsymbol{\Delta}\boldsymbol{\psi} - \ddot{\boldsymbol{\psi}} + \boldsymbol{f}_T &= 0 \end{aligned}$$

the right-hand sides $f_L, \boldsymbol{f}_T$ being obtained using decomposition (7.38) of $\boldsymbol{F}$ and the velocities c_L, c_T given by:

$$c_L^2 = \frac{\lambda + 2\mu}{\rho} \qquad c_T^2 = \frac{\mu}{\rho}$$

The irrotational wave $\boldsymbol{u}^L = \boldsymbol{\nabla}\phi$ ($\textbf{rot}\boldsymbol{u}^L = \boldsymbol{0}$, velocity c_L) and the equivoluminal wave $\boldsymbol{u}^T = \textbf{rot}\boldsymbol{\psi}$ ($\text{div}\boldsymbol{u}^T = 0$, velocity c_T) appearing in the Lamé decomposition (7.37) can be shown (for instance by considering the special case of plane waves) to have the physical nature of compressive (or 'longitudinal') and shear (or 'transverse') waves, respectively.

Reciprocity identity

Consider two elastodynamic states $(\boldsymbol{u}^1, \boldsymbol{\sigma}^1, \boldsymbol{F}^1)$ and $(\boldsymbol{u}^2, \boldsymbol{\sigma}^2, \boldsymbol{F}^2)$ over Ω, with initial conditions $(\boldsymbol{u}_0^1, \boldsymbol{v}_0^1)$ and $(\boldsymbol{u}_0^2, \boldsymbol{v}_0^2)$. The reciprocity identity (4.3), i.e.:

$$\int_{\partial\Omega} \left[\boldsymbol{T}^n(\boldsymbol{u}^1).\boldsymbol{u}^2 - \boldsymbol{T}^n(\boldsymbol{u}^2).\boldsymbol{u}^1\right] \,\mathrm{d}S = \int_{\Omega} \left[\text{div}\boldsymbol{\sigma}^1.\boldsymbol{u}^2 - \text{div}\boldsymbol{\sigma}^2.\boldsymbol{u}^1\right] \,\mathrm{d}V$$

implies, taking into account the fundamental equation of motion $(7.36)_2$:

$$\int_{\partial\Omega} \left[\boldsymbol{T}^n(\boldsymbol{u}^1).\boldsymbol{u}^2 - \boldsymbol{T}^n(\boldsymbol{u}^2).\boldsymbol{u}^1\right] \,\mathrm{d}S$$
$$= \int_{\Omega} \rho\left[(\ddot{\boldsymbol{u}}^1 - \boldsymbol{F}^1).\boldsymbol{u}^2 - (\ddot{\boldsymbol{u}}^1 - \boldsymbol{F}^1).\boldsymbol{u}^1\right] \,\mathrm{d}V$$

This relation still holds if the two elastodynamic states are considered at two different instants. Following the approach of Section 7.1, the states 1 and 2 are considered at instants $t_1 = \tau$ and $t_2 = t - \tau$, where $0 \leq \tau \leq t$. The above equation is then integrated over $0 \leq \tau \leq t$, which leads to a reciprocity identity applicable for any pair of elastodynamic states and any time instant $t \geq 0$ (Eringen and Suhubi [2]):

$$\begin{aligned}
&\int_{\partial\Omega} \left[(\boldsymbol{T}^n(\boldsymbol{u}^1) \star \boldsymbol{u}^2)(\boldsymbol{y},t) - (\boldsymbol{T}^n(\boldsymbol{u}^2) \star \boldsymbol{u}^1)(\boldsymbol{y},t)\right] \,\mathrm{d}S_y \\
&= \int_{\Omega} \rho\left[(\boldsymbol{F}^2 \star \boldsymbol{u}^1)(\boldsymbol{y},t) - (\boldsymbol{F}^1 \star \boldsymbol{u}^2)(\boldsymbol{y},t)\right] \,\mathrm{d}V_y \\
&\quad + \int_{\Omega} \rho\left[\boldsymbol{v}_0^2(\boldsymbol{y}).\boldsymbol{u}^1(\boldsymbol{y},t) - \boldsymbol{v}_0^1(\boldsymbol{y}).\boldsymbol{u}^2(\boldsymbol{y},t)\right] \,\mathrm{d}V_y \\
&\quad + \int_{\Omega} \rho\left[\boldsymbol{u}_0^2(\boldsymbol{y}).\dot{\boldsymbol{u}}^1(\boldsymbol{y},t) - \boldsymbol{u}_0^1(\boldsymbol{y}).\dot{\boldsymbol{u}}^2(\boldsymbol{y},t)\right] \,\mathrm{d}V_y
\end{aligned} \tag{7.39}$$

where the symbol $\star$ denotes the convolution product in time of two vector fields:

$$(\boldsymbol{a} \star \boldsymbol{b})(t) = \int_0^t \boldsymbol{a}(t-\tau).\boldsymbol{b}(\tau) \,\mathrm{d}\tau = \int_0^t \boldsymbol{a}(\tau).\boldsymbol{b}(t-\tau) \,\mathrm{d}\tau \tag{7.40}$$

7.7 ELASTODYNAMIC FUNDAMENTAL SOLUTIONS

Let $\boldsymbol{x} \in \mathbb{R}^3$ denote a fixed point and consider a domain $E \subseteq \mathbb{R}^3$. By 'elastodynamic fundamental solution', is meant any displacement $\boldsymbol{U}^k[\boldsymbol{x}, t, \boldsymbol{y}|f]$ and stress tensor $\mathbf{\Sigma}^k[\boldsymbol{x},t,\boldsymbol{y}|f]$ generated in E by a point force $\boldsymbol{x}$ applied at $\boldsymbol{x}$, of time-varying intensity $f(t)$ and along the $\boldsymbol{e}_k$ direction, that satisfy the fundamental equation of motion, i.e.:

$$\mathrm{div}\mathbf{\Sigma}^k - \rho\ddot{\boldsymbol{U}}^k + \delta(\boldsymbol{y}-\boldsymbol{x})f(t)\boldsymbol{e}_k = 0 \qquad (7.41)$$

and with vanishing initial conditions. The function $f(t)$ is assumed to be twice continuously differentiable and null for $t < 0$ (which implies the initial conditions $f(0) = \dot{f}(0) = 0$).

Stokes's fundamental solution

It is the elastodynamic response of an infinite medium to a point load: $E = \mathbb{R}^3$. Its expression is established (Eringen and Suhubi [2]) using the decomposition (7.37) into Lamé potentials. Recalling that $1/4\pi r$ (with $r = |\boldsymbol{y}-\boldsymbol{x}|$) is the infinite-space fundamental solution to the Laplace equation, one has:

$$\delta(\boldsymbol{y}-\boldsymbol{x})\boldsymbol{e}_k = -\frac{1}{4\pi}\Delta\frac{1}{r}\boldsymbol{e}_k = \frac{1}{4\pi}(\mathbf{rot\ rot} - \boldsymbol{\nabla}\mathrm{div})\left\{\frac{1}{r}\boldsymbol{e}_k\right\}$$

The point load (7.41) can then be recast in the form (7.38) as:

$$\rho\boldsymbol{F}(\boldsymbol{y},t) = \boldsymbol{\nabla}\mathrm{div}\left\{-\frac{f(t)}{4\pi r}\boldsymbol{e}_k\right\} + \mathbf{rot\ rot}\left\{\frac{f(t)}{4\pi r}\boldsymbol{e}_k\right\} \qquad (7.42)$$

The Stokes's solution is then sought in the form of a Lamé decomposition:

$$\boldsymbol{U}^k[\boldsymbol{x},t,\boldsymbol{y}|f] = \boldsymbol{\nabla}\mathrm{div}(A_L\boldsymbol{e}_k) - \mathbf{rot\ rot}(A_T\boldsymbol{e}_k) \qquad (7.43)$$

where, taking into account (7.37) and (7.42), the scalar functions A_α ($\alpha = L,T$) solve the following non-homogeneous wave equations:

$$c_\alpha^2\Delta A_\alpha - \ddot{A}_\alpha = \frac{f(t)}{4\pi r} \qquad (\alpha = L,T) \qquad (7.44)$$

and are given by:

$$A_\alpha(\boldsymbol{x},t,\boldsymbol{y}) = -\frac{r}{4\pi\rho c_\alpha^2}\left[\int_0^1 af(t-ar/c_\alpha)da + \int_1^{+\infty} f(t-ar/c_\alpha)da\right]$$

This result, upon substitution into the decomposition (7.43), yields the longitudinal and transverse parts of $\boldsymbol{U}[\boldsymbol{x},t,\boldsymbol{y}|f]$:

$$U_i^{(L)k}[\boldsymbol{x},t,\boldsymbol{y}|f] = \frac{\beta^2}{4\pi\mu r}\left[(\delta_{ik} - 3r_{,i}r_{,k})I_L + r_{,i}r_{,k}f(t-r/c_L)\right] \qquad (7.45)$$

$$U_i^{(T)k}[\boldsymbol{x},t,\boldsymbol{y}|f] = \frac{1}{4\pi\mu r}\left[(3r_{,i}r_{,k} - \delta_{ik})I_T + (\delta_{ik} - r_{,i}r_{,k})f(t-r/c_T)\right] \qquad (7.46)$$

where $I_{L,T}$ and the wave velocity ratio β are given by:

$$I_{L,T} = \int_0^1 af(t - ar/c_{L,T})da \tag{7.47}$$

$$\beta^2 = \frac{c_T^2}{c_L^2} = \frac{1-2\nu}{2(1-\nu)} = \frac{\mu}{\lambda + 2\mu} \tag{7.48}$$

Finally, the following expressions are obtained for the Stokes's fundamental solution:

$$U_i^k[\boldsymbol{x}, t, \boldsymbol{y}|f] = \frac{1}{4\pi\mu r}\left[Br_{,i}r_{,k} + (A + f(t - r/c_T))\delta_{ik}\right] \tag{7.49}$$

$$\Sigma_{ij}^k[\boldsymbol{x}, t, \boldsymbol{y}|f] = \frac{1}{4\pi r^2}\left[2Cr_{,i}r_{,k}r_{,j} + \delta_{ij}r_{,k}F_L + (\delta_{ik}r_{,j} + \delta_{jk}r_{,i})F_T\right] \tag{7.50}$$

where:

$$\begin{aligned}
A &= \int_1^\beta af(t - ar/c_T)da \\
B &= -3A - \Big(f(t - r/c_T) - \beta^2 f(t - r/c_L)\Big) \\
C &= -5B + \Big(f(t - r/c_T) + c_T \dot{f}(t - r/c_T)\Big) \\
&\qquad - \beta^2\Big(f(t - r/c_L) + c_L \dot{f}(t - r/c_L)\Big) \\
F_T &= 2B - \Big(f(t - r/c_T) + c_T \dot{f}(t - r/c_T)\Big) \\
F_L &= 2B - (1 - 2\beta^2)\Big(f(t - r/c_L) + c_L \dot{f}(t - r/c_L)\Big)
\end{aligned} \tag{7.51}$$

The Stokes solution depends on the time variable through the *retarded times* $t - r/c$, $c_T \le c \le c_L$: this is due to causality, i.e. the fact that waves propagate at finite velocities in the medium. In addition, the following symmetry properties hold:

$$\begin{aligned}
U_i^k[\boldsymbol{x}, t, \boldsymbol{y}|f] = U_k^i[\boldsymbol{y}, t, \boldsymbol{x}|f] = U_k^i[\boldsymbol{x}, t, \boldsymbol{y}|f] \\
U_{i,\bar{j}}^k[\boldsymbol{x}, t, \boldsymbol{y}|f] = -U_{i,j}^k[\boldsymbol{x}, t, \boldsymbol{y}|f]
\end{aligned} \tag{7.52}$$

The Stokes solution (7.49), (7.50) has some degree of similarity to the elastostatic Kelvin solution (4.8), (4.9); both solutions coincide for a static point load, i.e. $f(t) = 1$, a for sufficiently large time t.

Singular part of the Stokes solution

In view of eqs (7.49), (7.50), the Stokes solution is singular at $r = 0$:

$$U_i^k[\boldsymbol{x}, t, \boldsymbol{y}|f] = O(1/r) \qquad \Sigma_{ij}^k[\boldsymbol{x}, t, \boldsymbol{y}|f] = O(1/r^2) \tag{7.53}$$

In fact, a Taylor expansion in r about $r = 0$ of expressions (7.51) yields:

$$\begin{aligned} A + f(t - r/c_T) &= \frac{1+\beta^2}{2} f(t) - \frac{2+\beta^3}{3} \dot{f}(t) x_T + \frac{3+\beta^4}{8} \ddot{f}(t) x_T^2 + o(r^2) \\ B &= \frac{1-\beta^2}{2} f(t) - \frac{1-\beta^4}{8} \ddot{f}(t) x_T^2 + o(r^2) \\ C &= -\frac{3(1-\beta^2)}{2} f(t) + \frac{1-\beta^4}{8} \ddot{f}(t) x_T^2 + o(r^2) \\ F_T &= -\beta^2 f(t) + \frac{1+\beta^4}{4} \ddot{f}(t) x_T^2 + o(r^2) \\ F_L &= \beta^2 f(t) - \frac{1-2\beta^2+3\beta^4}{4} \ddot{f}(t) x_T^2 + o(r^2) \end{aligned} \tag{7.54}$$

where $x_T = r/c_T$. Then, substituting these expansions into eqs (7.49), (7.50) and recalling the definition (7.48) of β, one readily obtains:

$$\begin{aligned} U_i^k[\boldsymbol{x}, t, \boldsymbol{y}|f] &= f(t) U_i^k(\boldsymbol{x}, \boldsymbol{y}) + O(1) \\ \Sigma_{ij}^k[\boldsymbol{x}, t, \boldsymbol{y}|f] &= f(t) \Sigma_{ij}^k(\boldsymbol{x}, \boldsymbol{y}) + O(1) \end{aligned} \tag{7.55}$$

where (U_i^k, Σ_{ij}^k) are the components of the Kelvin elastostatic solution. In other words, the singular behaviour of the Stokes solution is identical to that of the Kelvin solution times $f(t)$. This property can also be shown without recourse to the Taylor expansions (7.54), by putting $U_i^k[f] = f(t) U_i^k + \bar{U}_i^k[f]$ in the Navier equation (7.36) with the point load (7.41). Taking into account the governing field equation for the Kelvin solution, one obtains:

$$\bar{\Sigma}_{ij,j}^k - \rho \ddot{\bar{U}}_i^k = \ddot{f} U_i^k$$

The right-hand side of the above equation is less singular than the point load (7.41), hence $\bar{U}_i^k[f]$ itself is less singular than $U_i^k[f]$. From expansions (7.54), $\bar{U}_i^k[f]$ is actually nonsingular.

Other fundamental solutions

Like in elastostatics, infinitely many fundamental solutions can be defined; they are characterized by the geometrical support E, the time modulation f of the point load and boundary conditions on ∂E. For instance, the two-dimensional Lamb solution (E: isotropic homogeneous half-plane with a point load applied on the otherwise stress-free boundary, see Eringen and Suhubi [2]) or the three-dimensional Kennett solution [4] (E: homogeneous or layered half-space with a stress-free boundary). Also, a fundamental solution for 2D and 3D anisotropic infinite media is available (Wang and Achenbach [7]).

All fundamental solutions exhibit at $r = |\boldsymbol{y} - \boldsymbol{x}| = 0$ a singular behaviour of the type (7.53) and identical to that of the corresponding static fundamental solution (same domain E and boundary conditions) and possess the symmetry property:

$$
\begin{aligned}
C_{ijab}U^{a}_{k,\bar{b}}[\boldsymbol{x},t,\boldsymbol{y}|f] &= \Sigma^{k}_{ij}[\boldsymbol{y},t,\boldsymbol{x}|f] \\
C_{ijpq}\Sigma^{p}_{kl,\bar{q}}[\boldsymbol{x},t,\boldsymbol{y}|f] &= C_{klab}\Sigma^{a}_{ij,b}[\boldsymbol{y},t,\boldsymbol{x}|f]
\end{aligned} \tag{7.56}
$$

Time-impulsive fundamental solution

It corresponds to the special case of a *time-impulsive* load applied at the point $\boldsymbol{x}$ and the time instant $t = 0$:

$$
f(t) = \delta(t)
$$

The time-impulsive fundamental solution, denoted:

$$
\begin{aligned}
U^{k}_{i}[\boldsymbol{x},t,\boldsymbol{y}|\delta] &\equiv U^{k}_{i}(\boldsymbol{x},t,\boldsymbol{y}) \\
\Sigma^{k}_{ij}[\boldsymbol{x},t,\boldsymbol{y}|\delta] &\equiv \Sigma^{k}_{ij}(\boldsymbol{x},t,\boldsymbol{y})
\end{aligned}
$$

satisfies at any $\boldsymbol{y} \neq \boldsymbol{x}$ the initial conditions $\boldsymbol{U}(\boldsymbol{x},0,\boldsymbol{y}) = \dot{\boldsymbol{U}}(\boldsymbol{x},0,\boldsymbol{y}) = 0$. One can easily show that:

$$
U^{k}_{i}[\boldsymbol{x},t,\boldsymbol{y}|f] = U^{k}_{i}(\boldsymbol{x},t,\boldsymbol{y}) \star f \tag{7.57}
$$

$$
\Sigma^{k}_{ij}[\boldsymbol{x},t,\boldsymbol{y}|f] = \Sigma^{k}_{ij}(\boldsymbol{x},t,\boldsymbol{y}) \star f \tag{7.58}
$$

In particular, the time-impulsive Stokes solution is given by eqs (7.49), (7.50) with $f(t) = \delta(t)$.

7.8 INTEGRAL REPRESENTATION OF AN ELASTODYNAMIC STATE

Consider a bounded elastic body Ω and let $\boldsymbol{x}$ be an arbitrarily chosen point *not on* $\partial\Omega$. The reciprocity identity (7.39) applied to the unknown elastodynamic state (state 1) and to the time-impulsive fundamental solution (state 2) yields:

$$
\begin{aligned}
u_k(\boldsymbol{x},t) &+ \int_{\partial\Omega} \left\{ T^{k}_{i}(\boldsymbol{x},t,\boldsymbol{y}) \star u_i(\boldsymbol{y},t) - U^{k}_{i}(\boldsymbol{x},t,\boldsymbol{y}) \star t_i(\boldsymbol{y},t) \right\} \,\mathrm{d}S_y \\
&= \int_{\Omega} \rho U^{k}_{i}(\boldsymbol{x},t,\boldsymbol{y}) \star F_i(\boldsymbol{y},t) \,\mathrm{d}V_y \\
&\quad + \int_{\Omega} \rho \left\{ v_{0i}(\boldsymbol{y}) U^{k}_{i}(\boldsymbol{x},t,\boldsymbol{y}) + u_{0i}(\boldsymbol{y}) \dot{U}^{k}_{i}(\boldsymbol{x},t,\boldsymbol{y}) \right\} \,\mathrm{d}V_y
\end{aligned}
$$

where $T^{k}_{i} = \Sigma^{k}_{ij} n_j$ are the components of the traction vector associated with the fundamental solution. The previous equation, rearranged using equations

(7.57)–(7.58), provides the integral representation formula for the displacement $\boldsymbol{u}(\boldsymbol{x},t)$:

$$\begin{aligned}\kappa u_k(\boldsymbol{x},t) = &\int_{\partial\Omega} \{U_i^k[\boldsymbol{x},t,\boldsymbol{y}|t_i(\boldsymbol{y},t)] - T_i^k[\boldsymbol{x},t,\boldsymbol{y}|u_i(\boldsymbol{y},t)]\}\,\mathrm{d}S_y \\ &+ \int_{\Omega} \rho U_i^k[\boldsymbol{x},t,\boldsymbol{y}|F_i(\boldsymbol{y},t)]\,\mathrm{d}V_y \\ &+ \int_{\Omega} \rho\left\{v_{0i}(\boldsymbol{y})U_i^k(\boldsymbol{x},t,\boldsymbol{y}) + u_{0i}(\boldsymbol{y})\dot{U}_i^k(\boldsymbol{x},t,\boldsymbol{y})\right\}\mathrm{d}V_y \qquad (7.59)\end{aligned}$$

where $\kappa = 1$ $(\boldsymbol{x} \in \Omega)$, $\kappa = 0$ $(\boldsymbol{x} \notin \Omega)$. This formula expresses $\boldsymbol{u}(\boldsymbol{x},t)$ as a function of the boundary traces of the elastodynamic variables and of the prescribed distributions of body forces and initial conditions. Since the integrals are indefinitely differentiable with respect to $\boldsymbol{x}$ for $\boldsymbol{x} \notin \partial\Omega$, the integral representation of the elastodynamic stress tensor stems from a direct application of Hooke's law $\sigma_{ij} = C_{ijk\ell}u_{k,\ell}$:

$$\begin{aligned}\kappa\sigma_{ij}(\boldsymbol{x},t) = &\int_{\partial\Omega} \{\Sigma_{ij}^k[\boldsymbol{y},t,\boldsymbol{x}|t_k(\boldsymbol{y},t)] - C_{abk\ell}\Sigma_{ij,\ell}^k[\boldsymbol{y},t,\boldsymbol{x}|u_a(\boldsymbol{y},t)]n_b(\boldsymbol{y})\}\,\mathrm{d}S_y \\ &+ \int_{\Omega} \rho\Sigma_{ij}^k[\boldsymbol{y},t,\boldsymbol{x}|F_k(\boldsymbol{y},t)]\,\mathrm{d}V_y \\ &+ \int_{\Omega} \rho\left\{v_{0k}(\boldsymbol{y})\Sigma_{ij}^k(\boldsymbol{y},t,\boldsymbol{x}) + u_{0k}(\boldsymbol{y})\dot{\Sigma}_{ij}^k(\boldsymbol{y},t,\boldsymbol{x})\right]\mathrm{d}V_y \qquad (7.60)\end{aligned}$$

in which the property (7.56) has been used. The formulas (7.59), (7.60) are applicable for any time instant $t \geq 0$ and any point $\boldsymbol{x} \notin \partial\Omega$. Any fundamental solution compatible with the domain Ω may be used. In practice, the most frequent choice is the Stokes fundamental solution. Taking into account the possibility to change $\boldsymbol{x}$-derivatives into $\boldsymbol{y}$-derivatives, eq. (7.52), the integral representation formula for the stress tensor becomes:

$$\begin{aligned}\kappa\sigma_{ij}(\boldsymbol{x},t) = C_{abk\ell} &\int_{\partial\Omega} \left\{\Sigma_{ij,\ell}^k[\boldsymbol{x},t,\boldsymbol{y}|u_a(\boldsymbol{y},t)]n_b(\boldsymbol{y}) - \Sigma_{ij}^k[\boldsymbol{x},t,\boldsymbol{y}|t_k(\boldsymbol{y},t)]\right\}\mathrm{d}S_y \\ &- \int_{\Omega} \rho\Sigma_{ij}^k[\boldsymbol{x},t,\boldsymbol{y}|F_k(\boldsymbol{y},t)]\,\mathrm{d}V_y \\ &- \int_{\Omega} \rho\left\{v_{0k}(\boldsymbol{y})\Sigma_{ij}^k(\boldsymbol{x},t,\boldsymbol{y}) + u_{0k}(\boldsymbol{y})\dot{\Sigma}_{ij}^k(\boldsymbol{x},t,\boldsymbol{y})\right\}\mathrm{d}V_y\end{aligned}$$

Then an integration by parts, using identity (A.8), gives:

$$\begin{aligned}&\int_{\partial\Omega} \Sigma_{ab,\ell}^k[\boldsymbol{x},t,\boldsymbol{y}|u_a(\boldsymbol{y},t)]n_b(\boldsymbol{y})\,\mathrm{d}S_y \\ &= \int_{\partial\Omega} D_{b\ell}\Sigma_{ab,\ell}^k[\boldsymbol{x},t,\boldsymbol{y}|u_a(\boldsymbol{y},t)]\,\mathrm{d}S_y + \int_{\partial\Omega} \Sigma_{ab,b}^k[\boldsymbol{x},t,\boldsymbol{y}|u_a(\boldsymbol{y},t)]n_\ell(\boldsymbol{y})\,\mathrm{d}S_y \\ &= \int_{\partial\Omega} \{\Sigma_{ab}^k[\boldsymbol{x},t,\boldsymbol{y}|D_{\ell b}u_a(\boldsymbol{y},t)] + \rho n_\ell(\boldsymbol{y})U_a^k[\boldsymbol{x},t,\boldsymbol{y}|\ddot{u}_a(\boldsymbol{y},t)]\}\,\mathrm{d}S_y\end{aligned}$$

where the field equation (7.41) has been invoked. One then obtains the representation formula of $\boldsymbol{\sigma}(\boldsymbol{x},t)$ in an integrated-by-parts form, which is the elastodynamic counterpart of eq. (4.25):

$$\begin{aligned}
\kappa\sigma_{ij}(\boldsymbol{x},t) = & \int_{\partial\Omega} \{C_{ijk\ell}\Sigma^k_{ab}[\boldsymbol{x},t,\boldsymbol{y}|D_{\ell b}u_a(\boldsymbol{y},t)] - \Sigma^k_{ij}[\boldsymbol{x},t,\boldsymbol{y}|t_k(\boldsymbol{y},t)]\}\,\mathrm{d}S_y \\
& + \rho C_{ijk\ell}\int_{\partial\Omega} U^k_a[\boldsymbol{x},t,\boldsymbol{y}|\ddot{u}_a(\boldsymbol{y},t)]n_\ell(\boldsymbol{y})\,\mathrm{d}S_y \\
& - \int_\Omega \rho\Sigma^k_{ij}[\boldsymbol{x},t,\boldsymbol{y}|F_k(\boldsymbol{y},t)]\,\mathrm{d}V_y \\
& - \int_\Omega \rho\left\{v_{0k}(\boldsymbol{y})\Sigma^k_{ij}(\boldsymbol{x},t,\boldsymbol{y}) + u_{0k}(\boldsymbol{y})\dot{\Sigma}^k_{ij}(\boldsymbol{x},t,\boldsymbol{y})\right\}\mathrm{d}V_y \qquad (7.61)
\end{aligned}$$

7.9 BOUNDARY INTEGRAL EQUATIONS IN ELASTODYNAMICS

Because of the singular behaviour of the fundamental solution at $\boldsymbol{y}=\boldsymbol{x}$, the representation formulas cannot be directly taken for $\boldsymbol{x}\in\partial\Omega$, and the boundary integral equations are established through the now usual limiting process, following the same lines as in the elastostatic case (Section 4.4). The equality of the singularities of the dynamic and static fundamental solutions at $\boldsymbol{y}=\boldsymbol{x}$ plays an essential role in the limiting process to be performed, and allows a straightforward extension to the elastodynamic case of the elastostatic regularization presented in Chapter 4. For simplicity, zero body forces and initial rest are assumed (the domain integrals which otherwise arise are unaffected by the regularization procedure and are valid for $\boldsymbol{x}\in\partial\Omega$). This approach first yields a result similar to the boundary integral equation (7.16) established for scalar wave problems: any elastodynamic state on Ω with zero body forces and initial conditions verifies the regularized displacement integral equation:

$$\begin{aligned}
& \int_{\partial\Omega} \left[T^k_i[\boldsymbol{x},t,\boldsymbol{y}|u_i(\boldsymbol{y},t)] - u_i(\boldsymbol{x},t)T^k_i(\boldsymbol{x},\boldsymbol{y})\right]\,\mathrm{d}S_y \\
& \qquad - \int_{\partial\Omega} U^k_i[\boldsymbol{x},t,\boldsymbol{y}|t_i(\boldsymbol{y},t)]\,\mathrm{d}S_y = 0 \qquad (7.62)
\end{aligned}$$

Regularized gradient integral equation

Assume that the Stokes solution is being used, so that the integration by parts followed by first-order regularization scheme can be followed. The gradient integral equation is thus sought as the limiting form when $\varepsilon\to 0$ of:

$$\begin{aligned}
& \int_{(\partial\Omega-e_\varepsilon)+s_\varepsilon} \{\Sigma^k_{ab}[\boldsymbol{x},t,\boldsymbol{y}|D_{\ell b}u_a(\boldsymbol{y},t)] + \rho n_\ell(\boldsymbol{y})U^k_a[\boldsymbol{x},t,\boldsymbol{y}|\ddot{u}_a(\boldsymbol{y},t)]\}\,\mathrm{d}S_y \\
& \qquad - \int_{(\partial\Omega-e_\varepsilon)+s_\varepsilon} U^k_{a,\ell}[\boldsymbol{x},t,\boldsymbol{y}|t_a(\boldsymbol{y},t)]\,\mathrm{d}S_y = 0 \qquad (7.63)
\end{aligned}$$

The static Kelvin solution is then introduced into this identity so that all singularities are transferred into elastostatic kernels:

$$
\begin{aligned}
0 = &\int_{(\partial\Omega - e_\varepsilon)+s_\varepsilon} \left\{ \Sigma^k_{ab}[\boldsymbol{x},t,\boldsymbol{y}|D_{\ell b}u_a(\boldsymbol{y},t)] - D_{\ell b}u_a(\boldsymbol{y},t)\Sigma^k_{ab}(\boldsymbol{x},\boldsymbol{y}) \right\} \mathrm{d}S_y \\
&+ \rho \int_{(\partial\Omega - e_\varepsilon)+s_\varepsilon} n_\ell(\boldsymbol{y}) U^k_a[\boldsymbol{x},t,\boldsymbol{y}|\ddot{u}_a(\boldsymbol{y},t)] \,\mathrm{d}S_y \\
&- \int_{(\partial\Omega - e_\varepsilon)+s_\varepsilon} \left[U^k_{a,\ell}[\boldsymbol{x},t,\boldsymbol{y}|t_a(\boldsymbol{y},t)] - t_a(\boldsymbol{y},t)U^k_{a,\ell}(\boldsymbol{x},\boldsymbol{y}) \right] \mathrm{d}S_y \\
&+ \int_{(\partial\Omega - e_\varepsilon)+s_\varepsilon} \left\{ D_{\ell b}u_a(\boldsymbol{y},t)\Sigma^k_{ab}(\boldsymbol{x},\boldsymbol{y}) - t_a(\boldsymbol{y},t)U^k_{a,\ell}(\boldsymbol{x},\boldsymbol{y}) \right\} \mathrm{d}S_y
\end{aligned}
$$

The first three integrals, in the limit $\varepsilon \to 0$, are weakly singular improper integrals over $\partial\Omega$: the first and third ones by virtue of the equality of the singularities of the dynamic and static fundamental solutions, eq. (7.55); the second one because the singularity of $U^k_a[t_a(\boldsymbol{y},t)]$ is integrable over $\partial\Omega$. Finally, invoking again the regularization identity (5.33) established in the elastostatic case, one has:

$$
\begin{aligned}
\lim_{\varepsilon \to 0} &\int_{(\partial\Omega - e_\varepsilon)+s_\varepsilon} \left\{ D_{\ell b}u_a(\boldsymbol{y},t)\Sigma^k_{ab}(\boldsymbol{x},\boldsymbol{y}) - t_a(\boldsymbol{y},t)U^k_{a,\ell}(\boldsymbol{x},\boldsymbol{y}) \right\} \mathrm{d}S_y \\
&= -\frac{1}{2}u_{k,\ell}(\boldsymbol{x},t) + D_{\ell b}u_a(\boldsymbol{x},t)A^k_{ab}(\boldsymbol{x},\partial\Omega) - t_a(\boldsymbol{x},t)B^k_{a\ell}(\boldsymbol{x},\partial\Omega) \\
&+ \int_{\partial\Omega} \left[D_{\ell b}u_a(\boldsymbol{y},t) - D_{\ell b}u_a(\boldsymbol{x},t) \right] \Sigma^k_{ab}(\boldsymbol{x},\boldsymbol{y}) \,\mathrm{d}S_y \\
&- \int_{\partial\Omega} \left[t_a(\boldsymbol{y},t) - t_a(\boldsymbol{x},t) \right] U^k_{a,\ell}(\boldsymbol{x},\boldsymbol{y}) \,\mathrm{d}S_y
\end{aligned}
$$

where the terms $A^k_{ab}(\boldsymbol{x},\partial\Omega)$, $B^k_{a\ell}(\boldsymbol{x},\partial\Omega)$ are given by eqs (5.29), (5.30). This analysis finally results in the integral representation formula for the displacement gradient at any point $\boldsymbol{x}$ of the boundary $\partial\Omega$ at which u and the unit normal $\boldsymbol{n}$ have $C^{1,\alpha}$ and $C^{0,\alpha}$ regularity, respectively:

$$
\begin{aligned}
\frac{1}{2}u_{k,\ell}(\boldsymbol{x},t) = &D_{\ell b}u_a(\boldsymbol{x},t)A^k_{ab}(\boldsymbol{x},\partial\Omega) - t_a(\boldsymbol{x},t)B^k_{a\ell}(\boldsymbol{x},\partial\Omega) \\
&+ \int_{\partial\Omega} \left[\Sigma^k_{ab}[\boldsymbol{x},t,\boldsymbol{y}|D_{\ell b}u_a(\boldsymbol{y},t)] - D_{\ell b}u_a(\boldsymbol{x},t)\Sigma^k_{ab}(\boldsymbol{x},\boldsymbol{y}) \right] \mathrm{d}S_y \\
&- \int_{\partial\Omega} \left[U^k_{a,\ell}[\boldsymbol{x},t,\boldsymbol{y}|t_a(\boldsymbol{y},t)] - t_a(\boldsymbol{x},t)U^k_{a,\ell}(\boldsymbol{x},\boldsymbol{y}) \right] \mathrm{d}S_y \\
&+ \rho \int_{\partial\Omega} U^k_a[\boldsymbol{x},t,\boldsymbol{y}|\ddot{u}_a(\boldsymbol{y},t)] n_\ell(\boldsymbol{y}) \,\mathrm{d}S_y
\end{aligned}
\tag{7.64}
$$

Comments

The transfer of the dynamical singularity into the static fundamental solution, eq. (7.55), plays like in scalar wave problems an important role, allowing a straightforward extension to elastodynamics of the regularization techniques developed in elastostatics.

The displacement integral equation (7.62) is valid for any fundamental solution, whereas the gradient integral equation (7.64), in this form, assumes that the Stokes solution is being used. The smoothness assumptions (in space) on $\boldsymbol{u}, \boldsymbol{t}$ which guarantee that the integrals in eqs (7.62), (7.64) are weakly singular are the same as in elastostatics.

7.10 EXTERIOR PROBLEMS AND CONDITIONS AT INFINITY

Behaviour at infinity of the Stokes solution

The asymptotic behaviour for $r = |\boldsymbol{y} - \boldsymbol{x}| \to \infty$ of the Stokes displacement solution is:

$$\boldsymbol{U}^{(L)k}.\boldsymbol{n} = \frac{\beta^2}{4\pi\mu r} f(t - r/c_L) n_k + o(r^{-2}) \qquad \boldsymbol{U}^{(L)k}.\boldsymbol{\tau} = o(r^{-2}) \tag{7.65}$$

$$\boldsymbol{U}^{(T)k}.\boldsymbol{\tau} = \frac{1}{4\pi\mu r} f(t - r/c_T) \tau_k + o(r^{-2}) \qquad \boldsymbol{U}^{(T)k}.\boldsymbol{n} = o(r^{-2}) \tag{7.66}$$

where $\boldsymbol{n}$ defines the radial direction with respect to origin $\boldsymbol{x}$:

$$\boldsymbol{n} = \boldsymbol{e}_r = (\boldsymbol{y} - \boldsymbol{x}) / |\boldsymbol{y} - \boldsymbol{x}| = r_{,i} \boldsymbol{e}_i$$

and $\boldsymbol{\tau}$ is any unit vector orthogonal to $\boldsymbol{n}$. In other words, far from the load point $\boldsymbol{x}$, the leading terms of the longitudinal and transverse parts of $\boldsymbol{U}^k[\boldsymbol{x}, t, \boldsymbol{y}|f]$ are spherical compressive and shear waves, respectively. These properties are established by using expressions (7.45), (7.46) and noticing that, for $r > ct$:

$$\int_0^1 af(t - ar/c)da = \int_0^{ct/r} af(t - ar/c)da = \frac{c^2t^2}{r^2} \int_0^1 bf((1-b)t)db$$

(a consequence of the assumption $f(t) = 0$ for $t \leq 0$), which implies (assuming f to be bounded):

$$\int_0^1 af(t - ar/c)da = o(r^{-1}) \qquad (r \to +\infty)$$

Moreover, the Stokes displacement and traction vectors satisfy the *radiation condition*:

$$\boldsymbol{T}^n(\boldsymbol{U}^{(\alpha)k}) + \rho c_\alpha \dot{\boldsymbol{U}}^{(\alpha)k} = O(r^{-2}) \qquad (\alpha = L, T) \tag{7.67}$$

Elastodynamic radiation conditions

To extend the validity of the integral representation (7.59) to unbounded domains of the form $\bar{\Omega} = \mathbb{R}^3 - \Omega$ (with Ω bounded), one must introduce certain restrictions on the behaviour at infinity of the solution $\boldsymbol{u}, \boldsymbol{t}$, following considerations similar to those invoked for scalar wave problems. Accordingly, the limiting process $R \to +\infty$ is again considered for the *finite* domain Ω_R bounded by $\partial\bar{\Omega}$ from within and the sphere S_R (centre $\boldsymbol{x}$, (large) radius R) from without. The displacement integral equation (7.62), applied to the domain Ω_R with $\boldsymbol{x}$ interior to S_R, reads:

$$\begin{aligned} u_k(\boldsymbol{x},t) &+ \int_{\partial\bar{\Omega}} \left[T_i^k[\boldsymbol{x},t,\boldsymbol{y}|u_i(\boldsymbol{y},t)] - T_i^k(\boldsymbol{x},\boldsymbol{y})u_i(\boldsymbol{y},t)\right] \, \mathrm{d}S_y \\ &+ \int_{\partial\bar{\Omega}} \{[u_i(\boldsymbol{y},t) - u_i(\boldsymbol{x},t)]T_i^k(\boldsymbol{x},\boldsymbol{y}) - U_i^k[\boldsymbol{x},t,\boldsymbol{y}|t_i(\boldsymbol{y},t)]\} \, \mathrm{d}S_y \\ &+ \int_{S_R} \{T_i^k[\boldsymbol{x},t,\boldsymbol{y}|u_i(\boldsymbol{y},t)] - U_i^k[\boldsymbol{x},t,\boldsymbol{y}|t_i(\boldsymbol{y},t)]\} \, \mathrm{d}S_y = 0 \end{aligned} \tag{7.68}$$

where use has been made of the identity (4.7) for the static fundamental solution, in the form:

$$\int_{S_R} T_i^k(\boldsymbol{x},\boldsymbol{y}) \, \mathrm{d}S_y = -\delta_{ik}$$

the unit normal $\boldsymbol{n}$ being exterior to Ω_R, hence in particular interior to Ω. At this point, sufficient conditions on $\boldsymbol{u}, \boldsymbol{t}$ ensuring that:

$$\lim_{R\to\infty} I_k = 0 \qquad I_k = \int_{S_R} \{U_i^k[\boldsymbol{x},t,\boldsymbol{y}|t_i(\boldsymbol{y},t)] - T_i^k[\boldsymbol{x},t,\boldsymbol{y}|u_i(\boldsymbol{y},t)]\} \, \mathrm{d}S_y$$

are sought. The asymptotic formulas (7.65), (7.67) and the properties:

$$U_i^k[f+g] = U_i^k[f] + U_i^k[g] \qquad \dot{U}_i^k[f] = U_i^k[\dot{f}]$$

verified by the Stokes solution allow to recast I_k as:

$$I_k = \int_{S_R} \left\{U_i^{(L)k}[t_i + \rho c_L \dot{u}_i] + U_i^{(T)k}[t_i + \rho c_T \dot{u}_i]\right\} R^2 \, \mathrm{d}\omega + o(R^{-1})$$

(having put $\mathrm{d}S_y = R^2 \, \mathrm{d}\omega$, where ω is the solid angle of origin $\boldsymbol{x}$) provided that:

$$\boldsymbol{u} = o(1) \qquad (R \to +\infty)$$

Then I_k can be rearranged to obtain:

$$\begin{aligned} I_k = \int_{S_R} &\left[U_i^{(L)k} + U_i^{(T)k}\right] \{(t_i^L + \rho c_L \dot{u}_i^L) + (t_i^T + \rho c_T \dot{u}_i^T)\} R^2 \, \mathrm{d}\omega \\ &+ \rho(c_L - c_T) \int_{S_R} \left\{U_i^{(L)k}[\dot{u}_i^T] - U_i^{(T)k}[\dot{u}_i^L]\right\} R^2 \, \mathrm{d}\omega \end{aligned}$$

Making use of the fact that $\boldsymbol{U}^k = O(1/R)$ and of the relations (7.65) and (7.67), one concludes that if $\boldsymbol{u}, \boldsymbol{t}$ satisfy the decay conditions (also called 'regularity conditions'):

$$\dot{\boldsymbol{u}}^L.\boldsymbol{n} = o(1) \qquad \dot{\boldsymbol{u}}^L.\boldsymbol{\tau} = o(R^{-1}) \tag{7.69}$$

$$\dot{\boldsymbol{u}}^T.\boldsymbol{n} = o(R^{-1}) \qquad \dot{\boldsymbol{u}}^T.\boldsymbol{\tau} = o(1) \tag{7.70}$$

and radiation conditions:

$$\boldsymbol{t}^L + \rho c_L \dot{\boldsymbol{u}}^L = o(R^{-1}) \qquad \boldsymbol{t}^T + \rho c_T \dot{\boldsymbol{u}}^T = o(R^{-1}) \tag{7.71}$$

the integral I_k vanishes in the limit $R \to \infty$. These conditions are known to ensure the uniqueness of the solution to an exterior problem (Eringen and Suhubi [2]). The condition (7.71) is the elastodynamic counterpart of the Sommerfeld condition (7.23) for scalar wave problems, and expresses the fact that the energy flux at infinity is outgoing; in other words, there is no source at infinity.

Displacement integral equation for an unbounded domain

The foregoing analysis shows that any elastodynamic state satisfying conditions (7.69), (7.70), (7.71) solves the integral equation

$$u_k(\boldsymbol{x},t) + \int_{\partial\Omega} \left[T_i^k[\boldsymbol{x},t,\boldsymbol{y}|u_i(\boldsymbol{y},t)] - u_i(\boldsymbol{x},t)T_i^k(\boldsymbol{x},\boldsymbol{y})\right] \mathrm{d}S_y$$

$$- \int_{\partial\Omega} U_i^k[\boldsymbol{x},t,\boldsymbol{y}|t_i(\boldsymbol{y},t)]\,\mathrm{d}S_y = 0 \tag{7.72}$$

which differs from eq. (7.62) only by the presence of a free term $u_k(\boldsymbol{x},t)$ and by the orientation of the normal $\boldsymbol{n}$, exterior to the domain $\bar{\Omega}$ and hence interior to $\partial\Omega$.

The limiting process $R \to \infty$ and the conditions at infinity (7.69), (7.70), (7.71) still hold in the presence of nonzero body forces $\boldsymbol{F}$ and initial conditions $\boldsymbol{u}_0, \boldsymbol{v}_0$ provided they satisfy:

$$\int_{\bar{\Omega}} \rho U_i^k[\boldsymbol{x},t,\boldsymbol{y}|F_i(\boldsymbol{y},t)]\,\mathrm{d}V_y$$

$$+ \int_{\bar{\Omega}} \rho \left\{ v_{0i}(\boldsymbol{y})U_i^k(\boldsymbol{x},t,\boldsymbol{y}) + u_{0i}(\boldsymbol{y})\dot{U}_i^k(\boldsymbol{x},t,\boldsymbol{y}) \right\} \mathrm{d}V_y < \infty \tag{7.73}$$

In that case, the right-hand side of the integral equation (7.72) is equal to the above limit (i.e. the integral over $\bar{\Omega}$, supposedly convergent) instead of zero.

Integral representation formulas

Equation (7.72) considered for $\boldsymbol{x} \notin \partial\Omega$ gives back, using identity (4.7), the representation formula (7.59). The latter is thus applicable to bounded as well as unbounded domains, provided the unit normal is always chosen exterior to the elastic body under study.

Any displacement field of the form (7.72) satisfies the conditions at infinity (7.69) to (7.71); this stems directly from the similar properties, eqs (7.65) to (7.67), of the Stokes solution. On the other hand, not all the elastodynamic states on unbounded domains admit representations of the form (7.59–7.60). In particular, a *plane wave*:

$$\boldsymbol{u}^I(\boldsymbol{y},t) = Ag(t - \boldsymbol{y}.\boldsymbol{p}/c)\boldsymbol{d} \tag{7.74}$$

where $\boldsymbol{d}, \boldsymbol{p}$ are two unit vectors defining the displacement and propagation directions, respectively, and such that either $\boldsymbol{d}.\boldsymbol{p} = 1$, $c = c_L$ (compressive wave) or $\boldsymbol{d}.\boldsymbol{p} = 0$, $c = c_T$ (shear wave), is a solution to the Navier equation (7.36) which does not satisfy conditions (7.69) to (7.71). In that case, or when condition (7.73) is violated, one should use the superposition principle.

For example, considering the scattering of a bounded stress-free obstacle (a cavity) by an incident wave $\boldsymbol{u}^I$ (e.g. a plane wave) which solves the homogeneous Navier equation (7.36), an integral representation for the total field $\boldsymbol{u} = \boldsymbol{u}^I + \boldsymbol{u}^S$ can be established in exactly the same way as in Section 7.4 for scalar wave problems and reads:

$$\kappa u_k(\boldsymbol{x},t) - u_k^I(\boldsymbol{x},t) = -\int_{\partial\Omega} T_i^k[\boldsymbol{x},t,\boldsymbol{y}|u_i(\boldsymbol{y},t)]\,\mathrm{d}S_y \tag{7.75}$$

REFERENCES

1. DAUBECHIES, I. *Ten lectures on wavelets.* SIAM Publications, Philadelphia, USA (1992).
2. ERINGEN, A. C., SUHUBI, E. S. *Elastodynamics (vol. II – linear theory).* Academic Press (1975).
3. KARABALIS, D. L. A simplified 3D time-domain BEM for dynamic soil-structure interaction problems. *Engng. Anal. with Bound. Elem.*, **8**, 139–145 (1991).
4. KENNETT, B. L. N. Seismic wave propagation in stratified media. In *Advances in Applied Mechanics*, vol. 21. Academic Press (1981).
5. METIVET, B. *Résolution spectrale des équations de Navier–Stokes par une méthode de sous-domaines courbes.* Ph.D. thesis, Université Paris VI, Paris, France (1987).
6. PEIRCE, A., SIEBRITS, E. Stability analysis and design of time-stepping schemes for general elastodynamic boundary element models. *Int. J. Num. Meth. in Eng.*, **40**, 319–342 (1997).
7. WANG, C. Y., ACHENBACH, J. D., HIROSE, S. Two-dimensional time-domain BEM for scattering of elastic waves in solids of general anisotropy. *Int. J. Solids Struct.*, **33**, 3843–3864 (1996).

8

Scalar and elastic wave problems in the frequency domain

This chapter deals with dynamical problems for which the field variables are time-harmonic with a fixed angular frequency ω, i.e. of the form:

$$u(\boldsymbol{x}, t) = \Re[\hat{u}(\boldsymbol{x})\mathrm{e}^{-\mathrm{i}\omega t}] \tag{8.1}$$

where $\hat{u}(\boldsymbol{x})$ is a complex-valued (scalar or vector) function which encodes amplitude and phase information. Solving for $u(\boldsymbol{x}, t)$ thus reduces to the determination of $\hat{u}(\boldsymbol{x})$. The representation (8.1) may be motivated by any of the following interpretations:

- Forced response to steady-state excitations $F(\boldsymbol{y})\mathrm{e}^{-\mathrm{i}\omega t}$, $\boldsymbol{t}^D(\boldsymbol{y})\mathrm{e}^{-\mathrm{i}\omega t}$ ... assuming that a permanent regime has been reached, i.e. that the influence of initial conditions is negligible;

- Resolution in the Fourier domain of a transient dynamical problem, ω being the transformed variable associated to physical time t:

  $$u(\boldsymbol{y}, t) = \frac{1}{\sqrt{2\pi}} \int_{\mathbb{R}} \hat{u}(\boldsymbol{y}, \omega)\mathrm{e}^{-\mathrm{i}\omega t} \, \mathrm{d}\omega$$

  For each value of ω, the transformed unknown $\hat{u}(\boldsymbol{y}, \omega)$ solves a boundary-value problem with prescribed frequency;

- Resolution in the Laplace domain of a transient dynamical problem, $p = -\mathrm{i}\omega$ being the transformed variable associated to physical time t.

In the sequel, following the traditional convention, the factor $\mathrm{e}^{\mathrm{i}\omega t}$ is systematically omitted and the notation u is used instead of $\hat{u}$.

8.1 HELMHOLTZ EQUATION AND FUNDAMENTAL SOLUTIONS

For a time-harmonic unknown, eq. (8.1), the wave equation (7.1) reduces to:

$$\Delta u + k^2 u + F = 0 \tag{8.2}$$

i.e. the Helmholtz equation, where $F(\boldsymbol{y})$ is the body source distribution and $k = \omega/c$ the wave number. A well-posed problem must include one boundary condition at any point of $\partial\Omega$ as well as conditions at infinity, but initial conditions are meaningless in this context and are no longer needed.

Fundamental solutions

The fundamental solutions $G(\boldsymbol{x},\boldsymbol{y};\omega)$ in the frequency domain correspond to a point source of unit amplitude located at the fixed point $\boldsymbol{x}$, i.e. must solve:

$$\Delta G + k^2 G + \delta(\boldsymbol{y}-\boldsymbol{x}) = 0 \tag{8.3}$$

They can be viewed as transient solutions (7.4) for a specific type of source modulation in time, namely $f(t) = \mathrm{e}^{-\mathrm{i}\omega t}$:

$$G(\boldsymbol{x},\boldsymbol{y};\omega) = G[\boldsymbol{x},t,\boldsymbol{y}|\mathrm{e}^{-\mathrm{i}\omega t}] \tag{8.4}$$

and rejecting the initial conditions back to $t=-\infty$. The transposition of the general properties (singular behaviour (7.53), equality of static and dynamic singularity (7.55), symmetry (7.56) ...) of time-domain fundamental solutions to their frequency-domain counterparts is thus immediate. When $G[f]$ is given by (7.5), the correspondence (8.4) gives the fundamental solution of (8.2) for the three-dimensional infinite space:

$$G(\boldsymbol{x},\boldsymbol{y};\omega) = \frac{\mathrm{e}^{\mathrm{i}kr}}{4\pi r} \qquad G_{,i}(\boldsymbol{x},\boldsymbol{y};\omega) = -\frac{r_{,i}}{4\pi r^2}(1-\mathrm{i}kr)\mathrm{e}^{\mathrm{i}kr} \tag{8.5}$$

The following series expansion is absolutely convergent for any x:

$$\mathrm{e}^{\mathrm{i}x} = \sum_{p\geq 0} \frac{(\mathrm{i}x)^p}{p!} \tag{8.6}$$

and allows us to establish that:

$$G(\boldsymbol{x},\boldsymbol{y};\omega) - G(\boldsymbol{x},\boldsymbol{y}) = \frac{ik}{4\pi}\sum_{p\geq 0}\frac{1}{(p+1)!}(\mathrm{i}kr)^p \tag{8.7}$$

$$G_{,i}(\boldsymbol{x},\boldsymbol{y};\omega) - G_{,i}(\boldsymbol{x},\boldsymbol{y}) = -\frac{k^2}{4\pi}r_{,i}\sum_{p\geq 0}\frac{p+1}{(p+2)!}(\mathrm{i}kr)^p \tag{8.8}$$

where $G(\boldsymbol{x},\boldsymbol{y}) = 1/(4\pi r)$ is the corresponding static fundamental solution. This result leads to a direct proof of the fact that the singular behaviours at

$\boldsymbol{y} = \boldsymbol{x}$ of the dynamic and static fundamental solutions are identical:

$$(r \to 0) \qquad (\forall \omega > 0) \qquad \left\{ \begin{array}{l} G(\boldsymbol{x}, \boldsymbol{y}; \omega) - G(\boldsymbol{x}, \boldsymbol{y}) = O(1) \\ \boldsymbol{\nabla} G(\boldsymbol{x}, \boldsymbol{y}; \omega) - \boldsymbol{\nabla} G(\boldsymbol{x}, \boldsymbol{y}) = O(1) \end{array} \right. \tag{8.9}$$

$$(\omega \to 0) \qquad (\forall r > 0) \qquad \left\{ \begin{array}{l} G(\boldsymbol{x}, \boldsymbol{y}; \omega) - G(\boldsymbol{x}, \boldsymbol{y}) = O(\omega) \\ \boldsymbol{\nabla} G(\boldsymbol{x}, \boldsymbol{y}; \omega) - \boldsymbol{\nabla} G(\boldsymbol{x}, \boldsymbol{y}) = O(\omega^2) \end{array} \right. \tag{8.10}$$

Fundamental solution for the infinite plane

Integrating the three-dimensional solution (8.5) for $-\infty \leq x_3 \leq +\infty$ leads to the fundamental solution for the infinite (y_1, y_2)-plane:

$$G(\tilde{x}, \tilde{y}; \omega) = \frac{\mathrm{i}}{4} H_0^1(kr) \qquad G_{,i}(\tilde{x}, \tilde{y}; \omega) = -\frac{\mathrm{i}k}{4} r_{,i} H_1^1(kr) \tag{8.11}$$

where $\tilde{x} = (x_1, x_2), \tilde{y} = (y_1, y_2)$ and $r = |\tilde{y} - \tilde{x}|$; $H_{0,1}^1$ are the Hankel functions of first kind and order 0,1 (see Abramowitz and Stegun [1], Gradshteyn and Ryzhik [7]). The following series expansions hold:

$$H_0^1(z) = 1 + \frac{2\mathrm{i}}{\pi} \left(\gamma + \ln \frac{z}{2} \right) + z^2 \sum_{p \geq 0} a_p^0 \left(\frac{z}{2} \right)^{2p} \tag{8.12}$$

$$H_1^1(z) = -\frac{2\mathrm{i}}{\pi z} + \frac{\mathrm{i}z}{\pi} \left(\gamma - \frac{1}{2} + \ln \frac{z}{2} \right) + \frac{z}{2} + \frac{z^3}{2} \sum_{p \geq 0} a_p^1 \left(\frac{z}{2} \right)^{2p} \tag{8.13}$$

where $\gamma = 0.57721\ldots$ is the Euler constant and

$$a_p^0 = \frac{1}{4} \left[1 + \frac{2\mathrm{i}}{\pi} \left(\gamma + \ln \frac{z}{2} - b_{p+1} \right) \right] \frac{(-1)^{p+1}}{(p+1)!^2}$$

$$a_p^1 = \frac{1}{4} \left[1 + \frac{2\mathrm{i}}{\pi} \left(\gamma + \ln \frac{z}{2} - b_{p+1} - \frac{1}{2(p+2)} \right) \right] \frac{(-1)^{p+1}}{(p+1)!(p+2)!}$$

$$b_p = \sum_{m=1}^{p} \frac{1}{m}$$

Using the above expansions, one can show that:

$$\begin{aligned} & G(\boldsymbol{x}, \boldsymbol{y}; \omega) - G(\boldsymbol{x}, \boldsymbol{y}) \\ & \quad = 1 - \frac{1}{2\pi} \left(\gamma + \ln \frac{k}{2} \right) + \frac{\mathrm{i}}{4} (kr)^2 \sum_{p \geq 0} a_p^0 \left(\frac{z}{2} \right)^{2p} \\ & G_{,i}(\boldsymbol{x}, \boldsymbol{y}; \omega) - G_{,i}(\boldsymbol{x}, \boldsymbol{y}) \\ & \quad = -\frac{kr}{4\pi} \left(\gamma - 1 + \ln \frac{z}{2} \right) + \frac{\mathrm{i}kr}{8} \left(1 + (kr)^2 \sum_{p \geq 0} a_p^0 \left(\frac{z}{2} \right)^{2p} \right) \end{aligned}$$

where $G(\boldsymbol{x},\boldsymbol{y}) = -1/(2\pi)\ln r$ is the static fundamental solution for the infinite plane. The two-dimensional dynamic fundamental solution thus possesses the following asymptotic properties:

$$(r \to 0) \qquad (\forall \omega > 0) \qquad \left\{ \begin{array}{r} G(\boldsymbol{x},\boldsymbol{y};\omega) - G(\boldsymbol{x},\boldsymbol{y}) = O(1) \\ \boldsymbol{\nabla} G(\boldsymbol{x},\boldsymbol{y};\omega) - \boldsymbol{\nabla} G(\boldsymbol{x},\boldsymbol{y}) = O(1) \end{array} \right. \tag{8.14}$$

$$(\omega \to 0) \qquad (\forall r > 0) \qquad \left\{ \begin{array}{r} G(\boldsymbol{x},\boldsymbol{y};\omega) - G(\boldsymbol{x},\boldsymbol{y}) = O(\ln\omega) \\ \boldsymbol{\nabla} G(\boldsymbol{x},\boldsymbol{y};\omega) - \boldsymbol{\nabla} G(\boldsymbol{x},\boldsymbol{y}) = O(\omega\ln\omega) \end{array} \right. \tag{8.15}$$

8.2 INTEGRAL EQUATIONS AND REPRESENTATIONS

Making the substitutions:

$$u(\boldsymbol{y},t) \to u(\boldsymbol{y})\mathrm{e}^{-\mathrm{i}\omega t} \qquad q(\boldsymbol{y},t) \to u(\boldsymbol{y})\mathrm{e}^{-\mathrm{i}\omega t}$$

and rejecting the initial conditions to $t = -\infty$ in the integral statements for time-domain problems (Sections 7.2, 7.3), one obtains at once the corresponding integral statements for frequency-domain problems.

Integral representation

Any solution $u(\boldsymbol{x})$ of the Helmholtz equation is given at any point $\boldsymbol{x}$ *not on the boundary* $\partial\Omega$ by the integral representation:

$$\begin{aligned} \kappa u_k(\boldsymbol{x}) = & \int_{\partial\Omega} \{q(\boldsymbol{y})G(\boldsymbol{x},\boldsymbol{y};\omega) - u(\boldsymbol{y})H(\boldsymbol{x},\boldsymbol{y};\omega)\}\,\mathrm{d}S_y \\ & + \int_\Omega F(\boldsymbol{y})G(\boldsymbol{x},\boldsymbol{y};\omega)\,\mathrm{d}V_y \end{aligned} \tag{8.16}$$

$$\begin{aligned} \kappa u_{,i}(\boldsymbol{x}) = & \int_{\partial\Omega} \{q(\boldsymbol{y})G_{,\bar{i}}(\boldsymbol{x},\boldsymbol{y};\omega) - u(\boldsymbol{y})n_j(\boldsymbol{y})G_{,\bar{i}j}(\boldsymbol{x},\boldsymbol{y};\omega)\}\,\mathrm{d}S_y \\ & + \int_\Omega F(\boldsymbol{y})G_{,\bar{i}}(\boldsymbol{x},\boldsymbol{y};\omega)\,\mathrm{d}V_y \end{aligned} \tag{8.17}$$

where $\kappa = 1$ ($\boldsymbol{x} \in \Omega$) or $\kappa = 0$ ($\boldsymbol{x} \notin \Omega$). The above equation holds for any fundamental solution compatible with the domain Ω. When the infinite-space solution (8.5) is used, an integration by parts leads to the following alternative integral representation of the gradient:

$$\begin{aligned} \kappa u_{,i}(\boldsymbol{x}) = & \int_{\partial\Omega} \left\{D_{ij}u(\boldsymbol{y})G_{,j}(\boldsymbol{y},\boldsymbol{x};\omega) - k^2u(\boldsymbol{y})n_i(\boldsymbol{y})G(\boldsymbol{x},\boldsymbol{y};\omega)\right\}\,\mathrm{d}S_y \\ & - \int_{\partial\Omega} q(\boldsymbol{y})G_{,i}(\boldsymbol{y},\boldsymbol{x};\omega)\,\mathrm{d}S_y - \int_\Omega F(\boldsymbol{y})G_{,i}(\boldsymbol{y},\boldsymbol{x};\omega)\,\mathrm{d}V_y \end{aligned} \tag{8.18}$$

Regularized boundary integral equations

The following regularized boundary integral equation holds, irrespective of the particular fundamental solution used, for any solution of the Helmholtz equation over a bounded domain Ω:

$$\int_{\partial\Omega} \{u(\boldsymbol{y})H(\boldsymbol{x},\boldsymbol{y};\omega) - u(\boldsymbol{x})H(\boldsymbol{x},\boldsymbol{y})\}\,\mathrm{d}S_y - \int_{\partial\Omega} q(\boldsymbol{y})U_i^k(\boldsymbol{x},\boldsymbol{y};\omega)\,\mathrm{d}S_y \\ - \int_{\Omega} F(\boldsymbol{y})G(\boldsymbol{y},\boldsymbol{x};\omega)\,\mathrm{d}V_y = 0 \quad (8.19)$$

The gradient of u at a boundary point $\boldsymbol{x} \in \partial\Omega$ has the integral representation:

$$\begin{aligned} \frac{1}{2}u_{,i}(\boldsymbol{x}) = & \int_{\partial\Omega} \{D_{ij}u(\boldsymbol{y})G_{,j}(\boldsymbol{x},\boldsymbol{y};\omega) - D_{ij}u(\boldsymbol{x})G_{,j}(\boldsymbol{x},\boldsymbol{y})\}\,\mathrm{d}S_y \\ & - \int_{\partial\Omega} \{q(\boldsymbol{y})G_{,i}(\boldsymbol{x},\boldsymbol{y};\omega) - q(\boldsymbol{x})G_{,i}(\boldsymbol{x},\boldsymbol{y})\}\,\mathrm{d}S_y \\ & - k^2 \int_{\partial\Omega} u(\boldsymbol{y})n_i(\boldsymbol{y})G(\boldsymbol{x},\boldsymbol{y};\omega)\,\mathrm{d}S_y \\ & - \int_{\Omega} F(\boldsymbol{y})G_{,i}(\boldsymbol{y},\boldsymbol{x};\omega)\,\mathrm{d}V_y \\ & + D_{ij}u(\boldsymbol{x})I_j(\boldsymbol{x},\Gamma) - q(\boldsymbol{x})I_i(\boldsymbol{x},\Gamma) \end{aligned} \quad (8.20)$$

where the $I_i(\boldsymbol{x},\Gamma)$ are given by eq. (5.31).

Conditions at infinity and exterior problems

The sufficient conditions under which the integral representations and equations are applicable to unbounded domains $\bar{\Omega} = \mathbb{R}^3 - \Omega$ (Ω bounded) were given in Section 8.3 for time-domain problems. Their transposition to problems in the frequency domain is immediate. As a result, the following sufficient conditions at infinity:

$$u = o(1) \qquad q - iku = o(R^{-1}) \qquad (R \to \infty) \quad (8.21)$$

(respectively decay and radiation conditions) ensure the uniqueness of the solution to an exterior problem. Then, the following integral equation holds true for any function u satisfying the Helmholtz equation (8.2) and the conditions at infinity (8.21) above:

$$u(\boldsymbol{x}) + \int_{\partial\Omega} \{u(\boldsymbol{y})H(\boldsymbol{x},\boldsymbol{y};\omega) - u(\boldsymbol{x})H(\boldsymbol{x},\boldsymbol{y})\}\,\mathrm{d}S_y \\ - \int_{\partial\Omega} q(\boldsymbol{y})G(\boldsymbol{x},\boldsymbol{y};\omega)\,\mathrm{d}S_y - \int_{\bar{\Omega}} F(\boldsymbol{y})G(\boldsymbol{y},\boldsymbol{x};\omega)\,\mathrm{d}V_y = 0 \quad (8.22)$$

When $\boldsymbol{x} \notin \partial\Omega$, this integral equation reduces to the integral representation (8.16), provided the domain integral is convergent. Conversely, any u of the form (8.16) satisfies the conditions at infinity (8.21).

Comments

The ability to reduce exterior problems to the solution of an integral equation over a bounded geometrical support (the boundary $\partial\Omega$) is of great practical value. The regularity and radiation conditions at infinity are exactly accounted for, without having to resort to the kind of additional treatment needed in FEMs or FDMs like fictitious 'absorbing' boundaries for the simulation of radiation conditions (Halpern [8], Thompson and Pinski [14], Harari *et al.* [9]). The difficulties associated with the latter are thus avoided.

These considerations, which apply to any exterior problem, are of particular importance in steady-state dynamics. Introducing artificial boundaries has a greater impact on the solution in dynamics than in statics: the artificial boundary (non-physically) traps the outgoing waves in the artificially bounded domain instead of letting the energy escape to infinity. In the time domain, assuming initial rest, the artificial boundary is reached by the waves only after a finite elapsed time, and does not affect the solution at previous time instants. For a given time interval of study, reflection by artificial boundaries is thus avoidable by truncating the unbounded domain at a sufficiently remote distance. In contrast, steady-state situations affect the entire domain.

Integral representations and equations are thus particularly well adapted to the numerical solution of wave propagation problems in unbounded media.

8.3 INDIRECT INTEGRAL FORMULATIONS AND THE FICTITIOUS EIGENVALUE PROBLEM

Indirect integral representations on the boundary

As in Section 2.4, consider simultaneously a bounded domain Ω and its unbounded complement $\bar{\Omega} = \mathbb{R}^3 - \Omega$. Let $u, \bar{u}$ denote solutions of the Helmholtz equation over Ω and $\bar{\Omega}$ respectively (*interior* and *exterior* problems, with $F = \bar{F} = 0$).

Regularized integral representations for single-layer potentials. Assume that u is continuous across $\partial\Omega$ (i.e. $u = \bar{u}$ on $\partial\Omega$) and denote by $\psi = \bar{q} + q$ the flux jump across $\partial\Omega$. Then u admits at a boundary point $\boldsymbol{x}$ the integral representation formula:

$$\bar{u}(\boldsymbol{x}) = \int_{\partial\Omega} \psi(\boldsymbol{y}) G(\boldsymbol{x}, \boldsymbol{y}; \omega) \, \mathrm{d}S_y \quad = \quad u(\boldsymbol{x}) \tag{8.23}$$

Moreover, the integral representation of the derivatives $u_{,i}$ at $\boldsymbol{x} \in \partial\Omega$ is:

$$\frac{1}{2}(u_{,i} + \bar{u}_{,i})(\boldsymbol{x}) = -\int_{\partial\Omega} \{\psi(\boldsymbol{y}) G_{,i}(\boldsymbol{x}, \boldsymbol{y}; \omega) - \psi(\boldsymbol{x}) G_{,i}(\boldsymbol{x}, \boldsymbol{y})\} \, \mathrm{d}S_y - \psi(\boldsymbol{x}) I_i(\boldsymbol{x}, \partial\Omega) \tag{8.24}$$

Regularized integral representations for double-layer potentials. Assume that $u_{,n}$ is continuous across $\partial\Omega$ (i.e. $q + \bar{q} = 0$ on $\partial\Omega$) and denote by $\phi = \bar{u} - u$ the potential jump across $\partial\Omega$. Then u admits at a boundary point $\boldsymbol{x}$ the integral representation formula:

$$\bar{u}(\boldsymbol{x}) = \int_{\partial\Omega} \{\phi(\boldsymbol{y})H(\boldsymbol{x},\boldsymbol{y};\omega) - \phi(\boldsymbol{x})H(\boldsymbol{x},\boldsymbol{y})\} \, \mathrm{d}S_y = u(\boldsymbol{x}) + \phi(\boldsymbol{x}) \tag{8.25}$$

Moreover, the integral representation of the derivatives $u_{,i}$ at $\boldsymbol{x} \in \partial\Omega$ is:

$$\frac{1}{2}(\bar{u}_{,i} + u_{,i})(\boldsymbol{x}) = -\int_{\partial\Omega} \{D_{ij}\phi(\boldsymbol{y})G_{,j}(\boldsymbol{x},\boldsymbol{y};\omega) - D_{ij}\phi(\boldsymbol{x})G_{,j}(\boldsymbol{x},\boldsymbol{y})\} \, \mathrm{d}S_y \\ - D_{ij}\phi(\boldsymbol{x})I_j(\boldsymbol{x},\partial\Omega) \tag{8.26}$$

Indirect integral equations for fundamental boundary-value problems

Equations (8.23) to (8.26) provide the basis for indirect boundary integral formulations of the fundamental boundary-value problems.

Dirichlet problem. The potential is prescribed over the boundary: $u = u^D$ on $\partial\Omega$. From eqs (8.23) and (8.25), the Dirichlet problem admits the following indirect integral equation formulations, of unknown ψ:

$$u^D = \int_{\partial\Omega} \psi(\boldsymbol{y})G(\boldsymbol{x},\boldsymbol{y};\omega) \, \mathrm{d}S_y \tag{8.27}$$

$$u^D(\boldsymbol{x}) = (\kappa - 1)\phi(\boldsymbol{x}) + \int_{\partial\Omega} \{\phi(\boldsymbol{y})H(\boldsymbol{x},\boldsymbol{y};\omega) - \phi(\boldsymbol{x})H(\boldsymbol{x},\boldsymbol{y})\} \, \mathrm{d}S_y \tag{8.28}$$

(where $\kappa = 1$ or $\kappa = 1$ for exterior and interior problems, respectively).

Neumann problem. The flux is prescribed over the boundary: $q = q^D$ on $\partial\Omega$. From eqs (8.24) and (8.26) the Neumann problem admits the following indirect integral equation formulations, of unknown ψ:

$$q^D(\boldsymbol{x}) = \left(\frac{1}{2} - I_i(\boldsymbol{x},\partial\Omega)n_i(\boldsymbol{x})\right)\psi(\boldsymbol{x}) \\ - n_i(\boldsymbol{x}) \int_{\partial\Omega} \{\psi(\boldsymbol{y})G_{,i}(\boldsymbol{x},\boldsymbol{y};\omega) - \psi(\boldsymbol{x})G_{,i}(\boldsymbol{x},\boldsymbol{y})\} \, \mathrm{d}S_y \tag{8.29}$$

$$q^D(\boldsymbol{x}) = -D_i\phi(\boldsymbol{x})I_i(\boldsymbol{x},\partial\Omega) \\ - n_i(\boldsymbol{x}) \int_{\partial\Omega} \{D_{ij}\phi(\boldsymbol{y})G_{,j}(\boldsymbol{x},\boldsymbol{y};\omega) - D_{ij}\phi(\boldsymbol{x})G_{,j}(\boldsymbol{x},\boldsymbol{y})\} \, \mathrm{d}S_y \tag{8.30}$$

Fictitious eigenvalues

Concerning the uniqueness of the solution u to a boundary-value problem for the Helmholtz equation (8.2), the following facts are known:

- An interior problem (i.e. over a bounded domain Ω) has a unique solution except when ω belongs to the infinite sequence of *eigenvalues* $0 \leq \omega_1 \leq \ldots \omega_n \leq \ldots$, for which the Helmholtz equation supplemented with *homogeneous* boundary data following the same boundary condition structure as in the original problem admits nontrivial *eigensolutions* u_n. Each ω_n has *finite multiplicity* k, i.e. corresponds to a finite number k of linearly independent eigensolutions.
- Any exterior problem (over the unbounded complement $\bar{\Omega} = \mathbb{R}^3 - \Omega$ of a bounded domain Ω) has, for any ω, a unique solution satisfying the conditions at infinity (7.23).

However, it turns out for reasons explained below that the boundary integral equation formulations for exterior problems suffer loss of uniqueness for certain specific values of ω, called *fictitious eigenvalues*, which are eigenvalues of a certain interior problem, even though the exterior boundary-value problem itself is known to admit a unique solution.

Loss of solution uniqueness in BIE formulations: the exterior Dirichlet problem. The direct regularized integral equations (8.19) for the exterior (ED) and homogeneous interior (ID) Dirichlet problems (unknown: q) read:

$$\int_{\partial\bar{\Omega}} q(\boldsymbol{y})G(\boldsymbol{x},\boldsymbol{y};\omega)\,\mathrm{d}S_y = u^D(\boldsymbol{x}) + \int_{\partial\bar{\Omega}} \{u^D(\boldsymbol{y})H(\boldsymbol{x},\boldsymbol{y};\omega) - u^D(\boldsymbol{x})H(\boldsymbol{x},\boldsymbol{y})\}\,\mathrm{d}S_y$$

$$\int_{\partial\Omega} q(\boldsymbol{y})G(\boldsymbol{x},\boldsymbol{y};\omega)\,\mathrm{d}S_y = 0$$

The same integral operator appears in the left-hand side of the integral equations (ED) and (ID). Since problem (ID) admits eigenmodes for certain values $\omega_n^D, n \geq 1$ of ω, it follows that the integral equation (8.19) for problem (ED) has infinitely many solutions for any eigenvalue $\omega = \omega_n^D$ of problem (ID).

Following a similar argument, the formulation of the exterior Dirichlet problem using the indirect integral equation (8.28) of unknown ϕ, and of the homogeneous interior Neumann problem usinq the direct integral equation (8.19) give respectively:

$$\int_{\partial\bar{\Omega}} \{\phi(\boldsymbol{y})H(\boldsymbol{x},\boldsymbol{y};\omega) - \phi(\boldsymbol{x})H(\boldsymbol{x},\boldsymbol{y})\}\,\mathrm{d}S_y = u^D(\boldsymbol{x})$$

$$\int_{\partial\Omega} \{u(\boldsymbol{y})H(\boldsymbol{x},\boldsymbol{y};\omega) - u(\boldsymbol{x})H(\boldsymbol{x},\boldsymbol{y})\}\,\mathrm{d}S_y = 0$$

The same integral operator, acting on different variables, appears in the left-hand side of both integral equations; the reasoning is not altered by the fact that the unit normals $\boldsymbol{n}$ have opposite signs. Since problem (IN) admits eigenmodes for certain values $\omega_n^N, n \geq 1$ of ω, it follows that the indirect integral equation (8.28) for problem (ED) has infinitely many solutions for any eigenvalue $\omega = \omega_n^N$ of problem (IN).

Loss of solution uniqueness in BIE formulations: the exterior Neumann problem. The formulation of the exterior Neumann problem (EN) using the direct integral equation (8.19), and of the homogeneous Dirichlet interior problem usinq the indirect integral equation (8.28) of unknown ϕ give respectively:

$$
\begin{aligned}
u(\boldsymbol{x}) + \int_{\partial\bar{\Omega}} \{u(\boldsymbol{y})H(\boldsymbol{x},\boldsymbol{y};\omega) - u(\boldsymbol{x})H(\boldsymbol{x},\boldsymbol{y})\} \,\mathrm{d}S_y \\
= \int_{\partial\bar{\Omega}} q^D(\boldsymbol{y})G(\boldsymbol{x},\boldsymbol{y};\omega) \,\mathrm{d}S_y \\
\phi(\boldsymbol{x}) - \int_{\partial\Omega} \{\phi(\boldsymbol{y})H(\boldsymbol{x},\boldsymbol{y};\omega) - \phi(\boldsymbol{x})H(\boldsymbol{x},\boldsymbol{y})\} \,\mathrm{d}S_y = 0
\end{aligned}
$$

The same integral operator, acting on different variables, appears in the left-hand side of both integral equations (wherein the unit normal $\boldsymbol{n}$ have opposite signs). The indirect integral equation (8.19) for problem (EN) has thus infinitely many solutions for any eigenvalue $\omega = \omega_n^D$ of problem (ID).

Finally, considering the formulation of problem (EN) using the integral equations (8.20), it is easy to see that the homogeneous problem (IN) is governed by the same integral operator. It follows that the indirect integral equation (8.20) applied to problem (EN) has infinitely many solutions for any eigenvalue $\omega = \omega_n^N$ of problem (IN).

8.4 THE BOUNDARY ELEMENT METHOD FOR FREQUENCY-DOMAIN PROBLEMS

The integral equations are discretized following the approach developed for static problems (Chapter 3), except for the fact that the unknowns (u, q) are represented in terms of complex, instead of real, nodal values. Likewise, the matrices in the resulting linear system of equation have complex coefficients. The treatment of the linear system is very much the same as in statics; in particulier, the Gaussian elimination technique is still applicable [5].

Influence of the wavelength

The mesh size (i.e. the typical diameter of a boundary element) is, generally speaking, chosen so as to provide a good approximation of the domain boundary and to match the expected spatial variability of the unknown fields.

In frequency-domain acoustics, the latter consideration is strongly influenced by the wavelength $\lambda = 2\pi/k$, which defines a characteristic length scale for the spatial variations of the field variables. Indeed, the spatial interpolation should provide a good representation of spatial variations over distances

Table 8.1 Dynamic severity indicator and number of Gauss points: nonsingular integrations (from [11]).

OSI	Number of Gauss points on the element quadrilaterals ($N_e = 8$)	triangles ($N_e = 6$)
$0.0 \le OSI < 0.5$	3×3	7
$0.5 \le OSI < 1.0$	4×4	16
$1.0 \le OSI < 1.5$	5×5	19
$1.5 \le OSI < 2.0$	6×6	28
$2.0 \le OSI < 2.5$	$4 \times (4 \times 4)$	4×16
$2.5 \le OSI < 3.0$	$4 \times (5 \times 5)$	4×19
$3.0 \le OSI < 3.5$	$4 \times (6 \times 6)$	4×28

of one wavelength, which implies that the mesh size be chosen as at most a fraction of a wavelength. For example, using eight-noded quadrilateral elements, we observed [2] that numerical results deteriorate strongly if the typical element edge size was less than about one-third of the wavelength (i.e. less than seven nodes along a line one wavelength long); this observation seems representative of what is usually observed.

Moreover, the presence of spatially oscillating quantities (behaving like e^{ikr}) in the field variables induces new constraints on the optimal choice of Gauss points for the numerical evaluation of element integrals (see Rezayat et al. [11]). These authors thus define, in addition to the 'static severity' indicator SSI reported in Chapter 3, an 'oscillation severity' indicator OSI:

$$\text{OSI} = \omega \Delta r / (2\pi c_T)$$

where $\Delta r = r_{max} - r_{min}$, r_{max} and r_{min} denoting the maximum and minimum distances, respectively, between $\boldsymbol{x}$ and a node of the boundary element under

Table 8.2 Dynamic severity indicator and number of Gauss points: singular integrations (from [11]).

OSI	Number of Gauss points per subdivision quadrilaterals ($N_e = 8$) corner	midside	triangles ($N_e = 6$) corner	midside
$0.0 \le OSI < 1.0$	4×3	3×3	4×3	3×4
$1.0 \le OSI < 1.5$	5×4	4×4	5×3	4×5
$1.5 \le OSI < 2.0$	6×4	5×5	6×4	5×6
$2.0 \le OSI < 2.5$	(2×4)×5	6×6	(2×4)×5	6×(2×4)
$2.5 \le OSI < 3.0$	(2×5)×6	(2×4)×(2×4)	(2×5)×6	(2×4)×(2×4)
$3.0 \le OSI < 3.5$	(2×6)×6	(2×5)×(2×5)	(2×6)×(2×4)	(2×5)×(2×6)

consideration. The relationship between their recommended number of Gauss points and the OSI is summarized in Table 8.1.

It is assumed that the singular integrations are performed using polar coordinates (ρ, α) in the parent element (Section 3.5), the latter being subdivided into triangles which share as a common vertex the antecedent $\boldsymbol{\eta}$ of $\boldsymbol{x}$. Table 8.2 displays, in the form $N_\rho \times N_\alpha$, the number of Gauss points for each triangular subelement.

Transfer of singularity into the static fundamental solution

The integral equations feature integrals like

$$\int_{\partial\Omega} \Big\{ u(\boldsymbol{y}) H(\boldsymbol{x}, \boldsymbol{y}; \omega) - u(\boldsymbol{x}) H(\boldsymbol{x}, \boldsymbol{y}) \Big\} \, \mathrm{d}S_y$$

The difference $u(\boldsymbol{y})H(\boldsymbol{x}, \boldsymbol{y}; \omega) - u(\boldsymbol{x})H(\boldsymbol{x}, \boldsymbol{y})$ is known to be nonsingular. It is useful to make this fact explicit before performing the numerical integrations, using series expansions like (8.6) for scalar problems or (8.38) for elastodynamic problems. Moreover, this manipulation is better circumscribed to a small portion $S(\boldsymbol{x})$ of $\partial\Omega$ such that $\boldsymbol{x}$ is interior to $S(\boldsymbol{x})$, for example by recasting equation (8.19) into:

$$\begin{aligned} \kappa u(\boldsymbol{x}) &+ \int_{\Gamma - S(x)} [u(\boldsymbol{y}) H(\boldsymbol{x}, \boldsymbol{y}; \omega) - (\boldsymbol{x}) H(\boldsymbol{x}, \boldsymbol{y})] \, \mathrm{d}S_y \\ &- \int_{\partial\Omega} q(\boldsymbol{y}) G(\boldsymbol{x}, \boldsymbol{y}; \omega) \, \mathrm{d}S_y \\ &+ \int_{S(x)} \Big\{ u(\boldsymbol{y}) H(\boldsymbol{x}, \boldsymbol{y}; \omega) - u(\boldsymbol{x}) H(\boldsymbol{x}, \boldsymbol{y}) \Big\} \, \mathrm{d}S_y = 0 \end{aligned} \tag{8.31}$$

(in practice, $S(\boldsymbol{x})$ is taken as the set of boundary elements containing $\boldsymbol{x}$). This procedure allows us to limit the use of series expansions to a neighbourhood of the singular point, thereby lowering the number of terms actually needed.

8.5 FREQUENCY-DOMAIN ELASTODYNAMICS

Only isotropic (homogeneous and linear) elasticity is addressed here. The local partial differential governing system of equations (7.36), which must be verified by any elastodynamic state $(\boldsymbol{u}, \boldsymbol{\sigma}, \boldsymbol{F})$, takes the following form for time-harmonic problems:

$$\mathrm{div}\boldsymbol{\sigma} + \rho\omega^2 \boldsymbol{u} + \rho \boldsymbol{F} = 0 \qquad \text{in } \Omega \tag{8.32}$$

($\boldsymbol{F}(\boldsymbol{y})$: specific body force distribution). For a well-posed problem, the above local equation must be supplemented by boundary conditions on $\partial\Omega$ and (if appropriate) conditions at infinity; initial conditions are obviously irrelevant.

For a given angular frequency ω, the wavelengths λ_L, λ_T of compressive (also termed 'longitudinal') and shear (also termed 'transversal') elastic waves, respectively, are:

$$\lambda_L = \frac{2\pi}{\omega}\sqrt{(\lambda + 2\mu)/\rho} \qquad \lambda_T = \frac{2\pi}{\omega}\sqrt{\mu/\rho} \qquad (\lambda_L > \lambda_T)$$

λ_T defines a length scale of spatial variability of the elastic fields.

Fundamental solutions in frequency-domain elastodynamics

They are defined as elastodynamic states (displacement $\boldsymbol{U}^k(\boldsymbol{x}, \boldsymbol{y}; \omega)$, stress $\boldsymbol{\Sigma}^k(\boldsymbol{x}, \boldsymbol{y}; \omega)$) generated in an auxiliary domain E by a time-harmonic point force of unit amplitude applied at a fixed point $\boldsymbol{x}$ and in a fixed coordinate direction k:

$$\rho \boldsymbol{F}(\boldsymbol{y}) = \delta(\boldsymbol{y} - \boldsymbol{x})\boldsymbol{e}_k \tag{8.33}$$

As usual, such states exist for any given auxiliary domain E and boundary conditions over ∂E. In the simplest case $E = \mathbb{R}^3$, the fundamental solution is known in closed form (see later). In the case of the elastic half-space (either homogeneous or layered) with traction-free boundary conditions, the fundamental solution has been derived by Kennett [10]; in contrast to the Mindlin solution of elastostatics, it is not known in closed form.

The fundamental solutions of frequency-domain elastodynamics can be considered as time-domain solutions (7.41) with the special choice $f(t) = \mathrm{e}^{-\mathrm{i}\omega t}$ for the source function, i.e.:

$$\left\{U_i^k, \Sigma_{ij}^k\right\}(\boldsymbol{x}, \boldsymbol{y}; \omega) = \left\{U_i^k, \Sigma_{ij}^k\right\}[\boldsymbol{x}, t, \boldsymbol{y}|\mathrm{e}^{-\mathrm{i}\omega t}] \tag{8.34}$$

where, in addition, initial conditions are shifted from $t = 0$ to $t = -\infty$. The general properties of the time-domain fundamental solution (singular behaviour (7.53,7.55), symmetry (7.56) ...) immediately translate into similar properties for the frequency-domain counterpart as defined above.

Helmholtz fundamental solution

It is the fundamental solution for an infinite elastic body $E = \mathbb{R}^3$. Its expression is readily obtained by putting $f(t) = \mathrm{e}^{-\mathrm{i}\omega t}$ into the time-domain Stokes solution (7.49), (7.50):

$$U_i^k(\boldsymbol{x}, \boldsymbol{y}; \omega) = \frac{1}{4\pi\mu r}\left[A\delta_{ik} + B r_{,i} r_{,k}\right] \tag{8.35}$$

$$\Sigma_{ij}^k(\boldsymbol{x}, \boldsymbol{y}; \omega) = \frac{1}{4\pi r^2}\left[2C r_{,i} r_{,k} r_{,j} + (\delta_{ik} r_{,j} + \delta_{jk} r_{,i})D + \delta_{ij} r_{,k} E\right] \tag{8.36}$$

where:

$$
\begin{aligned}
A &= \left(1 + \frac{\mathrm{i}}{x_T} - \frac{1}{x_T^2}\right) \mathrm{e}^{\mathrm{i}x_T} - \beta^2 \left(\frac{\mathrm{i}}{x_L} - \frac{1}{x_L^2}\right) \mathrm{e}^{\mathrm{i}x_L} \\
B &= \left(\frac{3}{x_T^2} - \frac{3\mathrm{i}}{x_T} - 1\right) \mathrm{e}^{\mathrm{i}x_T} - \beta^2 \left(\frac{3}{x_L^2} - \frac{3\mathrm{i}}{x_L} - 1\right) \mathrm{e}^{\mathrm{i}x_L} \\
C &= \left(-\frac{15}{x_T^2} + \frac{15\mathrm{i}}{x_T} + 6 - \mathrm{i}x_T\right) \mathrm{e}^{\mathrm{i}x_T} - \beta^2 \left(-\frac{15}{x_L^2} + \frac{15\mathrm{i}}{x_L} + 6 - \mathrm{i}x_L\right) \mathrm{e}^{\mathrm{i}x_L} \\
D &= (\mathrm{i}x_T - 1)\mathrm{e}^{\mathrm{i}x_T} + 2B \\
E &= (1 - 2\beta^2)(\mathrm{i}x_L - 1)\mathrm{e}^{\mathrm{i}x_L} + 2B
\end{aligned}
\tag{8.37}
$$

and using the notations $k_L r = x_L$, $k_T r = x_T$; the ratio $\beta = c_T/c_L$ is given by (7.48). The factors of $\exp(\mathrm{i}k_L r)$ and $\exp(\mathrm{i}k_T r)$ are respectively the longitudinal and transversal parts of the Helmholtz solution.

Singular behaviour of the three-dimensional Helmholtz solution. A substitution of the expansion (8.6) into expressions (8.37) results in the expansions:

$$
\begin{aligned}
A &= \frac{1+\beta^2}{2} + \sum_{p\geq 1} \frac{1+p+\beta^{p+2}}{p!(p+2)}(\mathrm{i}x_T)^p \\
B &= \frac{1-\beta^2}{2} + \sum_{p\geq 1} \frac{1-\beta^{p+2}(1-p)}{p!(p+2)}(\mathrm{i}x_T)^p \\
C &= -\frac{3(1-\beta^2)}{2} + x_T^2 \sum_{p\geq 0} \frac{1-\beta^{p+4}(p-1)}{p!(p+2)(p+4)}(\mathrm{i}x_T)^p \\
D &= -\beta^2 - x_T^2 \sum_{p\geq 0} \frac{p+2+2\beta^{p+4}}{p!(p+2)(p+4)}(\mathrm{i}x_T)^p \\
E &= \beta^2 + x_T^2 \sum_{p\geq 0} \frac{2-(p+4)\beta^{p+2}+2(p+3)\beta^{p+4}}{p!(p+2)(p+4)}(\mathrm{i}x_T)^p
\end{aligned}
\tag{8.38}
$$

The above expansions are absolutely convergent for any x_T, given that $\beta^2 \leq 1/2 < 1$. They establish the nonsingular character of the terms $A, B, \ldots$, which was not apparent in expressions (8.37). From the definition (7.48) of β, the leading terms in eqs (8.38) correspond to the Kelvin elastostatic solution. The Helmholtz and Kelvin solutions thus have identical singularities:

$$
(r \to 0) \qquad (\forall \omega > 0) \qquad \left\{ \begin{aligned} &\boldsymbol{U}(\boldsymbol{x}, \boldsymbol{y}; \omega) - \boldsymbol{U}(\boldsymbol{x}, \boldsymbol{y}) = O(1) \\ &\boldsymbol{\Sigma}(\boldsymbol{x}, \boldsymbol{y}; \omega) - \boldsymbol{\Sigma}(\boldsymbol{x}, \boldsymbol{y}) = O(1) \end{aligned} \right. \tag{8.39}
$$

$$
(\omega \to 0) \qquad (\forall r > 0) \qquad \left\{ \begin{aligned} &\boldsymbol{U}(\boldsymbol{x}, \boldsymbol{y}; \omega) - \boldsymbol{U}(\boldsymbol{x}, \boldsymbol{y}) = O(\omega) \\ &\boldsymbol{\Sigma}(\boldsymbol{x}, \boldsymbol{y}; \omega) - \boldsymbol{\Sigma}(\boldsymbol{x}, \boldsymbol{y}) = O(\omega^2) \end{aligned} \right. \tag{8.40}
$$

Helmholtz solution in plane strain

Integration of the three-dimensional solution (8.35) over $-\infty \le x_3 \le +\infty$ yields the plane-strain fundamental solution for the infinite plane:

$$U_i^k(\boldsymbol{x}, \boldsymbol{y}; \omega) = \frac{\mathrm{i}}{4\mu}\left[A\delta_{ik} + Br_{,i}r_{,k}\right] \tag{8.41}$$

$$\Sigma_{ij}^k(\boldsymbol{x}, \boldsymbol{y}; \omega) = \frac{\mathrm{i}}{4r}\left[2Cr_{,i}r_{,k}r_{,j} + (\delta_{ik}r_{,j} + \delta_{jk}r_{,i})D + \delta_{ij}r_{,k}E\right] \tag{8.42}$$

where:

$$\begin{aligned}
A &= H_0^1(k_T r) - \left[H_1^1(k_T r) - \beta H_1^1(k_L r)\right] \\
B &= -2A - \left[H_0^1(k_T r) - \beta H_0^1(k_L r)\right] \\
C &= -4B + k_T r\left[H_1^1(k_T r) - \beta^3 H_1^1(k_L r)\right] \\
D &= k_T r H_1^1(k_T r) + 2B \\
E &= -(1-2\beta^2)k_L r H_1^1(k_L r) + 2B
\end{aligned} \tag{8.43}$$

(the range of indices i, j, k being $\{1, 2\}$).

Singular behaviour of the two-dimensional Helmholtz solution. The series expansions (8.12)–(8.13) of the Hankel functions allow us to reformulate eqs (8.43) into the following series expansions, absolutely convergent for any $r > 0$:

$$\begin{aligned}
A &= -\frac{\mathrm{i}}{\pi}(1-\beta^2)\ln r - \frac{1+\beta^2}{2}\left(1+\frac{2\mathrm{i}}{\pi}\gamma\right) - \frac{\mathrm{i}}{\pi}(1-\beta^2)\left(1+\ln\frac{k_T}{2}\right) \\
&\quad + \frac{\mathrm{i}}{\pi}\beta^2\ln\beta - \frac{1}{2}x_T^2\left[S_1(x_T) - \beta^4 S_1(x_L)\right] \\
B &= -\frac{\mathrm{i}}{\pi}(1-\beta^2) + x_T^2\left[S_1(x_T) - \beta^4 S_1(x_L) - S_0(x_T) - \beta^4 S_0(x_L)\right] \\
C &= \frac{2\mathrm{i}}{\pi}(1-\beta^2) + x_T^2\left[\frac{1}{2}(1-\beta^4)\left(1+\frac{2\mathrm{i}}{\pi}(\delta-1)\right)\right. \\
&\quad + 4(S_0(x_T) - \beta^4 S_0(x_L)) - S_1(x_T) - \beta^4 S_1(x_L) \\
&\quad \left. + x_T^2(S_1(x_T) - \beta^6 S_1(x_L)) + \frac{2\mathrm{i}}{\pi}(1-\beta^2)\ln\frac{x_T}{2} - \frac{2\mathrm{i}}{\pi}\beta^2\ln\beta\right] \\
D &= \frac{2\mathrm{i}}{\pi}\beta^2 - x_T^2\left[2(S_1(x_T) - \beta^4 S_1(x_L)) - S_0(x_T) - \beta^4 S_0(x_L)\right. \\
&\quad \left. + \frac{2\mathrm{i}}{\pi}\left(\gamma - 1 + \ln\frac{x_T}{2}\right) + \frac{1}{2}\beta^2\left(1 + x_T^2 S_1(x_T)\right)\right] \\
E &= \frac{2\mathrm{i}}{\pi}\beta^2 - x_T^2\left[2(S_1(x_T) - \beta^4 S_1(x_L)) - S_0(x_T) - \beta^4 S_0(x_L)\right. \\
&\quad \left. + \frac{2\mathrm{i}}{\pi}\left(\gamma - 1 + \ln\frac{x_L}{2}\right) - \beta^2\frac{1-2\beta^2}{2}\left(1 + x_T^2 S_1(x_T)\right)\right]
\end{aligned} \tag{8.44}$$

where

$$S_0(z) = \sum_{p \geq 0} a_p^0 \left(\frac{z}{2}\right)^{2p} \qquad S_1(z) = \sum_{p \geq 0} a_p^1 \left(\frac{z}{2}\right)^{2p}$$

and the a_p^0, a_p^1 are as given in eqs (8.13). The Helmholtz and Kelvin two-dimensional solutions are readily seen from these expansions to verify the asymptotic properties:

$$(r \to 0) \qquad (\forall \omega > 0) \qquad \left\{ \begin{array}{l} \boldsymbol{U}(\boldsymbol{x}, \boldsymbol{y}; \omega) - \boldsymbol{U}(\boldsymbol{x}, \boldsymbol{y}) = O(1) \\ \boldsymbol{\Sigma}(\boldsymbol{x}, \boldsymbol{y}; \omega) - \boldsymbol{\Sigma}(\boldsymbol{x}, \boldsymbol{y}) = O(r \ln r) \end{array} \right.$$

$$(\omega \to 0) \qquad (\forall r > 0) \qquad \left\{ \begin{array}{l} \boldsymbol{U}(\boldsymbol{x}, \boldsymbol{y}; \omega) - \boldsymbol{U}(\boldsymbol{x}, \boldsymbol{y}) = O(\ln \omega) \\ \boldsymbol{\Sigma}(\boldsymbol{x}, \boldsymbol{y}; \omega) - \boldsymbol{\Sigma}(\boldsymbol{x}, \boldsymbol{y}) = O(\omega^2 \ln \omega) \end{array} \right.$$

Notice in particular that, contrarily to the three-dimensional case, the Helmholtz displacement does not yield the Kelvin displacement in the limit $\omega \to 0$.

8.6 INTEGRAL EQUATIONS AND REPRESENTATIONS FOR FREQUENCY-DOMAIN ELASTODYNAMICS

The integral formulations for frequency-domain problems are obtained by transposing the formulations for time-domain problems (Section 7.9).

Integral representations

The displacement $\boldsymbol{u}(\boldsymbol{x})$ at a given point $\boldsymbol{x} \in E$ *not located on* $\partial\Omega$ is given by the integral representation formula::

$$\begin{aligned} \kappa u_k(\boldsymbol{x}) = & \int_{\partial\Omega} \left\{ t_i(\boldsymbol{y}) U_i^k(\boldsymbol{x}, \boldsymbol{y}; \omega) - u_i(\boldsymbol{y}) T_i^k(\boldsymbol{x}, \boldsymbol{y}; \omega) \right\} \, \mathrm{d}S_y \\ & + \int_{\Omega} \rho F_i(\boldsymbol{y}) U_i^k(\boldsymbol{x}, \boldsymbol{y}; \omega) \, \mathrm{d}V_y \end{aligned} \tag{8.45}$$

The integral representation for the stress tensor $\boldsymbol{\sigma}(\boldsymbol{x})$ then follows by virtue of Hooke's law:

$$\begin{aligned} \kappa \sigma_{ij}(\boldsymbol{x}) = & \int_{\partial\Omega} \left\{ u_a(\boldsymbol{y}) n_b(\boldsymbol{y}) C_{abk\ell} \Sigma_{ij,\ell}^k(\boldsymbol{x}, \boldsymbol{y}; \omega) - t_k(\boldsymbol{y}) \Sigma_{ij}^k(\boldsymbol{x}, \boldsymbol{y}; \omega) \right\} \, \mathrm{d}S_y \\ & - \int_{\Omega} \rho F_k(\boldsymbol{y}) \Sigma_{ij}^k(\boldsymbol{x}, \boldsymbol{y}; \omega) \, \mathrm{d}V_y \end{aligned} \tag{8.46}$$

where $\kappa = 1$ $(\boldsymbol{x} \in \Omega)$ or $\kappa = 0$ $(\boldsymbol{x} \notin \Omega)$. These formulas can be employed in conjunction with any fundamental solution, provided $\Omega \subseteq E$. If the Helmholtz solution is specifically used in eq. (8.46), an integration by parts leads to the following variant:

$$\begin{aligned}\kappa\sigma_{ij}(\boldsymbol{x}) = C_{ijab}\int_{\partial\Omega} D_{\ell b}u_a(\boldsymbol{y})\Sigma^k_{ab}(\boldsymbol{x},\boldsymbol{y};\omega)\,\mathrm{d}S_y \\ - \rho\omega^2 C_{ijab}\int_{\partial\Omega} u_a(\boldsymbol{y})n_\ell(\boldsymbol{y})U^k_a(\boldsymbol{x},\boldsymbol{y};\omega)\,\mathrm{d}S_y \\ - \int_{\partial\Omega} t_k(\boldsymbol{y})\Sigma^k_{ij}(\boldsymbol{x},\boldsymbol{y};\omega)\,\mathrm{d}S_y - \int_{\Omega}\rho F_k(\boldsymbol{y})\Sigma^k_{ij}(\boldsymbol{x},\boldsymbol{y};\omega)\,\mathrm{d}V_y \quad (8.47)\end{aligned}$$

Regularized boundary integral equations

The displacement boundary integral equation:

$$\begin{aligned}\int_{\partial\Omega}\{u_i(\boldsymbol{y})T^k_i(\boldsymbol{x},\boldsymbol{y};\omega) - u_i(\boldsymbol{x})T^k_i(\boldsymbol{x},\boldsymbol{y})\}\,\mathrm{d}S_y \\ - \int_{\partial\Omega} t_i(\boldsymbol{y})U^k_i(\boldsymbol{x},\boldsymbol{y};\omega)\,\mathrm{d}S_y = 0 \quad (8.48)\end{aligned}$$

holds true for any elastodynamic state $(\boldsymbol{u},\boldsymbol{\sigma},\boldsymbol{F}=\boldsymbol{0})$ over a bounded domain Ω and any fundamental solution with $E \supseteq \Omega$.

Next, the displacement gradient at $\boldsymbol{x}\in\partial\Omega$ is given by the integral representation:

$$\begin{aligned}\frac{1}{2}u_{k,\ell}(\boldsymbol{x},t) = D_{\ell b}u_a(\boldsymbol{x},t)A^k_{ab}(\boldsymbol{x},\Gamma) - t_a(\boldsymbol{x},t)B^k_{a\ell}(\boldsymbol{x},\Gamma) \\ + \int_{\partial\Omega}\{D_{\ell b}u_a(\boldsymbol{y})\Sigma^k_{ab}(\boldsymbol{x},\boldsymbol{y};\omega) - D_{\ell b}u_a(\boldsymbol{x})\Sigma^k_{ab}(\boldsymbol{x},\boldsymbol{y})\}\,\mathrm{d}S_y \\ - \int_{\partial\Omega}\{t_a(\boldsymbol{y})U^k_{a,\ell}(\boldsymbol{x},\boldsymbol{y};\omega) - t_a(\boldsymbol{x})U^k_{a,\ell}(\boldsymbol{x},\boldsymbol{y})\}\,\mathrm{d}S_y \\ - \rho\omega^2\int_{\partial\Omega} u_a(\boldsymbol{y})n_\ell(\boldsymbol{y})U^k_a(\boldsymbol{x},\boldsymbol{y};\omega)\,\mathrm{d}S_y \quad (8.49)\end{aligned}$$

with $A^k_{ab}(\boldsymbol{x},\Gamma)$, $B^k_{a\ell}(\boldsymbol{x},\Gamma)$ given by eqs (5.29), (5.30).

Conditions at infinity and exterior problem

The analysis presented in Section 8.3 of the conditions under which integral representations and equations for time-domain elastodynamic problems over unbounded domains of the form $\bar{\Omega} = \mathbb{R}^3 - \Omega$ (Ω bounded) are valid is immediately transposable to frequency-domain problems. The regularity condition:

$$\begin{aligned}&\boldsymbol{u}^L.\boldsymbol{n} = o(1) \qquad \boldsymbol{u}^T.\boldsymbol{\tau} = o(R^{-1}) \\ &\boldsymbol{u}^T.\boldsymbol{n} = o(R^{-1}) \qquad \boldsymbol{u}^T.\boldsymbol{\tau} = o(1)\end{aligned} \qquad (R\to\infty) \qquad (8.50)$$

and radiation condition:

$$\begin{aligned} \boldsymbol{t}^L - \mathrm{i}\omega\rho c_L \boldsymbol{u}^L &= o(R^{-1}) \\ \boldsymbol{t}^T - \mathrm{i}\omega\rho c_T \boldsymbol{u}^T &= o(R^{-1}) \end{aligned} \qquad (R \to \infty) \tag{8.51}$$

ensure the uniqueness of the solution of any exterior elastodynamic problem. Any elastodynamic state over the unbounded domain $\bar{\Omega}$ and compatible with the conditions at infinity (8.50, 8.51) solves the integral equation:

$$u_k(\boldsymbol{x}) + \int_{\partial\Omega} \{u_i(\boldsymbol{y})T_i^k(\boldsymbol{x},\boldsymbol{y};\omega) - u_i(\boldsymbol{x})T_i^k(\boldsymbol{x},\boldsymbol{y})\}\ \mathrm{d}S_y \\ - \int_{\partial\Omega} t_i(\boldsymbol{y})U_i^k(\boldsymbol{x},\boldsymbol{y};\omega)\,\mathrm{d}S_y = 0 \tag{8.52}$$

which reduces to the representation formula (8.45) when $\boldsymbol{x} \notin \partial\Omega$. Also, the conditions at infinity (8.50, 8.51) are known to be sufficient conditions for an exterior problem to have a unique solution (Eringen and Suhubi [6]).

Conversely, from the asymptotic behaviour at infinity of the Helmholtz solution, any pair $(\boldsymbol{u}(\boldsymbol{x}), \boldsymbol{\sigma}(\boldsymbol{x}))$, $\boldsymbol{x} \notin \partial\Omega$ given by the integral representation formulas (8.45), (8.47) satisfy the conditions at infinity (8.50, 8.51). This property obviously suggests that integral formulations are very attractive for solving exterior problems.

Indirect boundary integral equations and fictitious eigenvalues

The derivation of indirect integral representations for frequency-domain elastodynamics follows the same steps as in Section 8.3. The various integral formulations for exterior problems suffer from the existence of fictitious eigenvalues for exactly the same reasons as in acoustics.

8.7 NUMERICAL EXAMPLES IN TIME-HARMONIC ELASTODYNAMICS

Scattering of an elastic plane wave by a spherical cavity (Bonnet [2])

A spherical cavity (radius R) is embedded in an unbounded elastic body Ω; the cavity boudary is traction-free: ($\boldsymbol{t} = \boldsymbol{0}$ on $\partial\Omega$). It is illuminated by an incident longitudinal plane wave $\boldsymbol{u}^I$ which propagates along $\boldsymbol{e}_3$:

$$\boldsymbol{u}^I(y) = \boldsymbol{u}_0 \mathrm{e}^{\mathrm{i}k_L y_3} \boldsymbol{e}_3 \tag{8.53}$$

The incident field $\boldsymbol{u}^I$ does not satisfy the conditions at infinity (8.50, 8.51). In order to compute the disturbance created by the presence of the cavity, the total displacement $\boldsymbol{u}^{total}$ is split:

$$\boldsymbol{u}^{total} = \boldsymbol{u}^I + \boldsymbol{u}$$

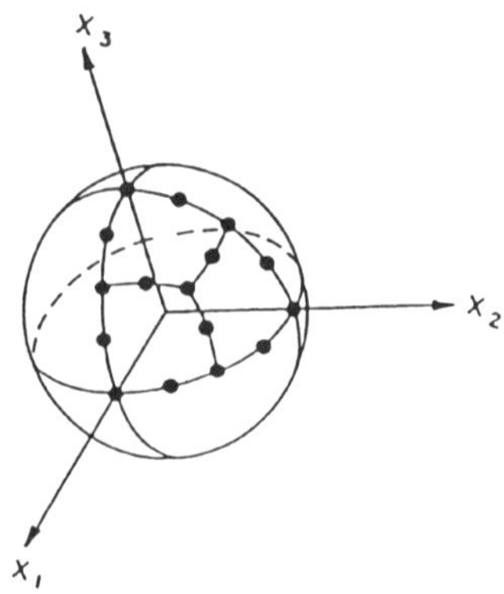

Figure 8.1 Mesh M_1.

Since $\partial\Omega$ is traction-free, the scattered field $\boldsymbol{u}$, which is the new unknown, satisfies a Neumann condition:

$$\boldsymbol{T}^n(\boldsymbol{u}) = -\boldsymbol{T}^n(\boldsymbol{u}^I) \qquad \text{on } \partial\Omega \tag{8.54}$$

The problem defined by eqs (8.32), (8.53), (8.54) has an analytical solution (Eringen and Suhubi [6]), and thus provides a good test for elastodynamic boundary element programs (see e.g. Rizzo et al. [12]). The numerical results shown below were obtained using the displacement boundary integral equation (8.48), in discretized form. Two types of errors were computed between the numerical and analytical solutions: the maximum relative error

$$\Delta u_{max} = \max_{\boldsymbol{y}\in\partial\Omega} \frac{|u - u_{exact}|}{|u_{exact}|} \times 100(\%)$$

and the relative mean-squared error

$$\Delta u_{av}^2 = \frac{\int_{\partial\Omega} |u - u_{exact}|^2 \,\mathrm{d}S_y}{\int_{\partial\Omega} |u_{exact}|^2 \,\mathrm{d}S_y} \times 100(\%)$$

Owing to the symmetry of the mesh with respect to the three coordinate planes, only one octant of the sphere is actually modelled. Results are displayed here for two meshes, respectively denoted M_1 (3 elements, 16 nodes) and M_2 (12 elements, 49 nodes); both employ the quadrilateral, eight-noded boundary element (Appendix B).

Results: displacement on the boundary. The relative errors, expressed in percentages, are displayed in Table 8.3 for several values of the non-dimensional wavenumber $k_L r = \omega R/c_L$:

Results: far-field displacement. Figure 8.2 shows values computed using mesh M_2 of the moduli $|u_r|, |u_\theta|$ on the cavity surface ($r/R = 1$) and in far-field

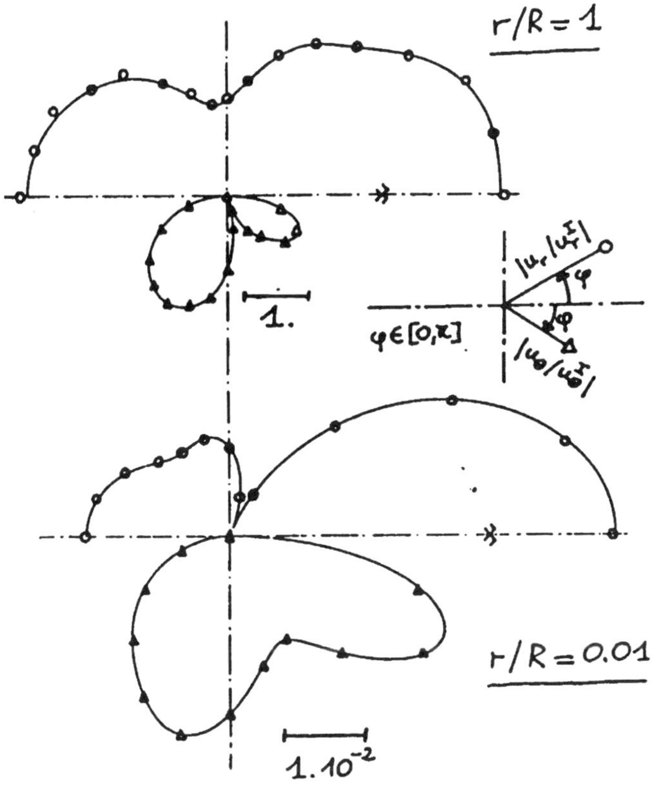

Figure 8.2 Diffraction of a plane L elastic wave by a spherical cavity (from [2]).

conditions ($r/R = 100$), where (r, θ, ϕ) denote spherical coordinates originating at the cavity centre and of axis e_3. The good agreement between the analytical solution (lines) and the computed values (symbols) is apparent.

Table 8.3 Displacement on the boundary: percentage errors for several values of the non-dimensional wavenumber $k_L r = \omega R/c_L$

$k_L R$	Δu_{max}	Δu_{moy}
1.	3.50 (M_1)	2.00 (M_1)
1.	0.43 (M_2)	0.19 (M_2)
2.	1.50 (M_2)	0.68 (M_2)
2.8	5.40 (M_2)	2.20 (M_2)
3.	7.70 (M_2)	3.50 (M_2)

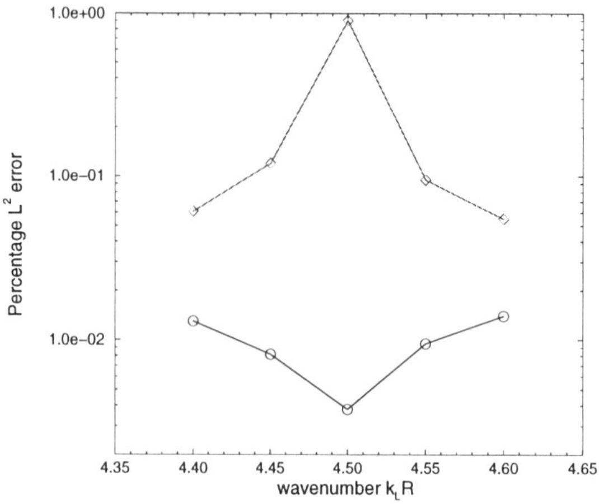

Figure 8.3 Avoidance of the fictitious eigenvalue $k_L R \approx 4.493$ using the Schenck method (from [2]).

Fictitious eigenvalues. The integral equation (8.48) for the Neumann exterior problem has a non-unique solution if ω is an eigenvalue of the Dirichlet interior problem (Section 8.3). For the spherical cavity, the lowest such eigenvalue is [6] $k_L R \approx 4.493$. To illustrate the effect of the Schenck method in avoiding fictitious eigenvalues, the error Δu_{moy} (using mesh M_4: 48 elements, 169 nodes) has been computed for several wave numbers $k_L R$ close to the above critical value, with and without using the Schenck method, see Figure 8.3. The numerical results are clearly divergent at the eigenvalue when no measure is taken to avoid it.

Relationship between mesh size and wavelength. The effect of ω (or, equivalently, of the wavelength) on the numerical results is illustrated in Table 8.4, for $k_T R \approx 8/3$ (using mesh M_1) and $k_T R \approx 4/3$ (using mesh M_2). Denoting

Table 8.4 Relationship between mesh size and wavelength (from [2]).

n	Mesh M_1			Mesh M_2		
	Δu_{max}	Δu_{av}	$[\text{Cond}K]^{-1}$	Δu_{max}	Δu_{av}	$[\text{Cond}K]^{-1}$
3.34	2.1	1.8	0.11	1.4	0.79	0.032
3.	2.1	1.49	0.10	2.7	1.59	0.025
2.73	2.4	1.72	0.094	6.2	3.02	0.003
2.5	5.7	3.42	0.10	10.2	5.64	0.014
2.31	11.9	6.98	0.10	19.	10.4	0.004

by n the number of element edges per transverse wavelength along a diameter of the sphere, it is apparent that below approx. $n = 3$ the accuracy of the numerical results substantially deteriorates (using eight-noded quadrilaterals). For this choice of boundary element, the mesh size should thus be determined so as to provide at least three element edges, or seven nodes, per transverse wave length.

Dirichlet and Neumann problems: performance comparison [2]

The spherical cavity embedded in an infinite elastic body and submitted to a (time-harmonic) uniform pressure is another classical problem for which an exact solution is known, namely [6]:

$$
\begin{aligned}
u_r &= -\frac{pR^3}{4\pi r^2}\frac{4\beta^2(\mathrm{i}x_L - 1)}{4\beta^2(1-\mathrm{i}k_L r) - (k_L R)^2}\mathrm{e}^{\mathrm{i}k_L(r-R)} \\
\sigma_{rr} &= -\frac{pR^3}{r^3}\frac{4\beta^2(1-\mathrm{i}x_L) - x_L^2}{4\beta^2(1-\mathrm{i}k_L r) - (k_L R)^2}\mathrm{e}^{\mathrm{i}k_L(r-R)} \\
\sigma_{\theta\theta} &= \frac{pR^3}{r^3}\frac{4\beta^2(1-\mathrm{i}x_L) + (2-4\beta^2)x_L^2}{4\beta^2(1-\mathrm{i}k_L r) - (k_L R)^2}\mathrm{e}^{\mathrm{i}k_L(r-R)} = \sigma_{\phi\phi}
\end{aligned}
$$

where p is the applied pressure and (r, θ, ϕ) is a spherical coordinate system emanating from the cavity centre. Here this solution is used to compare the accuracy of the boundary element method based on the displacement integral equation (8.48) in solving a Dirichlet or a Neumann problem, with the prescribed boundary data $\boldsymbol{u}^D$ (Dirichlet) or $\boldsymbol{t}^D$ (Neumann) given by:

$$
\boldsymbol{t}^D = p\boldsymbol{e}_r \qquad \boldsymbol{u}^D = -\frac{pR}{4\pi}\frac{4\beta^2(\mathrm{i}k_L r - 1)}{4\beta^2(1-\mathrm{i}k_L r) - (k_L R)^2}
$$

The results obtained for both boundary-value problems [2] are displayed in Table 8.5. It is seen that the numerical results obtained using the displacement integral equation (8.48) are much better for the Neumann problem than for the Dirichlet problem. A possible explanation is that the boundary integral equations take the respective forms:

$$
\begin{aligned}
\frac{1}{2}u_k(\boldsymbol{x}) + \int_{\partial\Omega}^{\star} u_i(\boldsymbol{y})T_i^k(\boldsymbol{x},\boldsymbol{y};\omega)\,\mathrm{d}S_y &= \int_{\partial\Omega} t_i^D(\boldsymbol{y})U_i^k(\boldsymbol{x},\boldsymbol{y};\omega)\,\mathrm{d}S_y \\
\int_{\partial\Omega} t_i^D(\boldsymbol{y})U_i^k(\boldsymbol{x},\boldsymbol{y};\omega)\,\mathrm{d}S_y &= \frac{1}{2}u_k^D(\boldsymbol{x}) + \int_{\partial\Omega}^{\star} u_i^D(\boldsymbol{y})T_i^k(\boldsymbol{x},\boldsymbol{y};\omega)\,\mathrm{d}S_y
\end{aligned}
$$

i.e. are (singular) Fredholm integral equations of the second and first kinds, respectively.[1] It is known from the classical theory of nonsingular integral equations that the solutions to Fredholm equations of the second kind are stable

[1]Putting the integral equations in CPV form allows to emphasize the Fredholm equation structure.

Table 8.5 Pressurized spherical cavity: performances for Neumann (top) and Dirichlet (bottom) problems (errors expressed in percentage value), from [2].

$k_T R$	mesh M_1: Δu_{max}	Δu_{moy}	mesh M_2: Δu_{max}	Δu_{moy}
0.	2.8	2.4	.20	0.19
0.3162	2.8	2.4	.20	0.19
1.581	2.5	2.5	.26	0.20
2.8	2.2	1.5	.22	0.10
3.162	5.1	3.0	.20	0.10
5.6	—	—	6.40	3.4

$k_T R$	mesh M_1: Δu_{max}	Δu_{moy}	mesh M_2: Δu_{max}	Δu_{moy}
0.	2.1	1.3	10.0	4.9
0.3162	2.5	2.3	14.0	6.6
1.581	18.0	11.0	10.0	4.9
2.8	—	—	8.3	4.1
3.162	—	—	8.3	4.9
5.6	—	—	18.0	11.0

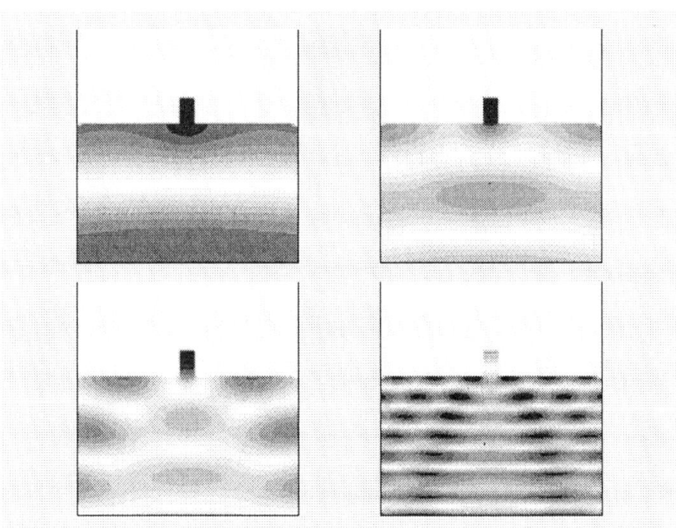

Figure 8.4 Building on a half-space: vertically incident plane SH shear wave. The figure shows antiplane displacement at various frequencies (from Semblat [13]).

with respect to right-hand-side perturbations whereas solutions to first-kind equations are unstable (Wing [15]). The above numerical results suggest that the singular first-kind equation is unstable as well; in particular, a mesh refinement ($M_1 \rightarrow M_2$) appears to cause the numerical solution of the Dirichlet problem to *deteriorate*.

This example suggests we can solve the Dirichlet problem using a traction integral equation like (8.49), which has a second-kind structure. Boundary element methods based on a traction BIE are not yet very common, however, except for crack problems, where they are necessary. The fact that their regularization has been understood only in recent years, together with the implementation difficulties associated with the $C^{1,\alpha}$ regularity requirement at the collocation points for $\boldsymbol{u}$ (Chapter 5), probably explain why this is so.

Elastic wave propagation in semi-infinite media (Semblat [13])

Figures 8.4 to 8.5 show some numerical results communicated by Semblat [13]. They were obtained using the boundary element module implemented in the computer code CESAR of the Laboratoire Central des Ponts et Chaussées (Civil Engineering Central Laboratory), Paris. Elastic wave propagation in semi-infinite media is treated by using a full-space fundamental solution, which entails the use of infinite elements on the free surface and the interfaces (in that case developed by Dangla [4]).

Figure 8.5 Diffraction of an SH shear wave by cylindrical cavities for two different values of the inter-cavity spacing. The figure shows isovalues for the antiplane displacement (from Semblat [13]).

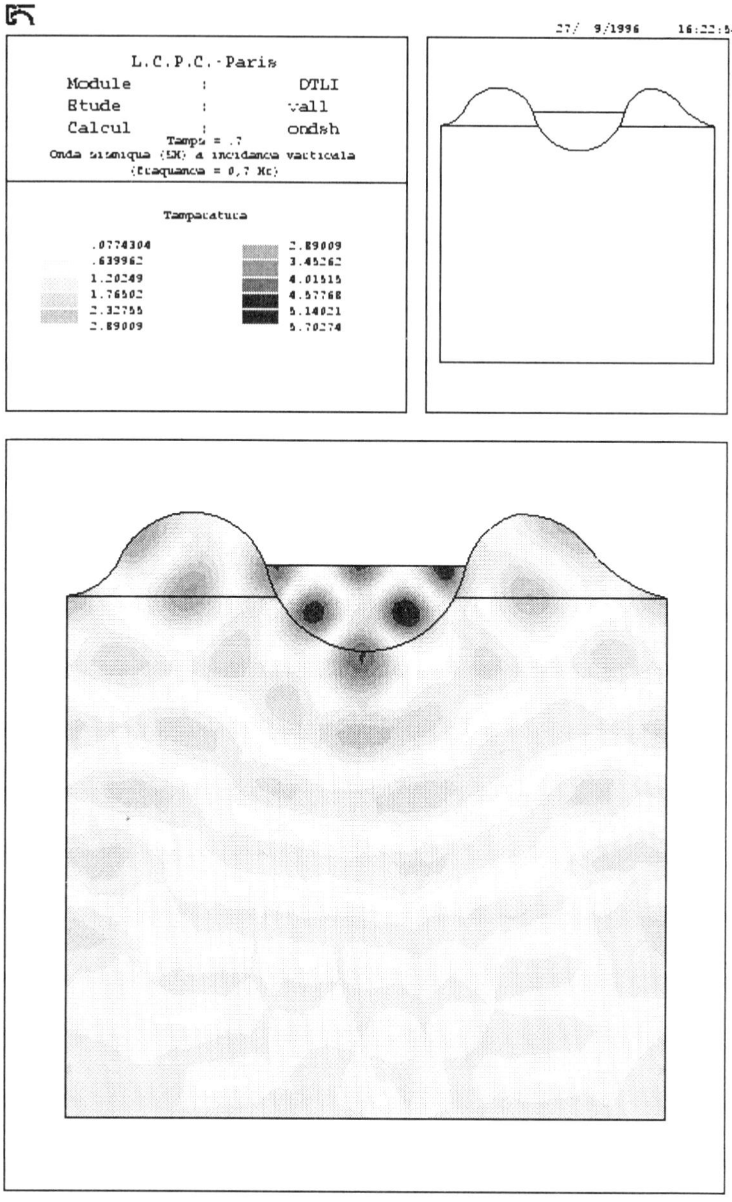

Figure 8.6 Valley in a non-planar substratum: vertically incident plane SH shear wave. The figure shows antiplane displacement for a certain frequency corresponding to a seismic amplification factor (from Semblat [13]).

Wave propagation in semi-infinite media (Clouteau [3])

In this example [3], an ellipsoidal valley (bounded by the surface of equation $0.25y_1^2 + y_2^2 + y_3^2 = 1$ in non-dimensional coordinates) is embedded in a homogeneous elastic half-space (free surface $y_3 = 0$) such that $c_L/c_T = 2$.The valley is either empty or filled with an elastic material such that $c_L/c'_L = 2$, $c_T/c'_T = 2$.

Due to the presence of the valley, the propagation of elastic waves is modified ('site effect'). The fundamental solution for an elastic half-space with free surface (see Kennett [10] for the stratified half-space) is used, so that the boundary element discretization is reduced to only the half-ellipsoidal surface/interface of the valley (Figure 8.7). Figure 8.8 shows some numerical results obtained by Clouteau [3].

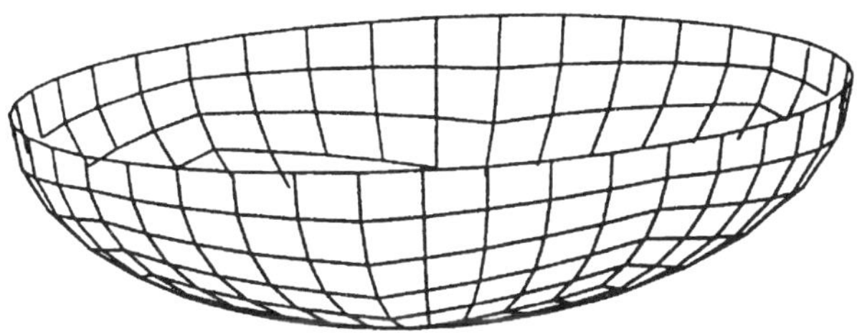

Figure 8.7 Boundary element mesh for an ellipsoidal valley (from Clouteau [3]).

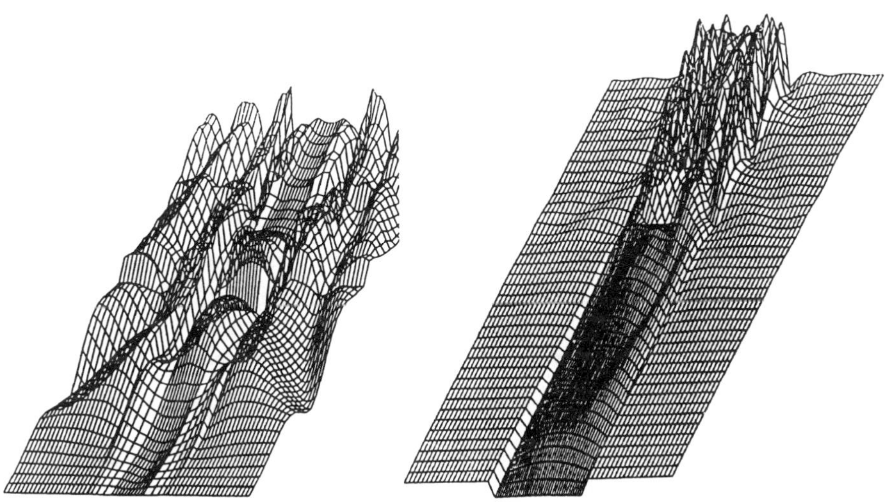

Figure 8.8 Site effect in an empty (left) or filled (right) ellipsoidal valley: vertical displacement along the major axis of the ellipsoid (vertically incident compressive wave), from Clouteau [3].

REFERENCES

1. ABRAMOWITZ, M., STEGUN, I. *Handbook of mathematical functions.* Dover Publications, New York (1972).

2. BONNET, M. *Méthode des équations intégrales régularisées en élastodynamique tridimensionnelle.* Doctoral thesis, Ecole Nationale des Ponts et Chaussées, Paris (1986).

3. CLOUTEAU, D. *Propagation d'ondes dans les milieux hétérogènes. Application à la tenue des ouvrages sous séismes.* Ph.D. thesis, Ecole Centrale de Paris, France (1990).

4. DANGLA, P. *Couplage éléments finis – équations intégrales en élastodynamique et interaction sol-structure.* Doctoral thesis, Ecole Nationale des Ponts et Chaussées, Paris, France (1989).

5. DONGARRA, J. J., BUNCH, J. R., MOLER, C. B., STEWART, G. B. LINPACK *users'guide.* Society for Industrial and Applied Mathematics (SIAM), Philadelphia, USA (1979).

6. ERINGEN, A. C., SUHUBI, E. S. *Elastodynamics (vol. II – linear theory).* Academic Press (1975).

7. GRADSHTEYN, I. S., RYZHIK, I. M. *Tables of integrals, series and products.* Academic Press (1980).

8. HALPERN, L. *Etude des conditions aux limites absorbantes pour des schémas numériques relatifs à des équations hyperboliques linéaires.* Ph.D. thesis, Université Paris VI (1987).

9. HARARI, I., BARBONE, P. E., MONTGOMERY, J. M. Finite element formulations for exterior problems: application to hybrid methods, non-reflecting boundary conditions, and infinite elements. *Int. J. Num. Meth. in Eng.*, **40**, 2791–2805 (1997).

10. KENNETT, B. L. N. Seismic wave propagation in stratified media. In *Advances in Applied Mechanics*, vol. 21. Academic Press (1981).

11. REZAYAT, M., SHIPPY, D. J., RIZZO, F. J. On time-harmonic elastic wave analysis by the boundary element method for moderate to high frequencies. *Comp. Meth. in Appl. Mech. Engng.*, **55**, 349–367 (1986).

12. RIZZO, F. J., SHIPPY, D. J., REZAYAT, M. A boundary integral equation method for radiation and scattering of elastic waves in three dimensions. *Int. J. Num. Meth. in Eng.*, **21**, 115–129 (1985).

13. SEMBLAT, J. F. Private communication. Laboratoire Central des Ponts et Chaussées, Paris (1997).

14. THOMPSON, L. L., PINSKI, P. M. A space-time finite element method for the exterior structural acoustics problem: time-dependent radiation boundary conditions in two space dimensions. *Int. J. Num. Meth. in Eng.*, **39**, 1635–1657 (1996).

15. WING, G. M. *A primer on integral equations of the first kind.* SIAM Publications, Philadelphia, USA (1991).

9

Diffusion, fluid flow

9.1 DIFFUSION EQUATION

One seeks a boundary integral formulation for the scalar diffusion equation (unknown time-dependent field $u(\boldsymbol{y}, t)$):

$$k\Delta u - \dot{u} = 0 \qquad \forall(\boldsymbol{y}, t) \in \Omega \times [0, T] \tag{9.1}$$

where k is a known physical constant of the medium under consideration. For simplicity, distributed sources over Ω are not considered here (their introduction would eventually lead to additional integrals over the domain, like in potential theory or elasticity). As usual, a well-posed problem must include initial conditions (conventionally set at $t = 0$):

$$u(\boldsymbol{y}, 0) = u_0(\boldsymbol{y})$$

as well as boundary conditions (prescription of either u, $q = \boldsymbol{\nabla} u.\boldsymbol{n}$ or some relation between u and q at any point of the boundary $\partial\Omega$); in addition, conditions at infinity should be introduced if Ω is unbounded.

Reciprocity identity

Following the general approach introduced in chapter 7 for the wave equation in the time domain, applying the Green's formula (2.4) to two fields u^1, u^2, solution of the governing equation (9.1) and integrating the resulting equality in time over $\tau \in [0,t]$ leads to:

$$\int_{\partial\Omega} \left[(q^1 \star u^2)(\boldsymbol{y},t) - (q^2 \star u^1)(\boldsymbol{y},t)\right] \mathrm{d}S_y$$
$$= \int_{\Omega} \frac{1}{k} \left[u_0^2(\boldsymbol{y})u^1(\boldsymbol{y},t) - u_0^1(\boldsymbol{y})u^2(\boldsymbol{y},t)\right] \mathrm{d}V_y \quad (9.2)$$

where the symbol $\star$ denotes the *time convolution product* (7.3). One notes that the domain integral vanishes for homogeneous initial conditions u_0^1, u_0^2.

Fundamental solution

For given domain $E \subseteq \mathbb{R}^3$ and source point $\boldsymbol{x} \in E$, the fundamental solution $G(\boldsymbol{x},t,\boldsymbol{y})$ is the response in E to a unit time-impulsive point source applied at $\boldsymbol{x}$ at the instant $t = 0$ and for a homogeneous initial condition. The governing equation for $G(\boldsymbol{x},t,\boldsymbol{y})$ is thus:

$$k\Delta G - \dot{G} + \delta(\boldsymbol{y} - \boldsymbol{x})\delta(t) = 0 \quad (9.3)$$

together with homogeneous boundary conditions on ∂E. The infinite-space fundamental solution ($E = \mathbb{R}^3$) is given (see e.g. [2]) by:

$$G(\boldsymbol{x},t,\boldsymbol{y}) = \frac{1}{(4\pi kt)^{3/2}} \exp\left(-\frac{r^2}{4kt}\right) \quad (9.4)$$

and possesses the symmetry property:

$$G(\boldsymbol{x},t,\boldsymbol{y}) = G(\boldsymbol{y},t,\boldsymbol{x})$$

The singular behaviour of the fundamental solution (9.4) about $r = 0$ is not directly apparent. However, it is easy to show, for example by a direct calculation, that:

$$\int_0^t G(\boldsymbol{x},t-\tau,\boldsymbol{y})\, \mathrm{d}\tau = \frac{1}{4\pi r} + O(1) \qquad (r \sim 0) \quad (9.5)$$

Integral representation formula for the diffusion equation

Consider a bounded domain Ω and let $\boldsymbol{x}$ denote an arbitrary fixed point *located away from the boundary* $\partial\Omega$. The reciprocity identity (9.2) applied to the sought-for solution (state 1) and to the time-impulsive fundamental solution (state 2) leads to:

$$u(\boldsymbol{x},t) + \int_{\partial\Omega} \left\{H(\boldsymbol{x},t,\boldsymbol{y}) \star u(\boldsymbol{y},t) - G(\boldsymbol{x},t,\boldsymbol{y}) \star q(\boldsymbol{y},t)\right\} \mathrm{d}S_y$$
$$= \frac{1}{k} \int_{\Omega} u_0(\boldsymbol{y}) G(\boldsymbol{x},t,\boldsymbol{y})\, \mathrm{d}V_y \quad (9.6)$$

where $H = G_{,j}n_j$ denotes the normal derivative of the fundamental solution and with $\kappa = 1$ ($\boldsymbol{x} \in \Omega$) or $\kappa = 0$ ($\boldsymbol{x} \notin \Omega$). This formula is applicable for any time instant $t \geq 0$ and any observation point $\boldsymbol{x} \notin \partial\Omega$. Then, the integral representation of the gradient $\boldsymbol{\nabla} u(\boldsymbol{x})$ is obtained by means of a straightforward differentiation under the integral sign of eq. (9.6).

Boundary integral equation

The integral equation results from the usual limiting process $\varepsilon \to 0$ in the exterior representation formula for the domain $\Omega_\varepsilon = \Omega - v_\varepsilon$ obtained by removing from Ω an exclusion neighbourhood $v_\varepsilon(\boldsymbol{x})$ of $\boldsymbol{x}$. The property (9.5) (transfer of the singularity at $\boldsymbol{y} = \boldsymbol{x}$ from the time-dependent fundamental solution to its static counterpart) is used. Assuming $u_0 = 0$ for simplicity, the following regularized integral equation is obtained:

$$\int_{\partial\Omega} \{H(\boldsymbol{x}, t, \boldsymbol{y}) \star u(\boldsymbol{y}, t) - u(\boldsymbol{y}, t) H(\boldsymbol{x}, \boldsymbol{y})\} \, \mathrm{d}S_y \\ + \int_{\partial\Omega} [u(\boldsymbol{y}, t) - u(\boldsymbol{x}, t)] H(\boldsymbol{x}, \boldsymbol{y}) \, \mathrm{d}S_y - \int_{\partial\Omega} G(\boldsymbol{x}, t, \boldsymbol{y}) \star q(\boldsymbol{y}, t) \, \mathrm{d}S_y = 0 \tag{9.7}$$

Discretization

The approach presented in Chapter 7 for the wave equation in the time domain is applicable here, and the structure of the resulting discretized formulation is similar. In particular, the 'temporal translation' property is again verified provided a constant time step is being used. On the other hand, the concept of causality is no longer present.

Formulation in the Laplace domain

The Laplace transform:

$$f(t) \to \hat{f} = \mathcal{L}f(s) = \int_0^{+\infty} \mathrm{e}^{-st} f(t) \, \mathrm{d}t \tag{9.8}$$

applied to eq. (9.1) yields the transformed equation:

$$[k\Delta\hat{u} - s]\hat{u} = 0 \tag{9.9}$$

The infinite-space fundamental solution for this equation is:

$$G(\boldsymbol{x}, \boldsymbol{y}, s) = \frac{1}{4\pi k r} \mathrm{e}^{-r\sqrt{s/k}} \tag{9.10}$$

It is thus apparent that:

$$kG(\boldsymbol{x}, \boldsymbol{y}, s) - \frac{1}{4\pi r} = O(1) \qquad r \sim 0$$

so that, here again, a regularized boundary integral equation is obtained by transferring the singularity into the static fundamental solution:

$$\int_{\partial\Omega} [H(\boldsymbol{x},\boldsymbol{y},s) - H(\boldsymbol{x},\boldsymbol{y})]u(\boldsymbol{y},s)\,\mathrm{d}S_y + \int_{\partial\Omega} [u(\boldsymbol{y},s) - u(\boldsymbol{x},s)]H(\boldsymbol{x},\boldsymbol{y})\,\mathrm{d}S_y \\ - \int_{\partial\Omega} G(\boldsymbol{x},\boldsymbol{y},s) \star q(\boldsymbol{y},s)\,\mathrm{d}S_y = 0 \qquad (9.11)$$

where $H = kG_{,i}n_i$.

9.2 FLUID FLOW

Potential fluid flow

Irrotational flows of inviscid, incompressible fluids are formulated in terms of a harmonic velocity scalar potential. This class of fluid flow is thus especially well suited to boundary element solution techniques of the kind presented in Chapter 3.

This class of problems also includes inviscid fluid flow in porous media (Darcy law) [6] and free surface flows (gravity waves). Examples as well as many references can be found in Liggett [4].

Navier–Stokes equations

Flows of incompressible Newtonian fluids are governed by the Navier–Stokes equations:

$$\begin{cases} -\boldsymbol{\nabla} p + \mu\Delta\boldsymbol{U} = \rho\left(\dfrac{\partial\boldsymbol{U}}{\partial t} + \boldsymbol{\nabla}\boldsymbol{U}.\boldsymbol{U}\right) - \rho g \\ ,\ \mathrm{div}\boldsymbol{U} = 0 \end{cases} \qquad (9.12)$$

associated with boundary conditions, either kinematical (e.g. adherence to an obstacle ($\boldsymbol{U} = \boldsymbol{U}_{obstacle}$), given velocity at infinity) or dynamical (e.g. equilibrium with external pressure). The stress tensor depends linearly on the velocity through the constitutive equation of incompressible Newtonian fluids:

$$\sigma_{ij} = -p\delta_{ij} + \mu(U_{i,j} + U_{j,i}) \qquad (9.13)$$

The Navier–Stokes equations are nonlinear due to the presence of the convective term[1] $\boldsymbol{\nabla U.U}$. For this reason, the integral formulation is very complex and involves domain integrals. The reader is referred to Liggett [4], Wu and Wahbah [7].

[1]The acceleration is indeed the material derivative of the velocity, as defined by eq. (12.3).

Steady-state flows of viscous incompressible fluids at low velocities

The velocity $\boldsymbol{U}$ and the pressure p for a viscous incompressible fluid (viscosity μ) verify, for low-velocity steady-state flows and assuming zero gravity, the Stokes equations:

$$\begin{cases} -\boldsymbol{\nabla} p + \mu \Delta \boldsymbol{U} = \boldsymbol{0} \\ , \ \mathrm{div}\boldsymbol{U} = 0 \end{cases} \tag{9.14}$$

The low-velocity assumption is used to neglect the convective term $\boldsymbol{\nabla U . U}$ and allows us to linearize the flow equations.

The Stokes equations possess a fundamental solution, often called *stokeslet* (velocity $\boldsymbol{U}^k$, pressure P^k) associated with a point source applied at $\boldsymbol{x}$ along the $\boldsymbol{e}_k$-direction. Its expression for three-dimensional problems is:

$$U_i^k(\boldsymbol{x}, \boldsymbol{y}) = \frac{1}{8\pi r}[\delta_{ik} + r_{,i} r_{,k}] \qquad P^k(\boldsymbol{x}, \boldsymbol{y}) = -\frac{\mu}{4\pi}(\frac{1}{r})_{,k} \tag{9.15}$$

The components Σ_{ij}^k of the stress tensor assocated with $(\boldsymbol{U}^k, P^k)$ are given by:

$$\Sigma_{ij}^k(\boldsymbol{x}, \boldsymbol{y}) = -\frac{3\mu}{4\pi} \frac{r_{,i} r_{,j} r_{,k}}{r^2} \tag{9.16}$$

and satisfy:

$$\begin{cases} -\boldsymbol{\nabla} P^k + \mu \Delta \boldsymbol{U}^k + \delta(\boldsymbol{y} - \boldsymbol{x}) \boldsymbol{e}_k = 0 \\ , \ \mathrm{div}\boldsymbol{U} = 0 \end{cases} \tag{9.17}$$

One notices that U_i^k and (P^k, Σ_{ij}^k) have $|\boldsymbol{y} - \boldsymbol{x}|^{-1}$ and $|\boldsymbol{y} - \boldsymbol{x}|^{-2}$ singularities, respectively. Also, for a fluid domain Ω and fields $\boldsymbol{U}^1, \boldsymbol{V}^2, p^1$ sufficiently smooth, one has:

$$\int_\Omega \boldsymbol{U}^2 . (\mu \Delta \boldsymbol{U}^1 - \boldsymbol{\nabla} p^1) \, \mathrm{d}V$$
$$= \int_{\partial\Omega} \boldsymbol{U}^2 . \boldsymbol{\sigma}^1 . \boldsymbol{n} \, \mathrm{d}S - \int_\Omega (\mu \boldsymbol{\nabla U}^1 : \boldsymbol{\nabla U}^2 - p \mathrm{div} \boldsymbol{U}^2) \, \mathrm{d}V$$

which, for divergence-free velocity fields $\boldsymbol{U}^1, \boldsymbol{U}^2$, yields the reciprocity identity:

$$\int_\Omega \{\mu(\boldsymbol{U}^2 . \Delta \boldsymbol{U}^1 - \boldsymbol{U}^1 . \Delta \boldsymbol{U}^2) - (\boldsymbol{U}^2 . \boldsymbol{\nabla} p^1 - \boldsymbol{U}^1 . \boldsymbol{\nabla} p^2)\} \, \mathrm{d}V$$
$$= \int_{\partial\Omega} \{\boldsymbol{U}^2 . \boldsymbol{t}^1 - \boldsymbol{U}^1 . \boldsymbol{t}^2\} \, \mathrm{d}S$$

where $\boldsymbol{t} = \boldsymbol{\sigma} . \boldsymbol{n}$ denotes the traction vector.

Following the usual procedure, the reciprocity identity is applied to the unknown state (state 1) and the fundamental solution (9.15), (9.16) (state 2). This yields the representation formula of the velocity at a point $\boldsymbol{x}$ not on the boundary:

$$U_k(\boldsymbol{x}) = \int_{\partial\Omega} \{t_i(\boldsymbol{y})U_i^k(\boldsymbol{x},\boldsymbol{y}) - U_i(\boldsymbol{y})T_i^k(\boldsymbol{x},\boldsymbol{y})\}\ \mathrm{d}S_y$$

where $T_i^k = \Sigma_{ij}^k$ are the components of the fundamental traction vector. The Laplacian of $\boldsymbol{U}(\boldsymbol{x})$ is therefore represented by:

$$U_{k,\ell\ell}(\boldsymbol{x}) = \int_{\partial\Omega} \{t_i(\boldsymbol{y})U_{i,\ell\ell}^k(\boldsymbol{x},\boldsymbol{y}) - U_i(\boldsymbol{y})\Sigma_{ij,\ell\ell}^k(\boldsymbol{x},\boldsymbol{y})n_j(\boldsymbol{y})\}\ \mathrm{d}S_y \qquad (9.18)$$

(where we use the equality of second-order $\boldsymbol{x}$-derivatives and $\boldsymbol{y}$-derivatives of the fundamental solution (9.15)). Moreover, one notes that the fundamental solution (9.15) verifies:

$$U_{i,\ell\ell}^k(\boldsymbol{x},\boldsymbol{y}) = U_{i,\ell k}^\ell(\boldsymbol{x},\boldsymbol{y}) = -U_{i,\ell\bar{k}}^\ell(\boldsymbol{x},\boldsymbol{y})$$

$$P_{,\ell\ell}^k(\boldsymbol{x},\boldsymbol{y}) = P_{,\ell k}^\ell(\boldsymbol{x},\boldsymbol{y}) = -P_{,\ell\bar{k}}^\ell(\boldsymbol{x},\boldsymbol{y})$$

Since $\Delta\boldsymbol{U} = \boldsymbol{\nabla}p$ from (9.14), this property allows to establish the integral representation formula of the pression $p(\boldsymbol{x})$ by integrating eq. (9.18) with respect to x_k:

$$p(\boldsymbol{x}) - p_0 = \int_{\partial\Omega} \{U_i(\boldsymbol{y})\Sigma_{ij,\ell}^\ell(\boldsymbol{x},\boldsymbol{y})n_j(\boldsymbol{y}) - t_i(\boldsymbol{y})U_{i,\ell}^\ell(\boldsymbol{x},\boldsymbol{y})\}\ \mathrm{d}S_y \qquad (9.19)$$

(p_0 being an arbitrary integration constant).

Equation $(9.14)_1$ expresses in fact the equilibrium of the stresses given by the constitutive equation (9.13): div$\boldsymbol{\sigma} = \boldsymbol{0}$. One deduces at once the identity:

$$\int_{\partial\Omega} T_i^k(\boldsymbol{x},\boldsymbol{y})\,\mathrm{d}S_y + \kappa\delta_{ik} = 0$$

where $\kappa = 1$ ($\boldsymbol{x} \in \Omega$), $\kappa = 0$ ($\boldsymbol{x} \notin \Omega$). This identity allows us, like in elasticity, to formulate a regularized boundary integral equation:

$$\int_{\partial\Omega} \{[U_i(\boldsymbol{y}) - U_i(\boldsymbol{x})]T_i^k(\boldsymbol{x},\boldsymbol{y}) - t_i(\boldsymbol{y})U_i^k(\boldsymbol{x},\boldsymbol{y})\}\ \mathrm{d}S_y = 0 \qquad (9.20)$$

Numerical resolution

One notices the strong formal similarity between the integral equations and representations (9.15)–(9.20) for the Stokes problem and (4.8)–(4.32) for elastostatics. In particular, the discretization techniques introduced in chapter 4 are fully applicable here.

REFERENCES

1. ABOUSLEIMAN, Y., CHENG, A. H. D. Boundary element solution for steady and unsteady Stokes flow. *Comp. Meth. in Appl. Mech. Engng.*, **117**, 1–13 (1994).
2. CHANG, Y. P., KANG, C. S., CHEN, D. J. The use of fundamental Green's functions for the solutions of problems of heat conduction in anisotropic media. *Int. J. Heat Mass Transfer*, **16**, 1905–1918 (1973).
3. HU, H. Study of integral equation methods for transonic flow calculations. *Engng. Anal. with Bound. Elem.*, **11**, 101–107 (1993).
4. LIGGETT, J. A. Fluid mechanics. In D. E. Beskos (ed.), *Boundary element methods in mechanics.*, vol. 1, pp. 367–416. North-Holland (1987).
5. POWER, H., PARTRIDGE, P. W. The use of Stokes' fundamental solution for the boundary-only element formulation of the three-dimensional Navier–Stokes equations for moderate Reynolds numbers. *Int. J. Num. Meth. in Eng.*, **37**, 1825–1840 (1994).
6. VAN DER WEIDEN, R., DE HOOP, A. T. Boundary integral equations for the computational modeling of three-dimensional groundwater flow problems. *Comp. Mech.*, **4**, 283–291 (1989).
7. WU, J. C., WAHBAH, M. *Numerical solution of viscous flow equations using integral representations*, vol. 59 of *Lecture notes in Physics*, pp. 448–453. Springer-Verlag (1976).

Part III

Advanced topics

10

Symmetric Galerkin boundary integral formulations

The *collocation* approach to boundary integral equations, which is the most widely known, has been followed in the previous chapters. However, symmetric integral formulations of Galerkin type can also be constructed, following a weighted-residual approach. This chapter covers symmetric Galerkin boundary integral formulations for three-dimensional potential and elastic problems, with body forces ignored for brevity.

Galerkin integral formulations are the subject of many studies in applied mathematics, see e.g. Bendali [7], Cortey-Dumont [13], Nedelec [22], Wendland [29]. Their applications to engineering solid mechanics are still in an early stage of development compared to collocation techniques (see Sirtori et al. [26], Parreira and Guiggiani [25], Bécache [5]). However, the research activity in this area has noticeably increased during the last few years (see the survey paper [10]), and is expected to keep doing so in the future.

The boundary of the body under consideration is assumed smooth (see Section 1.6), except for the possibility of cracks. Following the usual convention, the unit normal vector points outwards from the body.

10.1 BASIC IDEA

Ignoring temporarily the difficulties related to the singular nature of fundamental solutions, the governing potential and flux integral equations for any solution (u, q) to the Laplace equation on a domain Ω read:

$$0 = \frac{1}{2}q(\boldsymbol{x}) + \int_{\partial\Omega} u(\boldsymbol{y})G_{,\bar{i}j}(\boldsymbol{x},\boldsymbol{y})n_i(\boldsymbol{x})n_j(\boldsymbol{y})\,\mathrm{d}S_y - \int_{\partial\Omega} q(\boldsymbol{y})H(\boldsymbol{y},\boldsymbol{x})\,\mathrm{d}S_y \quad \text{(a)}$$

$$0 = \frac{1}{2}u(\boldsymbol{x}) + \int_{\partial\Omega} u(\boldsymbol{y})H(\boldsymbol{x},\boldsymbol{y})\,\mathrm{d}S_y - \int_{\partial\Omega} q(\boldsymbol{y})G(\boldsymbol{x},\boldsymbol{y})\,\mathrm{d}S_y \quad \text{(b)}$$

The basic idea consists in multiplying equation (a) by a test function $\tilde{u}$ and equation (b) by a test function $\tilde{q}$, and then integrating the resulting equalities over $\partial\Omega$, in order to obtain integral equations defined in the weighted residuals sense:

$$\begin{aligned} 0 = &\int_{\partial\Omega}\int_{\partial\Omega} \tilde{u}(\boldsymbol{x})u(\boldsymbol{y})G_{,\bar{i}j}(\boldsymbol{x},\boldsymbol{y})n_i(\boldsymbol{x})n_j(\boldsymbol{y})\,\mathrm{d}S_y\,\mathrm{d}S_x \\ &+ \frac{1}{2}\int_{\partial\Omega} q(\boldsymbol{x})\tilde{u}(\boldsymbol{x})\,\mathrm{d}S_x - \int_{\partial\Omega}\int_{\partial\Omega} \tilde{u}(\boldsymbol{x})q(\boldsymbol{y})H(\boldsymbol{y},\boldsymbol{x})\,\mathrm{d}S_y\,\mathrm{d}S_x \\ 0 = &\frac{1}{2}\int_{\partial\Omega} u(\boldsymbol{x})\tilde{q}(\boldsymbol{x})\,\mathrm{d}S_x + \int_{\partial\Omega}\int_{\partial\Omega} \tilde{q}(\boldsymbol{x})u(\boldsymbol{y})H(\boldsymbol{x},\boldsymbol{y})\,\mathrm{d}S_y\,\mathrm{d}S_x \\ &- \int_{\partial\Omega}\int_{\partial\Omega} \tilde{q}(\boldsymbol{x})q(\boldsymbol{y})G(\boldsymbol{x},\boldsymbol{y})\,\mathrm{d}S_y\,\mathrm{d}S_x \end{aligned}$$

which possess a structure of the form:

$$\begin{aligned} \mathcal{B}_{uu}(u,\tilde{u}) + \mathcal{B}_{qu}(q,\tilde{u}) &= 0 \\ \mathcal{B}_{uq}(u,\tilde{q}) + \mathcal{B}_{qq}(q,\tilde{q}) &= 0 \end{aligned}$$

which is symmetric by virtue of the symmetry properties of the fundamental solution $G(\boldsymbol{x},\boldsymbol{y})$:

$$\begin{aligned} \mathcal{B}_{uu}(u,\tilde{u}) &= \mathcal{B}_{uu}(\tilde{u},u) \\ \mathcal{B}_{qq}(q,\tilde{q}) &= \mathcal{B}_{qq}(\tilde{q},q) \\ \mathcal{B}_{qu}(q,u) &= \mathcal{B}_{uq}(u,q) \end{aligned}$$

In practice, the formulation of a symmetric system of boundary integral equations needs a more sophisticated approach in order to take care of the singular behaviour of the fundamental solution and its derivatives and to incorporate boundary conditions explicitly.

10.2 DIRECT GALERKIN FORMULATION OF DIRICHLET AND NEUMANN PROBLEMS

Weighted residual for the strongly singular boundary integral equation

Recall the regularized integral equation (2.24):

$$0 = \kappa u(\boldsymbol{x}) + \int_{\partial\Omega} [u(\boldsymbol{y}) - u(\boldsymbol{x})]H(\boldsymbol{x},\boldsymbol{y})\,\mathrm{d}S_y - \int_{\partial\Omega} q(\boldsymbol{y})G(\boldsymbol{x},\boldsymbol{y})\,\mathrm{d}S_y \quad (10.1)$$

with $\kappa = 0$ (Ω bounded) and $\kappa = 1$ ($\mathbb{R}^3 - \Omega$ bounded), which holds true for any $\boldsymbol{x} \in \mathbb{R}^3$, including of course $\boldsymbol{x} \in \partial\Omega$. Upon multiplication by a test function $\tilde{q}(\boldsymbol{x})$ and integration of the result over $\partial\Omega$, one obtains:

$$\kappa \int_{\partial\Omega} \tilde{q}(\boldsymbol{x}) u(\boldsymbol{x}) \, \mathrm{d}S_x + \int_{\partial\Omega} \int_{\partial\Omega} \tilde{q}(\boldsymbol{x})[u(\boldsymbol{y}) - u(\boldsymbol{x})] H(\boldsymbol{x}, \boldsymbol{y}) \, \mathrm{d}S_y \, \mathrm{d}S_x \\ - \int_{\partial\Omega} \int_{\partial\Omega} \tilde{q}(\boldsymbol{x}) q(\boldsymbol{y}) G(\boldsymbol{x}, \boldsymbol{y}) \, \mathrm{d}S_y \, \mathrm{d}S_x = 0 \quad (10.2)$$

The inner integral is weakly singular while the outer one is nonsingular.

Symmetric Galerkin integral formulation of the Dirichlet problem

Let u be prescribed over the entire boundary: $u = u^D$ over $\partial\Omega$. Then, from eq. (10.2), the unknown flux q solves the symmetric Galerkin boundary integral equation:

$$\mathcal{B}_{qq}(q, \tilde{q}) = \mathcal{L}_q(\tilde{q}) \qquad \forall \tilde{q} \in \mathcal{V}_q \quad (10.3)$$

where $\mathcal{V}_q$ denotes the set of piecewise continuous functions $\tilde{q}$ on $\partial\Omega$, and:

$$\mathcal{B}_{qq}(q, \tilde{q}) = \int_{\partial\Omega} \int_{\partial\Omega} \tilde{q}(\boldsymbol{x}) q(\boldsymbol{y}) G(\boldsymbol{x}, \boldsymbol{y}) \, \mathrm{d}S_y \, \mathrm{d}S_x$$

$$\mathcal{L}_q(\tilde{q}) = \kappa \int_{\partial\Omega} \tilde{q}(\boldsymbol{x}) u^D(\boldsymbol{x}) \, \mathrm{d}S_x + \int_{\partial\Omega} \int_{\partial\Omega} \tilde{q}(\boldsymbol{x})[u^D(\boldsymbol{y}) - u^D(\boldsymbol{x})] H(\boldsymbol{x}, \boldsymbol{y}) \, \mathrm{d}S_y \, \mathrm{d}S_x$$

The symmetric character of the bilinear form $\mathcal{B}_{qq}$ stems from the symmetry property (2.9):

$$G(\boldsymbol{x}, \boldsymbol{y}) = G(\boldsymbol{y}, \boldsymbol{x}) \quad (10.4)$$

Weighted residual for the hypersingular boundary integral equation

This time, one seeks a weighted residual formulation for the integral representation (2.19) of the potential gradient:

$$0 = u_{,i}(\boldsymbol{z}) + \int_{\partial\Omega} u(\boldsymbol{y}) G_{,\bar{i}j}(\boldsymbol{z}, \boldsymbol{y}) n_j(\boldsymbol{y}) \, \mathrm{d}S_y - \int_{\partial\Omega} q(\boldsymbol{y}) G_{,\bar{i}}(\boldsymbol{z}, \boldsymbol{y}) \, \mathrm{d}S_y \quad (10.5)$$

which, in this non-regularized form, holds only for $\boldsymbol{z} \notin \partial\Omega$; the domaine Ω may be either bounded or unbounded. Following an idea used in Sirtori et al. [26], introduce a closed auxiliary surface $\tilde{S}$, interior to the domain Ω (in particular $\tilde{S} \cap \partial\Omega = \emptyset$), and such that there exists a one-to-one mapping $\boldsymbol{x} \in \partial\Omega \to \boldsymbol{z} \in \tilde{S}$. Given the (sufficiently smooth) boundary $\partial\Omega$, it is always possible to define a surface $\tilde{S}$ in such a way, for instance through:

$$\boldsymbol{x} \in \partial\Omega \to \boldsymbol{z} = \boldsymbol{x} - \eta \boldsymbol{n}(\boldsymbol{x}) \in \tilde{S} \quad (10.6)$$

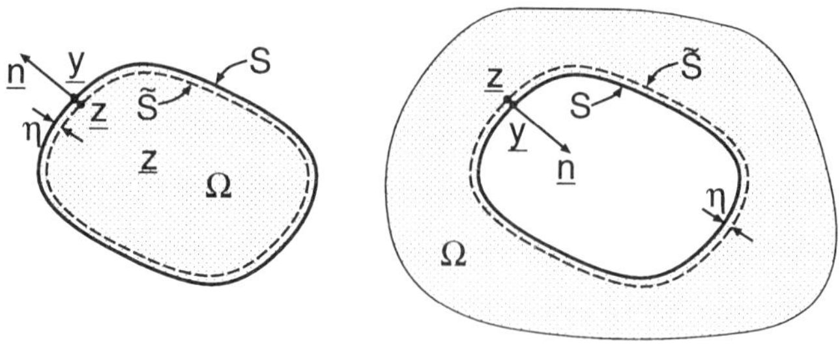

Figure 10.1 Auxiliary surface $\tilde{S}$.

with $\eta > 0$ sufficiently small. The unit normal on $\tilde{S}$ is oriented from $\tilde{S}$ to $\partial\Omega$.

Consider then a test function $\tilde{u}(\boldsymbol{z})$ defined over $\tilde{S}$. The inner product of (10.5) with $\tilde{u}(\boldsymbol{z})n_i(\boldsymbol{z})$ is taken and integrated over $\tilde{S}$ with respect to $\boldsymbol{z}$ (at this stage, the surfaces $\tilde{S}$ and $\partial\Omega$ are disconnected and the double surface integrals are thus nonsingular), which leads to the identity:

$$I_1(\tilde{S}, \partial\Omega) + I_2(\tilde{S}, \partial\Omega) = 0 \tag{10.7}$$

having put:

$$I_1(\tilde{S}, \partial\Omega) = \int_{\tilde{S}} \tilde{u}(\boldsymbol{z})q(\boldsymbol{z})\,\mathrm{d}S_z - \int_{\tilde{S}}\int_{\partial\Omega} \tilde{u}(\boldsymbol{z})q(\boldsymbol{y})H(\boldsymbol{y},\boldsymbol{z})\,\mathrm{d}S_y\,\mathrm{d}S_z \tag{10.8}$$

$$I_2(\tilde{S}, \partial\Omega) = \int_{\tilde{S}}\int_{\partial\Omega} \tilde{u}(\boldsymbol{z})u(\boldsymbol{y})G_{,\bar{i}j}n_i(\boldsymbol{z})n_j(\boldsymbol{y})\,\mathrm{d}S_y\,\mathrm{d}S_z \tag{10.9}$$

Analytical manipulations (namely regularization for I_1 and integrations by parts for I_2, see Appendix 10.A) are then performed, taking advantage of the fact that the two surfaces $\partial\Omega$ and $\tilde{S}$ are disconnected. As a result, the limiting forms $I_1(\partial\Omega, \partial\Omega)$ and $I_2(\partial\Omega, \partial\Omega)$ taken by $I_1(\tilde{S}, \partial\Omega)$ and $I_2(\tilde{S}, \partial\Omega)$ as the surface $\tilde{S}$ becomes arbitrarily close to $\partial\Omega$ (i.e. for $\eta \to 0$ in the mapping (10.6)) are:

$$\begin{aligned}\lim_{\tilde{S}\to\partial\Omega} I_1(\tilde{S}, \partial\Omega) = (1-\kappa)&\int_{\partial\Omega} q(\boldsymbol{y})\tilde{u}(\boldsymbol{y})\,\mathrm{d}S_y \\ &- \int_{\partial\Omega} q(\boldsymbol{y})\int_{\partial\Omega}[\tilde{u}(\boldsymbol{x}) - \tilde{u}(\boldsymbol{y})]H(\boldsymbol{y},\boldsymbol{x})\,\mathrm{d}S_x\,\mathrm{d}S_y\end{aligned} \tag{10.10}$$

$$\lim_{\tilde{S}\to\partial\Omega} I_2(\tilde{S}, \partial\Omega) = \int_{\partial\Omega} R_a\tilde{u}(\boldsymbol{x})\int_{\partial\Omega} B_{ab}(\boldsymbol{x},\boldsymbol{y})R_b u(\boldsymbol{y})\,\mathrm{d}S_y\,\mathrm{d}S_x \tag{10.11}$$

The inner (with respect to $\boldsymbol{x}$) and outer (with respect to $\boldsymbol{y}$) integrations are respectively weakly singular and regular. The differential operator $R_a u$ is defined by:

$$R_a u = e_{abc}n_b u_{,c} \tag{10.12}$$

To establish the result (10.11), use has been made of the fact that G verifies $G_{,ii}(\boldsymbol{y},\boldsymbol{z}) = G_{,\bar{i}\bar{i}}(\boldsymbol{y},\boldsymbol{z}) = 0$ $(\boldsymbol{y} \neq \boldsymbol{z})$, which implies the existence of a second-order tensor $\boldsymbol{B}(\boldsymbol{z},\boldsymbol{y})$ such that, for any $\boldsymbol{z} \neq \boldsymbol{y}$:

$$G_{,\bar{i}j}(\boldsymbol{y},\boldsymbol{z}) = e_{iap}e_{jbq}B_{ab,\bar{p}q}(\boldsymbol{y},\boldsymbol{z}) \tag{10.13}$$

The tensor $\boldsymbol{B}(\boldsymbol{z},\boldsymbol{y})$ has a r^{-1} weak singularity and possesses symmetry properties:

$$B_{ab}(\boldsymbol{z},\boldsymbol{y}) = B_{ba}(\boldsymbol{z},\boldsymbol{y}) = B_{ab}(\boldsymbol{y},\boldsymbol{z}) \tag{10.14}$$

In particular, if $G(\boldsymbol{y},\boldsymbol{z})$ is the infinite-space fundamental solution, the tensor $\boldsymbol{B}(\boldsymbol{y},\boldsymbol{z})$ is simply given by:

$$B_{ab}(\boldsymbol{z},\boldsymbol{y}) = -\delta_{ab}G(\boldsymbol{y},\boldsymbol{z}) = -\frac{\delta_{ab}}{4\pi r}$$

If $G(\boldsymbol{y},\boldsymbol{z})$ is the half-space fundamental solution with a Dirichlet or Neumann boundary condition (given respectively by eqs (2.11) and (2.12)), the tensor $\boldsymbol{B}(\boldsymbol{y},\boldsymbol{z})$ is given by:

$$B_{ab}(\boldsymbol{z},\boldsymbol{y}) = -\frac{1}{4\pi}\left(\frac{\delta_{ab}}{r} + \epsilon\frac{\delta_{ab} - 2n_a n_b}{\bar{r}}\right) \tag{10.15}$$

where $\boldsymbol{n}$ is the outward unit normal to the half-space and $\epsilon = -1$ (Dirichlet condition) or $\epsilon = 1$ (Neumann condition).

Taking into account the intermediary results (10.10) and (10.11), the limiting form of (10.7) when $\tilde{S} \to \partial\Omega$ reads:

$$(1-\kappa)\int_{\partial\Omega} q(\boldsymbol{y})\tilde{u}(\boldsymbol{y})\,\mathrm{d}S_y - \int_{\partial\Omega}\int_{\partial\Omega} q(\boldsymbol{y})[\tilde{u}(\boldsymbol{x}) - \tilde{u}(\boldsymbol{y})]H(\boldsymbol{y},\boldsymbol{x})\,\mathrm{d}S_x\,\mathrm{d}S_y$$
$$+ \int_{\partial\Omega}\int_{\partial\Omega} R_a\tilde{u}(\boldsymbol{x})R_b u(\boldsymbol{y})B_{ab}(\boldsymbol{x},\boldsymbol{y})\,\mathrm{d}S_y\,\mathrm{d}S_x = 0 \tag{10.16}$$

where the coefficient κ retains the same meaning as in eq. (10.1).

Each integral (inner and outer) in eq. (10.16) is convergent provided q and u be respectively piecewise continuous and $C^{0,\alpha}$-smooth at any point of $\partial\Omega$ (sufficient condition for convergence[1]). These conditions are identical to those invoked for the integral equation (10.2) or the displacement integral equation (10.1). In particular, the more restrictive and cumbersome to implement $C^{1,\alpha}$ smoothness requirement for u in collocation forms of the gradient boundary integral equation, e.g. eq. (10.5), is avoided.

[1]These requirements ensure that the singular integral is convergent, and thus put unequal emphasis on the two integrations. In fact, q and u should belong to the spaces $H^{-1/2}(\partial\Omega)$ and $H^{1/2}(\partial\Omega)$, respectively (see definition (10.30)), which define somewhat weaker smoothness requirements.

The differential operator $f \to R_a f$ has been introduced by Nedelec [21] under the name of *surface curl* of the scalar field f. Assume the surface has a parametric representation

$$\boldsymbol{\xi} = (\xi_1, \xi_2) \in \Delta \to \boldsymbol{y} \in \partial\Omega$$

in a neighbourhood of $\boldsymbol{y}$. Then one can show that:

$$R_a f \,\mathrm{d}S_y = e_{\alpha\beta 3} f_{,\alpha} (\boldsymbol{a}_\beta . \boldsymbol{e}_a) \,\mathrm{d}\xi_1 \mathrm{d}\xi_2 \tag{10.17}$$

In particular, $R_a f$ depends only on the values of f on $\partial\Omega$.

Galerkin integral formulation of the Neumann problem

Let the flux q be prescribed over the entire boundary: $q = q^D$ over $\partial\Omega$. Then, from eq. (10.16), the potential u solves the Galerkin integral equation:

$$\boldsymbol{\mathcal{B}}_{uu}(u, \tilde{u}) = \boldsymbol{\mathcal{L}}_u(\tilde{u}) \qquad \forall \tilde{u} \in \mathcal{V}_u \tag{10.18}$$

where $\mathcal{V}_u$ denotes the set of continuous functions $\tilde{u}$ on $\partial\Omega$, and:

$$\begin{aligned} \boldsymbol{\mathcal{B}}_{uu}(u, \tilde{u}) &= \int_{\partial\Omega}\int_{\partial\Omega} R_a \tilde{u}(\boldsymbol{x}) R_b u(\boldsymbol{y}) B_{ab}(\boldsymbol{x}, \boldsymbol{y}) \,\mathrm{d}S_y \,\mathrm{d}S_x \\ \boldsymbol{\mathcal{L}}_u(\tilde{u}) &= (\kappa - 1) \int_{\partial\Omega} q(\boldsymbol{y}) \tilde{u}(\boldsymbol{y}) \,\mathrm{d}S_y \\ &\quad - \int_{\partial\Omega}\int_{\partial\Omega} q(\boldsymbol{y}) [\tilde{u}(\boldsymbol{x}) - \tilde{u}(\boldsymbol{y})] H(\boldsymbol{y}, \boldsymbol{x}) \,\mathrm{d}S_x \,\mathrm{d}S_y \end{aligned}$$

The bilinear form $\boldsymbol{\mathcal{B}}_{uu}$ is symmetric due to the properties (10.14) of $\boldsymbol{B}(\boldsymbol{x}, \boldsymbol{y})$.

10.3 INDIRECT GALERKIN FORMULATION OF DIRICHLET AND NEUMANN PROBLEMS

Consider a bounded domain Ω and its complement $\bar{\Omega} = \mathbb{R}^3 \setminus \Omega$. Let the boundary variables (u, q) and $(\bar{u}, \bar{q})$ be respectively associated to solutions of the Laplace equation in Ω (*interior* problem) and $\bar{\Omega}$ (*exterior* problem). Let $\phi = \bar{u} - u$ and $\psi = \bar{q} + q$ denote the potential and flux[2] discontinuities across $\partial\Omega$; they must be continuous and piecewise continuous on $\partial\Omega$, respectively.

The potential u and its gradient can be expressed for any observation point $\boldsymbol{x}$ interior to Ω or $\bar{\Omega}$ using single-layer potentials, eqs (8.23), (8.24) of density ψ or double-layer potentials, eqs (8.25), (8.26) of density ϕ.

[2]The '+' sign in $\psi = \bar{q} + q$ stems from the fact that q and $\bar{q}$ are defined in terms of opposite normals.

Dirichlet problem. Consider interior and exterior Dirichlet problems having the same boundary data u^D: then $\phi = 0$. Taking equation (10.2) in turn for the interior and exterior problems and adding the resulting identities, one obtains the following formulation, of unknown ψ for the Dirichlet problem:

$$\mathcal{B}_{qq}(\psi, \tilde{q}) = \int_{\partial\Omega} \tilde{q}(\boldsymbol{x}) u^D(\boldsymbol{x})\, \mathrm{d}S_x \qquad \forall \tilde{q} \in \mathcal{V}_q \tag{10.19}$$

Neumann problem. Consider interior and exterior Neumann problems having the same boundary data q^D: then $\psi = 0$. Taking equation (10.16) in turn for the interior and exterior problems and adding the resulting identities, one obtains the following formulation, of unknown ϕ for the Neumann problem:

$$\mathcal{B}_{uu}(\phi, \tilde{u}) = \int_{\partial\Omega} q^D(\boldsymbol{y}) \tilde{u}(\boldsymbol{y})\, \mathrm{d}S_y \qquad \forall \tilde{u} \in \mathcal{V}_u \tag{10.20}$$

The right-hand sides of the indirect formulations have a much simpler form than those of the corresponding direct formulations (10.3), (10.18). However, once the density ψ or ϕ is computed, one still has to evaluate a representation formula at every node of $\partial\Omega$ for q and u. This task is very similar, in terms of complexity and computational costs, to that of setting up directly the right-hand sides of eqs (10.3), (10.18).

10.4 DIRICHLET AND NEUMANN SCALAR WAVE PROBLEMS

The foregoing approach is applicable to linear acoustics in the frequency domain, i.e. to problems governed by the Helmholtz equation:

$$\Delta u + k^2 u = 0 \qquad (k = \frac{\omega}{c}):\ \text{wavenumber}$$

The main difference from static problems lies in the double integration by parts used for the Galerkin formulation of the Neumann problem. The infinite-space fundamental solution is:

$$G(\boldsymbol{x}, \boldsymbol{y}; \omega) = \frac{\mathrm{e}^{ikr}}{4\pi r}$$

and verifies the relation

$$G_{,\bar{i}j}(\boldsymbol{x}, \boldsymbol{y}; \omega) = \delta_{ij} k^2 G(\boldsymbol{x}, \boldsymbol{y}) - e_{iap} e_{jaq} G_{,\bar{p}q}(\boldsymbol{x}, \boldsymbol{y}; \omega) \tag{10.21}$$

Proceeding as in Section 10.2, the bilinear forms are found to be:

$$\mathcal{B}_{qq}(q, \tilde{q}; \omega) = \int_{\partial\Omega}\int_{\partial\Omega} \tilde{q}(\boldsymbol{x}) G(\boldsymbol{x}, \boldsymbol{y}; \omega) q(\boldsymbol{y})\, \mathrm{d}S_y\, \mathrm{d}S_x \tag{10.22}$$

$$\mathcal{B}_{uu}(u, \tilde{u}; \omega) = \int_{\partial\Omega}\int_{\partial\Omega} \Big[k^2 n_a(\boldsymbol{x}) n_a(\boldsymbol{y}) \tilde{u}(\boldsymbol{x}) u(\boldsymbol{y}) - R_a\tilde{u}(\boldsymbol{x}) R_a u(\boldsymbol{y})\Big] G(\boldsymbol{x}, \boldsymbol{y}; \omega)\, \mathrm{d}S_y\, \mathrm{d}S_x \tag{10.23}$$

The direct symmetric Galerkin formulations for the Dirichlet and Neumann problems (analogous to eqs (10.3) and (10.18) for static problems) are then:

$$\mathcal{B}_{qq}(q, \tilde{q}; \omega) = \mathcal{L}_q(\tilde{q}; \omega) \qquad \forall \tilde{q} \in \mathcal{V}_q \tag{10.24}$$

$$\mathcal{B}_{uu}(u, \tilde{u}; \omega) = \mathcal{L}_u(\tilde{u}; \omega) \qquad \forall \tilde{u} \in \mathcal{V}_u \tag{10.25}$$

where $\mathcal{V}_q$ (resp. $\mathcal{V}_u$) denotes the set of complex-valued test functions $\tilde{q}$ (resp. $\tilde{u}$) piecewise continuous (resp. continuous) on $\partial\Omega$, and the expressions of the linear forms $\mathcal{L}_q$ $\mathcal{L}_u$ are:

$$\begin{aligned}\mathcal{L}_q(\tilde{q}; \omega) = &\int_{\partial\Omega}\int_{\partial\Omega} \tilde{q}(\boldsymbol{x})\Big[H(\boldsymbol{x}, \boldsymbol{y}; \omega)u^D(\boldsymbol{y}) - H(\boldsymbol{x}, \boldsymbol{y})u^D(\boldsymbol{x})\Big]\,\mathrm{d}S_y\,\mathrm{d}S_x \\ &+ \kappa \int_{\partial\Omega} \tilde{q}(\boldsymbol{x})u^D(\boldsymbol{x})\,\mathrm{d}S_x \end{aligned}\tag{10.26}$$

$$\begin{aligned}\mathcal{L}_u(\tilde{u}; \omega) = &-\int_{\partial\Omega}\int_{\partial\Omega} q(\boldsymbol{y})\Big[H(\boldsymbol{y}, \boldsymbol{x}; \omega)\tilde{u}(\boldsymbol{x}) - H(\boldsymbol{y}, \boldsymbol{x})\tilde{u}(\boldsymbol{y})\Big]\,\mathrm{d}S_x\,\mathrm{d}S_y \\ &+ (1-\kappa) \int_{\partial\Omega} q(\boldsymbol{y})\tilde{u}(\boldsymbol{y})\,\mathrm{d}S_y \end{aligned}\tag{10.27}$$

The corresponding indirect Galerkin formulations (analogous to eqs (10.19) and (10.20) for static problems and with the same definitions of ϕ and ψ) are:

$$\mathcal{B}_{qq}(\psi, \tilde{q}; \omega) = \int_{\partial\Omega} \tilde{q}(\boldsymbol{x})u^D(\boldsymbol{x})\,\mathrm{d}S_x \qquad \forall \tilde{q} \in \mathcal{V}_q \tag{10.28}$$

for the Dirichlet problem, and

$$\mathcal{B}_{uu}(\phi, \tilde{u}; \omega) = \int_{\partial\Omega} q^D(\boldsymbol{y})\tilde{u}(\boldsymbol{y})\,\mathrm{d}S_y \qquad \forall \tilde{u} \in \mathcal{V}_u \tag{10.29}$$

for the Neumann problem.

10.5 SOME CONVERGENCE RESULTS

The mathematical properties of Galerkin integral formulations (symmetry, positive- or negative-definiteness) play a key role in the mathematical analysis of existence, uniqueness and convergence issues. Galerkin integral formulations for fundamental boundary-value problems have been analysed by Nedelec and co-workers [3], [20], [23] and Wendland [30]. Consider again the indirect Galerkin integral formulations for the Dirichlet problem (eq. (10.19), boundary data $u^D \in H^{1/2}(\partial\Omega)$, unknown density ψ) and the Neumann problem (eq. (10.20), boundary data $u^D \in H^{1/2}(\partial\Omega)$, unknown density ψ). The function spaces $\mathcal{V}_q$ and $\mathcal{V}_u$ actually are $H^{-1/2}(\partial\Omega)$ and $H^{1/2}(\partial\Omega)$, respectively, the space $H^s(\partial\Omega)$, $s > 0$ being defined by (see [28]):

$$H^s(\partial\Omega) = \left\{ f \middle| \, |f|^2_{H^s(\partial\Omega)} = \sum_{|\alpha|\leq[s]} \int_{\partial\Omega} |D^\alpha f|^2 \, \mathrm{d}S \right.$$
$$\left. + \sum_{|\alpha|=[s]} \int_{\partial\Omega}\int_{\partial\Omega} \frac{|D^\alpha f(\boldsymbol{y}) - D^\alpha f(\boldsymbol{x})|^2}{|\boldsymbol{y}-\boldsymbol{x}|^{n-1+2(s-[s])}} \, \mathrm{d}S_y \, \mathrm{d}S_x < +\infty \right\} \tag{10.30}$$

($n = 2$ or 3: dimension of the geometrical space) where $[s]$ is the integer part of s, $\boldsymbol{\alpha}$ is an n-tuple of positive integers, $|\boldsymbol{\alpha}| = \alpha_1 + \ldots \alpha_n$ and $D^\alpha f = (f_{,\alpha_1}, \ldots, f_{,\alpha_n})$; the space $H^{-s}(\Gamma)$ is the dual space of $H^s(\Gamma)$ with respect to the L^2 scalar product $(\boldsymbol{f}, \boldsymbol{g})_{L^2(\Gamma)} = \int_\Gamma \boldsymbol{f}^T \boldsymbol{g} \, \mathrm{d}\Gamma$. The smoothness of the functions of $H^s(\Gamma)$ increases with the value of s. It is interesting to note that for half-integer values of s a kernel function similar to the fundamental solutions used in BEMs appears in the definition of the norm $|f|_{H^s(\Gamma)}$.

The bilinear forms $\mathcal{B}_{qq}, \mathcal{B}_{uu}$ are shown to be *coercive*, i.e. there exist positive constants β_{qq}, β_{uu} such that:

$$\mathcal{B}_{qq}(q,q) \geq \beta_{qq} \, |q|^2_{H^{-1/2}(\partial\Omega)}$$
$$\mathcal{B}_{uu}(u,u) \geq \beta_{uu} \, |u|^2_{H^{1/2}(\partial\Omega)/\mathbb{R}}$$

The existence and uniqueness of the solution to problems (10.19) and (10.20) follows by virtue of the Lax–Milgram theorem. Moreover, the following convergence results are given in [23], where the unknown density ψ or ϕ and the boundary $\partial\Omega$ interpolations are assumed to use P_k and P_m Lagrangian (two-dimensional) finite elements, respectively (h: element diameter):

$$\begin{aligned} &\int_{\partial\Omega} \left| q(\boldsymbol{\xi}) - q_h(\mathcal{P}^{-1}(\boldsymbol{\xi})\right|^2 \, \mathrm{d}\Gamma_\xi \leq C_1 h^m + C_2 h^{k+1} \\ &\int_{\partial\Omega} \left| u(\boldsymbol{\xi}) - u_h(\mathcal{P}^{-1}(\boldsymbol{\xi})\right|^2 \, \mathrm{d}\Gamma_\xi \leq C_1 [h^{k+1} + h^{m+1}] + C_2 h^m \end{aligned} \tag{10.31}$$

(the orthogonal projection $\mathcal{P}$ onto Γ is used in order to 'transport' onto the exact boundary Γ the approximate solution q_h or u_h defined on the approximate boundary Γ_h). They suggest the choice of interpolations so that $m = k+1$ (e.g. flat triangles with piecewise constant unknowns) for the Dirichlet problem, and $m = k$ (e.g. flat triangles with piecewise linear unknowns) for the Neumann problem. Convergence results of a similar nature are known for symmetric Galerkin integral formulations of fundamental boundary-value problems associated with e.g. Helmholtz or linear elasticity equations.

A more general approach to the mathematical study of Galerkin BIE formulations is found e.g. in Wendland [30, 28]. It is based on the fact that the various integral operators $v \rightarrow \mathcal{G}v$:

$$\mathcal{G}v(\boldsymbol{\xi}) \equiv \int_\Gamma \boldsymbol{G}(\boldsymbol{\xi}, \boldsymbol{x}) \boldsymbol{v}(\boldsymbol{x}) \, \mathrm{d}\Gamma_x + (\text{free term})$$

involved in integral formulations of elliptic problems are particular instances of *pseudo-differential operators*, which are characterized by an *index* α related to the degree of singularity of the kernel $K(\boldsymbol{x}, \boldsymbol{y})$ ($\alpha = -1, 0, 1$ for weakly singular, strongly singular and hypersingular kernels, respectively).

A general convergence result is available for boundary element discretization of Galerkin formulations associated with strongly elliptic integral operators of order α, in connection with the use of the generic family of so-called $S_h^{d,r}$ finite element approximations systems in $(n-1)$ dimensions (Babuska and Aziz [2]) where the polynomial degree d of the interpolation functions and the regularity degree $r < d - 1$ are non-negative integers, and h denotes the maximum element size. This family of interpolations includes the usual C^0 Lagrange interpolations P_m (for $d = m+1, r = 1$) and piecewise constant interpolations on triangles (for $d = 1, r = 0$). The result states that, for $\alpha \leq 2r+1$ $(n=2)$ or $\alpha \leq 2r$ $(n=3)$ and for any real numbers t, s such that $\alpha \leq t \leq \alpha/2 \leq s \leq d$, the approximation $\boldsymbol{v}_h$ to the true solution $\boldsymbol{v}$ of the Galerkin formulation for the integral equation $\mathcal{G}\boldsymbol{v} = \boldsymbol{b}$ is such that

$$|\boldsymbol{v}_h - \boldsymbol{v}|_{H^t(\Gamma)} \leq c h^{s-t} |\boldsymbol{v}|_{H^s(\Gamma)} \qquad (10.32)$$

(the H^s norm being defined in eq. (10.30)). Estimates (10.31), neglecting the effect of geometrical modelling, are particular instances of this result.

In contrast, convergence results concerning collocation BEM formulations are scarce, and mostly address spline discretizations for 2D problems [28, 1].

10.6 NUMERICAL IMPLEMENTATION

To fix our ideas, this section presents the boundary element treatment for the Galerkin formulation (10.20) of the interior Neumann problem.

Discretization of boundary and unknowns

The concepts introduced in Chapter 3 still apply. The boundary $\partial\Omega$ is subdivided into a mesh of N_E elements $E_1, \ldots, E_{N_E}$ of triangular or quadrilateral shape. The element E_e $(1 \leq e \leq N_E)$ is mathematically described as the interpolation of N_e geometrical nodes $\boldsymbol{y}^q$ by shape functions N_q:

$$\boldsymbol{\xi} \in \Delta_e \to \boldsymbol{y} = \sum_{q=1}^{N_e} N_q(\boldsymbol{\xi}) \boldsymbol{y}^q$$

Let N_N denote the total number of nodes on the mesh. The functions $\phi, \tilde{\phi}$ are approximated using nodal values $\phi^q, \tilde{\phi}^q$ and interpolation functions $N_q(\boldsymbol{\xi})$:

$$(\phi, \tilde{\phi})(\boldsymbol{y}) = \sum_{q=1}^{N_i} N_q(\boldsymbol{\xi})(\phi, \tilde{\phi})^q \qquad (\boldsymbol{\xi} \in \Delta_e)$$

Upon substitution of these interpolations into formulation (10.20), the discretized Galerkin formulation is:

$$\forall\{\tilde{\phi}\} \qquad \{\tilde{\phi}\}^T[\boldsymbol{K}]\{\phi\} = \{\tilde{\phi}\}^T\{\boldsymbol{f}\} \qquad \Big(\{\phi\} = \{\phi^1, \ldots, \phi^{N_N}\}\Big)$$

where:

$$K_{\tilde{q}q} = \int_{D_q}\int_{D_{\tilde{q}}} \{e_{\tilde{\alpha}\tilde{\beta}3} N_{\tilde{q},\tilde{\alpha}}(\tilde{\boldsymbol{\xi}})(\boldsymbol{a}_{\tilde{\beta}}(\tilde{\boldsymbol{\xi}}).\boldsymbol{e}_a)\} B_{ab}(\boldsymbol{x}(\tilde{\boldsymbol{\xi}}), \boldsymbol{y}(\boldsymbol{\xi}))$$
$$\{e_{\alpha\beta 3} N_{q,\alpha}(\boldsymbol{\xi})(\boldsymbol{a}_\beta(\boldsymbol{\xi}).\boldsymbol{e}_b)\} \,\mathrm{d}\tilde{\boldsymbol{\xi}}\,\mathrm{d}\boldsymbol{\xi} \tag{10.33}$$

$$f_{\tilde{q}} = \int_{D_{\tilde{q}}} q^D(\boldsymbol{x}(\tilde{\boldsymbol{\xi}})) N_{\tilde{q}}(\tilde{\boldsymbol{\xi}}) J(\tilde{\boldsymbol{\xi}})\,\mathrm{d}\tilde{\boldsymbol{\xi}} \tag{10.34}$$

and $D_{\tilde{q}}, D_q$ denote the respective geometrical supports in the $\tilde{\boldsymbol{\xi}}$- and $\boldsymbol{\xi}$-spaces of the interpolation functions $N_{\tilde{q}}(\tilde{\boldsymbol{\xi}})$ and $N_q(\boldsymbol{\xi})$. The governing linear system of equations is then:

$$[\boldsymbol{K}]\{\phi\} = \{f\} \tag{10.35}$$

Element integrations

The symmetric Galerkin boundary element method involves *double* surface integrations over pairs of elements E, E'. Three different situations arise: singular coincident integration ($E = E'$), singular adjacent integration (E, E' are distinct but share a common edge or vertex) and regular integration (E, E' are not connected). The singular adjacent case is specific to Galerkin boundary element methods and does not arise in collocation-based implementations.

Nonsingular integrations. They are performed using Gauss points (see Section 3.4). The double element integrals, of the form:

$$I_{e\tilde{e}} = \int_{\Delta_{\tilde{e}}}\int_{\Delta_e} f(\tilde{\boldsymbol{\xi}}, \boldsymbol{\xi})\,\mathrm{d}\tilde{\boldsymbol{\xi}}\,\mathrm{d}\boldsymbol{\xi}$$

are evaluated by means of a product formula involving the Gauss points and weights $(\boldsymbol{\xi}_i, w_i)$ and $(\tilde{\boldsymbol{\xi}}_j, \tilde{w}_j)$ on the reference elements Δ_e and $\tilde{\Delta}_e$:

$$I_{e\tilde{e}} \approx \sum_{i=1}^{N} w'_i \int_{\Delta_e} f(\tilde{\boldsymbol{\xi}}_i, \boldsymbol{\xi})\,\mathrm{d}\boldsymbol{\xi} \approx \sum_{i=1}^{N}\sum_{j=1}^{N} w'_i w_j f(\tilde{\boldsymbol{\xi}}_i, \tilde{\boldsymbol{\xi}}_j)$$

Singular integrations: coincident case. Thanks to the regularization approach, all integrals are integrable over a *single* element (their order of singularity being r^{-1}). Hence, the double integral can be treated as an inner singular integral over $\boldsymbol{y} \in E$ followed by an outer nonsingular integral over $\boldsymbol{x} \in E$, hence computed using Gauss points. This in turn implies that the inner integrals must be computed with the singular points $\boldsymbol{x}$ taken as the Gauss points used for the outer integral.

The double surface integration thus takes the form:

$$I_{e,e}^{sing} = \int_{\Delta_e}\int_{\Delta_e} f(\tilde{\boldsymbol{\xi}},\boldsymbol{\xi})\,\mathrm{d}\tilde{\boldsymbol{\xi}}\,\mathrm{d}\boldsymbol{\xi} \approx \sum_{i=1}^{N} \tilde{w}_i \int_{\Delta_e} f(\tilde{\boldsymbol{\xi}}_i,\boldsymbol{\xi})\,\mathrm{d}\boldsymbol{\xi}$$

where $f(\boldsymbol{\xi},\tilde{\boldsymbol{\xi}})$ has a $|\tilde{\boldsymbol{\xi}}-\boldsymbol{\xi}|^{-1}$ singularity. The inner integral is then evaluated using polar coordinates (ρ,α) in the reference element as in Section 3.5, taking as origin, in turn, each Gauss point $\tilde{\boldsymbol{\xi}}_i$ of the outer integration:

$$(\xi_1,\xi_2) = (\tilde{\xi}_{i1},\tilde{\xi}_{i2}) + (\rho\cos\alpha,\rho\sin\alpha)$$

One can then define $\hat{f}(\rho,\alpha;\tilde{\boldsymbol{\xi}}_i)$, nonsingular at $\rho=0$, so that:

$$f(\tilde{\boldsymbol{\xi}}_i,\boldsymbol{\xi}) = \frac{1}{\rho}\hat{f}(\rho,\alpha;\tilde{\boldsymbol{\xi}}_i)$$

This leads to the following formula:

$$\int_E\int_E R_a\psi(\boldsymbol{x})R_b\phi(\boldsymbol{y})B_{ab}(\boldsymbol{x},\boldsymbol{y})\,\mathrm{d}S_y\,\mathrm{d}S_x$$
$$= \sum_{\tilde{q}=1}^{N_e}\tilde{\phi}^{\tilde{q}}\sum_{q=1}^{N_e}\phi^q\sum_{i=1}^{N}\tilde{w}_i\int_E \hat{B}_{ab}(\rho,\alpha;\boldsymbol{\xi}_i)e_{\alpha\beta 3}N_{q,\alpha}(\boldsymbol{\xi})(\boldsymbol{a}_\beta.\boldsymbol{e}_b)\,\mathrm{d}\rho\mathrm{d}\alpha$$

The other singular double integrals appearing in Galerkin formulations like (10.37) are computed in the same manner:

$$\int_E\int_E \tilde{q}(\boldsymbol{x})[u(\boldsymbol{y})-u(\boldsymbol{x})]H(\boldsymbol{x},\boldsymbol{y})\,\mathrm{d}S_y\,\mathrm{d}S_x$$
$$= \sum_{\tilde{q}=1}^{N_e}\tilde{q}^{\tilde{q}}\sum_{q=1}^{N_e}u^q\sum_{i=1}^{N}\tilde{w}_i\int_E \hat{H}(\rho,\alpha;\boldsymbol{\xi}_i)\hat{N}_q(\rho,\alpha;\boldsymbol{\xi}_i)J(\boldsymbol{\xi})\,\mathrm{d}\rho\mathrm{d}\alpha$$

$$\int_E\int_E \tilde{\psi}(\boldsymbol{y})\psi(\boldsymbol{y})G(\boldsymbol{x},\boldsymbol{y})\,\mathrm{d}S_y\,\mathrm{d}S_x$$
$$= \sum_{\tilde{q}=1}^{N_e}\tilde{\psi}^{\tilde{q}}\sum_{q=1}^{N_e}\psi^q\sum_{i=1}^{N}\tilde{w}_i\int_E \hat{G}(\rho,\alpha;\boldsymbol{\xi}_i)N_q(\boldsymbol{\xi})J(\boldsymbol{\xi})\,\mathrm{d}\rho\mathrm{d}\alpha$$

The regularization is thus used to full advantage.

Singular integrations: adjacent case. They correspond to a weaker type of singularity. The two-dimensional case is relatively easy to handle by introducing *relative intrinsic coordinates* (see e.g. Frangi and Novati [15]):

$$u = \frac{1}{2}(\xi+\eta) \qquad v = \frac{1}{2}(\xi-\eta)$$

Alternatively, analytical integrations are available for the simpler types of boundary elements (straight elements with linear interpolation), see Carini et al. [12].

The issue becomes quite complex for three-dimensional problems, and general numerical integration algorithms are still under research. In some cases (especially triangular elements with linear interpolation), analytical evaluation is feasible, see Hamdi [17], Barbier [4].

Solution of the linear system

The symmetric character of the Galerkin formulation (10.37) entails significant gains in storage and computer time over collocation methods with the same mesh and unknowns. Indeed, equation (10.33) highlights the symmetry (resting upon property (10.14) of the kernel $\boldsymbol{B}(\boldsymbol{x},\boldsymbol{y})$) of the matrix $[\boldsymbol{K}]$ in the linear system of equations (10.35). The Galerkin boundary integral formulation thus has similarities with usual finite element methods. However, the matrix $[\boldsymbol{K}]$ is fully populated (instead of banded): the presence of a kernel function like $\boldsymbol{B}(\boldsymbol{x},\boldsymbol{y})$ means that the influence coefficients $\int_{E_{\tilde{e}}}\int_{E_e}\{\ldots\}\,\mathrm{d}\tilde{\boldsymbol{\xi}}\,\mathrm{d}\boldsymbol{\xi}$ are nonzero for any pair $(e,\tilde{e})$, i.e. not only for pairs of elements $E_e, E_{\tilde{e}}$ sharing common nodes. The linear system of equations (10.35) can be solved e.g. using the Choleski factorization.

10.7 GALERKIN INTEGRAL FORMULATION FOR MIXED BOUNDARY-VALUE PROBLEMS

A symmetric Galerkin formulation for mixed boundary-value problems of the form:

$$\Delta u = 0 \qquad u = u^D \text{ on } S_u \quad q = q^D \text{ on } S_q \tag{10.36}$$

can be obtained from the identities established in Section 10.2. In order to do so, the geometrical support of the test functions $\tilde{q}$ in eq. (10.2) and $\tilde{u}$ in eq. (10.16) are restricted to S_q and S_u, respectively, i.e.:

$$\tilde{u} \in \mathcal{V}_u = \left\{ \tilde{u} \mid \tilde{u} \in C^{0,\alpha}(\partial\Omega) \text{ and } \tilde{u}(\boldsymbol{x}) = 0 \forall \boldsymbol{x} \in S_u \right\}$$

$$\tilde{q} \in \mathcal{V}_q = \left\{ \tilde{q} \mid \tilde{q} \text{ piecewise continuous over } \partial\Omega \text{ and } \tilde{q}(\boldsymbol{x}) = 0 \forall \boldsymbol{x} \in S_q \right\}$$

Also, let $\widehat{u^D}$ denote a continuation of u^D, i.e. a continuous function defined on the entire boundary $\partial\Omega$ and such that $\widehat{u^D} = u^D$ on S_u. A new unknown $\hat{u} = u - \widehat{u^D}$ is then introduced. From the definition of u^D, one has $\hat{u} = 0$ on ∂S_u, and the new unknown $\hat{u}$ belongs to the same space $\mathcal{V}_u$ as the test functions $\tilde{u}$. This change of unknown is not required when the data u^D vanishes on the edge ∂S_u: in that case, $\hat{u} = u, \widehat{u^D} = 0$ over S_q.

These definitions and choices lead to the following symmetric Galerkin boundary integral equation formulation for the problem (10.36), with unknowns $\hat{u}$ (on S_q) and q (on S_u):

$$\left\{\begin{aligned} \mathcal{B}_{uu}(\hat{u},\tilde{u}) + \mathcal{B}_{qu}(q,\tilde{u}) &= \mathcal{L}_u(\tilde{u}) \\ \mathcal{B}_{uq}(\hat{u},\tilde{q}) + \mathcal{B}_{qq}(q,\tilde{q}) &= \mathcal{L}_q(\tilde{q}) \end{aligned}\right. \qquad \forall \tilde{u} \in \mathcal{V}_u, \forall \tilde{q} \in \mathcal{V}_q \tag{10.37}$$

where the bilinear forms are given by:

$$\begin{aligned} \mathcal{B}_{uq}(\hat{u},\tilde{q}) &= -\int_{S_u}\int_{S_q} \tilde{q}(\boldsymbol{x})\hat{u}(\boldsymbol{y})H(\boldsymbol{x},\boldsymbol{y}) \,\mathrm{d}S_y \,\mathrm{d}S_x \\ \mathcal{B}_{qq}(q,\tilde{q}) &= \int_{S_u}\int_{S_u} \tilde{q}(\boldsymbol{x})q(\boldsymbol{y})G(\boldsymbol{x},\boldsymbol{y}) \,\mathrm{d}S_y \,\mathrm{d}S_x \\ \mathcal{B}_{qu}(q,\tilde{u}) &= -\int_{S_u}\int_{S_q} q(\boldsymbol{y})\tilde{u}(\boldsymbol{x})H(\boldsymbol{y},\boldsymbol{x}) \,\mathrm{d}S_x \,\mathrm{d}S_y \\ \mathcal{B}_{uu}(\hat{u},\tilde{u}) &= \int_{S_q}\int_{S_q} R_a\tilde{u}(\boldsymbol{x})B_{ab}(\boldsymbol{x},\boldsymbol{y})R_b\hat{u}(\boldsymbol{y}) \,\mathrm{d}S_y \,\mathrm{d}S_x \end{aligned}$$

and the linear forms (which incorporate the boundary data u^D and q^D) by:

$$\begin{aligned} \mathcal{L}_q(\tilde{q}) = \kappa \int_{S_u} \tilde{q}(\boldsymbol{x})u^D(\boldsymbol{x}) \,\mathrm{d}S_x &- \int_{S_u}\int_{S_q} \tilde{q}(\boldsymbol{x})q^D(\boldsymbol{y})G(\boldsymbol{x},\boldsymbol{y}) \,\mathrm{d}S_y \,\mathrm{d}S_x \\ &+ \int_{S_u}\int_{\partial\Omega} \tilde{q}(\boldsymbol{x})[\widehat{u^D}(\boldsymbol{y}) - \widehat{u^D}(\boldsymbol{x})]H(\boldsymbol{x},\boldsymbol{y}) \,\mathrm{d}S_y \,\mathrm{d}S_x \\ \mathcal{L}_u(\tilde{u}) = (\kappa - 1) \int_{S_q} q^D(\boldsymbol{y})\tilde{u}(\boldsymbol{y}) \,\mathrm{d}S_y & \\ &- \int_{S_q}\int_{S_u} R_a\tilde{u}(\boldsymbol{x})B_{ab}(\boldsymbol{x},\boldsymbol{y})R_b\hat{u}(\boldsymbol{y}) \,\mathrm{d}S_y \,\mathrm{d}S_x \\ &+ \int_{S_q}\int_{S_q} q^D(\boldsymbol{y})[\tilde{u}(\boldsymbol{x}) - \tilde{u}(\boldsymbol{y})]H(\boldsymbol{y},\boldsymbol{x}) \,\mathrm{d}S_x \,\mathrm{d}S_y \\ &- \int_{S_q}\int_{S_u} q^D(\boldsymbol{y})\tilde{u}(\boldsymbol{y})H(\boldsymbol{y},\boldsymbol{x}) \,\mathrm{d}S_x \,\mathrm{d}S_y \end{aligned}$$

The formulation (10.37) is symmetric by virtue of the properties (10.4) and (10.14) of the fundamental solution:

$$\forall(\hat{u},\tilde{u}) \in \mathcal{V}_u, \forall(q,\tilde{q}) \in \mathcal{V}_q \qquad \left\{\begin{aligned} \mathcal{B}_{uu}(\hat{u},\tilde{u}) &= \mathcal{B}_{uu}(\tilde{u},\hat{u}) \\ \mathcal{B}_{qq}(q,\tilde{q}) &= \mathcal{B}_{qq}(\tilde{q},q) \\ \mathcal{B}_{uq}(\hat{u},\tilde{q}) &= \mathcal{B}_{qu}(\tilde{q},\hat{u}) \end{aligned}\right.$$

It is worth noting that for most choices of the fundamental solution the explicit expression of the tensor $\boldsymbol{B}$ defined by (10.13) is not *a priori* known. Also, without introducing the continuation $\widehat{u^D}$ of u^D and the new unknown $\hat{u}$, the

Galerkin formulation (10.37) would not in general be symmetric because the solution u and the test functions $\tilde{u}$ do not belong to the same set of functions when $u^D \neq 0$ on ∂S_u.

Interpretation as a stationarity condition

The Galerkin boundary integral equation formulation (10.37), obtained as a combination of the strongly singular and hypersingular integral equations in weighted residual form, is in fact the stationarity condition for the potential energy $E(v)$ of the system:

$$E(v) = \int_{\Omega} \boldsymbol{\nabla}(v).\boldsymbol{\nabla}(v)\,\mathrm{d}V - \int_{S_q} q^D v\,\mathrm{d}S - \int_{S_u} q(v - u^D)\,\mathrm{d}S$$

where the last term is the Dirichlet boundary condition over S_u expressed as a constraint.

The functional $E(v)$ is stationary, i.e. has a zero gradient, at the solution u to problem (10.36). Using the augmented form given above of $E(v)$ and putting $v = u + \delta u$, the stationarity condition reads:

$$\delta E(u).\delta u = \int_{\Omega} \boldsymbol{\nabla}(u).\boldsymbol{\nabla}(\delta u)\,\mathrm{d}V - \int_{S_q} q^D \delta u\,\mathrm{d}S - \int_{S_u} q\delta u\,\mathrm{d}S = 0 \qquad (10.38)$$

where, due to the fact that the boundary constraint on u has been introduced explicitly in $E(v)$, δu is unconstrained. In fact, since:

$$\int_{\Omega} \boldsymbol{\nabla}(v).\boldsymbol{\nabla}(\delta u)\,\mathrm{d}V = \int_{\partial\Omega} \frac{\partial v}{\partial n}\delta u\,\mathrm{d}S - \int_{\Omega} \Delta v \delta u\,\mathrm{d}V$$

one shows easily that u solves (10.36) and that the Lagrange multiplier q, defined on S_u, is equal to $\partial u/\partial n$.

Now, let $E(v)$ be restricted to *harmonic* functions v (i.e. which verify $\Delta v = 0$). Since u is also harmonic, the stationarity condition (10.38) can then be restricted to variations $\delta u = v - u$ such that $\Delta \delta u = 0$. Integrating by parts the domain integral in eq. (10.38) gives:

$$\int_{S_u} u^D \frac{\partial \delta u}{\partial n}\,\mathrm{d}S + \int_{S_T} u \frac{\partial \delta u}{\partial n}\,\mathrm{d}S - \int_{S_T} q^D \delta u\,\mathrm{d}S - \int_{S_u} q\delta u\,\mathrm{d}S = 0 \qquad (10.39)$$

Construction of the test functions. The construction of harmonic test functions utilizes the fact that *any* harmonic function has an integral epresentation (for $\boldsymbol{x}$ interior to Ω) of the form:

$$\delta u(\boldsymbol{x}) = -\int_{\partial\Omega} \delta u(\boldsymbol{y}) H(\boldsymbol{x},\boldsymbol{y})\,\mathrm{d}S_y + \int_{\partial\Omega} \delta q(\boldsymbol{y}) G(\boldsymbol{x},\boldsymbol{y})\,\mathrm{d}S_y \qquad (10.40)$$

where δq denotes the normal derivative of δu. In the same way, any function δu harmonic in the *exterior* domain $\bar{\Omega} = \mathbb{R}^3 - \Omega$ verifies at any point $\boldsymbol{x}$ interior to Ω the exterior (with respect to $\bar{\Omega}$) representation formula:

$$0 = -\int_{\partial\Omega} \delta\bar{u}(\boldsymbol{y})H(\boldsymbol{x},\boldsymbol{y})\,\mathrm{d}S_y + \int_{\partial\Omega} \delta\bar{q}(\boldsymbol{y})G(\boldsymbol{x},\boldsymbol{y})\,\mathrm{d}S_y \tag{10.41}$$

where $\delta\bar{q}$ denotes the normal derivative of $\delta\bar{u}$. The unit normal $\boldsymbol{n}$ exterior to Ω is used in both equations (10.40) and (10.41). Then, subtracting eq. (10.41) from (10.40), a representation formula for $u(\boldsymbol{x})$ in terms of $\tilde{u} = \delta\bar{u} - \delta u$ and $\tilde{q} = \delta\bar{q} - \delta q$, the potential and flux discontinuities across $\partial\Omega$, is obtained:

$$\delta u(\boldsymbol{x}) = \int_{\partial\Omega} \tilde{u}(\boldsymbol{y})H(\boldsymbol{x},\boldsymbol{y})\,\mathrm{d}S_y - \int_{\partial\Omega} \tilde{q}(\boldsymbol{y})G(\boldsymbol{x},\boldsymbol{y})\,\mathrm{d}S_y$$

One can restrict this formula to the case $\delta\bar{u} = \delta u$, i.e. $\tilde{u} = 0$ over S_u and $\delta\bar{q} = -\delta q$, i.e. $\tilde{t} = 0$ over S_T, which gives:

$$\delta u(\boldsymbol{x}) = \int_{S_q} \tilde{u}(\boldsymbol{y})H(\boldsymbol{x},\boldsymbol{y})\,\mathrm{d}S_y - \int_{S_u} \tilde{q}(\boldsymbol{y})G(\boldsymbol{x},\boldsymbol{y})\,\mathrm{d}S_y \tag{10.42}$$

In this way, any harmonic test functions δu can be represented in a form which reflect the boundary condition structure of the original problem (10.36). Similarly, the gradient of δu is expressed from eq. (10.42) by the representation:

$$\delta u_{,i}(\boldsymbol{x}) = \int_{S_q} \tilde{u}(\boldsymbol{y})G_{,\bar{i}\bar{j}}(\boldsymbol{x},\boldsymbol{y})n_j(\boldsymbol{y})\,\mathrm{d}S_y - \int_{S_u} \tilde{q}(\boldsymbol{y})G_{,\bar{i}}(\boldsymbol{x},\boldsymbol{y})\,\mathrm{d}S_y \tag{10.43}$$

Integral variational formulation. A substitution of (10.42) and (10.43) into the stationarity equation (10.39) then yields the Galerkin boundary integral formulation (10.37). To establish this result, one has to replace the boundary $\partial\Omega$ by the auxiliary surface $\tilde{S}$ in the definition (10.42) of δu, regularize the integrals and take their limiting form for $\tilde{S} \to \partial\Omega$, like in Section 10.2. The terms containing $\tilde{q}$ and $\tilde{u}$ arising from this manipulation are none other than the weighted residual forms of the boundary integral equations (2.24) and (2.19), so that equations (10.2) and (10.16) are obtained.

The formulation (10.37) is thus seen to express the stationarity of the (augmented) potential energy functional, and hence is of a *variational* nature.

10.8 FLUID–STRUCTURE COUPLING IN ACOUSTICS

Galerkin boundary integral equations allow us to establish symmetrical formulations for coupling problems while retaining the possibility to use boundary elements. To illustrate this idea, consider an elastic structure (specific mass ρ_e) occupying a bounded region $\Omega_e \subset \mathbb{R}^3$, the complementary (infinite) region $\Omega_f = \mathbb{R}^3 \setminus \Omega_e$ being filled by an acoustical fluid (specific mass ρ_f). One wants to compute the mechanical state of the coupled system (displacement $\boldsymbol{u}$ in Ω_e, total pressure p^T in Ω_f) resulting from a known incident wave p^I solution to the Helmholtz equation. The scattered pressure $p = p^T - p^I$, which verifies decay and radiation conditions at infinity (which is not necessarily true for P^T, p^I), is introduced.

The transmission conditions at the interface Γ separating the solid and fluid regions are:

$$\boldsymbol{t} = -(p + p^I)\boldsymbol{n} \qquad \frac{1}{\rho_f \omega^2}\left(\frac{\partial p}{\partial n} + \frac{\partial p^I}{\partial n}\right) = w \tag{10.44}$$

where w denotes the normal deplacement: $w = \boldsymbol{u}.\boldsymbol{n}$. The governing weak formulation for the elastodynamic state, incorporating the transmission condition $(10.44)_1$, reads:

$$\int_{\Omega_e} [\boldsymbol{\sigma}(\boldsymbol{u}) : \boldsymbol{\varepsilon}(\delta \boldsymbol{u}) - \rho_e \omega^2 \boldsymbol{u}.\delta \boldsymbol{u}]\,\mathrm{d}V + \int_\Gamma (p + p^I)\delta w\,\mathrm{d}S = 0 \qquad (\forall \delta \boldsymbol{u}) \tag{10.45}$$

where $\delta \boldsymbol{u}$ is the displacement variation (test function) and $\delta w = \delta \boldsymbol{u}.\boldsymbol{n}$. Since the fluid region is unbounded, the scattered pressure p verifies for $\boldsymbol{x} \in \Gamma$ the integral equation:

$$p(\boldsymbol{x}) = \int_\Gamma \Big\{ G(\boldsymbol{y}, \boldsymbol{x}; \omega) \frac{\partial p}{\partial n}(\boldsymbol{y}) - [H(\boldsymbol{x}, \boldsymbol{y}; \omega) p(\boldsymbol{y}) - H(\boldsymbol{x}, \boldsymbol{y}) p(\boldsymbol{x})] \Big\}\,\mathrm{d}S_y$$

Substitution of this relation into (10.45) and elimination of $\partial p/\partial n$ by means of the transmission condition $(10.44)_2$ gives:

$$E(\boldsymbol{u}, \delta \boldsymbol{u}) + \rho_f \omega^2 \mathcal{B}_{qq}(w, \delta w) - \mathcal{B}_{uq}(p, \delta w) = \mathcal{L}_q(\delta w) \tag{10.46}$$

where $\mathcal{B}_{qq}$ is given by eq. (10.22) and

$$E(\boldsymbol{u}, \delta \boldsymbol{u}) = \int_{\Omega_e} [\boldsymbol{\sigma}(\boldsymbol{u}) : \boldsymbol{\varepsilon}(\delta \boldsymbol{u}) - \rho_e \omega^2 \boldsymbol{u}.\delta \boldsymbol{u}]\,\mathrm{d}V$$

$$\mathcal{B}_{uq}(p, \delta w) = \int_\Gamma \int_\Gamma \delta w(\boldsymbol{x}) \Big[H(\boldsymbol{x}, \boldsymbol{y}; \omega) p(\boldsymbol{y}) - H(\boldsymbol{x}, \boldsymbol{y}) p(\boldsymbol{x}) \Big]\,\mathrm{d}S_y\,\mathrm{d}S_x$$

$$\mathcal{L}_q(\delta w) = -\int_\Gamma p^I(\boldsymbol{x}) \delta w(\boldsymbol{x})\,\mathrm{d}S_x - \int_\Gamma \int_\Gamma \frac{\partial p^I}{\partial n}(\boldsymbol{y}) G(\boldsymbol{y}, \boldsymbol{x}; \omega) \delta w(\boldsymbol{x})\,\mathrm{d}S_y\,\mathrm{d}S_x$$

Also, the Galerkin boundary integral formulation (10.25) for the acoustic exterior Neumann problem gives the following relation between p and w:

$$\mathcal{B}_{uu}(p, \delta p) + \mathcal{B}_{qu}(w, \delta p) = \mathcal{L}_u(\delta p) \tag{10.47}$$

where $\mathcal{B}_{uu}$ is given by (10.23) and

$$\mathcal{B}_{qu}(w, \delta p) = \int_\Gamma \int_\Gamma w(\boldsymbol{y}) \Big[H(\boldsymbol{y}, \boldsymbol{x}; \omega) \delta p(\boldsymbol{x}) - H(\boldsymbol{y}, \boldsymbol{x}) \delta p(\boldsymbol{y}) \Big]\,\mathrm{d}S_x\,\mathrm{d}S_y$$

$$\mathcal{L}_{qu}(\delta p) = \frac{1}{\rho_f \omega^2} \int_\Gamma \int_\Gamma \frac{\partial p^I}{\partial n}(\boldsymbol{y}) \Big[H(\boldsymbol{y}, \boldsymbol{x}; \omega) \delta p(\boldsymbol{x}) - H(\boldsymbol{y}, \boldsymbol{x}) \delta p(\boldsymbol{y}) \Big]\,\mathrm{d}S_x\,\mathrm{d}S_y$$

In particular, one has $\mathcal{B}_{qu}(w, \delta p) = \mathcal{B}_{uq}(\delta p, w)$. From this property and the known symmetry of the bilinear forms $E, \mathcal{B}_{qq}, \mathcal{B}_{uu}$, the governing equations (10.46), (10.47) of the coupled problem constitute a symmetric formulation.

The discretized formulation of the coupled problem has the form:

$$\begin{bmatrix} \mathcal{B}_{uu} & \mathcal{B}_{qu} & 0 \\ \mathcal{B}_{uq} & \boldsymbol{E}_{ww} + \rho_f \omega^2 \mathcal{B}_{qq} & \boldsymbol{E}_{wv} \\ 0 & \boldsymbol{E}_{vw} & \boldsymbol{E}_{vv} \end{bmatrix} \begin{Bmatrix} p \\ w \\ v \end{Bmatrix} = \begin{Bmatrix} \mathcal{L}_u \\ \mathcal{L}q \\ 0 \end{Bmatrix}$$

where $\{\boldsymbol{w}\}$ and $\{\boldsymbol{v}\}$ gather the degrees of freedom associated with the normal displacements on Γ and all other displacements, respectively (i.e. $\{\boldsymbol{u}\} = \{\boldsymbol{w}; \boldsymbol{v}\}$). The matrices $\mathcal{B}_{uu} = \mathcal{B}_{qq}$ are symmetric and fully populated; the matrix $\mathcal{B}_{qu} = \mathcal{B}_{qu}^T$ is full; the matrix $\boldsymbol{E}$ (impedance matrix of the finite element model) is symmetric and banded.

This coupled formulation has been used e.g. by Hamdi and Jean [18]; see also Zeng and Bielak [32] for a similar development.

10.9 LINEAR ELASTICITY

Mixed boundary-value problem

The derivation of Galerkin formulations for elastic problems essentially follows the steps described in Section 10.7 for potential problems. The typical elastostatic boundary-value problem is (assuming isotropic and homogeneous elastic properties):

$$C_{ijab} u_{a,bj} = 0 \qquad \boldsymbol{u} = \boldsymbol{u}^D \text{ on } S_u \quad \boldsymbol{t} = \boldsymbol{t}^D \text{ on } S_T \tag{10.48}$$

The augmented potential energy functional $E(\boldsymbol{v})$ is now defined by:

$$E(\boldsymbol{v}) = \frac{1}{2} \int_\Omega \boldsymbol{\varepsilon}(\boldsymbol{v}) : \boldsymbol{C} : \boldsymbol{\varepsilon}(\boldsymbol{v}) \, \mathrm{d}V - \int_{S_T} \boldsymbol{t}^D . \boldsymbol{v} \, \mathrm{d}S - \int_{S_u} \boldsymbol{t}.(\boldsymbol{v} - \boldsymbol{u}^D) \, \mathrm{d}S \tag{10.49}$$

Following the same steps as in Section 10.7, the stationarity condition for $E(\boldsymbol{v})$ reads:

$$\delta E(\boldsymbol{u}).\delta \boldsymbol{u} = \int_{S_u} \boldsymbol{u}^D . \boldsymbol{T}^n(\delta \boldsymbol{u}) \, \mathrm{d}S + \int_{S_T} \boldsymbol{u}.\boldsymbol{T}^n(\delta \boldsymbol{u}) \, \mathrm{d}S \\ - \int_{S_T} \boldsymbol{t}^D . \delta \boldsymbol{u} \, \mathrm{d}S - \int_{S_u} \boldsymbol{t}.\delta \boldsymbol{u} \, \mathrm{d}S = 0 \tag{10.50}$$

where the test functions $\delta \boldsymbol{u}$ are solutions of the homogeneous local elastic equilibrium equations:

$$C_{ijab} \delta u_{a,bj} = 0$$

and thus admit integral representations of the form:

$$\delta u_k(\boldsymbol{x}) = \int_{S_T} \tilde{u}_i(\boldsymbol{y}) T_i^k(\boldsymbol{x},\boldsymbol{y})\,\mathrm{d}S_y - \int_{S_u} \tilde{t}_i(y) U_i^k(\boldsymbol{x},\boldsymbol{y})\,\mathrm{d}S_y$$

$$\delta\sigma_{ij}(\boldsymbol{x}) = C_{ijab} \int_{S_T} \tilde{u}_k(\boldsymbol{y}) n_\ell(\boldsymbol{y}) \Sigma^a_{k\ell,b}(\boldsymbol{y},\boldsymbol{x})\,\mathrm{d}S_y - \int_{S_u} \tilde{t}_k(y) \Sigma^k_{ij}(\boldsymbol{y},\boldsymbol{x})\,\mathrm{d}S_y$$

Galerkin boundary integral formulation for elastostatics

Still following the same path as in Section 10.7 (see [9] for a detailed presentation), one finally establishes that the boundary unknowns of problem (10.48) (displacement $\boldsymbol{u}$ on S_T and traction $\boldsymbol{t}$ on S_u) solve the Galerkin formulation:

Find $\hat{\boldsymbol{u}}^D \in \mathcal{V}_u, \boldsymbol{t} \in \mathcal{V}_T$ such that

$$\begin{cases} \boldsymbol{\mathcal{B}}_{uu}(\hat{\boldsymbol{u}},\tilde{\boldsymbol{u}}) + \boldsymbol{\mathcal{B}}_{tu}(\boldsymbol{t},\tilde{\boldsymbol{u}}) &= \boldsymbol{\mathcal{L}}_u(\tilde{\boldsymbol{u}}) \\ \boldsymbol{\mathcal{B}}_{ut}(\hat{\boldsymbol{u}},\tilde{\boldsymbol{t}}) + \boldsymbol{\mathcal{B}}_{tt}(\boldsymbol{t},\tilde{\boldsymbol{t}}) &= \boldsymbol{\mathcal{L}}_t(\tilde{\boldsymbol{t}}) \end{cases} \qquad \forall \tilde{\boldsymbol{u}} \in \mathcal{V}_u, \forall \tilde{\boldsymbol{t}} \in \mathcal{V}_T \tag{10.51}$$

where the bilinear forms are given by:

$$\boldsymbol{\mathcal{B}}_{uu}(\hat{\boldsymbol{u}},\tilde{\boldsymbol{u}}) = \int_{S_T} R_q \hat{u}_i(\boldsymbol{x}) \int_{S_T} B_{ikqs}(\boldsymbol{y},\boldsymbol{x}) R_s \tilde{u}_k(\boldsymbol{y})\,\mathrm{d}S_y\,\mathrm{d}S_x$$

$$\boldsymbol{\mathcal{B}}_{tu}(\boldsymbol{t},\tilde{\boldsymbol{u}}) = -\int_{S_u}\int_{S_T} t_k(\boldsymbol{x}) \tilde{u}_i(\boldsymbol{y}) T_i^k(\boldsymbol{x},\boldsymbol{y})\,\mathrm{d}S_y\,\mathrm{d}S_x$$

$$\boldsymbol{\mathcal{B}}_{ut}(\hat{\boldsymbol{u}},\tilde{\boldsymbol{t}}) = -\int_{S_u}\int_{S_T} \tilde{t}_k(\boldsymbol{y}) u_i(\boldsymbol{x}) T_i^k(\boldsymbol{y},\boldsymbol{x})\,\mathrm{d}S_x\,\mathrm{d}S_y$$

$$\boldsymbol{\mathcal{B}}_{tt}(\boldsymbol{t},\tilde{\boldsymbol{t}}) = \int_{S_u}\int_{S_u} t_k(\boldsymbol{x}) \tilde{t}_i(\boldsymbol{y}) U_i^k(\boldsymbol{x},\boldsymbol{y})\,\mathrm{d}S_y\,\mathrm{d}S_x$$

and the linear forms are given by:

$$\begin{aligned}\boldsymbol{\mathcal{L}}_u(\tilde{\boldsymbol{u}}) = (\kappa-1)\int_{S_T} t_k^D(\boldsymbol{x}) \tilde{u}_k(\boldsymbol{x})\,\mathrm{d}S_x - \int_{S_T} t_k^D(\boldsymbol{x}) \tilde{u}_i(\boldsymbol{x}) \int_{S_u} T_i^k(\boldsymbol{x},\boldsymbol{y})\,\mathrm{d}S_y\,\mathrm{d}S_x \\ + \int_{S_T}\int_{S_T} t_k^D(\boldsymbol{x})[\tilde{u}_i(\boldsymbol{y}) - \tilde{u}_i(\boldsymbol{x})] T_i^k(\boldsymbol{x},\boldsymbol{y})\,\mathrm{d}S_y\,\mathrm{d}S_x \\ - \int_{\partial\Omega} R_q \hat{\bar{u}}_i(\boldsymbol{x}) \int_{S_u} B_{iqks}(\boldsymbol{y},\boldsymbol{x}) R_s \tilde{u}_k(\boldsymbol{y})\,\mathrm{d}S_y\,\mathrm{d}S_x\end{aligned}$$

$$\begin{aligned}\boldsymbol{\mathcal{L}}_t(\tilde{\boldsymbol{t}}) = \kappa \int_{S_u} \tilde{t}_k(\boldsymbol{x}) \bar{u}_k(\boldsymbol{x})\,\mathrm{d}S_x - \int_{S_u}\int_{S_T} t_k^D(\boldsymbol{x}) \tilde{t}_i(\boldsymbol{y}) U_i^k(\boldsymbol{x},\boldsymbol{y})\,\mathrm{d}S_y\,\mathrm{d}S_x \\ + \int_{S_u}\int_{\partial\Omega} \tilde{t}_k(\boldsymbol{y})[\hat{\bar{u}}_i(\boldsymbol{x}) - \hat{\bar{u}}_i(\boldsymbol{y})] T_i^k(\boldsymbol{y},\boldsymbol{x})\,\mathrm{d}S_x\,\mathrm{d}S_y\end{aligned}$$

The sets $\mathcal{V}_u, \mathcal{V}_T$ of admissible functions are:

$$\mathcal{V}_u = \left\{ \tilde{\boldsymbol{u}} \mid \tilde{\boldsymbol{u}} \in C^{0,\alpha}(S_T) \text{ and } \tilde{\boldsymbol{u}} = \boldsymbol{0} \text{ on } \partial S_T \right\}$$

$$\mathcal{V}_T = \left\{ \tilde{\boldsymbol{t}} \mid \tilde{\boldsymbol{t}} \text{ piecewise continuous on } S_T \right\}$$

Like in Section 10.7, an arbitrarily chosen continuation $\hat{\boldsymbol{u}}^D$ of $\boldsymbol{u}^D$ over $\partial\Omega$ has been introduced, in order to define the new unknown:

$$\hat{\boldsymbol{u}} = \boldsymbol{u} - \hat{\boldsymbol{u}}^D \qquad \text{on } S_T$$

Note that $(\boldsymbol{t}, \tilde{\boldsymbol{t}})$ can take arbitrary values on ∂S_T but $(\hat{\boldsymbol{u}}, \tilde{\boldsymbol{u}})$ must be continuous on S_u and vanish on ∂S_u. The latter provision ensures in particular:

$$\forall(\hat{\boldsymbol{u}}^D, \tilde{\boldsymbol{u}}), \forall(\boldsymbol{t}, \tilde{\boldsymbol{t}}) \qquad \begin{cases} \mathcal{B}_{uu}(\hat{\boldsymbol{u}}^D, \tilde{\boldsymbol{u}}) & = \mathcal{B}_{uu}(\tilde{\boldsymbol{u}}, \hat{\boldsymbol{u}}^D) \\ \mathcal{B}_{tt}(\boldsymbol{t}, \tilde{\boldsymbol{t}}) & = \mathcal{B}_{tt}(\tilde{\boldsymbol{t}}, \boldsymbol{t}) \\ \mathcal{B}_{ut}(\hat{\boldsymbol{u}}^D, \tilde{\boldsymbol{t}}) & = \mathcal{B}_{tu}(\tilde{\boldsymbol{t}}, \hat{\boldsymbol{u}}^D) \end{cases}$$

and hence that the formulation (10.51) is symmetric. Finally, a double integration by parts has been performed. Following an argument put forward by Nishimura and Kobayashi [24] and as a consequence of the symmetry property (4.16) of the elastic fundamental solution, one has for $\boldsymbol{z} \neq \boldsymbol{y}$:

$$0 = \frac{\partial}{\partial z_\ell}\left\{C_{ijab}\frac{\partial}{\partial y_b}\Sigma^a_{k\ell}(\boldsymbol{y}, \boldsymbol{z})\right\} = C_{ijab}\frac{\partial}{\partial y_b}\left\{\frac{\partial}{\partial z_\ell}\Sigma^a_{k\ell}(\boldsymbol{y}, \boldsymbol{z})\right\}$$

$$0 = \frac{\partial}{\partial y_j}\left\{C_{ijab}\frac{\partial}{\partial y_b}\Sigma^a_{k\ell}(\boldsymbol{y}, \boldsymbol{z})\right\} = C_{k\ell ab}\frac{\partial}{\partial y_b}\left\{\frac{\partial}{\partial y_j}\Sigma^a_{ij}(\boldsymbol{z}, \boldsymbol{y})\right\}$$

which implies the existence of a fourth-order tensor $\boldsymbol{A}$ such that for $\boldsymbol{z} \neq \boldsymbol{y}$:

$$C_{ijab}\frac{\partial}{\partial z_b}\Sigma^a_{k\ell}(\boldsymbol{y}, \boldsymbol{z}) = e_{iep}e_{jfq}e_{kgr}e_{\ell hs}\frac{\partial}{\partial z_e}\frac{\partial}{\partial z_f}\frac{\partial}{\partial y_g}\frac{\partial}{\partial y_h}A_{pqrs}(\boldsymbol{y}, \boldsymbol{z}) \tag{10.52}$$

which is the counterpart of identity (10.13).

The tensor $\boldsymbol{A}$ associated with the Kelvin fundamental solution has been given by Nedelec [22]. It can be found by introducing the Galerkin tensor $G^k_i = F\delta_{ik}$ of the Kelvin displacement, eqs (4.48)–(4.49), so that one has:

$$\begin{aligned} \frac{1}{\mu^2}C_{ijab}\Sigma^a_{k\ell,b}(\boldsymbol{y}, \boldsymbol{z}) &= -4\nu[F_{,ijk\ell} - \delta_{ij}F_{,ppk\ell} - \delta_{k\ell}F_{,ppij}] \\ &- 2(1-\nu)[2F_{,ijk\ell} - \delta_{ik}F_{,ppj\ell} - \delta_{j\ell}F_{,ppik} - \delta_{i\ell}F_{,ppjk} - \delta_{jk}F_{,ppi\ell}] \end{aligned} \tag{10.53}$$

Then, δ being the Kronecker symbol and e the Ricci permutation tensor, using the algebraic identity $e_{aep}e_{cfp} = \delta_{ac}\delta_{ef} - \delta_{af}\delta_{be}$, one obtains

$$e_{iep}e_{jfq}e_{kgr}e_{\ell hs}\delta_{pq}\delta_{rs}F_{,efgh} = F_{,ijk\ell} - \delta_{ij}F_{,ppk\ell} - \delta_{k\ell}F_{,ppij} + \delta_{ij}\delta_{k\ell}\Delta^2 F$$

and similar relations under permutations of i, j, k, ℓ. Finally, since $\Delta^2 F = 0$ for any $\boldsymbol{y} \neq \boldsymbol{z}$, substitution of the above identities into eq. (10.53) yield eq. (10.52) with:

$$A_{pqrs}(\boldsymbol{y}, \boldsymbol{z}) = \frac{\mu}{8\pi}\left[\delta_{pr}\delta_{qs} + \delta_{ps}\delta_{qr} + \frac{2\nu}{1-\nu}\delta_{pq}\delta_{rs}\right]|\boldsymbol{z} - \boldsymbol{y}|$$

One then puts:

$$B_{ikqs}(\boldsymbol{y}, \boldsymbol{z}) = e_{iep}e_{kgr}\frac{\partial}{\partial z_e}\frac{\partial}{\partial y_g}A_{pqrs}(\boldsymbol{y}, \boldsymbol{z}) \tag{10.54}$$

Crack in an infinite medium

Consider a crack (described by the open surface Γ) embedded in an infinite elastic body, its faces subjected to a symmetric loading. This situation is the limiting form of a Neumann problem for an infinitely thin cavity of boundary $\Gamma^+ \cup \Gamma^-$. The Galerkin formulation (10.51) becomes in this simple case:

Find $\varphi \in \mathcal{V}_u$ such that

$$\int_\Gamma \int_\Gamma R_q \psi_i(\boldsymbol{x}) B_{ikqs}(\boldsymbol{y}, \boldsymbol{x}) R_s \phi_k(\boldsymbol{y}) \, \mathrm{d}S_y \, \mathrm{d}S_x = -\int_\Gamma \boldsymbol{\psi}.\boldsymbol{t}^D \, \mathrm{d}S \qquad (10.55)$$

where $\boldsymbol{\phi}$ is the *crack opening displacement*, i.e. the displacement discontinuity, across the crack surface Γ. Formulations of this type have been investigated by e.g. Bonnet [8], Bui [11] Gu, Yew [16], Xu, Argon, Ortiz and Bower [31]).

The Galerkin formulation for cracks embedded in finite bodies can be established in a straightforward way, for instance by splitting the solid Ω into two subdomains by means of a (fictitious) surface containing the crack surface(s) and then combining the formulation (10.51) for each subdomain (i.e. following and adapting the 'displacement discontinuity approach' presented in Section 13.5). The usual unknowns on the external boundary are then supplemented by the unknown crack opening displacement $\boldsymbol{\phi}$.

In a recent paper, Li, Mear and Xiao [19] report the first numerical implementation of a Galerkin integral formulation for three-dimensional fracture mechanics applicable to arbitrary solids of finite dimensions.

Elastodynamics

Galerkin boundary integral formulations in elastodynamics can be established in much the same way as for potential or elastostatic problems. The regularization of terms containing the strongly singular kernel $\boldsymbol{\Sigma}(\boldsymbol{x}, \boldsymbol{y}; \omega)$ uses again the properties of the Kelvin fundamental stress tensor. Using the Helmholtz fundamental solution for the isotropic infinite space, the hypersingular kernel can be recast into a form similar to (10.52) by invoking the representation of $\boldsymbol{U}(\boldsymbol{x}, \boldsymbol{y}; \omega)$ in terms of a *Somigliana potential* $F(\boldsymbol{x}, \boldsymbol{y}; \omega)$ [14], the elastodynamic counterpart to the Galerkin tensor representation:

$$U_i^k = 2(1-\nu)[\Delta F + k_L^2 F]\delta_{ik} - F_{,ik} \qquad (10.56)$$

with

$$F(\boldsymbol{x}, \boldsymbol{y}; \omega) = \frac{1}{4\pi\mu k_T^2}(\mathrm{e}^{\mathrm{i}k_L r} - \mathrm{e}^{\mathrm{i}k_T r})\frac{1}{r} \qquad (10.57)$$

and $k_L^2 = \rho\omega^2/(\lambda + 2\mu), k_T^2 = \rho\omega^2/\mu$. One then finds after some calculations:

$$C_{ijab}\frac{\partial}{\partial y_b}\Sigma_{k\ell}^a(\boldsymbol{y}, \boldsymbol{x}; \omega) = e_{iep}e_{jfq}e_{kgr}e_{\ell hs}\frac{\partial}{\partial x_e}\frac{\partial}{\partial x_f}\frac{\partial}{\partial y_g}\frac{\partial}{\partial y_h}A_{pqrs}(\boldsymbol{y}, \boldsymbol{x}; \omega) + \rho\omega^2 D_{ijk\ell}(\boldsymbol{y}, \boldsymbol{x}; \omega) \qquad (10.58)$$

with

$$A_{pqrs}(\boldsymbol{y},\boldsymbol{x};\omega) = \left[\delta_{pr}\delta_{qs} + \delta_{ps}\delta_{qr} + \frac{2\nu}{1-\nu}\delta_{pq}\delta_{rs}\right] F(\boldsymbol{y},\boldsymbol{x};\omega)$$
$$D_{ijk\ell}(\boldsymbol{y},\boldsymbol{x};\omega) = -(1-2\nu)C_{ijab}C_{k\ell cd}\delta_{ac}F_{,bd} - \frac{\mu}{k_2^T}2(1-\nu)C_{ijk\ell}\Delta^2 F$$

Note that even though $\Delta^2 F$ does not vanish like in elastostatics, it is only weakly singular: one has:

$$\frac{\mu}{k_2^T}2(1-\nu)\Delta^2 F = \beta^2 \frac{\mathrm{e}^{\mathrm{i}k_L r}}{4\pi r} - \frac{\mathrm{e}^{\mathrm{i}k_T r}}{4\pi r}$$

where β is defined by eq. (7.48). The decomposition (10.58) appears in [24] and is generalized to anisotropic elasticity in [6]).

For instance, the elastodynamic response of an elastic body subjected to given time-harmonic tractions $\boldsymbol{t}^D$ on its external boundary S and containing a traction-free defect of boundary Γ is governed by the symmetric Galerkin integral equation:

$$\mathcal{B}(\boldsymbol{u},\boldsymbol{w}) = \mathcal{F}(\boldsymbol{w}) \qquad \forall \boldsymbol{w} \in [H^{1/2}(\partial\Omega)]^3 \tag{10.59}$$

where ($\partial\Omega = S + \Gamma$ denoting the entire boundary)

$$\mathcal{B}(\boldsymbol{u},\boldsymbol{v}) = \int_{\partial\Omega}\int_{\partial\Omega} \Big\{ R_q u_i(\boldsymbol{y}) R_s \bar{w}_k(\boldsymbol{x}) B_{ikqs}(\boldsymbol{x},\boldsymbol{y};\omega)$$
$$+ k_T^2 n_j(\boldsymbol{y}) n_\ell(\boldsymbol{x}) u_i(\boldsymbol{y}) \bar{w}_k(\boldsymbol{x}) D_{ijk\ell}(\boldsymbol{x},\boldsymbol{y};\omega)\Big\}\, \mathrm{d}S_y\, \mathrm{d}S_x$$

$$\mathcal{F}(\boldsymbol{w}) = \int_S\int_S [T_i^k(\boldsymbol{y},\boldsymbol{x})\bar{w}_k(\boldsymbol{x}) - T_i^{(0)k}(\boldsymbol{y},\boldsymbol{x})\bar{w}_k(\boldsymbol{y})] t_a^D(\boldsymbol{y})\, \mathrm{d}S_y\, \mathrm{d}S_x$$
$$+ \int_\Gamma\int_S T_i^k(\boldsymbol{y},\boldsymbol{x})\bar{w}_k(\boldsymbol{x}) t_a^D(\boldsymbol{y})\, \mathrm{d}S_y\, \mathrm{d}S_x - \int_S t_a^D(\boldsymbol{y})\bar{w}_k(\boldsymbol{y})\, \mathrm{d}S_y$$

Depending on whether the defect Γ is a cavity or a crack, $\boldsymbol{u}$ on Γ denotes the displacement or the displacement jump.

Symmetric Galerkin integral formulations can also be extended to problems in the time domain. Up to now, computational efforts have been limited to problems with simple geometries (e.g. cracks in infinite media, see Bécache [5] and Barbier [4]). As for collocation methods in the time domain, one of the main obstacles to overcome lies in the evaluation of element integrals involving retarded times (see Section 7.5).

10.10 NUMERICAL EXAMPLES

Acoustical directivity of a vibrating tube (Hamdi [17])

The doctoral thesis of Hamdi [17], in which the Galerkin integral formulation is established and three-dimensional numerical examples in exterior acoustics are presented, is a pioneering work in the area of symmetric Galerkin integral equation formulations. In addition to being one of the earliest implementations of SGBEM to deal with mixed boundary conditions, it also incorporates the double integration by parts regularization for the hypersingular kernel.

One of the numerical examples presented therein concerns the comparison of computed and experimental values for the acoustical radiation of a cylindrical tube (radius $a = 1$ m, length $6a$). The excitation consists of point sources located on a circle of radius $0.9a$ (Figure 10.2). Three meshes of the tube (12, 16 or 20 rings of 36 linear three-noded triangular elements, for a total of 234, 316 or 388 unknown nodal values, respectively) have been used. A sample result of [17], where computed and experimental directivity diagrams are compared, is reproduced in Figure 10.3.

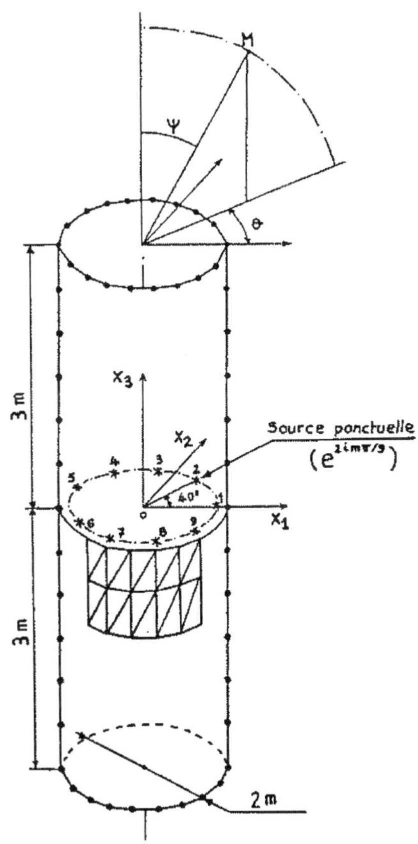

Figure 10.2 Acoustical directivity of a vibrating tube: geometrical configuration (from Hamdi [17]).

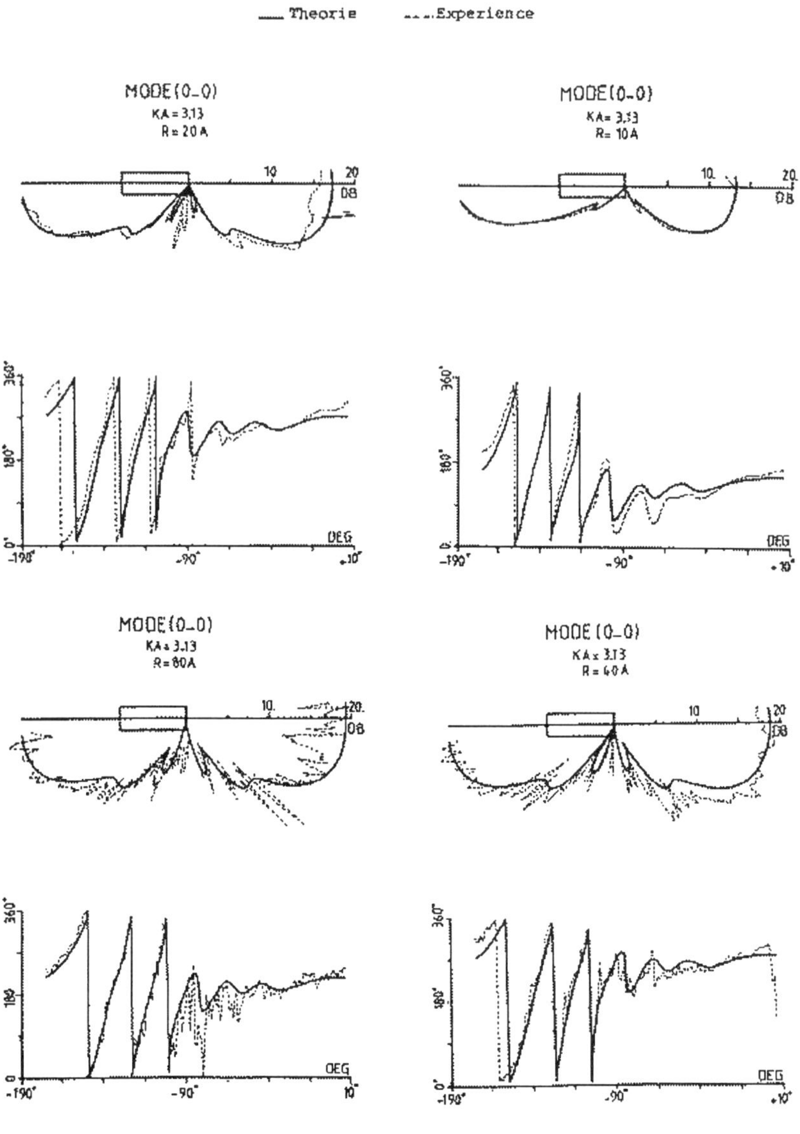

Figure 10.3 Acoustical directivity of a vibrating tube: comparison between numerical and experimental results, with $ka = 3,13$ (from Hamdi [17]).

Elliptical crack under uniform loading

The problem of an elliptical plane crack Γ (centre O, axes $a \geq b$) embedded in an infinite elastic body and subjected to an arbitrary uniform loading has

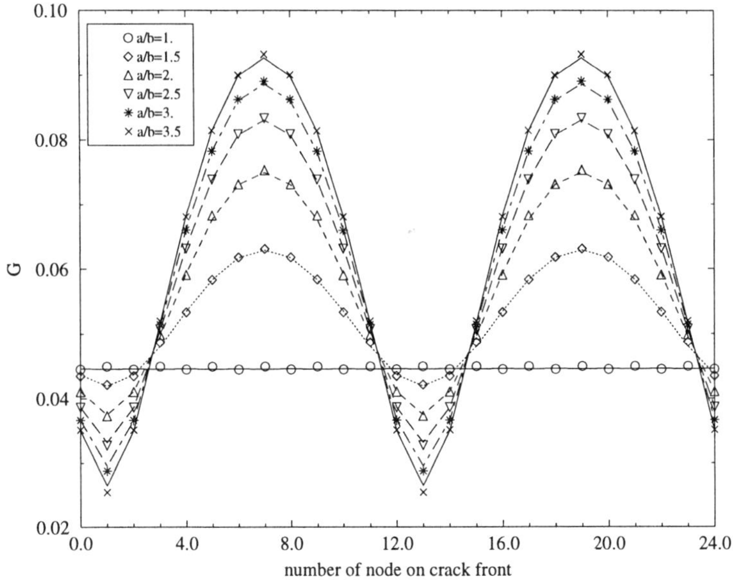

Figure 10.4 Example 1: computed and exact distributions of G along the crack front for several aspect ratio values (mode I loading).

an exact solution (see e.g. [27]). For the particular case of uniform tension ($\bar{t} = p e_3$, mode I problem[3]), analytical expressions for the normal opening displacement ϕ and energy release rate G are available:

$$\phi_3 = \frac{(1-\nu)p}{\mu} \frac{b}{E(k)} \left(1 - \frac{x_1^2}{a^2} - \frac{x_2^2}{b^2}\right)^{1/2}$$

$$G = p^2 \frac{1-\nu^2}{E} \frac{\pi b}{E^2(k)} \left(1 - k^2 \cos^2 \alpha\right)^{1/2}$$

where $E(k)$ denotes the complete elliptic integral of first kind:

$$E(k) = \int_0^{\pi/2} \left(1 - k^2 \cos^2 \alpha\right)^{1/2} \, \mathrm{d}\alpha \qquad k^2 = 1 - \frac{b^2}{a^2}$$

The symmetric Galerkin boundary element method was applied to this test problem, using a 9-noded interpolation for ϕ on the crack surface.

Table 10.1 displays mean-square errors on ϕ for several meshes (circular crack, i.e. $a/b = 1$) and also for several values of the aspect ratio a/b ranging

[3]The reader is referred to Sections 13.6 and 13.8 for the definitions of the concepts from fracture mechanics appearing in this example.

Table 10.1 Elliptical crack under uniform loading: relative mean-square error on the normal crack opening displacement ϕ for various meshes (top) and for various values of the aspect ratio a/b with a 30-element mesh (bottom).

number of elements	6	12	18	30	42	56
rel. error ($\times 10^{-3}$)	98.7	8.58	7.01	3.41	2.25	1.46

a/b	1.	2.	4.	6.	8.
rel. error ($\times 10^{-3}$)	3.40	3.35	6.23	13.9	33.1

between 1 and 8, using a 30-element mesh. Even though the latter mesh is not very fine, the errors (which increase with a/b) are acceptable even for the larger aspect ratios, i.e. the most distorted meshes. In addition, computed and exact distributions of G along the crack front are plotted in Figure 10.4 for various values of a/b.

Appendix 10.A REGULARIZED EXPRESSIONS OF I_1 AND I_2

In this appendix, the expressions (10.10) and (10.11) for the limiting forms $I_1(\partial\Omega, \partial\Omega)$ and $I_2(\partial\Omega, \partial\Omega)$ of $I_1(\tilde{S}, \partial\Omega)$ and $I_2(\tilde{S}, \partial\Omega)$ as $\tilde{S} \to \partial\Omega$ (i.e. for $\eta \to 0$ in the construction (10.6)) are established.

Limiting process for I_1. The term I_1, eq. (10.8), can be transformed into:

$$I_1(\tilde{S}, \partial\Omega) = \int_{\tilde{S}} \tilde{u}(\boldsymbol{z})q(\boldsymbol{z})\,\mathrm{d}S_z - \int_{\partial\Omega} q(\boldsymbol{y}) \int_{\tilde{S}} [\tilde{u}(\boldsymbol{z}) - \tilde{u}(\boldsymbol{y})]H(\boldsymbol{y}, \boldsymbol{z})\,\mathrm{d}S_z\,\mathrm{d}S_y$$
$$- \int_{\partial\Omega} q(\boldsymbol{y})\tilde{u}(\boldsymbol{y}) \left\{ \int_{\tilde{S}} H(\boldsymbol{y}, \boldsymbol{z})\,\mathrm{d}S_z \right\} \mathrm{d}S_y \tag{10.A.1}$$

by transposing integrations over $\partial\Omega$ and $\tilde{S}$ and then adding and subtracting the same quantity. Next, identity (4.7) gives, taking into account the orientation convention for $\boldsymbol{n}$ on $\tilde{S}$:

$$\int_{\tilde{S}} H(\boldsymbol{y}, \boldsymbol{z})\,\mathrm{d}S_z = \kappa \tag{10.A.2}$$

Indeed, $\boldsymbol{y}$ is exterior to $\tilde{S}$ if Ω is bounded ($\kappa = 0$) and interior to $\tilde{S}$ if $\mathbb{R}^3 - \Omega$ is bounded ($\kappa = 1$); in the latter case, the normal $\boldsymbol{n}$ to $\tilde{S}$ is *interior* to $\tilde{S}$. At this point, it is possible to take the limit $\tilde{S} \to \partial\Omega$: the quantity

$$\int_{\tilde{S}} [\tilde{u}(\boldsymbol{z}) - \tilde{u}(\boldsymbol{y})]H(\boldsymbol{y}, \boldsymbol{z})\,\mathrm{d}S_z$$

considered as a function of $\boldsymbol{y}$ is continuous across $\tilde{S}$. The limiting expression of $I_1(\tilde{S}, \partial\Omega)$ when $\tilde{S} \to \partial\Omega$ thus merely results from substituting $\tilde{S}$ by $\partial\Omega$ and $\boldsymbol{z} \in \tilde{S}$ by $\boldsymbol{x} \in \partial\Omega$ into eq. (10.A.1) and using eq. (10.A.2):

$$\lim_{\tilde{S}\to\partial\Omega} I_1(\tilde{S}, \partial\Omega) = (1-\kappa)\int_{\partial\Omega} q(\boldsymbol{y})\tilde{u}(\boldsymbol{y})\,\mathrm{d}S_y - \int_{\partial\Omega} q(\boldsymbol{y})\int_{\partial\Omega}[\tilde{u}(\boldsymbol{x}) - \tilde{u}(\boldsymbol{y})]H(\boldsymbol{y},\boldsymbol{x})\,\mathrm{d}S_x\,\mathrm{d}S_y$$

The inner (with respect to $\boldsymbol{x}$) and outer (with respect to $\boldsymbol{y}$) integrations are respectively weakly singular and nonsingular.

Limiting process for I_2. The term $I_2(\tilde{S}, \partial\Omega)$, eq. (10.9), involves the hypersingular kernel $G_{,\bar{i}j}$. Substituting eq. (10.13) into eq. (10.9) and applying twice the Stokes formula (A.5) for the closed surfaces $\partial\Omega$ and $\tilde{S}$ yields:

$$\begin{aligned} I_2(\tilde{S}, \partial\Omega) &= \int_{\tilde{S}} \tilde{u}(\boldsymbol{z})e_{iap}n_i(\boldsymbol{z})\int_{\partial\Omega} e_{jbq}n_j(\boldsymbol{y})u(\boldsymbol{y})B_{ab,\bar{p}q}(\boldsymbol{z},\boldsymbol{y})\,\mathrm{d}S_y\,\mathrm{d}S_z \\ &= -\int_{\tilde{S}} \tilde{u}(\boldsymbol{z})e_{iab}n_i(\boldsymbol{z})\int_{\partial\Omega} R_b u(\boldsymbol{y})B_{ab,\bar{p}}(\boldsymbol{z},\boldsymbol{y})\,\mathrm{d}S_y\,\mathrm{d}S_z \\ &= \int_{\tilde{S}} R_a\tilde{u}(\boldsymbol{z})\int_{\partial\Omega} R_b u(\boldsymbol{y})B_{ab}(\boldsymbol{z},\boldsymbol{y})\,\mathrm{d}S_y\,\mathrm{d}S_z \end{aligned} \tag{10.A.3}$$

using the operator $R_a f$, eq. (10.12). Due to the weakly singular character of the tensor $\boldsymbol{B}(\boldsymbol{y},\boldsymbol{z})$, the expression:

$$\int_{\partial\Omega} B_{ab}(\boldsymbol{z},\boldsymbol{y})R_b u(\boldsymbol{y})\,\mathrm{d}S_y$$

considered as a function of $\boldsymbol{z}$ is continuous across $\partial\Omega$. The limiting expression of $I_2(\tilde{S}, \partial\Omega)$ when $\tilde{S} \to \partial\Omega$ thus results again from substuting $\tilde{S}$ by $\partial\Omega$ and $\boldsymbol{z} \in \tilde{S}$ by $\boldsymbol{x} \in \partial\Omega$ into eq. (10.A.3):

$$\lim_{\tilde{S}\to\partial\Omega} I_2(\tilde{S}, \partial\Omega) = \int_{\partial\Omega} R_a\tilde{u}(\boldsymbol{x})\int_{\partial\Omega} B_{ab}(\boldsymbol{x},\boldsymbol{y})R_b u(\boldsymbol{y})\,\mathrm{d}S_y\,\mathrm{d}S_x$$

The inner (with respect to $\boldsymbol{x}$) and outer (with respect to $\boldsymbol{y}$) integrations are respectively weakly singular and nonsingular.

REFERENCES

1. ARNOLD, D. N., WENDLAND, W. L. On the asymptotic convergence of collocation methods. *Math. Comp.*, **41**, 349–381 (1983).
2. BABUSKA, I, AZIZ, A K. Survey lectures on the mathematical foundations of the finite element method. In A K Aziz (ed.), *The mathematical foundations of*

the finite element method with application to partial differential equations, pp. 3–359. Academic Press (1972).

3. BAMBERGER, A., HA DUONG, T. Formulation variationnelle espace-temps pour le calcul par potentiel retardé de la diffraction d'une onde acoustique (I). *Math. Meth. in Appl. Sc.*, **8**, 405–435 (1986).
4. BARBIER, D. *Méthode des potentiels retardés pour la simulation de la diffraction d'ondes élastodynamiques par une fissure tridimensionnelle*. Ph.D. thesis, Ecole Polytechnique, France (1999).
5. BÉCACHE, E.. A variational boundary integral equation method for an elastodynamic antiplane crack. *Int. J. Num. Meth. in Eng.*, **36**, 969–993 (1993).
6. BÉCACHE, E., NEDELEC, J.-C., NISHIMURA, N.. Regularization in 3D for anisotropic elastodynamic crack and obstacle problems. *J. Elast.*, **31**, 25–46 (1993).
7. BENDALI, A. Problème aux limites extérieur et intérieur pour le système de Maxwell en régime harmonique. Internal report no. 59, CMAP, Ecole Polytechnique, Palaiseau, France (1980).
8. BONNET, M. Equations intégrales variationnelles pour le problème en vitesse de propagation de fissures en élasticité linéaire. *C.R. Acad. Sci. Paris, série II,* **318**, 429–434 (1994).
9. BONNET, M. Regularized direct and indirect symmetric variational BIE formulations for three-dimensional elasticity. *Engng. Anal. with Bound. Elem.*, **15**, 93–102 (1995).
10. BONNET, M., MAIER, G., POLIZZOTTO, C. On symmetric Galerkin boundary element method. *Appl. Mech. Rev.*, **51**, 669–704 (1998).
11. BUI, H. D. An integral equation method for solving the problem of a plane crack of arbitrary shape. *J. Mech. Phys. Solids*, **25**, 29–39 (1977).
12. CARINI, A, DILIGENTI, M, MARANESI, P., ZANELLA, M. Analytical integrations for two-dimensional elastic analysis by the symmetric Galerkin boundary element method. *Comp. Mech.* (1997). (soumis).
13. CORTEY-DUMONT, PH. *Simulation numérique de problèmes de diffraction d'ondes par une fissure*. Ph.D. thesis, Université Paris 6, France (1985).
14. ERINGEN, A. C., SUHUBI, E. S. *Elastodynamics (vol II. – linear theory)*. Academic Press (1975).
15. FRANGI, A., NOVATI, G. Symmetric BE method in two-dimensional elasticity: evaluation of double integrals for curved elements. *Comp. Mech.*, **19**, 58–68 (1996).
16. GU, H., YEW, C. H. Finite element solution of a boundary integral equation for mode I embedded three-dimensional fractures. *Int. J. Num. Meth. in Eng.*, **26**, 1525–1540 (1988).
17. HAMDI, M. A. *Formulation variationnelle par équations intégrales pour le calcul de champs acoustiques linéaires proches et lointains*. Ph.D. thesis, Université Technologique de Compiègne, France (1982).
18. HAMDI, M. A., JEAN, PH. Résolution numérique des problèmes couplés par une méthode d'éléments de frontière. In *Tendances actuelles en calcul des structures*, pp. 449–464. Pluralis (1985).

19. LI, S., MEAR, M.E., XIAO, L. Symmetric weak-form integral equation method for three-dimensional fracture analysis. *Comp. Meth. in Appl. Mech. Engng.*, **151**, 435–459 (1998).

20. NEDELEC, J-C. Curved finite element methods for the solution of singular integral equations on surfaces in $\mathbb{R}^3$. *Comp. Meth. in Appl. Mech. Engng.*, **8**, 61–80 (1976).

21. NEDELEC, J. C. Approximation des équations intégrales en physique et en mécanique. Course notes, Ecole Polytechnique, Palaiseau, France (1977).

22. NEDELEC, J. C. Integral equations with non integrable kernels. *Integral equations and operator theory*, **5**, 562–572 (1982).

23. NEDELEC, J-C. Approximation des équations intégrales par éléments finis. Etude d'erreur. In J. L. Lions, R. Dautray (ed.), *Analyse mathématique et calcul scientifique pour les sciences et les techniques (chapter 13)*, pp. 953–968. Masson, Paris (1985).

24. NISHIMURA, N., KOBAYASHI, S. A regularized boundary integral equation method for elastodynamic crack problems. *Comp. Mech.*, **4**, 319–328 (1989).

25. PARREIRA, P., GUIGGIANI, M. On the implementation of the Galerkin approach in the boundary element method. *Computers and Structures*, **33**, 269–279 (1989).

26. SIRTORI, S., MAIER, G., NOVATI, G., MICCOLI, S. A Galerkin symmetric boundary element method in elasticity: formulation and implementation. *Int. J. Num. Meth. in Eng.*, **35**, 255–282 (1992).

27. TADA, H., PARIS, P., IRWIN, G. *The stress analysis of cracks handbook.* Technical report, Del. Research Corporation, Hellertown, Pennsylvania, USA (1973).

28. WENDLAND, W L. On the asymptotic convergence of boundary integral methods. In C. A. Brebbia (ed.), *Boundary element methods*, pp. 412–430. Springer-Verlag, Berlin, Heidelberg (1981).

29. WENDLAND, W. L. Asymptotic accuracy and convergence for point collocation methods. In C. A. Brebbia (ed.), *Topics in boundary element research.*, vol. 2. Springer-Verlag (1984).

30. WENDLAND, W. L. Boundary element methods for elliptic problems. Mathematical theory of finite and boundary element methods: lecture notes, Dusseldorf (Germany) (Jun. 1987).

31. XU, G., ARGON, A. S., ORTIZ, M., BOWER, A. Development of a variational boundary integral method for the analysis of fully three dimensional crack advance problems. In S. N. Atluri, G. Yagawa, T. A. Cruse (eds), *Computational Mechanics '95*, vol. 2, pp. 2874–2889. Springer-Verlag (1995).

32. ZENG, X, BIELAK, J. Stable symmetric finite element–boundary integral coupling methods for fluid–structure interface problems. *Engng. Anal. with Bound. Elem.*, **15**, 79–91 (1995).

11

Exploitation of geometrical symmetries

11.1 INTRODUCTION

A lot of problems arising in computational engineering deal with bodies Ω which have some geometrical symmetry, which can be described by a *symmetry group* G of n isometries under which Ω is invariant:

$$(\forall \boldsymbol{s} \in G) \qquad \boldsymbol{s}(\Omega) = \Omega \tag{11.1}$$

In such cases, linear boundary-value problems can be reduced to n *subproblems* stated on a *symmetry cell* C, i.e. a part, n times smaller, of the original body Ω which generates the whole body under the action of all the elements of G:

$$\Omega = \bigcup_{\boldsymbol{s} \in G} \boldsymbol{s}(C) \tag{11.2}$$

Such a procedure allows substantial computational savings, since the computational cost increases much faster than the size of the problem considered.

It is easy to reduce a problem when the loading (source, boundary conditions) and the geometry share the same symmetry properties, but far less easy to obtain and implement in the general case the maximal reduction (i.e. on the smallest possible symmetry cell) when the loading has no, or weaker, symmetry. A systematic approach of the use of geometrical symmetry in solving boundary-value problems, based on the use of *group representation theory*, has been carried out by Bossavit [3] in the framework of variational formulations of elliptic boundary-value problems. As a result Bossavit's work provides by means of algebraic formulas a systematic and automatic way (a 'recipe', according to the author) to *decompose* the initial problem into n subproblems and then *reconstruct* the global solution from the results of the n subproblems.

To our best knowledge, very few publications address systematic treatments allowing one to exploit geometrical symmetries in boundary integral equation (BIE) methods (see [2] and Allgower et al. [1], Guy and Mangeot [5] Lobry and Broche [6]), although the use of geometrical symmetries is obviously desirable. The framework and results of [3] are oriented towards variational formulations and finite-element methods. The present chapter shows their adaptation to boundary integral equation (BIE) methods.

For the sake of exposition, the developments to follow are presented for three-dimensional elastostatics. Their adaptation to other linear boundary-value problems is straightforward. Accordingly, the classical elastic equilibrium boundary-value problem for an elastic body Ω with boundary $\partial\Omega$ is considered:

$$(\mathrm{P}) \qquad \begin{cases} \mathrm{div}\boldsymbol{\sigma}(\boldsymbol{u}) = \boldsymbol{0} & \\ \boldsymbol{\sigma}\cdot\boldsymbol{n} = \boldsymbol{t}^D & \text{on } S_T \\ \boldsymbol{u} = \boldsymbol{u}^D & \text{on } S_u \end{cases}$$

where $S_T \cap S_u = \emptyset$ and $S_T \cup S_u = \partial\Omega$. The problem (P) can equivalently be stated in a variational (or *weak*) form:

$$(\mathrm{P}) \qquad \text{Find } \boldsymbol{u} \in \mathcal{V} = \{\boldsymbol{w} \in H^1(\Omega),\ \boldsymbol{w}\,|_{S_u} = \boldsymbol{u}^D\},$$
$$\int_\Omega (\boldsymbol{\nabla} u : \boldsymbol{C} : \boldsymbol{\nabla} v)\,\mathrm{d}V = \int_{S_T} (\boldsymbol{t}^D.\boldsymbol{v})\,\mathrm{d}S$$
$$(\forall \boldsymbol{v} \in \mathcal{V}_0 = \{\boldsymbol{w} \in H^1(\Omega),\ \boldsymbol{w}\,|_{S_u} = \boldsymbol{0}\}) \qquad (11.3)$$

In this chapter, the domain Ω is assumed to be invariant under a symmetry group G, eq. (11.1). Also, the complementary components S_u and S_T of $\partial\Omega$ are assumed to possess this invariance, so that symmetrical portions of the boundary are subjected to boundary conditions *of the same type*. On the other hand, no symmetry assumption is made regarding the *values* of the boundary data ($\boldsymbol{u}^d, \boldsymbol{t}^d$: the geometrical symmetry must be respected by the type of the boundary conditions but not by the values. One wishes to make the best computational use of the *geometrical* symmetry of the domain Ω.

11.2 GEOMETRICAL SYMMETRY IN BOUNDARY-VALUE PROBLEMS

This section is devoted to a brief review of the essential concepts, notations and results used for the approach of geometrical symmetry in boundary-value problems in [3].

For the geometrical symmetry to be of use, the differential operator governing the boundary-value problem (here $\mathrm{div}\boldsymbol{\sigma}(\boldsymbol{u})$, which is understood as a second-order elliptic operator acting on $\boldsymbol{u}$) must be invariant under the symmetry group G. This rather vague statement can be given a precise form using the variational formulation (11.3), stating that the bilinear form

$$a(\boldsymbol{u}, \boldsymbol{v}) = \int_{\Omega} (\boldsymbol{\nabla} u : \boldsymbol{C} : \boldsymbol{\nabla} v) \, \mathrm{d}V$$

which appears in the left-hand side of (11.3) must satisfy the following *equivariance property*:

$$\forall \boldsymbol{s} \in G \qquad a(\boldsymbol{S}\boldsymbol{u}, \boldsymbol{S}\boldsymbol{v}) = a(\boldsymbol{u}, \boldsymbol{v}) \tag{11.4}$$

using the notation:

$$\boldsymbol{S}\boldsymbol{u}(\boldsymbol{x}) = \boldsymbol{s}\boldsymbol{u}(\boldsymbol{s}^{-1}\boldsymbol{x})$$

In the present context of linear elasticity, equation (11.4) is obviously true if the elastic coefficients are constant on Ω, but could be violated in the case of heterogeneous elastic moduli. To put it another way, G must be the group of *geometrical and material* symmetry of the structure considered. We restrict ourselves to *Abelian* finite symmetry groups, i.e. for which $\boldsymbol{s}_1\boldsymbol{s}_2 = \boldsymbol{s}_2\boldsymbol{s}_1$ ($\boldsymbol{s}_1, \boldsymbol{s}_2 \in G$); let n denote the *order* (i.e. the number of elements) of G.

The basic idea consists in replacing the initial problem (P) on Ω by n *subproblems* on a *symmetry cell* C; then a linear combination (with adequate weighting coefficients, see later) of the solutions of the subproblems produces the global solution to (P), using the linearity of the boundary-value problem with respect to its data (superposition principle of linear elasticity). A symmetry cell is a subdomain of Ω, n times smaller than Ω, such that

$$\Omega = \boldsymbol{s}_1(C) \cup \ldots \cup \boldsymbol{s}_n(C)$$

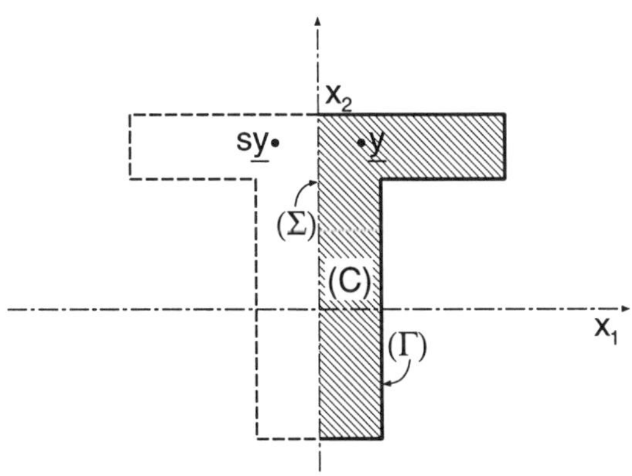

Figure 11.1 Symmetrical domain Ω and symmetry cell C: case $G = \{\boldsymbol{1}, \boldsymbol{s}\}$, $\Omega = C \cup \boldsymbol{s}(C)$.

(Figure 11.1). A more abstract and precise definition of a symmetry cell is given in [3], the 'naive' one being sufficient for the present purposes. Let $\Gamma = \partial\Omega \cap \partial C$ denote the part of the cell boundary lying on the boundary of Ω and let $\Sigma = \partial C \setminus \Gamma$ be the additional boundary (see Figure 11.1). In fact [3]:

$$\Sigma = \{\boldsymbol{x} \in C | \exists \boldsymbol{s} \in G \ (\boldsymbol{s} \neq \mathbf{1}), \ \boldsymbol{sx} \in \Sigma\} \tag{11.5}$$

and for almost all $\boldsymbol{x} \in \Sigma$ this $\boldsymbol{s}$ is unique.

11.2.1 Elementary analysis for plane symmetry

In the simple situation $G = \{\mathbf{1}, \boldsymbol{s}\}$ ($\boldsymbol{s}$ being a plane symmetry), one builds the even and odd parts of the boundary data:

$$\begin{aligned} (\boldsymbol{u}_E^D, \boldsymbol{t}_E^D)(\boldsymbol{x}) &= \left\{ \frac{1}{2} \left[\boldsymbol{u}^D(\boldsymbol{x}) + \boldsymbol{s}\boldsymbol{u}^D(\boldsymbol{sx}) \right], \frac{1}{2} \left[\boldsymbol{t}^D(\boldsymbol{x}) + \boldsymbol{s}\boldsymbol{t}^D(\boldsymbol{sx}) \right] \right\} \\ (\boldsymbol{u}_O^D, \boldsymbol{t}_O^D)(\boldsymbol{x}) &= \left\{ \frac{1}{2} \left[\boldsymbol{u}^D(\boldsymbol{x}) - \boldsymbol{s}\boldsymbol{u}^D(\boldsymbol{sx}) \right], \frac{1}{2} \left[\boldsymbol{t}^D(\boldsymbol{x}) - \boldsymbol{s}\boldsymbol{t}^D(\boldsymbol{sx}) \right] \right\} \end{aligned} \tag{11.6}$$

Then $\boldsymbol{u}$ is the sum of its odd and even parts: $\boldsymbol{u} = \boldsymbol{u}_O + \boldsymbol{u}_E$, solutions of (P) with boundary data given by (11.6). But since $\boldsymbol{u}_O$ and $\boldsymbol{u}_E$ have known symmetry properties:

$$\forall \boldsymbol{x} \in \boldsymbol{s}(C) \quad \left\{ \begin{aligned} \boldsymbol{u}_E(\boldsymbol{sx}) &= \boldsymbol{s}\boldsymbol{u}_E(\boldsymbol{x}) \\ \boldsymbol{u}_O(\boldsymbol{sx}) &= \boldsymbol{s}\boldsymbol{u}_O(\boldsymbol{x}) \end{aligned} \right.$$

the problem (P) with data (11.6) can be formulated and solved on the *symmetry cell* of Figure 11.1, provided additional boundary conditions are given on the additional boundary Σ, Σ being here the intersection of Ω with the plane of symmetry:

$$\begin{aligned} \boldsymbol{n}.\boldsymbol{u}_E &= 0 \quad \text{and} \quad (\boldsymbol{\sigma}.\boldsymbol{n} - (\boldsymbol{n}.\boldsymbol{\sigma}.\boldsymbol{n})\boldsymbol{n})_E = \mathbf{0} \qquad \text{on } \Sigma \\ (\boldsymbol{u} - (\boldsymbol{u}.\boldsymbol{n})\boldsymbol{n})_O &= \mathbf{0} \quad \text{and} \quad (\boldsymbol{n}.\boldsymbol{\sigma}.\boldsymbol{n})_O = 0 \qquad \text{on } \Sigma \end{aligned} \tag{11.7}$$

They are consequences of the symmetry properties of the solution $\boldsymbol{u}_E, \boldsymbol{u}_O$ and as such should be regarded as *constraints* rather than *conditions*.

Equations (11.6) and (11.7) thus define the *decomposition* of problem (P) into two subproblems (P_E), (P_O) for the respective unknowns ($\tilde{\boldsymbol{u}}_E, \tilde{\boldsymbol{u}}_O$), restrictions to C of the symmetric and antisymmetric parts $\boldsymbol{u}_E$ and $\boldsymbol{u}_O$ of $\boldsymbol{u}$ (the notation $\tilde{()}$ indicates fields defined on the symmetry cell C).

The reconstruction step consists in defining a continuation of ($\tilde{\boldsymbol{u}}_E, \tilde{\boldsymbol{u}}_O$) to the entire domain Ω by invoking the known symmetry properties:

$$\forall \boldsymbol{x} \in C \qquad \left\{ \begin{aligned} \tilde{\boldsymbol{u}}_E(\boldsymbol{sx}) &= \boldsymbol{s}\boldsymbol{u}_E(\boldsymbol{x}) \\ \tilde{\boldsymbol{u}}_O(\boldsymbol{sx}) &= -\boldsymbol{s}\boldsymbol{u}_O(\boldsymbol{x}) \end{aligned} \right. \tag{11.8}$$

The solution $\boldsymbol{u}$ to the original problem is finally given by:

$$(\forall \boldsymbol{x} \in \Omega) \qquad \boldsymbol{u}(\boldsymbol{x}) = \boldsymbol{u}_O(\boldsymbol{x}) + \boldsymbol{u}_E(\boldsymbol{x}) \tag{11.9}$$

The operations (11.8) and (11.9) can be merged into a single formula giving $\boldsymbol{u}$ in Ω from $(\tilde{\boldsymbol{u}}_E, \tilde{\boldsymbol{u}}_O)$ in C:

$$(\forall \boldsymbol{x} \in C) \qquad \begin{cases} \boldsymbol{u}(\boldsymbol{x}) = \tilde{\boldsymbol{u}}_E(\boldsymbol{x}) + \tilde{\boldsymbol{u}}_O(\boldsymbol{x}) \\ \boldsymbol{u}(\boldsymbol{s}\boldsymbol{x}) = \boldsymbol{s}\tilde{\boldsymbol{u}}_E(\boldsymbol{x}) - \boldsymbol{s}\tilde{\boldsymbol{u}}_O(\boldsymbol{x}) \end{cases} \tag{11.10}$$

Such an intuitive way of deriving these steps works nicely in the simplest case $G = \{\mathbf{1}, \boldsymbol{s}\}$, but becomes very cumbersome for more complicated symmetry groups, to say nothing of *nonAbelian* symmetry groups, see [3]. Thus the usefulness of a systematic treatment, which gives the algebraic decomposition and reconstruction formulas like (11.6), (11.7), (11.8) and (11.10) in the general case.

11.2.2 General case

The systematic treatment in [3] uses as a key tool the group-theoretic concept of *irreducible representation* of a finite group G. For the sake of simplicity, the presentation is here limited to the case of commutative, or *Abelian*, symmetry groups, for which:

$$\forall \boldsymbol{s}_1, \boldsymbol{s}_2 \in G, \boldsymbol{s}_1\boldsymbol{s}_2 = \boldsymbol{s}_2\boldsymbol{s}_1$$

However, Bossavit's analysis includes the possibility of non-Abelian groups as well. This situation, which is outside the scope of this chapter, arises e.g. for bodies possessing both plane and cyclic symmetries.

An *irreducible representation* is a complex-valued function ρ on G which preserves the group structure, and hence has the following properties:

$$\rho(\mathbf{1}) = 1 \qquad \text{and} \qquad (\forall \boldsymbol{s}_1, \boldsymbol{s}_2 \in G) \quad \rho(\boldsymbol{s}_1\boldsymbol{s}_2) = \rho(\boldsymbol{s}_1)\rho(\boldsymbol{s}_2) \tag{11.11}$$

A finite group G of order n possesses exactly n irreducible representations $\{\rho_\nu, \nu = 0, \ldots, n-1\}$. Such irreducible representations are tabulated for many finite groups; for example:

- *Cyclic symmetry.* Let G denote the cyclic group $G = \{\mathbf{1}, \boldsymbol{r}, \ldots, \boldsymbol{r}^{n-1}\}$ ($\boldsymbol{r}$: rotation of angle $2\pi/n$). Its irreducible representation is defined by:

$$\rho_\nu(\boldsymbol{r}) = \exp(2\mathrm{i}\pi\nu/n) \qquad (\nu = 0, \ldots, n-1) \tag{11.12}$$

- *Plane symmetry.* Let $\boldsymbol{s}_1$, $\boldsymbol{s}_2$ and $\boldsymbol{s}_3$ denote the plane symmetries with respect to the coordinate planes $(x_1 = 0)$, $(x_2 = 0)$ and $(x_3 = 0)$. The symmetry groups:

$$D_1 = \{\mathbf{1}, \boldsymbol{s}_1\}$$
$$D_2 = \{\mathbf{1}, \boldsymbol{s}_1, \boldsymbol{s}_2, \boldsymbol{q} = \boldsymbol{s}_1\boldsymbol{s}_2\}$$
$$D_3 = \{1, \boldsymbol{s}_1, \boldsymbol{s}_2, \boldsymbol{s}_3, \boldsymbol{p}_1 = \boldsymbol{s}_2\boldsymbol{s}_3, \boldsymbol{p}_2 = \boldsymbol{s}_3\boldsymbol{s}_1, \boldsymbol{p}_3 = \boldsymbol{s}_1\boldsymbol{s}_2, \boldsymbol{q} = \boldsymbol{s}_1\boldsymbol{s}_2\boldsymbol{s}_3\}$$

Table 11.1 Irreducible representations for plane symmetries with respect to one, two and three coordinate planes (respective orders $n = 2, 4, 8$).

D_1	$\mathbf{1}$	s_1
ρ_0	+1	+1
ρ_1	+1	-1

D_2	$\mathbf{1}$	s_1	s_2	q
ρ_0	+1	+1	+1	+1
ρ_1	+1	-1	+1	-1
ρ_2	+1	+1	-1	-1
ρ_3	+1	-1	-1	+1

D_3	$\mathbf{1}$	s_1	s_2	s_3	p_1	p_2	p_3	q
ρ_0	+1	+1	+1	+1	+1	+1	+1	+1
ρ_1	+1	+1	+1	-1	-1	-1	+1	-1
ρ_2	+1	-1	+1	+1	+1	-1	-1	-1
ρ_3	+1	-1	+1	-1	-1	+1	-1	+1
ρ_4	+1	+1	-1	+1	-1	+1	-1	-1
ρ_5	+1	+1	-1	-1	+1	-1	-1	+1
ρ_6	+1	-1	-1	+1	-1	-1	+1	+1
ρ_7	+1	-1	-1	-1	+1	+1	+1	-1

respectively associated to plane symmetries with respect to one, two and three coordinate planes and of respective orders $n = 2, 4, 8$, admit the irreducible representations defined by the coefficients given in Table 11.1.

Bossavit has shown in [3] that the irreducible representations give the coefficients of the decomposition and reconstruction steps in the general case,

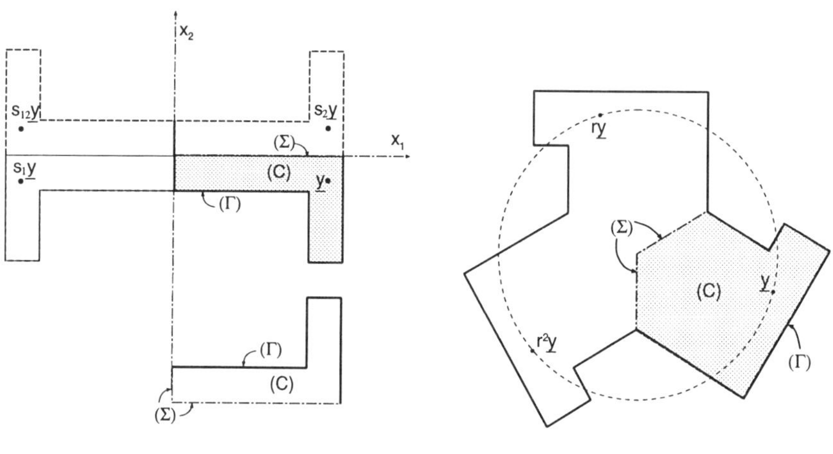

Figure 11.2 Symmetrical domain Ω and symmetry cell C:
$G = \{1, s_1, s_2, s_1 s_2\}$, $\Omega = C \cup s_1(C) \cup s_2(C) \cup s_1 s_2(C)$ (left)
$G = \{1, r, r^2\}$, $\Omega = C \cup r(C) \cup r^2(C)$ (right).

as follows (the reader is referred to [3] for any further explanation about the theory and should concentrate on the result and its subsequent use). The exploitation of geometrical symmetries in the weak formulation (P) require three steps.

1 - Choice of a symmetry cell. A symmetry cell C is chosen as an adequate subset of Ω such that:

$$|C| = \frac{1}{n}|\Omega| \qquad \text{and} \qquad \Omega = \boldsymbol{s}_1(C) \cup \ldots \cup \boldsymbol{s}_n(C) \quad (\boldsymbol{s}_1, \ldots, \boldsymbol{s}_n \in G)$$

We will denote by Γ_u and Γ_T the respective restrictions of S_u and S_T to C ($\Gamma_T \cap \Gamma_u = \emptyset$) and put $\Gamma_T \cup \Gamma_u = \Gamma$. Then the boundary of C is $\partial C = \Sigma \cup \Gamma$ (Figure 11.2), $\Sigma = \partial C \cap \Omega$ being the additional boundary. Thus S_u and S_T are generated by the action of G on Γ_u and Γ_T respectively.

2 - Reduction to subproblems stated on C. The following sequence $(P_\nu)_{\nu=1,\ldots,n}$ of *uncoupled* subproblems on Ω is introduced:

$$\operatorname{div}\boldsymbol{\sigma}(\boldsymbol{u}_\nu) = \boldsymbol{0} \qquad \text{in } \Omega$$

$$(P_\nu): \quad \boldsymbol{T}^n(\boldsymbol{u}^\nu)(\boldsymbol{x}) = \boldsymbol{t}^{D(\nu)}(\boldsymbol{x}) = \frac{1}{n}\sum_{\boldsymbol{s}\in G} \rho_\nu(\boldsymbol{s})\boldsymbol{s}^{-1}\boldsymbol{t}^D(\boldsymbol{s}\boldsymbol{x}) \qquad \text{on } S_T \tag{11.13}$$

$$\boldsymbol{u}^\nu(\boldsymbol{x}) = \boldsymbol{u}^{D(\nu)}(\boldsymbol{x}) = \frac{1}{n}\sum_{\boldsymbol{s}\in G} \rho_\nu(\boldsymbol{s})\boldsymbol{s}^{-1}\boldsymbol{u}^D(\boldsymbol{s}\boldsymbol{x}) \qquad \text{on } S_u$$

which generalize the 'odd' and 'even' boundary data defined by eq. (11.6), the subscript ν playing a role similar to the subscripts O, E. Subproblems (P_ν) are mixed symmetric/antisymmetric problems (e.g. symmetric with respect to the plane $x_1 = 0$ and antisymmetric with respect to the plane $x_2 = 0$ for a domain admitting plane symmetries).

Moreover, it may occur that the boundary data $(\boldsymbol{u}^{\nu(d)}, \boldsymbol{t}^{\nu(d)})$ appearing in (11.13) vanish for certain values of ν; this fact arises if the original boundary data $(\boldsymbol{u}^{(d)}, \boldsymbol{t}^{(d)})$ of (P) is itself invariant under some subgroup G' of G. In particular:

- $G' = \{1\}$ (data with maximum symmetry): the subproblem (P_1) only has nonzero data.
- $G' = G$ (no data symmetry): the n subproblems are to be considered.

We will denote by n_0 the number of nontrivial (i.e. with nonzero boundary data $(\boldsymbol{u}^{\nu(d)}, \boldsymbol{t}^{\nu(d)})$) subproblems associated to a given problem (P).

By construction, the solutions $\boldsymbol{u}^\nu$ to the subproblems (P_ν) verify, for any chosen element $\boldsymbol{s}_0$ of G:

$$\boldsymbol{u}^\nu(\boldsymbol{s}_0\boldsymbol{y}) = \rho_\nu(\boldsymbol{s}_0^{-1})\boldsymbol{s}_0\boldsymbol{u}^\nu(\boldsymbol{y}) \tag{11.14}$$

which generalizes (11.8); this can be seen e.g. by calculating $\boldsymbol{t}^{D(\nu)}(\boldsymbol{s}_0\boldsymbol{x})$ as defined by (11.13) with the change of variable $\boldsymbol{s} = \boldsymbol{t}\boldsymbol{s}_0^{-1}$ and using the properties of the representation ρ_ν. The subproblems (11.13) thus can be reformulated as problems on the symmetry cell C:

$$\begin{cases} \operatorname{div}\boldsymbol{\sigma}(\tilde{\boldsymbol{u}}_\nu) = \boldsymbol{0} & \text{in } C \\ \boldsymbol{T}^n(\boldsymbol{u}^\nu)(\boldsymbol{x}) = \tilde{\boldsymbol{t}}^{D(\nu)}(\boldsymbol{x}) = \dfrac{1}{n}\displaystyle\sum_{\boldsymbol{s}\in G} \rho_\nu(\boldsymbol{s})\boldsymbol{s}^{-1}\tilde{\boldsymbol{t}}(\boldsymbol{s}\boldsymbol{x}) & \text{on } \Gamma_T \\ \tilde{\boldsymbol{u}}^\nu(\boldsymbol{x}) = \tilde{\boldsymbol{u}}^{D(\nu)}(\boldsymbol{x}) = \dfrac{1}{n}\displaystyle\sum_{\boldsymbol{s}\in G} \rho_\nu(\boldsymbol{s})\boldsymbol{s}^{-1}\boldsymbol{u}^D(\boldsymbol{s}\boldsymbol{x}) & \text{on } \Gamma_u \\ \gamma_\nu\tilde{\boldsymbol{u}}^\nu = \boldsymbol{0} & \text{on } \Sigma \end{cases} \tag{11.15}$$

The last equation expresses the constraint on Σ induced by the symmetry property (11.14) of $\boldsymbol{u}^\nu$. In [3] it is shown that for any point $\boldsymbol{x}$ of Σ there exists a fixed element $\boldsymbol{s}_x \in G$ such that $\boldsymbol{s}_x\boldsymbol{x} \in \Sigma$ and that $\boldsymbol{s}_x$ is unique for almost all points $\boldsymbol{x} \in \Sigma$. Then, one puts:

$$\gamma_\nu\tilde{\boldsymbol{u}}^\nu = \tilde{\boldsymbol{u}}^\nu(\boldsymbol{x}) - \rho_\nu(\boldsymbol{s}_x^{-1})\boldsymbol{s}_x\boldsymbol{u}^\nu(\boldsymbol{s}_x\boldsymbol{x}) \tag{11.16}$$

One can readily show that the conditions defined by $(11.15)_4$ and (11.16) reduce to (11.7) for the special case of plane symmetry.

3 - Reconstruction of the global solution. The subproblems (P_ν) are solved for the unknowns $\tilde{\boldsymbol{u}}^\nu$. Then the global solution of the problem of interest (P) is reconstructed using:

$$\boldsymbol{u}(\boldsymbol{s}\boldsymbol{y}) = \sum_{\nu=0}^{n-1} \rho_\nu(\boldsymbol{s}^{-1})\boldsymbol{s}\boldsymbol{u}^\nu(\boldsymbol{y}) \qquad (\forall \boldsymbol{y} \in C, \forall \boldsymbol{s} \in G) \tag{11.17}$$

which reduces to (11.10) for the special case of plane symmetry.

11.2.3 Case of complex-valued irreducible representations

An irreducible representation ρ of a finite Abelian group G is basically a complex-valued function over G (see eq. (11.12) for cyclic groups). In some particular cases, including the most commonly used one (domains with orthogonal planes of symmetry), the irreducible representations ρ_ν are real-valued (see Table 11.1).

The appearance of complex-valued ρ_ν may cause additional difficulties when one considers a real-valued boundary-value problem (as in elastostatics), because equations (11.15) imply that the subproblems (P_ν) have complex-valued boundary data. However, it is known [3] that the complex-valued ρ_ν can be gathered into *conjugate pairs* $(\rho_{\nu_1}, \rho_{\nu_2})$ such that:

$$\overline{\rho}_{\nu_1}(\boldsymbol{s}) = \rho_{\nu_2}(\boldsymbol{s}) \qquad (\forall \boldsymbol{s} \in G)$$

Then, if a real-valued problem is considered, the pairs $(\boldsymbol{u}^{\nu_1}, \boldsymbol{u}^{\nu_2})$ and $(\boldsymbol{t}^{\nu_1}, \boldsymbol{t}^{\nu_2})$ are conjugate owing to (11.15). Putting $\tilde{\boldsymbol{u}}^{\nu_1} = \boldsymbol{v} + \mathrm{i}\boldsymbol{w}$, $\tilde{\boldsymbol{u}}^{\nu_2} = \boldsymbol{v} - \mathrm{i}\boldsymbol{w}$, the subproblems (P_{ν_1}), (P_{ν_2}) involve the same condition $(11.15)_4$ on the complementary boundary Σ. This condition has the explicit form:

$$\begin{Bmatrix} \boldsymbol{v}(\boldsymbol{x}) \\ \boldsymbol{w}(\boldsymbol{x}) \end{Bmatrix} - \begin{bmatrix} \mathrm{Re}(\rho_\nu(\boldsymbol{s}_x)) & \mathrm{Im}(\rho_\nu(\boldsymbol{s}_x)) \\ -\mathrm{Im}(\rho_\nu(\boldsymbol{s}_x)) & \mathrm{Re}(\rho_\nu(\boldsymbol{s}_x)) \end{bmatrix} \begin{Bmatrix} \boldsymbol{s}_x \boldsymbol{v}(\boldsymbol{x}) \\ \boldsymbol{s}_x^{-1} \boldsymbol{w}(\boldsymbol{x}) \end{Bmatrix} = \begin{Bmatrix} 0 \\ 0 \end{Bmatrix} \tag{11.18}$$

at any point $\boldsymbol{x} \in \Sigma$ such that there exists $\boldsymbol{s}_x \in G$ verifying $\boldsymbol{s}_x \boldsymbol{x} \in \Sigma$.

Thus, complex-valued irreducible representations lead to pairs of subproblems on real-valued fields, *coupled by the boundary constraints* $(11.15)_4$ *and (11.16)*. This is to be taken into account when considering the implementation of cyclic symmetry with loadings having no, or weaker, cyclic symmetry.

11.3 BOUNDARY INTEGRAL EQUATIONS AND GEOMETRICAL SYMMETRY

The exploitation of geometrical symmetry in boundary integral equation formulations is now investigated, using as a model situation the regularized displacement elastostatic integral equation (4.32):

$$\int_{\partial\Omega} \{[u_i(\boldsymbol{y}) - u_i(\boldsymbol{x})]T_i^k(\boldsymbol{x}, \boldsymbol{y}) - t_i(\boldsymbol{y})U_i^k(\boldsymbol{x}, \boldsymbol{y})\} \, \mathrm{d}S_y = 0 \tag{11.19}$$

To make use of geometrical symmetry in a BEM context one has to define a symmetry cell C, build the subproblems (P_ν) and apply the reconstruction formula (11.17).

A possible way of doing this is simply to translate into BEM language the subproblems (P_ν) defined by (11.15), that is, to apply (11.19) to the symmetry cell C of boundary $\partial C = \Sigma \cup \Gamma$ and to the boundary conditions and constraints (11.15). However, in a BEM context, the computational efficiency of using geometrical symmetry is not related to the ratio $|C|/|\Omega|$ of the cell and domain volumes as in an FEM context but rather to the relative sizes of the *boundaries*, i.e. the ratio $|\partial C|/|\partial\Omega|$. If the additional boundary Σ of the symmetry cell is of similar size to the trace Γ of $\partial\Omega$, very few will be gained by using symmetry. For instance, if Ω is a sphere and C an octant defined by the coordinate planes, one has $|C|/|\Omega| = 1/8$ but $|\partial C|/|\partial\Omega| = 5/16$ only.

It is thus better to define subproblems that use *integral equations stated on* Γ *only*, because the ratios $|\Gamma|/|\partial\Omega|$ and $|C|/|\Omega|$ are equal. In other words, Γ must play the role of the symmetry cell with respect to the boundary $\partial\Omega$. In order to do so, let us revert to the definition (11.13) of the subproblem (P_ν) on the entire domain Ω. The boundary fields $(\boldsymbol{u}^\nu, \boldsymbol{t}^\nu)$ are then governed by the integral equation (11.19):

$$\int_{\partial\Omega} \{[u_i^\nu(\boldsymbol{y}) - u_i^\nu(\boldsymbol{x})]T_i^k(\boldsymbol{x}, \boldsymbol{y}) - t_i^\nu(\boldsymbol{y})U_i^k(\boldsymbol{x}, \boldsymbol{y})\} \, \mathrm{d}S_y = 0$$

The change of variable ($\boldsymbol{y} \rightarrow \boldsymbol{s}^{-1}\boldsymbol{y}$) then transforms the above relation into a sum of integrals over Γ:

$$\sum_{\boldsymbol{s}\in G} \int_\Gamma \{[u_i^\nu(\boldsymbol{sy}) - u_i^\nu(\boldsymbol{x})]T_i^k(\boldsymbol{x}, \boldsymbol{sy}) - t_i^\nu(\boldsymbol{sy})U_i^k(\boldsymbol{x}, \boldsymbol{sy})\} \, \mathrm{d}S_y = 0 \quad (11.20)$$

The collocation point $\boldsymbol{x}$ is fixed, so that $\boldsymbol{s} \in G$ acts only on the integration point bfy in (11.20). Moreover, since every $\boldsymbol{s} \in G$ is an isometry, the mapping $\boldsymbol{y} \rightarrow \boldsymbol{sy}$ preserves the differential element $\mathrm{d}S_y$.

Now comes the equivalent of the *equivariance property* (11.4): the Kelvin tensors satisfy the symmetry properties

$$\begin{aligned} U_i^k(\boldsymbol{x}, \boldsymbol{sy}) &= s_{ip}s_{kq}U_p^q(\boldsymbol{s}^{-1}\boldsymbol{x}, \boldsymbol{y}) \\ T_i^k(\boldsymbol{x}, \boldsymbol{sy}) &= s_{ip}s_{kq}T_p^q(\boldsymbol{s}^{-1}\boldsymbol{x}, \boldsymbol{y}) \end{aligned} \qquad (\forall \boldsymbol{s} \in G) \qquad (11.21)$$

where an element $\boldsymbol{s} \in G$ (or, more exactly, the linear mapping between vector spaces associated to the *affine* mapping $\boldsymbol{s}$) is represented by its matrix $[s_{ij}]_{1\leq i,j\leq 3}$. Indeed, for equation (11.21) as written here to be correct, the origin $\boldsymbol{O}$ of the (affine) geometrical space must be chosen such as to be *invariant under all* $\boldsymbol{s} \in G$.

The equivariance (11.21) and the symmetry property (11.14) verified by $(\boldsymbol{u}^\nu, \boldsymbol{t}^\nu)$ are then inserted into the integral equation (11.20), which becomes:

$$\sum_{\boldsymbol{s}\in G} \rho_\nu(\boldsymbol{s}^{-1})s_{ip}s_{kq} \int_\Gamma [s_{im}\tilde{u}_m^\nu(\boldsymbol{y}) - \tilde{u}_i^\nu(\boldsymbol{x})\rho_\nu(\boldsymbol{s})]T_p^q(\boldsymbol{s}^{-1}\boldsymbol{x}, \boldsymbol{y}) \, \mathrm{d}S_y$$
$$- \sum_{\boldsymbol{s}\in G} \rho_\nu(\boldsymbol{s}^{-1})s_{ip}s_{kq}s_{im} \int_\Gamma t_m^\nu(\boldsymbol{y})U_p^q(\boldsymbol{s}^{-1}\boldsymbol{x}, \boldsymbol{y}) \, \mathrm{d}S_y = 0$$

Now, keeping in mind that $\boldsymbol{s}$ is an isometry, one has $\boldsymbol{s}^T\boldsymbol{s} = \boldsymbol{ss}^T = \mathbf{1}$, that is, $s_{im}s_{ip} = \delta_{mp}$:

$$\sum_{\boldsymbol{s}\in G} \rho_\nu(\boldsymbol{s}^{-1})s_{kq} \int_\Gamma [\tilde{u}_p^\nu(\boldsymbol{y}) - s_{ip}\tilde{u}_i^\nu(\boldsymbol{x})\rho_\nu(\boldsymbol{s})]T_p^q(\boldsymbol{s}^{-1}\boldsymbol{x}, \boldsymbol{y}) \, \mathrm{d}S_y$$
$$- \sum_{\boldsymbol{s}\in G} \rho_\nu(\boldsymbol{s}^{-1})s_{kq} \int_\Gamma t_p^\nu(\boldsymbol{y})U_p^q(\boldsymbol{s}^{-1}\boldsymbol{x}, \boldsymbol{y}) \, \mathrm{d}S_y = 0 \quad (11.22)$$

Finally, one notes that eq. (11.14) with $\boldsymbol{s}$ replaced by $\boldsymbol{s}^{-1} = \boldsymbol{s}^T$ gives:

$$\rho_\nu(\boldsymbol{s})\boldsymbol{s}^{-1}\boldsymbol{u}^\nu(\boldsymbol{x}) = \boldsymbol{u}^\nu(\boldsymbol{s}^{-1}\boldsymbol{x}) \qquad (\forall \boldsymbol{x} \in C, \forall \boldsymbol{s} \in G)$$

or, in component notation:

$$\rho_\nu(\boldsymbol{s})s_{ip}\tilde{u}_i^\nu(\boldsymbol{x}) = \tilde{u}_p^\nu(\boldsymbol{sx}) \qquad (\forall \boldsymbol{x} \in C, \forall \boldsymbol{s} \in G)$$

which implies:

$$\tilde{u}_p^\nu(\boldsymbol{y}) - \tilde{u}_i^\nu(\boldsymbol{x})\rho_\nu(\boldsymbol{s})s_{ip} = \tilde{u}_p^\nu(\boldsymbol{y}) - \tilde{u}_p^\nu(\boldsymbol{s}^{-1}\boldsymbol{x})$$

This relation is substituted into eq. (11.22). As a result, the governing integral equation for the subproblem (P_ν) of unknowns ($\tilde{\boldsymbol{u}}^\nu, \tilde{\boldsymbol{t}}^\nu$) is:

$$\sum_{s \in G} \rho_\nu(\boldsymbol{s}^{-1}) s_{kq} \int_\Gamma [\tilde{u}_p^\nu(\boldsymbol{y}) - \tilde{u}_p^\nu(\boldsymbol{s}^{-1}\boldsymbol{x})] T_p^q(\boldsymbol{s}^{-1}\boldsymbol{x}, \boldsymbol{y}) \, \mathrm{d}S_y$$
$$- \sum_{s \in G} \rho_\nu(\boldsymbol{s}^{-1}) s_{kq} \int_\Gamma t_p^\nu(\boldsymbol{y}) U_p^q(\boldsymbol{s}^{-1}\boldsymbol{x}, \boldsymbol{y}) \, \mathrm{d}S_y = 0 \quad (11.23)$$

Its geometrical support is Γ, intersection of C with $\partial\Omega$; Γ is hence in that sense a symmetry cell for $\partial\Omega$. Equation (11.23) is applicable in connection with plane or cyclic symmetries, and more generally with any Abelian finite symmetry group G. Once all subproblems are solved on the boundary, the solution $(\boldsymbol{u}, \boldsymbol{t}) \,|_{\partial\Omega}$ to the initial integral equation (11.19) is reconstructed using eq. (11.17).

11.4 DISCRETIZATION OF THE INTEGRAL EQUATION

Here the surface Γ (*not* the entire boundary $\partial\Omega$) is divided into N_E boundary elements totalting N_N nodes. Moreover, let $\Gamma(\boldsymbol{x})$ denote the union of those boundary elements which contain the collocation point $\boldsymbol{x}$. The discretized form of equation (11.23) is, using notations similar to those of Chapter 3:

$$0 = \sum_{s \in G} \rho_\nu(\boldsymbol{s}^{-1}) s_{kq} \Big\{ \sum_{e \in I(s^{-1}x)} \sum_{q=1}^{N_e} AS_{iq}(\boldsymbol{s}^{-1}\boldsymbol{x}, e, q) \tilde{u}_i^{m(e,q)\nu}$$
$$+ \sum_{e \notin I(s^{-1}x)} \sum_{q=1}^{N_e} AR_{iq}(\boldsymbol{s}^{-1}\boldsymbol{x}, e, q) \tilde{u}_i^{m(e,q)\nu}$$
$$- \sum_{e=1}^{N_E} \sum_{q=1}^{N_e} B_{iq}(\boldsymbol{s}^{-1}\boldsymbol{x}, e, q) \tilde{t}_i^{m(e,q)\nu}$$
$$- \Big[\sum_{e \notin I(s^{-1}x)} \widehat{AR}_{iq}(\boldsymbol{s}^{-1}\boldsymbol{x}, e) \Big] \tilde{u}_i^\nu(\boldsymbol{x}) \Big\} \quad (11.24)$$

where the element matrices, given by:

$$AS_{ir}(\boldsymbol{z}, e, q) = \int_{E_e} \hat{N}_q(\rho, \alpha; \boldsymbol{z}) T_i^r(\rho, \alpha) J(\boldsymbol{\xi}) \, \mathrm{d}\rho \mathrm{d}\alpha$$
$$AR_{ir}(\boldsymbol{z}, e, q) = \int_{E_e} N_q(\boldsymbol{\xi}) T_i^r(\boldsymbol{z}, \boldsymbol{y}) J(\boldsymbol{\xi}) \, \mathrm{d}\boldsymbol{\xi}$$
$$\widehat{AR}_{ir}(\boldsymbol{z}, e, q) = \int_{E_e} T_i^r(\boldsymbol{z}, \boldsymbol{y}) J(\boldsymbol{\xi}) \, \mathrm{d}\boldsymbol{\xi}$$
$$B_{ir}(\boldsymbol{z}, e, q) = \int_{E_e} N_q(\boldsymbol{\xi}) U_i^r(\boldsymbol{z}, \boldsymbol{y}) J(\boldsymbol{\xi}) \, \mathrm{d}\boldsymbol{\xi}$$

do not depend on the subproblem index ν. Each element matrix $AS(\boldsymbol{s}^{-1}\boldsymbol{x}, e, q)$, $AR(\boldsymbol{s}^{-1}\boldsymbol{x}, e, q)$, $B(\boldsymbol{s}^{-1}\boldsymbol{x}, e, q)$, once computed, contributes to *all* subproblems. Only the weights of the element matrices depend on the subproblem index, through the coefficients $\rho_\nu(\boldsymbol{s}^{-1})$. This essential remark shows in which way the exploitation of geometrical symmetry provides computational savings.

Indeed, the dependence on ν of the subproblem (P_ν) appears only through the boundary conditions and constraints, the governing operator (either a local differential operator, a bilinear form or an integral operator depending on the viewpoint considered) being the same for all subproblems. Hence, in a discrete BEM formulation, the subproblem (P_ν) depends on ν only through the nodal values. Hence no dependence on ν must appear in the element integrals, and the previous result was to be expected.

Reduction of computational costs

A closer examination of (11.24) allows an approximate operation count. The boundary element modelling is restricted to the symmetry cell G, which is n times smaller than the original boundary $\partial\Omega$. The element integrals $AS(\boldsymbol{s}^{-1}\boldsymbol{x}, e, q)$, $AR(\boldsymbol{s}^{-1}\boldsymbol{x}, e, q)$ and $J(\boldsymbol{s}^{-1}\boldsymbol{x}, e, q)$ have to be evaluated for all boundary elements on Γ and all images $\boldsymbol{s}^{-1}\boldsymbol{x}$ of the collocation points $\boldsymbol{x}$ on Γ. In other words, the building loop scans the number of collocation points which would have been introduced for the entire boundary $\partial\Omega$ but for a number of boundary elements n times smaller. As a result, the overall number of element integrals to be computed using symmetry is expected to be n times smaller than without using symmetry. However, one should bear in mind that the setting up of the systems of equations (11.24) involves sums of element integrals weighted by the coefficients $\rho_\nu(\boldsymbol{s})$. An additional amount of arithmetic operations, not required in the absence of symmetry exploitation, is thus entailed and the ideal computational gain of $1/n$ is degraded.

Similarly, the solution of a linear system of equations of size $nN \times nN$ is replaced by that of n linear systems of size $N \times N$ (actually, due to the presence of constraints on Σ, the ratio n of the matrix sizes is approximate and becomes exact as the mesh is refined). Since the matrices are full and non-symmetric, the solution of an $(N \times N)$ system requires $O(N^3)$ operations. As a consequence, the expected ratio of the respective operation counts is:

$$\frac{\text{operation count without symmetry}}{\text{operation count with symmetry}} = \frac{(nN)^3}{n(N)^3} = n^2$$

11.5 INTEGRAL REPRESENTATIONS AT INTERIOR POINTS

The boundary integral equation (11.23) allows us to solve (P) for the surface fields, which is the usual first step of all boundary element methods. Next comes the post-processing step, in which one uses the well-known integral representation formulas of $\boldsymbol{u}(\boldsymbol{x})$ and $\boldsymbol{\sigma}(\boldsymbol{x})$ at user-specified interior points $\boldsymbol{x}$. Obviously much is to be gained by also using geometrical symmetry in this step.

Integral representation of the displacement

Let us consider first the displacement integral representation formula (4.21) of $\boldsymbol{u}(\boldsymbol{x})$:

$$u_k(\boldsymbol{x}) = \int_{\partial\Omega} \{t_i(\boldsymbol{y})U_i^k(\boldsymbol{x},\boldsymbol{y}) - u_i(\boldsymbol{y})T_i^k(\boldsymbol{x},\boldsymbol{y})\} \, \mathrm{d}S_y \tag{11.25}$$

which is next reduced to a sum of integrals over the symmetry cell Γ:

$$u_k(\boldsymbol{x}) = \sum_{s\in G} \int_{\Gamma} \{t_i(\boldsymbol{s}\boldsymbol{y})U_i^k(\boldsymbol{x},\boldsymbol{s}\boldsymbol{y}) - u_i(\boldsymbol{s}\boldsymbol{y})T_i^k(\boldsymbol{x},\boldsymbol{s}\boldsymbol{y})\} \, \mathrm{d}S_y \tag{11.26}$$

Then the equivariance property (11.21) and the identity (11.17) are inserted in eq. (11.26). This gives:

$$\begin{aligned} u_k(\boldsymbol{x}) = &\sum_{s\in G}\sum_{\nu=0}^{n-1} \rho_\nu(\boldsymbol{s}^{-1}) s_{ip} s_{kq} s_{im} \int_\Gamma t_m^\nu(\boldsymbol{y}) U_p^q(\boldsymbol{s}^{-1}\boldsymbol{x},\boldsymbol{y}) \, \mathrm{d}S_y \\ &- \sum_{s\in G}\sum_{\nu=0}^{n-1} \rho_\nu(\boldsymbol{s}^{-1}) s_{ip} s_{kq} s_{im} \int_\Gamma \tilde{u}_m^\nu(\boldsymbol{y}) T_p^q(\boldsymbol{s}^{-1}\boldsymbol{x},\boldsymbol{y}) \, \mathrm{d}S_y \end{aligned}$$

or, since $s_{im}s_{ip} = \delta_{mp}$:

$$u_k(\boldsymbol{x}) = \sum_{s\in G}\sum_{\nu=0}^{n-1} \rho_\nu(\boldsymbol{s}^{-1}) s_{kq} \int_\Gamma \{t_i^\nu(\boldsymbol{y}) U_i^q(\boldsymbol{s}^{-1}\boldsymbol{x},\boldsymbol{y}) - \tilde{u}_i^\nu(\boldsymbol{y}) T_i^q(\boldsymbol{s}^{-1}\boldsymbol{x},\boldsymbol{y})\} \, \mathrm{d}S_y \tag{11.27}$$

The discretized version of the representation formula (11.27) thus reads:

$$\begin{aligned} u_k(\boldsymbol{x}) = &\sum_{s\in G}\sum_{\nu=0}^{n-1} \rho_\nu(\boldsymbol{s}^{-1}) s_{rk} \\ &\times \sum_{q=1}^{N_e} \sum_{E_e\subset\Gamma} \left\{ \tilde{t}_i^{(q)\nu} B_{ir}(\boldsymbol{s}^{-1}\boldsymbol{x},e,q) - \tilde{u}_i^{(q)\nu} A_{ir}(\boldsymbol{s}^{-1}\boldsymbol{x},e,q) \right\} \end{aligned} \tag{11.28}$$

where the element matrices, which do not depend on the subproblem index, are given by:

$$\begin{aligned} A(\boldsymbol{z},e,q)_{ir} &= \int_E N_q(\boldsymbol{\xi}) T_i^r(\boldsymbol{z},\boldsymbol{y}) J(\boldsymbol{\xi}) \, \mathrm{d}\boldsymbol{\xi} \\ B(\boldsymbol{z},e,q)_{ir} &= \int_E N_q(\boldsymbol{\xi}) U_i^r(\boldsymbol{z},\boldsymbol{y}) J(\boldsymbol{\xi}) \, \mathrm{d}\boldsymbol{\xi} \end{aligned}$$

Reduction of computational costs

An examination of (11.28) reveals that the computation of $\boldsymbol{u}(\boldsymbol{x})$ requires the evaluation of $n \times N_E$ element integrals for each kernel $\boldsymbol{U}, \boldsymbol{\Sigma}$. This is exactly the same amount as if the representation (4.21) were used directly without any attempt at incorporating symmetry. Thus the representation formula (11.27) seems to offer no computational advantage over the conventional formula (4.21). However, let us apply the discretized formula (11.28) to the interior point $\boldsymbol{s}_0\boldsymbol{x}$, $\boldsymbol{s}_0\boldsymbol{x}$ being a fixed element of G:

$$u_k(\boldsymbol{s}_0\boldsymbol{x}) = \sum_{\boldsymbol{s}\in G}\sum_{\nu=0}^{n-1} \rho_\nu(\boldsymbol{s}^{-1}) s_{rk} \times \sum_{q=1}^{N_e}\sum_{E_e\subset\Gamma} \left\{ \tilde{t}_i^{(q)\nu} B_{ir}(\boldsymbol{s}^{-1}\boldsymbol{s}_0\boldsymbol{x}, e, q) - \tilde{u}_i^{(q)\nu} A_{ir}(\boldsymbol{s}^{-1}\boldsymbol{s}_0\boldsymbol{x}, e, q) \right\} \tag{11.29}$$

A classical result of group theory states that the mapping $(\boldsymbol{t} \in G) \rightarrow (\boldsymbol{s} = \boldsymbol{s}_0\boldsymbol{t} \in G)$ is one-to-one. Upon introducing the substitution $\boldsymbol{s} = \boldsymbol{s}_0\boldsymbol{t}$, eq. (11.29) thus takes the equivalent form:

$$u_k(\boldsymbol{s}_0\boldsymbol{x}) = \sum_{\boldsymbol{t}\in G}\sum_{\nu=0}^{n-1} \rho_\nu(\boldsymbol{t}^{-1})\rho_\nu(\boldsymbol{s}_0)(\boldsymbol{s}_0\boldsymbol{t})_{rk} \sum_{q=1}^{N_e}\sum_{E_e\subset\Gamma} \left\{ \tilde{u}_i^{(q)\nu} A_{ir}(\boldsymbol{t}^{-1}\boldsymbol{x}, e, q) - \tilde{t}_i^{(q)\nu} B_{ir}(\boldsymbol{t}^{-1}\boldsymbol{x}, e, q) \right\} \tag{11.30}$$

having used the fact that, from the definition of ρ_ν:

$$\rho_\nu(\boldsymbol{s}^{-1}) = \rho_\nu(\boldsymbol{s}^{-1}\boldsymbol{s}_0^{-1}\boldsymbol{s}_0) = \rho_\nu(\boldsymbol{s}^{-1}\boldsymbol{s}_0^{-1})\rho_\nu(\boldsymbol{s}_0)$$

The consequence of this analysis is as follows: once the element integrals $A(\boldsymbol{t}^{-1}\boldsymbol{x}, e, q)$, $B(\boldsymbol{t}^{-1}\boldsymbol{x}, e, q)$ $(\boldsymbol{t} \in G, 1 \leq e \leq N_E, 1 \leq q \leq N_e)$ are evaluated, they allow the computation of $\boldsymbol{u}(\boldsymbol{s}_0\boldsymbol{x})$ for *all* the n images of $\boldsymbol{x}$ under the action of G by means of eq. (11.30), rather than just one single value $\boldsymbol{u}(\boldsymbol{x})$. Indeed (11.30) shows that the computation of all the $\boldsymbol{u}(\boldsymbol{s}_0\boldsymbol{x})$, $\boldsymbol{s}_0 \in G$ uses the same element integrals, weighted by coefficients which depend on $\boldsymbol{s}_0$.

Integral representation of the stress tensor

The same considerations apply for the stress tensor. The integral representation formula is obtained by applying Hooke's law to eq. (11.29).

11.6 EXPLOITATION OF SYMMETRY: RECAPITULATION

As a prerequisite, for any symmetry group which is to be made available to the user, the BEM code must contain several pieces of information:

- The *multiplication table* of G, i.e. an $n \times n$ integer array TABLE such that $k = \text{TABLE}(i, j)$ is the index of the composition $\boldsymbol{s}_i \boldsymbol{s}_j$ of $\boldsymbol{s}_i \in G$ and $\boldsymbol{s}_j \in G$, i.e. such that $\boldsymbol{s}_i \boldsymbol{s}_j = \boldsymbol{s}_k \in G$.

- The table of *irreducible representations* of G, i.e. an $n \times n$ table containing the weighting coefficients $\rho_\nu(\boldsymbol{s}_i)$ $(0 \leq \nu \leq n-1,\ 1 \leq i \leq n)$.

- The definition of the action of a symmetry element $\boldsymbol{s} \in G$ on a point (or vector) $\boldsymbol{x}$, i.e. the n matrices $[s_{ij}]_{1 \leq i,j \leq 3})$ associated to the elements of G.

Then the solution procedure is as follows:

1. **Initialization**. This preliminary step consists in:

   - Select G from the possible choices implemented in the code.
   - Input mesh data, i.e. nodes, connectivity table, ... This mesh is assumed to be the discrete *symmetry cell* (in a boundary element sense) Γ.
   - Input boundary conditions data. Each point $\boldsymbol{y} \in \Gamma$ must be associated to n boundary data (n being the order of G): if (say) $\boldsymbol{y}$ supports a given traction $\boldsymbol{t}^{(d)}(\boldsymbol{y})$, then the user must provide the value of $\boldsymbol{t}^{(d)}(\boldsymbol{s}_1\boldsymbol{y}), \ldots, \boldsymbol{t}^{(d)}(\boldsymbol{s}_n\boldsymbol{y})$, $(G = \{\boldsymbol{s}_1, \ldots \boldsymbol{s}_n\}$.
   - List the subproblems (P_ν) needed for the case under consideration and, from the previous step, compute the boundary data $(\boldsymbol{u}^{\nu(d)}, \boldsymbol{t}^{\nu(d)})$ (on Γ) for each (P_ν) according to (11.13). Among these boundary data, isolate those which are nonzero and list the corresponding *nontrivial subproblems*.
   - For each $\boldsymbol{s} \in G$ in turn, identify the nodes $\boldsymbol{x} \in \Gamma$ such that $\boldsymbol{s}\boldsymbol{x} \in \Gamma$ and work out the additional boundary constraints (11.15). Use of these constraints ensures the adequate behaviour of the subproblems and allows a further reduction of the number of unknowns of each subproblem.

2. **Build and solve the subproblems (P_ν)**:

   - Build the discretized boundary integral equation (11.23) *for each nontrivial subproblem (P_ν)*.
   - Insert the boundary data $(\boldsymbol{u}^{\nu(d)}, \boldsymbol{t}^{\nu(d)})$.
   - Take into account the boundary constraints (11.15) (build during the initialization phase), either by eliminating unknowns and equations (no coupling between subproblems) or adding new equations (in case of coupled subproblems, e.g. when using cyclic symmetry).
   - Solve these subproblems for the unknown parts of $(\boldsymbol{u}^\nu, \boldsymbol{t}^\nu)\,|_\Gamma$.

3. Reconstruct the solution of (P) on $\partial\Omega$ by using the reconstruction identity (11.17).

4. Compute $\boldsymbol{u}$ and $\boldsymbol{\sigma}$ at interior points if requested. For every user-specified interior point, $\boldsymbol{u}(\boldsymbol{s}\boldsymbol{x})$, $\boldsymbol{\sigma}(\boldsymbol{s}\boldsymbol{x})$ are computed for all $\boldsymbol{s} \in G$ using the discretized representation formula (11.30).

11.7 NUMERICAL EXAMPLE IN ELASTODYNAMICS

The example is taken from [2]. A spherical cavity of unit radius is embedded in an infinite elastic medium. The cavity surface is loaded by tractions $\boldsymbol{t}^D = (A + B\sin\theta - C\sin\theta e^{i\phi})e^{-i\omega t}$ (A, B, C: complex constants, θ, ϕ: spherical angles). The elastic constants, the angular frequency ω and the mass density are such that $k_L = 0.5345, k_T = 1$. The loading, which does not possess plane symmetries, is a linear combination of spherical modes (0,0), (1,0), (1,1), which are known analytically (Eringen and Suhubi [4]).

The symmetry cell Γ is one octant of the sphere. Two boundary element meshes M_1 (3 elements, 16 nodes) and M_2 (12 elements, 49 nodes), using eight-noded quadrilateral elements, are considered. The eight subproblems associated with the group of plane symmetries (of order $n = 8$) have, after application of the constraints defined by eqs $(11.15)_4$ and (11.16), 33, 30, 30, 30, 26, 26, 26 and 21 scalar unknowns (mesh M_1) and 120, 113, 113, 113, 105, 105, 105 and 96 scalar unknowns (mesh M_2).

To solve the same problem without exploiting geometrical symmetries, two meshes M_3 (24 elements, 74 nodes, 222 scalar unknowns) and M_4 (96 elements, 290 nodes, 870 scalar unknowns) of the entire sphere have been used.

The main purpose of this example is of course the study of the relative computational gains brought by taking advantage of symmetries. Most of the

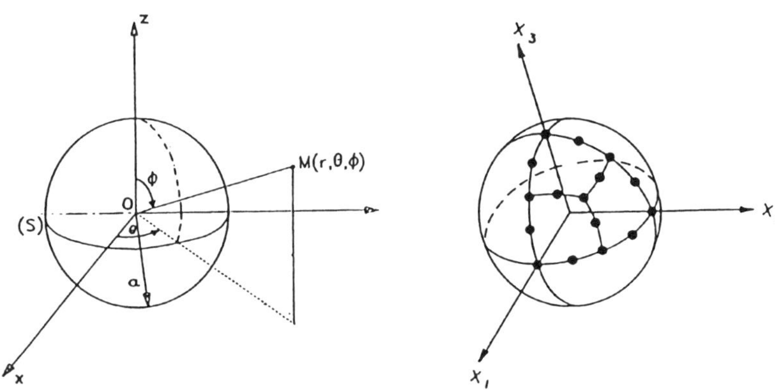

Figure 11.3 Spherical cavity and spherical coordinates (left); mesh M_1 (right).

Table 11.2 Ratios of computing times without and with symmetry, and relative errors in L^2-norm for the computed surface displacement.

	Assembly	Solution	Postprocessing	L^2 relative error
M_3 to M_1 ratio	2.97	35.1	0.58	6.5×10^{-3}
M_4 to M_2 ratio	3.66	51.3	0.58	5.2×10^{-4}

computer time is spent in the assembly and solution stages. Table 11.2 displays the computer time ratios spent in the assembly, solution and evaluation of integral representations at a single observation point ('postprocessing'). The relative errors in L^2-norm obtained for the computed displacement on the cavity surface are also displayed.

The actual ratio of assembly computer times with and without symmetry is roughly 1/3 to 1/4 instead of the ideal value 1/8. This is due to two factors: the presence of additional arithmetic operations (weighted sums of element matrices, the weights being different for each subproblem) and the increased complexity of the assembly loop.

The actual ratio of solution computer times with and without symmetry (roughly 1/35 for meshes M_1, M_3 and 1/51 for meshes (M_2, M_4) gets closer to the ideal value 1/64 as the mesh is refined.

The post-processing computer times is greater for M_1 than for M_3 (and similarly with M_2 and M_4). When symmetries are used (meshes M_1, M_2), $(\boldsymbol{u}(\boldsymbol{x}), \boldsymbol{\sigma}(\boldsymbol{x}))$ are in fact computed simultaneously at the 8 images of $\boldsymbol{x}$ under the symmetry group, while they are computed at $\boldsymbol{x}$ only without using symmetry (M_3, M_4).

Geometrical symmetries are also exploited in the numerical example shown at the end of Chapter 8 (scattering of an elastic wave by a spherical cavity).

REFERENCES

1. Allgower, E. L., Böhmer, K., Georg, K., Miranda, R. Exploiting symmetry in boundary element methods. *SIAM J. Numer. Anal.*, **29**, 534–552 (1992).
2. Bonnet, M. On the use of geometrical symmetry in the boundary element methods for 3D elasticity. In C.A. Brebbia (ed.), *Boundary element technology VI*, pp. 185–201. Computational Mechanics Publications and Elsevier (1991).
3. Bossavit, A. Symmetry, groups and boundary-value problems : a progressive introduction to noncommutative harmonic analysis of partial differential equations in domains with geometrical symmetry. *Comp. Meth. in Appl. Mech. Engng.*, **56**, 167–215 (1986).
4. Eringen, A. C., Suhubi, E. S. *Elastodynamics (vol. II – linear theory)*. Academic Press (1975).

5. Guy, J., Mangeot, B. Use of group theory in various integral equations. *SIAM J. Appl. Math.*, **40**, 390–399 (1981).

6. Lobry, J., Broche, Ch. Geometrical symmetry in the boundary element method. *Engng. Anal. with Bound. Elem.*, **9**, 233–245 (1994).

12

Domain differentiation of boundary integral equations

12.1 MOTIVATION

The consideration of first and second variations of integral functionals with respect to a geometrical domain arises in the study of many situations where a geometrical domain plays a major role. These include shape optimization processes with respect to mechanical constraints, usually expressed in terms of integrals involving the mechanical quantities (displacement, stress ...) which themselves are shape-dependent. In the related area of geometrical inverse problems, a part of the domain boundary is unknown (e.g. in crack or defect identification), its determination being attempted by minimizing a distance between computed (for a given domain configuration) and known (for the actual, unknown, domain configuration) values of some mechanical quantity. In addition, free-boundary problems can be dealt with using similar considerations. Finally, evolution problems in fracture or damage mechanics often involve geometrical fronts which propagate according to a Griffith-like criterion, where the second derivative of the potential energy at equilibrium with respect to front perturbations plays an essential role. All these situations involve integral functionals having both explicit (through the geometrical support of the integral) and implicit (through the solution to a boundary-value problem) dependence on the geometrical domain under consideration.

For several reasons, including computational efficiency, it is essential to be able to evaluate first and second variations of such functionals. Most usual optimization algorithms use first-order derivatives, while Newton-type algorithms also use second-order derivatives. Moreover, accurate evaluation of second-order derivatives allows a practical checking of optimality conditions.

Finally, as mentioned before, second-order derivatives of potential energy govern evolution problems involving energy release rates. Also, it is a known fact that finite-difference estimation of gradients (which would require here finite perturbations of the geometrical domain) is both computationally expensive and prone to inaccuracies due to its mathematical ill-posedness. Hence it is natural to consider analytical differentiation with respect to a variable domain. This concept has been studied and used by many authors (see e.g. Haug et al. [7], Petryk and Mroz [13]), up to now mostly in FEM-oriented contexts. However, since in such problems the domain (and hence its boundary) is a primary unknown, it is natural to consider boundary integral formulations, because they offer in this context the 'minimal' modelling.

The formulation of shape sensitivities, in a BIE context or otherwise, may result from the direct differentiation approach (DDA) or the adjoint state method (ASM), applied either before or after discretization. This chapter presents both approaches for continuous formulations (i.e. before discretization). Domain perturbations are mathematically formulated using diffeomorphisms, in the fashion of the Lagrangian description of continuum kinematics. These diffeomorphisms depend on a finite number of parameters, so that all domain evolutions ultimately reduce to variations of those parameters. These parameters may either be actual design variables (for shape optimization) or result from the discretization of infinite-dimensional domain perturbations (e.g. crack front propagation in 3D fracture mechanics). This approach also allows us to define and compute directional derivatives associated to any (sufficiently regular) boundary perturbation. As a result, first- and second-order shape sensitivities are defined in terms of transformation 'velocities' and 'accelerations', which are vector fields on the boundary (Section 12.2).

The direct differentiation approach consists in applying to the relevant governing boundary integral equation the material differentiation concept of continuum kinematics, the (possibly non-material) domain changes being described in terms of a finite number of parameters (e.g. design parameters in shape optimization) instead of the physical time variable. This leads to governing boundary integral equations for the shape sensitivities of elastic boundary variables. The latter may be considered for their own interest or as auxiliary variables which ultimately allow the computation of the shape sensitivities of domain- and state-dependent integral functionals. References about DDA include e.g. Nishimura and Kobayashi [12] for crack identification, Mukherjee and Chandra [11] for design sensitivities in nonlinear solid mechanics, Bonnet [3] for crack front stability, among others. A Taylor series approach having some similarity with the DDA is given by Zeng and Saigal [18] for crack identification problems in potential theory. Also, the direct differentiation of discretized BEM formulations is treated in e.g. Kane et al. [8]. A difficulty with the DDA lies in the singular character of the boundary integral equation. Barone and Yang [1] apply the material derivative concept to the Somigliana identity (4.33), involving integrals defined in the Cauchy principal value (CPV) sense. However, since the exclusion neighbourhood used in the limiting process which defines a CPV integral is distorted by a domain perturbation, this procedure raises the difficulty of whether the material derivative of a CPV integral equals the CPV of the material derivative of the integrand. The answer to this question is assumed affirmative in [1], although in other instances erroneous results induced by misapplication of general techniques (change of variables, integration by parts ...) to CPV integrals are well-known. Zhang and Mukherjee [19] circumvent this potential objection by using a 2D elastic BIE formulated in terms of tangential gradient of displacements ('derivative boundary integral equation') and thus of a weakly singular nature; they obtain weakly singular integral equations for first- and second-order elastic shape sensitivities. However, the generalization of the 'derivative boundary integral equation' to 3D situations is not easy. In the present chapter (Sections 12.4 to 12.8), boundary integral equation formulations are established for first- and second-order elastic shape sensitivities in 3D situations, starting from regularized versions of the conventional boundary integral equations where only weakly singular integrals occur (see Chapter 2 for scalar problems, Chapter 4 for elastostatics, Chapters 7, 9 for scalar and elastic wave problems). Applying the material derivatives to such formulations is mathematically sound, and results in weakly singular integral equation formulations for first- and second-order shape sensitivities.

The adjoint state method (Section 12.9) consists in introducing the solution to another boundary-value problem, termed the *adjoint problem*, defined so as to cancel all terms containing the sensitivities of the primal variables (and thus eliminate the need to compute them). Continuous formulations using the adjoint state method to shape sensitivity in a BIE context have been considered e.g. by Choi and Kwak [6] for 2D elasticity, Bonnet [2] for 3D geometrical inverse problems in acoustics, Burczyński and Fedelinski for transient elastodynamics [5], Meric [9] for thermal problems. For both the DDA and the ASM, the presentation in this chapter is restricted for ease of exposition to 3D scalar equilibrium problems, but can be extended in a rather straightforward way to other situations: elastostatics, acoustics, elastodynamics ...

12.2 DIFFERENTIATION IN A DOMAIN PERTURBATION

A governing formulation for the perturbation of a physical state induced by a perturbation of its geometrical support Ω is sought. The adequate, and natural, mathematical setting allowing us to describe the domain perturbations is an adaptation of the classical concepts of continuum kinematics (Salençon [14]), whereby domain changes are mathematically represented by means of *transformations.* A domain transformation is a family of one-to-one mappings $\boldsymbol{\Phi}$ depending on a *kinematical parameter* (fictitious time) p (Lagrangian-type description):

$$\boldsymbol{X} \in \Omega \rightarrow \boldsymbol{x} = \boldsymbol{\Phi}(\boldsymbol{X}, p) \in \Omega(p) \tag{12.1}$$

such that, in addition, $\boldsymbol{\Phi}(\boldsymbol{X},0) = \boldsymbol{X}$. The *transformation velocity* $\boldsymbol{\theta}(\boldsymbol{x},p)$ associated to the transformation $\boldsymbol{\Phi}$ is defined as:

$$\boldsymbol{\theta}(\boldsymbol{x},p) = \boldsymbol{\Phi}_{,p}(\boldsymbol{X},p) \qquad \boldsymbol{x} = \boldsymbol{\Phi}(\boldsymbol{X},p) \tag{12.2}$$

In other words, $\boldsymbol{\theta}(\boldsymbol{x},p)$ is the 'velocity' (p being the 'time' variable) at point $\boldsymbol{x}$, in Eulerian representation. Note that the adjectives 'Lagrangian' and 'Eulerian' refer throughout this chapter to the *fictitious* time p.

Differentiation of field variables in the domain transformation

Let $f(\boldsymbol{x},p)$ denote a scalar or tensor field, in Eulerian representation. The Lagrangian, or material, derivative $\overset{\star}{f}$ of f, which describes the variations of f at a point moving with the transformation (12.1), is defined by:

$$\overset{\star}{f}(\boldsymbol{x},p) = \lim_{h\to 0} \frac{1}{h}\left[f(\boldsymbol{\Phi}(\boldsymbol{X},p+h),p+h) - f(\boldsymbol{\Phi}(\boldsymbol{X},p),p)\right] \tag{12.3}$$

and is connected to the partial 'time' derivative $f_{,p}$ by:

$$\overset{\star}{f}(\boldsymbol{x},p) = f_{,p}(\boldsymbol{x},p) + \boldsymbol{\nabla} f(\boldsymbol{x},p).\boldsymbol{\theta}(\boldsymbol{x},p) \tag{12.4}$$

The relations (12.3)–(12.4) are standard concepts of continuum kinematics, applied to material transformations of deformable bodies (in eq. (12.3), $\boldsymbol{\nabla} f$ denotes the Eulerian gradient of f with respect to $\boldsymbol{x}$). Moreover, the Lagrangian derivative of the gradient $\boldsymbol{\nabla} f$ of a scalar or tensor field f is given by:

$$(\boldsymbol{\nabla} f)^{\star} = \boldsymbol{\nabla} \overset{\star}{f} - \boldsymbol{\nabla} f.\boldsymbol{\nabla}\theta \tag{12.5}$$

while the Lagrangian derivative of a *material* vector $\boldsymbol{a}$ attached to the moving point $\boldsymbol{x} = \boldsymbol{\Phi}(\boldsymbol{X};p)$ is given by:

$$\overset{\star}{\boldsymbol{a}} = \boldsymbol{\nabla}\boldsymbol{\theta}.\boldsymbol{a} \tag{12.6}$$

Derivative of volume, area and length differential elements

The Lagrangian derivative of differential elements $\mathrm{d}V$, $\mathrm{d}S$, $\mathrm{d}s$ associated with volumes $\Omega(p)$, surfaces $S(p)$ and curves $\Gamma(p)$ undergoing the transformation $\boldsymbol{\Phi}$, at 'time' p, are given by:

$$\overset{\star}{\mathrm{d}V} = \mathrm{div}\boldsymbol{\theta}\,\mathrm{d}V \qquad \overset{\star}{\mathrm{d}S} = \mathrm{div}_S\boldsymbol{\theta}\,\mathrm{d}S \qquad \overset{\star}{\mathrm{d}s} = \boldsymbol{\tau}.\boldsymbol{\nabla}\boldsymbol{\theta}.\boldsymbol{\tau}\,\mathrm{d}s \tag{12.7}$$

where div_S denotes the surface divergence (Appendix A) and $\boldsymbol{\tau}$ is the unit tangent to the material curve $\Gamma(p)$. In similar fashion, one has:

$$\overset{\star}{\boldsymbol{n}} = -\boldsymbol{n}.\boldsymbol{\nabla}_S\boldsymbol{\theta} \qquad \overset{\star}{\boldsymbol{\tau}} = \boldsymbol{\nabla}\boldsymbol{\theta}.\boldsymbol{\tau} - (\boldsymbol{\tau}.\boldsymbol{\nabla}\boldsymbol{\theta}.\boldsymbol{\tau})\boldsymbol{\tau} \tag{12.8}$$

where $\boldsymbol{n}$ is the unit normal to the material surface $S(p)$ and $\boldsymbol{\nabla}_S$ denotes the surface gradient (Appendix A). Some of the previous identities are established in Appendix 12.A. Note that the vectors $\boldsymbol{n}, \boldsymbol{\tau}$ are, in general, not material.

Derivatives of integrals

Consider integrals over volumes $\Omega(p)$, surfaces $S(p)$ or lines $\Gamma(p)$ 'moving' according to the transformation $\boldsymbol{\Phi}$:

$$I(p) = \int_{\Omega(p)} f(\boldsymbol{y}, p)\,\mathrm{d}V$$

$$J(p) = \int_{S(p)} g(\boldsymbol{y}, p)\,\mathrm{d}S$$

$$H(p) = \int_{\Gamma(p)} h(\boldsymbol{y}, p)\,\mathrm{d}s$$

Their derivatives with respect to the variable p are expressible by means of classical identities of continuum kinematics. First, $\overset{\star}{I}(p)$ can be expressed in terms of either the Lagrangian or the Eulerian derivative of f:

$$\overset{\star}{I}(p) = \int_{\Omega(p)} [\overset{\star}{f} + f\mathrm{div}\boldsymbol{\theta}]\,\mathrm{d}V \tag{12.9}$$

$$= \int_{\Omega(p)} f_{,p}\,\mathrm{d}V + \int_{\partial\Omega(p)} f[\boldsymbol{\theta}.\boldsymbol{n}]\,\mathrm{d}S \tag{12.10}$$

where $\boldsymbol{n}$ denotes the outward unit normal to $\partial\Omega$. Next, two equivalent expressions are also available for $\overset{\star}{J}(p)$:

$$\overset{\star}{J}(p) = \int_{S(p)} \left[\overset{\star}{g} + g\mathrm{div}_S\boldsymbol{\theta}\right]\mathrm{d}S \tag{12.11}$$

$$= \int_{S(p)} \left[g_{,p} + (g_{,n} - 2Kg)(\boldsymbol{\theta}.\boldsymbol{n})\right]\mathrm{d}S + \int_{\partial S(p)} g(\boldsymbol{\theta}.\boldsymbol{\nu})\,\mathrm{d}s \tag{12.12}$$

where K is the mean curvature at $\boldsymbol{y} \in S$:

$$K = \frac{1}{2}\left(\frac{1}{R_1} + \frac{1}{R_2}\right) = -\frac{1}{2}\mathrm{div}_S\boldsymbol{n}$$

R_1, R_2 are the principal radii of curvature at $\boldsymbol{y}$ and, if S is an open surface with edge ∂S, $\boldsymbol{\nu}$ denotes the outward unit normal to ∂S (Figure 12.1). Finally the derivative $\overset{\star}{H}(p)$ is given by:

$$\overset{\star}{H}(p) = \int_{\Gamma(p)} \left[\overset{\star}{h} + h\left(\frac{\partial\boldsymbol{\theta}}{\partial\boldsymbol{\tau}}.\boldsymbol{\tau}\right)\right]\mathrm{d}s \tag{12.13}$$

where $\boldsymbol{\tau}$ is the unit tangent to $\Gamma(p)$. When $\Gamma(p)$ is a *closed* curve of algebraic radius of curvature $R(s)$, one has:

$$\overset{\star}{H}(p) = \int_{C} \left[\overset{\star}{h} + h\left(\frac{\mathrm{d}}{\mathrm{d}s}(\boldsymbol{\theta}.\boldsymbol{\tau}) - \frac{1}{R}\theta_\nu\right)\right]\mathrm{d}s \tag{12.14}$$

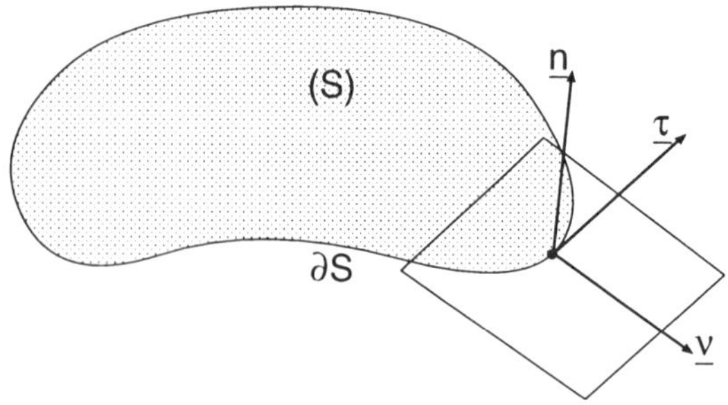

Figure 12.1 Frame (n, ν, τ) attached to ∂S.

Expressions (12.9–12.14) for $\overset{\star}{I}, \overset{\star}{J}, \overset{\star}{H}$ are derived in several references, see e.g., Petryk and Mroz [13], Mialon [10]. They provide the mathematical tool allowing us to investigate the shape sensitivity of integral functionals that depend on a geometrical domain both explicitly (through their geometrical support) and implicitly (through the solution to a boundary integral equation).

12.3 DIFFERENTIATION OF AN OBJECTIVE FUNCTION

Consider the following situation: an objective function J is defined as an integral over (a part of) the boundary of a variable domain $\Omega = \Omega(p)$, in terms of the solution to a boundary-value problem on $\Omega(p)$ which may be formulated in boundary integral equation form. To fix our ideas, consider on $\Omega(p)$ a boundary-value problem for the Laplacian, and denote by $S_u(p)$ and $S_q(p)$ the Dirichlet and Neumann parts of the boundary $\partial\Omega(p)$ of the variable domain:

$$\begin{cases} \Delta u = 0 & \text{in } \Omega(p) \\ q \equiv \dfrac{\partial u}{\partial n} = \bar{q} & \text{on } S_q(p) \\ u = \bar{u} & \text{on } S_u(p) \end{cases} \tag{12.15}$$

The boundary data $\bar{u}$ and $\bar{q}$ may have a known dependence on p. The domain transformation $\Omega(p = 0) \to \Omega(p)$ is assumed to transform $S_u(0)$ into $S_u(p)$ and $S_q(0)$ into $S_q(p)$ (this is not *a priori* obvious). Then, consider the generic objective function $\mathcal{J}(\Omega)$:

$$\mathcal{J}(\Omega) = J(u_\Omega, q_\Omega, \Omega) = \int_{S_q} \varphi(u_\Omega, \boldsymbol{y}) \, \mathrm{d}S_y + \int_{S_u} \psi(q_\Omega, \boldsymbol{y}) \, \mathrm{d}S_y \tag{12.16}$$

where (u_Ω, q_Ω) denotes the solution to the problem (12.15) for a given domain Ω. The distinction between the notations J and $\mathcal{J}$ emphasizes the fact that the objective function $\mathcal{J}$ depends on Ω both explicitly and implicitly through the solution (u_Ω, q_Ω). The derivative of $\mathcal{J}(\Omega)$ in a domain perturbation is *a priori* given by eq. (12.12), i.e.:

$$\overset{\star}{\mathcal{J}} = \int_{S_q} \left[\varphi_{,u} \overset{\star}{u} + \varphi_{,y}.\boldsymbol{\theta} + \varphi \mathrm{div}_S \boldsymbol{\theta} \right] \mathrm{d}S_y + \int_{S_u} \left[\psi_{,q} \overset{\star}{q} + \psi_{,y}.\boldsymbol{\theta} + \psi \mathrm{div}_S \boldsymbol{\theta} \right] \mathrm{d}S_y \quad (12.17)$$

The *direct differentiation method* consists in formulating a governing problem for the derivatives $(\overset{\star}{u}, \overset{\star}{q})$ of the state variables (u, q). This can be accomplished either in boundary integral equation form (see Section 12.4), in a finite element-oriented weak form, or in local field equation form (see e.g. Sokolowski and Zolesio [16]).

The *adjoint state method* (Section 12.9) consists in introducing the solution $(\hat{u}, \hat{q})$ to another boundary-value problem, termed the *adjoint problem*, defined so as to cancel all terms containing the derivatives $(\overset{\star}{u}, \overset{\star}{q})$ (and thus eliminate the need to compute them) in eq. (12.17).

12.4 GOVERNING BOUNDARY INTEGRAL EQUATION FOR THE FIRST-ORDER DOMAIN DERIVATIVE OF THE STATE

For expository purposes, consider a physical state (u, q) solving the Poisson equation over some domain Ω, and hence governed by the boundary integral equation (2.24). The latter reads:

$$I_1(\boldsymbol{x}, u) - I_2(\boldsymbol{x}, q) = F(\boldsymbol{x}) \quad (12.18)$$

in terms of abbreviated notations introduced for later convenience:

$$I_1(\boldsymbol{x}, u) = \bar{\kappa} u(\boldsymbol{x}) + \int_{\partial\Omega} [u(\boldsymbol{y}) - u(\boldsymbol{x})] H(\boldsymbol{x}, \boldsymbol{y}) \, \mathrm{d}S_y \quad (12.19)$$

$$I_2(\boldsymbol{x}, q) = \int_{\partial\Omega} q(\boldsymbol{y}) G(\boldsymbol{x}, \boldsymbol{y}) \, \mathrm{d}S_y \quad (12.20)$$

$$F(\boldsymbol{x}) = \int_{\Omega} G(\boldsymbol{x}, \boldsymbol{y}) b(\boldsymbol{y}) \, \mathrm{d}V_y \quad (12.21)$$

with $\bar{\kappa} = 0$ (Ω bounded) or $\bar{\kappa} = 1$ ($\mathbb{R}^3 - \Omega$ bounded) and where G is the infinite-space fundamental solution: $G(\boldsymbol{x}, \boldsymbol{y}) = 1/(4\pi r)$. Equation (12.18) holds for any $\boldsymbol{x} \in \mathbb{R}^3$ and is equivalent to the field equation $\Delta u + b = 0$ over $\Omega(p)$. For $\boldsymbol{x} \in \partial\Omega(p)$, equation (12.18) governs the values on the boundary of the potential u and the flux $q = \boldsymbol{\nabla} u.\boldsymbol{n}$.

Formulation of the derivative boundary integral equation

The integral equation (12.18) allows us to investigate the effect on the solution (u, q) of a variation of the domain $\Omega(p)$, i.e. of the fictitious time p, under a given geometrical transformation $\mathbf{\Phi}$. Suppose that a boundary-value problem is defined over each $\Omega(p)$ ($p \geq 0$), and denote by (u_p, q_p) the corresponding family of solutions (the distributed source b being also allowed to vary with p). Then, (u_p, q_p) solves the integral equation (12.18), with $\omega \equiv \Omega(p)$. One can then differentiate this p-family of integral equations with respect to p, assuming that the collocation point $\boldsymbol{x}$ 'moves' according to the transformation of the domain. It is essential at this point to remark that, since eq. (12.18) involves only weakly singular integrals, the mathematical validity of its forthcoming differentiation with respect to p is assured.

In short, one now seeks the explicit form of:

$$\overset{\star}{I_1}(\boldsymbol{x}, u) - \overset{\star}{I_2}(\boldsymbol{x}, q) = \overset{\star}{F}(\boldsymbol{x}) \tag{12.22}$$

First, taking the Lagrangian derivative of the surface integrals $I_1(\boldsymbol{x}, u)$ and $I_2(\boldsymbol{x}, q)$ by means of identity (12.11) yields:

$$\frac{\mathrm{d}}{\mathrm{d}p} I_1(\boldsymbol{x}, u) = I_1(\boldsymbol{x}, \overset{\star}{u}) + J_1(\boldsymbol{x}, u; \boldsymbol{\theta}) \tag{12.23}$$

$$\frac{\mathrm{d}}{\mathrm{d}p} I_2(\boldsymbol{x}, q) = I_2(\boldsymbol{x}, \overset{\star}{q}) + J_2(\boldsymbol{x}, q; \boldsymbol{\theta}) \tag{12.24}$$

with:

$$\begin{aligned} J_1(\boldsymbol{x}, u; \boldsymbol{\theta}) = & \int_{\partial\Omega} [u(\boldsymbol{y}) - u(\boldsymbol{x})] \left\{ \overset{\star}{n_j}(\boldsymbol{y}) G_{,j}(\boldsymbol{x}, \boldsymbol{y}) - n_j(\boldsymbol{y}) (G_{,j})^\star(\boldsymbol{x}, \boldsymbol{y}) \right\} \mathrm{d}S_y \\ & + \int_{\partial\Omega} [u(\boldsymbol{y}) - u(\boldsymbol{x})] H(\boldsymbol{x}, \boldsymbol{y}) \mathrm{div}_S \boldsymbol{\theta}(\boldsymbol{y}) \, \mathrm{d}S_y \end{aligned} \tag{12.25}$$

$$J_2(\boldsymbol{x}, q; \boldsymbol{\theta}) = \int_{\partial\Omega} q(\boldsymbol{y}) \left\{ \overset{\star}{G}(\boldsymbol{x}, \boldsymbol{y}) + G(\boldsymbol{x}, \boldsymbol{y}) \mathrm{div}_S \boldsymbol{\theta}(\boldsymbol{y}) \right\} \mathrm{d}S_y \tag{12.26}$$

(the fact that I_1, I_2 are linear in u, q respectively has been used). Formulas (12.23), (12.25) and (12.24), (12.26), with $\Omega \equiv \Omega(p)$ and $(u, q) \equiv (u_p, q_p)$, are valid for any p such that the geometrical transformation $\mathbf{\Phi}(\cdot, p)$ is defined.

More explicit expressions are now sought for J_1, J_2. First, formulas (12.7), (12.8) in component form, using the notations of Appendix A, read:

$$\overset{\star}{n_j} = -n_r D_j \theta_r \qquad \mathrm{div}_S \boldsymbol{\theta} = D_r \theta_r \, \mathrm{d}S$$

Also, the fundamental solution G is a function of the *two* material points $\boldsymbol{x}, \boldsymbol{y}$ and does not depend explicitly on p. Its Lagrangian derivatives, obtained from (12.3) and using the symmetry property (4.13), are thus given by:

$$\begin{aligned} \overset{\star}{G}(\boldsymbol{x}, \boldsymbol{y}) &= \theta_r(\boldsymbol{y}) G_{,r}(\boldsymbol{x}, \boldsymbol{y}) + \theta_r(\boldsymbol{x}) G_{,\bar{r}}(\boldsymbol{x}, \boldsymbol{y}) \\ &= [\theta_r(\boldsymbol{y}) - \theta_r(\boldsymbol{x})] \, G_{,r}(\boldsymbol{x}, \boldsymbol{y}) \end{aligned} \tag{12.27}$$

$$\begin{aligned} (G_{,j})^\star(\boldsymbol{x}, \boldsymbol{y}) &= \theta_r(\boldsymbol{y}) G_{,jr}(\boldsymbol{x}, \boldsymbol{y}) + \theta_r(\boldsymbol{x}) G_{,j\bar{r}}(\boldsymbol{x}, \boldsymbol{y}) \\ &= [\theta_r(\boldsymbol{y}) - \theta_r(\boldsymbol{x})] \, G_{,jr}(\boldsymbol{x}, \boldsymbol{y}) \end{aligned} \tag{12.28}$$

These expressions show that, if $\boldsymbol{\theta}$ is sufficiently smooth, the quantities

$$[u(\boldsymbol{y}) - u(\boldsymbol{x})](G_{,j})^{\star}\,, \qquad \overset{\star}{G}$$

have weak (i.e. integrable) singularities. Taking eqs (12.27), (12.28) into account, J_2 becomes:

$$J_2(\boldsymbol{x}, q; \boldsymbol{\theta}) = \int_{\partial\Omega} q(\boldsymbol{y}) \left\{ [\theta_r(\boldsymbol{y}) - \theta_r(\boldsymbol{x})]\, G_{,r}(\boldsymbol{x}, \boldsymbol{y}) + G(\boldsymbol{x}, \boldsymbol{y}) D_r \theta_r(\boldsymbol{y}) \right\} \mathrm{d}S_y$$

Moreover, J_1 takes the form:

$$\begin{aligned} J_1&(\boldsymbol{x}, u; \boldsymbol{\theta}) \\ &= \int_{\partial\Omega} [u(\boldsymbol{y}) - u(\boldsymbol{x})]\, [\theta_r(\boldsymbol{y}) - \theta_r(\boldsymbol{x})]\, n_j(\boldsymbol{y}) G_{,jr}(\boldsymbol{x}, \boldsymbol{y})\, \mathrm{d}S_y \\ &\quad + \int_{\partial\Omega} [u(\boldsymbol{y}) - u(\boldsymbol{x})] G_{,j}(\boldsymbol{x}, \boldsymbol{y}) \left\{ n_j(\boldsymbol{y}) D_r \theta_r(\boldsymbol{y}) - n_r(\boldsymbol{y}) D_j \theta_r(\boldsymbol{y}) \right\} \mathrm{d}S_y \\ &= \int_{\partial\Omega} [u(\boldsymbol{y}) - u(\boldsymbol{x})]\, [\theta_r(\boldsymbol{y}) - \theta_r(\boldsymbol{x})]\, n_j G_{,jr}(\boldsymbol{x}, \boldsymbol{y})\, \mathrm{d}S_y \\ &\quad + \int_{\partial\Omega} [u(\boldsymbol{y}) - u(\boldsymbol{x})] G_{,j}(\boldsymbol{x}, \boldsymbol{y}) D_{jr} \theta_r(\boldsymbol{y}) dSy \end{aligned} \tag{12.29}$$

using the tangential differential operator D_{ij}, defined by eq. (A.9). Then, using the fact that G solves the field equation $G_{,jj} = 0$ $(\boldsymbol{y} \neq \boldsymbol{x})$, one has:

$$\begin{aligned} n_j(\boldsymbol{y}) G_{,jr}(\boldsymbol{x}, \boldsymbol{y}) &= [n_j(\boldsymbol{y}) G_{,jr}(\boldsymbol{x}, \boldsymbol{y}) - n_r(\boldsymbol{y}) G_{,jj}(\boldsymbol{x}, \boldsymbol{y})] \\ &\qquad + n_r(\boldsymbol{y}) G_{,jj}(\boldsymbol{x}, \boldsymbol{y})(\boldsymbol{x}, \boldsymbol{y}) \\ &= D_{jr} G_{,j}(\boldsymbol{x}, \boldsymbol{y}) \end{aligned} \tag{12.30}$$

so that the right-hand side of eq. (12.29) becomes:

$$\begin{aligned} J_1(\boldsymbol{x}, u; \boldsymbol{\theta}) &= \int_{\partial\Omega} [u(\boldsymbol{y}) - u(\boldsymbol{x})]\, [\theta_r(\boldsymbol{y}) - \theta_r(\boldsymbol{x})]\, D_{jr} G_{,j}(\boldsymbol{x}, \boldsymbol{y})\, \mathrm{d}S_y \\ &\quad + \int_{\partial\Omega} [u(\boldsymbol{y}) - u(\boldsymbol{x})] G_{,j}(\boldsymbol{x}, \boldsymbol{y}) D_{jr} \theta_r(\boldsymbol{y})\, \mathrm{d}S_y \\ &= \int_{\partial\Omega} [u(\boldsymbol{y}) - u(\boldsymbol{x})] D_{jr} \left\{ [\theta_r(\boldsymbol{y}) - \theta_r(\boldsymbol{x})]\, G_{,j}(\boldsymbol{x}, \boldsymbol{y}) \right\} \mathrm{d}S_y \end{aligned}$$

An integration by parts using identity (A.8) then yields:

$$\begin{aligned} J_1(\boldsymbol{x}, u; \boldsymbol{\theta}) &= \int_{\partial\Omega} \left\{ D_{rj} [u(\boldsymbol{y}) - u(\boldsymbol{x})] \right\} [\theta_r(\boldsymbol{y}) - \theta_r(\boldsymbol{x})]\, G_{,j}(\boldsymbol{x}, \boldsymbol{y})\, \mathrm{d}S_y \\ &= \int_{\partial\Omega} D_{rj} u(\boldsymbol{y})\, [\theta_r(\boldsymbol{y}) - \theta_r(\boldsymbol{x})]\, G_{,j}(\boldsymbol{x}, \boldsymbol{y})\, \mathrm{d}S_y \end{aligned}$$

Finally, the Lagrangian derivative of the right-hand side $F(\boldsymbol{x})$ (eq. (12.21)) in the domain transformation is obtained using formula (12.9):

$$\overset{\star}{F}(\boldsymbol{x};\boldsymbol{p}) = \int_{\Omega} \overset{\star}{b}_i(\boldsymbol{y})G(\boldsymbol{x},\boldsymbol{y})\,\mathrm{d}V_y + \int_{\Omega} b(\boldsymbol{y})\left\{G_{,j}(\boldsymbol{x},\boldsymbol{y})\,[\theta_j(\boldsymbol{y})-\theta_j(\boldsymbol{x})] + G(\boldsymbol{x},\boldsymbol{y})\theta_{j,j}(\boldsymbol{y})\right\}\mathrm{d}V_y$$

and can then be recast into a more convenient form using the divergence formula:

$$\begin{aligned}
\overset{\star}{F}(\boldsymbol{x};\boldsymbol{p}) &= \int_{\Omega} \overset{\star}{b}_i(\boldsymbol{y})G(\boldsymbol{x},\boldsymbol{y})\,\mathrm{d}V_y + \int_{\Omega} b(\boldsymbol{y})\left\{G(\boldsymbol{x},\boldsymbol{y})\,[\theta_j(\boldsymbol{y})-\theta_j(\boldsymbol{x})]\right\}_{,j}\mathrm{d}V_y \\
&= \int_{\Omega}\left[\overset{\star}{b}_i(\boldsymbol{y}) - b_{,j}(\boldsymbol{y})\,[\theta_j(\boldsymbol{y})-\theta_j(\boldsymbol{x})]\right] G(\boldsymbol{x},\boldsymbol{y})\,\mathrm{d}V_y \\
&\quad + \int_{\partial\Omega} b(\boldsymbol{y})G(\boldsymbol{x},\boldsymbol{y})\,[\theta_j(\boldsymbol{y})-\theta_j(\boldsymbol{x})]\,n_j(\boldsymbol{y})\,\mathrm{d}S_y \\
&= \int_{\Omega} b_{,p}(\boldsymbol{y})G(\boldsymbol{x},\boldsymbol{y})\,\mathrm{d}V_y + \theta_j(\boldsymbol{x})\int_{\Omega} b_{,j}(\boldsymbol{y})G(\boldsymbol{x},\boldsymbol{y})\,\mathrm{d}V_y \\
&\quad + \int_{\partial\Omega} b(\boldsymbol{y})G(\boldsymbol{x},\boldsymbol{y})\,[\theta_j(\boldsymbol{y})-\theta_j(\boldsymbol{x})]\,n_j(\boldsymbol{y})\,\mathrm{d}S_y
\end{aligned} \tag{12.31}$$

Statement of the first-order derivative integral equation. One concludes from the previous calculations that for any p-family of potential–flux pairs (u_p, q_p) solving the integral equation (12.18) for the domain $\Omega = \Omega(p)$, any smooth perturbation of such a state in a geometrical transformation $\boldsymbol{\Phi}(\cdot,p)$ (having a $C^{0,1}$-continuous transformation velocity $\boldsymbol{\theta}$ at $\boldsymbol{y}=\boldsymbol{x}$) and for a smoothly changing source $b(\cdot,p)$ is such that the Lagrangian derivative pair $(\overset{\star}{u},\overset{\star}{q})\,|_{\partial\Omega}$ solve the integral equation:

$$I_1(\boldsymbol{x},\overset{\star}{u}) - I_2(\boldsymbol{x},\overset{\star}{q}) = \overset{\star}{F}(\boldsymbol{x}) - J_1(\boldsymbol{x},u;\boldsymbol{\theta}) + J_2(\boldsymbol{x},q;\boldsymbol{\theta}) \tag{12.32}$$

with:

$$J_1(\boldsymbol{x},u;\boldsymbol{\theta}) = \int_{\partial\Omega} D_{rj}u(\boldsymbol{y})\,[\theta_r(\boldsymbol{y})-\theta_r(\boldsymbol{x})]\,G_{,j}(\boldsymbol{x},\boldsymbol{y})\,\mathrm{d}S_y \tag{12.33}$$

$$\begin{aligned}
J_2(\boldsymbol{x},q;\boldsymbol{\theta}) &= \int_{\partial\Omega} q(\boldsymbol{y})\,[\theta_r(\boldsymbol{y})-\theta_r(\boldsymbol{x})]\,G_{,r}(\boldsymbol{x},\boldsymbol{y})\,\mathrm{d}S_y \\
&\quad + \int_{\partial\Omega} q(\boldsymbol{y})G(\boldsymbol{x},\boldsymbol{y})D_r\theta_r(\boldsymbol{y})\,\mathrm{d}S_y
\end{aligned} \tag{12.34}$$

$$\begin{aligned}
\overset{\star}{F}(\boldsymbol{x};\boldsymbol{p}) &= \int_{\Omega} b_{,p}(\boldsymbol{y})G(\boldsymbol{x},\boldsymbol{y})\,\mathrm{d}V_y + \theta_j(\boldsymbol{x})\int_{\Omega} b_{,j}(\boldsymbol{y})G(\boldsymbol{x},\boldsymbol{y})\,\mathrm{d}V_y \\
&\quad + \int_{\partial\Omega} b(\boldsymbol{y})G(\boldsymbol{x},\boldsymbol{y})\,[\theta_j(\boldsymbol{y})-\theta_j(\boldsymbol{x})]\,n_j(\boldsymbol{y})\,\mathrm{d}S_y
\end{aligned} \tag{12.35}$$

The above integral identity, which holds for any point $\boldsymbol{x} \in \mathbb{R}^3$, defines a boundary integral equation (if $\boldsymbol{x}$ is chosen on $\partial\Omega$) or an integral representation formula for $\overset{\star}{u}(\boldsymbol{x};p)$ (if $\boldsymbol{x}$ is chosen interior to Ω). The integral equation is weakly singular provided u and $\boldsymbol{\theta}$ have $C^{0,\alpha}$ and $C^{0,1}$ smoothness at $\boldsymbol{y} = \boldsymbol{x}$, respectively. Having started from the *regularized* integral equation (12.18) ensures, together with the previous smoothness assumptions on u and $\boldsymbol{\theta}$, that all terms in the final result (12.32) as well as in all intermediary steps are weakly singular. The integration by parts performed on eq. (12.25), although not mandatory, yields the more compact expression (12.33) for J_1.

Interpretation of the right-hand side of the derivative integral equation

It is at first sight surprising that the governing problem for $(\overset{\star}{u}, \overset{\star}{q})$ involves boundary conditions on $(\overset{\star}{u}, \overset{\star}{q})$ *and* a known right-hand side, made of surface integrals, which could also be thought of as incorporating information from boundary conditions. Actually, the latter terms happen to be related to a *body* source term, i.e similar to b. In order to show this, one first notes that:

$$\overset{\star}{q} = (\boldsymbol{\nabla} u.\boldsymbol{n})^\star = \boldsymbol{\nabla}\, \overset{\star}{u}\, .\boldsymbol{n} - \boldsymbol{\nabla} u.\boldsymbol{\nabla}\boldsymbol{\theta}.\boldsymbol{n} - \boldsymbol{\nabla} u.\, \overset{\star}{\boldsymbol{n}}$$

while the Lagrangian derivative $\overset{\star}{\boldsymbol{n}}$ (12.8) can be expressed as:

$$\overset{\star}{\boldsymbol{n}} = (\boldsymbol{n}.\boldsymbol{\theta}_{,n})\boldsymbol{n} - \boldsymbol{n}.\nabla\boldsymbol{\theta} = (\text{div}\boldsymbol{\theta} - \text{div}_S\boldsymbol{\theta})\boldsymbol{n} - \boldsymbol{n}.\boldsymbol{\nabla}\boldsymbol{\theta}$$

Then:

$$\begin{aligned} \overset{\star}{q} + (\text{div}_S\boldsymbol{\theta})q &= \overset{\star}{u}_{,n} - \boldsymbol{A}.\boldsymbol{n} \\ \boldsymbol{A} &= -(\text{div}\boldsymbol{\theta})\boldsymbol{\nabla} u + \boldsymbol{\nabla} u.\boldsymbol{\nabla}\boldsymbol{\theta} + \boldsymbol{\nabla}\boldsymbol{\theta}.\boldsymbol{\nabla} u \end{aligned}$$

Substituting this relation into $\overset{\star}{I}_2$ as expressed by (12.20) gives:

$$\begin{aligned} \overset{\star}{I}_2(\boldsymbol{x}, q; \boldsymbol{\theta}) &= \int_{\partial\Omega} \overset{\star}{u}_{,n}(\boldsymbol{y})G(\boldsymbol{x},\boldsymbol{y})\,\mathrm{d}S_y \\ &+ \int_{\partial\Omega} \{q(\boldsymbol{y})\,[\theta_j(\boldsymbol{y}) - \theta_j(\boldsymbol{x})]\,G_{,j}(\boldsymbol{x},\boldsymbol{y}) - A_j(\boldsymbol{y})n_j(\boldsymbol{y})G(\boldsymbol{x},\boldsymbol{y})\}\,\mathrm{d}S_y \end{aligned}$$

An analytical transformation using integrations by parts (Appendix 12.B) yields the identity:

$$\begin{aligned} &\int_{\partial\Omega} A_j(\boldsymbol{y})n_j(\boldsymbol{y})G(\boldsymbol{x},\boldsymbol{y})\,\mathrm{d}S_y \\ &= \int_{\Omega} G(\boldsymbol{x},\boldsymbol{y})\,\{A_j(\boldsymbol{y}) - b(\boldsymbol{y})\,[\theta_j(\boldsymbol{y}) - \theta_j(\boldsymbol{x})]\}_{,j}\,\mathrm{d}V_y \\ &\quad + \int_{\partial\Omega} [\theta_j(\boldsymbol{y}) - \theta_j(\boldsymbol{x})]\,\{q(\boldsymbol{y})G_{,j}(\boldsymbol{x},\boldsymbol{y}) + G(\boldsymbol{x},\boldsymbol{y})b(\boldsymbol{y})n_j(\boldsymbol{y})\}\,\mathrm{d}S_y \\ &\quad - \int_{\partial\Omega} [\theta_r(\boldsymbol{y}) - \theta_r(\boldsymbol{x})]\,G_{,j}(\boldsymbol{x},\boldsymbol{y})D_{rj}u_i(\boldsymbol{y})\,\mathrm{d}S_y \end{aligned} \tag{12.36}$$

Its substitution, together with the developed expressions of $\overset{\star}{I}_1$ and $\overset{\star}{F}$, into the derivative integral equation (12.22), leads to the identity:

$$I_2(\boldsymbol{x}, \overset{\star}{u}_{,n}) - I_1(\boldsymbol{x}, \overset{\star}{u}) \\ = \int_{\Omega} G(\boldsymbol{x}, \boldsymbol{y}) \left\{ (b(\boldsymbol{y})\theta_j(\boldsymbol{y}) - A_j(\boldsymbol{y}))_{,j} + b_{,p}(\boldsymbol{y}) \right\} \, \mathrm{d}V_y \qquad (12.37)$$

This result is interpreted as follows: the mathematical structure of the derivative integral equation is similar to that of the initial integral equation (12.18), with unknowns $(\overset{\star}{u}, \overset{\star}{u}_{,n})$ and body source $\mathrm{div}(b\boldsymbol{\theta} - \boldsymbol{A}) + b_{,p}$ (instead of (u, q) and b in (12.18)). When $b = 0$ (no body source in the initial problem), the body source for the derivative integral equation reduces to $-\mathrm{div}\boldsymbol{A}$ (but does not vanish); the corresponding domain integral thus happens to be analytically convertible into boundary integrals, giving rise to the right-hand side $J_2 - J_1$ in the derivative integral equation (12.32).

12.5 GOVERNING BOUNDARY INTEGRAL EQUATION FOR THE SECOND-ORDER DOMAIN DERIVATIVE OF THE STATE

In this section, the analysis is carried out one step further, to establish the governing integral equation for second-order derivatives of the solution (u, q).

Second-order differentiation in a domain transformation

In order to define mixed second-order derivatives, the geometrical transformation $\boldsymbol{\Phi}$ is assumed to depend on *two* shape parameters p_1, p_2:

$$\boldsymbol{X} \in \Omega(0) \rightarrow \boldsymbol{x} = \boldsymbol{\Phi}(\boldsymbol{X}; p_1, p_2)$$

The transformation velocities $\boldsymbol{\theta}$ and $\boldsymbol{\mu}$ with respect to p_1, p_2 and the 'acceleration' $\boldsymbol{\chi}$ are introduced accordingly:

$$\begin{aligned} \boldsymbol{\theta}(\boldsymbol{x}; p_1, p_2) &= \boldsymbol{\Phi}_{,p_1}(\boldsymbol{X}; p_1, p_2) \\ \boldsymbol{\mu}(\boldsymbol{x}; p_1, p_2) &= \boldsymbol{\Phi}_{,p_2}(\boldsymbol{X}; p_1, p_2) \qquad (\boldsymbol{x} = \boldsymbol{\Phi}(\boldsymbol{X}; p_1, p_2)) \\ \boldsymbol{\chi}(\boldsymbol{x}; p_1, p_2) &= \boldsymbol{\Phi}_{,p_1 p_2}(\boldsymbol{X}; p_1, p_2) \end{aligned}$$

Let $f(\boldsymbol{x}; p_1, p_2)$ denote a (sufficiently smooth) scalar or tensor field. Denote by $\overset{\star}{f}$ and $\overset{\vee}{f}$ the Lagrangian derivatives of f with respect to p_1, p_2 respectively. Define the second-order Lagrangian derivative $\overset{\star\vee}{f}$ by:

$$\overset{\star\vee}{f}(\boldsymbol{x}; p_1, p_2) = g_{,p_1 p_2}(\boldsymbol{X}; p_1, p_2) \qquad (12.38)$$

with

$$g(\boldsymbol{X};p_1,p_2) = f(\boldsymbol{\Phi}(\boldsymbol{X};p_1,p_2);p_1,p_2)$$

One remarks at once that these definitions imply the equality of mixed second-order derivatives, i.e.: $\boldsymbol{\chi} = \overset{\vee}{\boldsymbol{\theta}} = \overset{\star}{\boldsymbol{\mu}}$ and $\overset{\star\vee}{f} = \overset{\vee\star}{f} = (\overset{\vee}{f})^{\star} = (\overset{\star}{f})^{\vee}$.

Second-order derivative integral equation

Assume for simplicity the absence of body sources ($b = 0$). The governing integral equation (12.32) for the first-order Lagrangian derivatives of (u,q) reads:

$$I_1(\boldsymbol{x},\overset{\star}{u}) - I_2(\boldsymbol{x},\overset{\star}{q}) = -J_1(\boldsymbol{x},u;\boldsymbol{\theta}) + J_2(\boldsymbol{x},q;\boldsymbol{\theta}) \tag{12.39}$$

This holds true for any domain $\Omega = \Omega(p_1,p_2)$, any solution (u,q) to the Laplace equation and any Lagrangian derivative $(\overset{\star}{u},\overset{\star}{q})$ with respect to p_1 of (u,q). Also, as a consequence of the definitions previously introduced, taking the mixed second-order Lagrangian derivative of the initial integral equation (12.18) with respect to p_1,p_2 amounts to taking the first-order Lagrangian derivative of eq. (12.39) with respect to p_2. This step is now carried out.

First, the derivative of the left-hand side of (12.39) is:

$$\frac{\mathrm{d}}{\mathrm{d}p_2}\left\{I_1(\boldsymbol{x},\overset{\star}{u}) - I_2(\boldsymbol{x},\overset{\star}{q})\right\}$$
$$= I_1(\boldsymbol{x},\overset{\star\vee}{u}) - I_2(\boldsymbol{x},\overset{\star\vee}{q}) + J_1(\boldsymbol{x},\overset{\star}{q},\boldsymbol{\mu}) - J_2(\boldsymbol{x},\overset{\star}{u},\boldsymbol{\mu}) \tag{12.40}$$

Indeed, this result is a mere transposition (with (u,q) replaced by $(\overset{\star}{u},\overset{\star}{q})$) of the derivative of the left-hand side of eq. (12.32). The remaining task is to derive explicit expressions for the derivatives $\overset{\vee}{J}_1,\overset{\vee}{J}_2$ of J_1,J_2 with respect to p_2. The Lagrangian derivative $\overset{\vee}{J}_2$ is given, from eq. (12.11) and recalling that J_2 is linear with respect to q and $\boldsymbol{\theta}$, by:

$$\overset{\vee}{J}_2(\boldsymbol{x},q,\boldsymbol{\theta}) = J_2(\boldsymbol{x},\overset{\vee}{q},\boldsymbol{\theta}) + J_2(\boldsymbol{x},q,\boldsymbol{\chi}) + K_2(\boldsymbol{x},q;\boldsymbol{\theta},\boldsymbol{\mu}) \tag{12.41}$$

with (after some manipulations):

$$\begin{aligned}
K_2(\boldsymbol{x},q;\boldsymbol{\theta},\boldsymbol{\mu}) = &\int_{\partial\Omega} q(\boldsymbol{y})G^k_{,ij}(\boldsymbol{x},\boldsymbol{y})\,[\theta_i(\boldsymbol{y})-\theta_i(\boldsymbol{x})]\,[\mu_j(\boldsymbol{y})-\mu_j(\boldsymbol{x})]\,\mathrm{d}S_y \\
&+\int_{\partial\Omega} q(\boldsymbol{y})G_{,j}(\boldsymbol{x},\boldsymbol{y})\,[\theta_j(\boldsymbol{y})-\theta_j(\boldsymbol{x})]\,D_i\mu_i(\boldsymbol{y})\,\mathrm{d}S_y \\
&+\int_{\partial\Omega} q(\boldsymbol{y})G_{,j}(\boldsymbol{x},\boldsymbol{y})[\mu_j(\boldsymbol{y})-\mu_j(\boldsymbol{x})]D_i\theta_i(\boldsymbol{y})\,\mathrm{d}S_y \\
&+\int_{\partial\Omega} q(\boldsymbol{y})G(\boldsymbol{x},\boldsymbol{y})\,\{D_j\theta_j D_k\mu_k - D_j\mu_k D_k\theta_j\}\,(\boldsymbol{y})\,\mathrm{d}S_y \\
&+\int_{\partial\Omega} q(\boldsymbol{y})G(\boldsymbol{x},\boldsymbol{y})\,\{n_i n_j D_k\mu_i D_k\theta_j\}\,(\boldsymbol{y})\,\mathrm{d}S_y
\end{aligned} \tag{12.42}$$

Similarly, the Lagrangian derivative of $J_1(\boldsymbol{x}, u; \boldsymbol{\theta})$ is given by:

$$\begin{aligned}\overset{\vee}{J_1}(\boldsymbol{x}, u; \boldsymbol{\theta}) = J_1(\boldsymbol{x}, u; \boldsymbol{\chi}) &+ \int_{\partial\Omega} [\theta_i(\boldsymbol{y}) - \theta_i(\boldsymbol{x})]\, G_{,j}(\boldsymbol{x}, \boldsymbol{y})(D_{ij}u)^{\vee}(\boldsymbol{y})\, \mathrm{d}S_y \\ &+ \int_{\partial\Omega} [\theta_i(\boldsymbol{y}) - \theta_i(\boldsymbol{x})]\, G_{,j}(\boldsymbol{x}, \boldsymbol{y}) D_{ij}u(\boldsymbol{y}) D_k\mu_k(\boldsymbol{y})\, \mathrm{d}S_y \\ &+ \int_{\partial\Omega} [\theta_i(\boldsymbol{y}) - \theta_i(\boldsymbol{x})]\, [\mu_k(y) - \mu_k(x)] G_{,jk}(\boldsymbol{x}, \boldsymbol{y}) D_{ij}u(\boldsymbol{y})\, \mathrm{d}S_y\end{aligned}$$

It is shown (eq. (12.A.7) of Appendix 12.C) that:

$$(D_{ij}f)^{\star} = D_{ij}\overset{\star}{f} + D_j\theta D_{ki}f + D_i\theta_k D_{jk}f$$

which implies:

$$\overset{\vee}{J_1}(\boldsymbol{x}, u; \boldsymbol{\theta}) = J_1(\boldsymbol{x}, u; \boldsymbol{\chi}) + J_1(\boldsymbol{x}, \overset{\vee}{u}; \boldsymbol{\theta}) + K_1(\boldsymbol{x}, u; \boldsymbol{\theta}, \boldsymbol{\mu}) \tag{12.43}$$

where:

$$\begin{aligned}&K_1(\boldsymbol{x}, u; \boldsymbol{\theta}, \boldsymbol{\mu}) \\ &= \int_{\partial\Omega} [\theta_i(\boldsymbol{y}) - \theta_i(\boldsymbol{x})]\, \{D_j\mu_k D_{ki}u + D_i\mu_k D_{jk}u\}\, (\boldsymbol{y}) G_{,j}(\boldsymbol{x}, \boldsymbol{y})\, \mathrm{d}S_y \\ &\quad + \int_{\partial\Omega} [\theta_i(\boldsymbol{y}) - \theta_i(\boldsymbol{x})]\, [\mu_k(y) - \mu_k(x)] G_{,jk}(\boldsymbol{x}, \boldsymbol{y}) D_{ij}u(\boldsymbol{y})\, \mathrm{d}S_y \\ &\quad + \int_{\partial\Omega} [\theta_i(\boldsymbol{y}) - \theta_i(\boldsymbol{x})]\, D_k\mu_k(\boldsymbol{y}) G_{,j}(\boldsymbol{x}, \boldsymbol{y}) D_{ij}u(\boldsymbol{y})\, \mathrm{d}S_y\end{aligned} \tag{12.44}$$

Statement of the second-order derivative integral equation. One concludes from the previous calculations that for any p-family of potential–flux pairs (u_p, q_p) solving the integral equation (12.18) for the domain $\Omega = \Omega(p)$, any smooth perturbation of such a state in a geometrical transformation $\boldsymbol{\Phi}(\cdot, p)$ (having a $C^{0,1}$-continuous transformation velocity $\boldsymbol{\theta}$ at $\boldsymbol{y} = \boldsymbol{x}$) and for a smoothly changing source $b(\cdot, p)$ is such that the mixed second-order derivative pair $(\overset{\star}{u}, \overset{\star}{q})\mid_{\partial\Omega}$ solve the integral equation:

$$\begin{aligned}&I_1(\boldsymbol{x}, \overset{\star\vee}{u}) - I_2(\boldsymbol{x}, \overset{\star\vee}{q}) \\ &= J_2(\boldsymbol{x}, \overset{\star}{q}; \boldsymbol{\mu}) + J_2(\boldsymbol{x}, \overset{\vee}{q}; \boldsymbol{\theta}) - J_1(\boldsymbol{x}, \overset{\star}{u}; \boldsymbol{\mu}) - J_1(\boldsymbol{x}, \overset{\vee}{u}; \boldsymbol{\theta}) \\ &\quad + J_2(\boldsymbol{x}, q; \boldsymbol{\chi}) - J_1(\boldsymbol{x}, u; \boldsymbol{\chi}) + K_2(\boldsymbol{x}, q; \boldsymbol{\theta}, \boldsymbol{\mu}) - K_1(\boldsymbol{x}, u; \boldsymbol{\theta}, \boldsymbol{\mu})\end{aligned} \tag{12.45}$$

where J_1, J_2, K_1, K_2 are given by equations (12.33), (12.34), (12.44) and (12.42), respectively.

The above integral identity, which holds for any point $\boldsymbol{x} \in \mathbb{R}^3$, defines a boundary integral equation (if $\boldsymbol{x}$ is chosen on $\partial\Omega$) or an integral representation formula for $\overset{\star\vee}{u}(\boldsymbol{x};p)$ (if $\boldsymbol{x}$ is chosen interior to Ω). The integral equation is weakly singular provided u and $\boldsymbol{\theta}$ have $C^{0,\alpha}$ and $C^{0,1}$ smoothness at $\boldsymbol{y}=\boldsymbol{x}$, respectively.

Symmetry in $\boldsymbol{\theta}$, $\boldsymbol{\mu}$ of the integral equation (12.45)

The right-hand side of the integral equation (12.45) is symmetric with respect to $(\boldsymbol{\theta},\boldsymbol{\mu})$. This fact is not obvious, and needs to be established.

On one hand, the expression $J_2(\boldsymbol{x},\overset{\star}{q};\boldsymbol{\mu})+J_2(\boldsymbol{x},\overset{\vee}{q};\boldsymbol{\theta})-J_1(\boldsymbol{x},\overset{\star}{u};\boldsymbol{\mu})-J_1(\boldsymbol{x},\overset{\vee}{u};\boldsymbol{\theta})$ is, in view of Eqs (12.33)–(12.34), clearly symmetric with respect to $(\boldsymbol{\theta},\boldsymbol{\mu})$; likewise, $K_2(\boldsymbol{x},u;\boldsymbol{\theta},\boldsymbol{\mu})$ is also symmetric with respect to $(\boldsymbol{\theta},\boldsymbol{\mu})$ in view of eq. (12.42). Likewise, $J_2(\boldsymbol{x},q;\boldsymbol{\chi})-J_1(\boldsymbol{x},u;\boldsymbol{\chi})$ is symmetric in the sense that $\boldsymbol{\chi}=\overset{\vee}{\boldsymbol{\theta}}=\overset{\star}{\boldsymbol{\mu}}$. On the other hand, the symmetry of $K_1(\boldsymbol{x},u;\boldsymbol{\theta},\boldsymbol{\mu})$ in $(\boldsymbol{\theta},\boldsymbol{\mu})$ is not apparent. However, it is shown in Appendix 12.D that K_1 can be reformulated as:

$$
\begin{aligned}
K_1(\boldsymbol{x},u;\boldsymbol{\theta},\boldsymbol{\mu}) = &\int_{\partial\Omega}[u(\boldsymbol{y})-u(\boldsymbol{x})]n_j(\boldsymbol{y})(G_{,j})^{\star\vee}(\boldsymbol{x},\boldsymbol{y})\,\mathrm{d}S_y \\
&+\int_{\partial\Omega}[u(\boldsymbol{y})-u(\boldsymbol{x})](G_{,j})^{\star}(\boldsymbol{x},\boldsymbol{y})D_{jk}\mu_j(\boldsymbol{y})\,\mathrm{d}S_y \\
&+\int_{\partial\Omega}[u(\boldsymbol{y})-u(\boldsymbol{x})](G_{,j})^{\vee}(\boldsymbol{x},\boldsymbol{y})D_{jk}\theta_k(\boldsymbol{y})\,\mathrm{d}S_y \\
&-\int_{\partial\Omega}[u(\boldsymbol{y})-u(\boldsymbol{x})]G_{,j}(\boldsymbol{x},\boldsymbol{y})d_j(\boldsymbol{y})\,\mathrm{d}S_y
\end{aligned}
$$

with:

$$
\begin{aligned}
d_j = n_i\left(D_j\theta_i D_k\mu_k + D_j\mu_i D_k\theta_k - D_j\mu_k D_k\theta_i - D_j\theta_k D_k\mu_i\right) \\
+ n_j\left(D_k\theta_i D_i\mu_k - D_i\theta_i D_k\mu_k\right)
\end{aligned}
$$

and the above expression is symmetric in $(\boldsymbol{\theta},\boldsymbol{\mu})$.

12.6 COMMENTS ABOUT THE DERIVATIVE INTEGRAL EQUATIONS

The first-order derivative integral equation (12.32), for a given transformation velocity $\boldsymbol{\theta}$, does not define uniquely the pair $(\overset{\star}{u},\overset{\star}{q})$. In order to define a unique pair $(\overset{\star}{u},\overset{\star}{q})$, it is also necessary to prescribe the *evolution* in the domain transformation, i.e. as p varies, of the boundary conditions for the boundary-value problem at hand. One can for example choose the transformation $\boldsymbol{\Phi}(\cdot;p)$

describing a given domain evolution[1] such that S_u, S_T (the Dirichlet and Neumann parts of $\partial\Omega$) are material surfaces:

$$S_u(p) = \boldsymbol{\Phi}(S_u;p) \qquad S_T(p) = \boldsymbol{\Phi}(S_T;p)$$

In that case, $u, \overset{\star}{u}$ and $q, \overset{\star}{q}$ are known or unknown on the same portions of $\partial\Omega$, and consequently $(\overset{\star}{u}, \overset{\star}{q})$ depend linearly on $\boldsymbol{\theta}$, provided that the boundary conditions of the governing problem for $(\overset{\star}{u}, \overset{\star}{q})$ are themselves linear with respect to $\boldsymbol{\theta}$. Similar comments apply for the second-order derivative integral equation (12.45), which *a priori* does not define uniquely the pair $(\overset{\star\vee}{u}, \overset{\star\vee}{q})$ for given $\boldsymbol{\theta}, \boldsymbol{\mu}, \boldsymbol{\chi}$.

The pairs (u, q), $(\overset{\star}{u}, \overset{\star}{q})$ and $(\overset{\star\vee}{u}, \overset{\star\vee}{q})$, respective unknowns of the integral equations (12.18), (12.32) and (12.45), are governed by the same integral operators. More precisely, if the transformation $\boldsymbol{\Phi}(\cdot;p)$ is chosen so that S_u, S_T are material surfaces, the unknown parts of (u, q) are governed by the same integral operator. This fact has important practical implications: the discretized integral operator, which must in any case be set up and factored in order to solve numerically the initial integral equation (12.18), can be reused for solving the derivative integral equations (12.32), (12.45). The supplementary numerical burden entailed by the computation of the first- and second-order Lagrangian derivatives of the state (u, q) is thus limited to setting up the right-hand sides of eq. (12.32) for any $\boldsymbol{\theta}$ and of (12.45) for any pair $(\boldsymbol{\theta}, \boldsymbol{\mu})$ and then solving triangular systems of linear equations. In contrast, a finite-difference approach would involve k (resp. $k(k+1)/2$) complete integral equation solutions (including setting up and factoring a new matrix operator each time) for the evaluation of all first-order (resp. second-order) derivatives of (u, q), where k denotes the number of geometrical parameters.

Defining a second-order Lagrangian derivative entails some conceptual difficulties. For example, it seems natural to consider the action of one transformation with 'time' p_1 followed by another with 'time' p_2:

$$\boldsymbol{Y} \rightarrow \boldsymbol{z} = \boldsymbol{\Phi}^1(\boldsymbol{Y};p_1) \rightarrow \boldsymbol{y} = \boldsymbol{\Phi}^2(\boldsymbol{z};p_2)$$

the transformation velocities $\boldsymbol{\theta}, \boldsymbol{\mu}$ being associated respectively to $\boldsymbol{\Phi}^1, \boldsymbol{\Phi}^2$ according to (12.2). But then, a Lagrangian differentiation in the transformation $\boldsymbol{\Phi}^1$ followed by a differentiation of the result in the transformation $\boldsymbol{\Phi}^2$ would in general imply $\overset{\vee}{\boldsymbol{\theta}} \neq \overset{\star}{\boldsymbol{\mu}}$, $\overset{\star\vee}{f} \neq \overset{\vee\star}{f}$ and thus $\overset{\star\vee}{J} \neq \overset{\vee\star}{J}$. This natural-looking approach to second-order differentiation may then be inadequate. This is ultimately a consequence of the fact that the composition of two transformations generally does not commute:

$$\boldsymbol{\Phi}^2(\boldsymbol{\Phi}^1(\boldsymbol{Y}, p_1), p_2) \neq \boldsymbol{\Phi}^1(\boldsymbol{\Phi}^2(\boldsymbol{Y}, p_2), p_1)$$

[1] A given domain evolution can be described using infinitely many transformations $\mathbf{F}$, such that individual points evolve differently but the global evolution is the same.

Definition (12.38) does not suffer from this ambiguity because it applies to a *single* transformation $\mathbf{\Phi}$ which depends on two (or, more generally, on a finite number of) geometrical parameters. It is then applicable to second-order sensitivity analyses with finitely many geometrical (design) variables encountered in e.g. shape optimization.

In another approach (see e.g. Simon [15]), the *domain derivative* DJ of a functional $J(\Omega)$ is defined from the consideration of small perturbations $\boldsymbol{\theta}$ of the domain Ω, without reference to shape parameters:

$$\boldsymbol{y} \in \Omega \rightarrow \boldsymbol{y} + \delta \boldsymbol{y} \in \Omega + \delta\Omega \qquad \delta\boldsymbol{y} = \boldsymbol{\theta}(\boldsymbol{y})$$

where $\boldsymbol{\theta}$ belongs to an infinite-dimensional function space $\mathcal{V}$. Then, DJ is the linear operator on $\boldsymbol{\theta}$ such that:

$$\langle DJ(\Omega), \boldsymbol{\theta}\rangle = J(\Omega + \boldsymbol{\theta}) - J(\Omega) + o(|\boldsymbol{\theta}|_{\mathcal{V}}) \tag{12.46}$$

For surface integrals, the first-order derivative given by eq. (12.11), applicable to surfaces evolving according to (12.1) (and thus dependent on finitely many shape parameters), is equal to the domain derivative applied to the small variation $\boldsymbol{\theta}\,\mathrm{d}p$ of Ω (directional derivative) $\langle DJ(\Omega), \boldsymbol{\theta}\rangle\,\mathrm{d}p$ of J:

$$J(\Omega + \boldsymbol{\theta}\,\mathrm{d}p) - J(\Omega) = \overset{\star}{J}\ \mathrm{d}p + o(\,\mathrm{d}p)$$

This is not true any more for second-order derivatives. The *second-order domain derivative* $D^2 J(\Omega)$ (Simon [15]) is defined in terms of domain perturbations of the form:

$$\boldsymbol{y} \in \Omega \rightarrow \boldsymbol{y} + \boldsymbol{\theta}(\boldsymbol{y}) + \boldsymbol{\mu}(\boldsymbol{y})$$

and is thus a bilinear form over $\boldsymbol{\theta}, \boldsymbol{\mu}$ such that:

$$\langle \boldsymbol{\mu}, D^2 J(\Omega), \boldsymbol{\theta}\rangle = J(\Omega + (\boldsymbol{\theta} + \boldsymbol{\mu})) - J(\Omega) - \overset{\star}{J}\,(\Omega; \boldsymbol{\theta}) - \overset{\vee}{J}(\Omega; \boldsymbol{\mu}) + o(|\boldsymbol{\theta}|^2_{\mathcal{V}}\,, |\boldsymbol{\mu}|^2_{\mathcal{V}}) \tag{12.47}$$

for small $\boldsymbol{\theta}, \boldsymbol{\mu}$, following the usual way of defining the second-order variation of a functional. In contrast, definition (12.1) involves domain perturbations of the form:

$$\boldsymbol{y} \in \Omega_p \rightarrow \boldsymbol{y} + \boldsymbol{\theta}(\boldsymbol{y}; \boldsymbol{p})\,\mathrm{d}p_1 + \boldsymbol{\mu}(\boldsymbol{y}; p_1, p_2)\,\mathrm{d}p_2 + \frac{1}{2}\boldsymbol{\chi}(\boldsymbol{y}; p_1, p_2)\,\mathrm{d}p_1\,\mathrm{d}p_2$$

Thus, the second-order Lagrangian derivative does not in general coincide with the second-order domain derivative applied to $\boldsymbol{\theta}\,\mathrm{d}p_1, \boldsymbol{\mu}\,\mathrm{d}p_2$, except for geometrical transformations $\mathbf{\Phi}(\cdot; p_1, p_2)$ such that $\boldsymbol{\chi} = 0$:

$$\langle \boldsymbol{\mu}, D^2 J(\Omega), \boldsymbol{\theta}\rangle - \overset{\star\vee}{J} = \int_S \mathrm{div}_S \boldsymbol{\chi}\,\mathrm{d}S$$

Similarly, one notes that since $\Omega+(\boldsymbol{\theta}+\boldsymbol{\mu}) \neq (\Omega+\boldsymbol{\theta})+\boldsymbol{\mu}$, the definition (12.47) does not reflect two successive differentiations.

Here, the first-order derivative integral equation (12.32) can be considered as governing the domain derivative (in Simon's sense) of the initial state, without reference to a shape parameter. On the other hand, the interpretation of the second-order derivative integral equation (12.45) warrants more caution due to the fundamental distinction between second-order Lagrangian derivative and second-order domain derivative.

12.7 BOUNDARY ELEMENT DISCRETIZATION

Representation of variable surfaces

The construction of approximate surfaces using boundary elements has been discussed in Chapter 3. Each element is defined by means of N nodes $\boldsymbol{y}^k$ $(1 \leq k \leq N)$ and N shape functions $N^k(\boldsymbol{\xi})$:

$$\boldsymbol{y} = \sum_{k=1}^{N} N^k(\boldsymbol{\xi})\boldsymbol{y}^k \qquad (\boldsymbol{\xi} \in E_0) \tag{12.48}$$

For a given choice of shape functions, the surface approximating $\partial\Omega$ is then entirely determined by the nodes. A variable boundary $\partial\Omega$ can then be monitored by the evolution of its nodes, i.e. by representing the variable surface by an interpolation (12.48) in which the nodes move according to a given geometrical transformation:

$$\boldsymbol{y}^k = F(\mathbf{Y}^k; \boldsymbol{p}) \tag{12.49}$$

where $\boldsymbol{p} = (p_1, p_2, \ldots)$ is the set of shape parameters. Thus, the discretized transformation velocities $\boldsymbol{\theta}$, $\boldsymbol{\mu}$ and 'acceleration' $\boldsymbol{\chi}$ associated with two generic shape parameters p_1, p_2 are obtained from eqs (12.48)–(12.49) as:

$$\boldsymbol{\theta}(\boldsymbol{y}) = \sum_{k=1}^{N} \boldsymbol{\Phi}_{,p_1}(\boldsymbol{Y}_k; \boldsymbol{p}) N_k(\boldsymbol{\xi})$$

$$\boldsymbol{\mu}(\boldsymbol{y}) = \sum_{k=1}^{N} \boldsymbol{\Phi}_{,p_2}(\boldsymbol{Y}_k; \boldsymbol{p}) N_k(\boldsymbol{\xi})$$

$$\boldsymbol{\chi}(\boldsymbol{y}) = \sum_{k=1}^{N} \boldsymbol{\Phi}_{,p_1 p_2}(\boldsymbol{Y}_k; \boldsymbol{p}) N_k(\boldsymbol{\xi})$$

i.e. as interpolations of the form (12.48). One notes that this interpolation approach ensures the $C^{0,1}$ smoothness of $\boldsymbol{\theta}, \boldsymbol{\mu}, \boldsymbol{\chi}$ on the boundary.

Structure of the linear systems of equations

A boundary element discretization of the initial integral equation (12.18) and the derivative integral equations (12.32) and (12.45) using the techniques described in Chapter 3 leads (assuming the absence of body sources) to systems of linear equations of the form:

$$
\begin{aligned}
[\boldsymbol{A}]\,\{\boldsymbol{u}\} + [\boldsymbol{B}]\{\boldsymbol{q}\} &= \{0\} \\
[\boldsymbol{A}]\,\{\overset{\star}{\boldsymbol{u}}\} + [\boldsymbol{B}]\{\overset{\star}{\boldsymbol{q}}\} &= \{\boldsymbol{f}^1(\boldsymbol{u},\boldsymbol{q};\boldsymbol{\theta})\} \\
[\boldsymbol{A}]\,\{\overset{\star\vee}{\boldsymbol{u}}\} + [\boldsymbol{B}]\{\overset{\star\vee}{\boldsymbol{q}}\} &= \{\boldsymbol{f}^2(\boldsymbol{u},\boldsymbol{q},\overset{\star}{\boldsymbol{u}},\overset{\star}{\boldsymbol{q}};\boldsymbol{\theta},\boldsymbol{\mu},\boldsymbol{\chi})\}
\end{aligned}
$$

The same matrices $[\boldsymbol{A}], [\boldsymbol{B}]$ appear in the left-hand side of each of the three equations above. After appropriate column switching, matrix equations take the form:

$$
\begin{aligned}
[\boldsymbol{K}]\,\{\boldsymbol{v}\} &= \{\boldsymbol{g}^0\} \\
[\boldsymbol{K}]\,\{\overset{\star}{\boldsymbol{v}}\} &= \{\boldsymbol{g}^1(\boldsymbol{v};\boldsymbol{\theta})\} \\
[\boldsymbol{K}]\,\{\overset{\star\vee}{\boldsymbol{v}}\} &= \{\boldsymbol{g}^2(\boldsymbol{v},\overset{\star}{\boldsymbol{v}};\boldsymbol{\theta},\boldsymbol{\mu},\boldsymbol{\chi})\}
\end{aligned}
$$

where $\{\boldsymbol{v}\}$, $\{\overset{\star}{\boldsymbol{v}}\}$ and $\{\overset{\star\vee}{\boldsymbol{v}}\}$ gather the unknown degrees of freedom of the state and its Lagrangian derivatives. The three systems of equations are governed by the same matrix operator $[\boldsymbol{K}]$. This, however, is true only provided the domain evolution is described by a transformation $\boldsymbol{\Phi}(\,;\boldsymbol{p})$ chosen such that the images $\boldsymbol{\Phi}(S_u;\boldsymbol{p})$ and $\boldsymbol{\Phi}(S_T;\boldsymbol{p})$ of S_u, S_T (the initial Dirichlet and Neumann portions of the boundary) are the Dirichlet and Neumann portions for the boundary-value problem on the transformed domain $\boldsymbol{\Phi}(\Omega;\boldsymbol{p})$.

Discretization of the tangential differential operators

Equations (A.13), (A.14) and (A.17) of Appendix A provide expressions for the operators $\boldsymbol{\nabla}_S$, div_S and D_{ij} in terms of a parametric representation. They are thus applicable to the interpolation of the surface and unknowns by shape functions:

$$
\boldsymbol{\nabla}_S f = \sum_{k=1}^{N} \left\{ N_{k,\alpha} g^{\alpha\beta} \boldsymbol{e}_\beta \right\}(\boldsymbol{\xi}) f^k \tag{12.50}
$$

$$
\mathrm{div}_S \boldsymbol{u} = \sum_{k=1}^{N} \left\{ N_{k,\alpha} g^{\alpha\beta} \boldsymbol{e}_\beta \right\}(\boldsymbol{\xi}) . \boldsymbol{u}^k \tag{12.51}
$$

$$
D_{ij} f \,\mathrm{d}S_y = \varepsilon_{ijk} \sum_{k=1}^{N} \boldsymbol{e}_k . \left\{ N_{k,1} \boldsymbol{a}^2 - N_{k,2} \boldsymbol{a}_1 \right\}(\boldsymbol{\xi}) f^k \,\mathrm{d}\boldsymbol{\xi} \tag{12.52}
$$

where $\boldsymbol{a}_{1,2}$ denote the vectors of the natural basis associated with the mapping (12.48) of a boundary element and $g^{\alpha\beta}$ are the contravariant components of the metric tensor. The latter thus satisfy $g^{\alpha\gamma}(\boldsymbol{a}_\gamma.\boldsymbol{a}_\beta) = \delta^\alpha_\beta$. Formulas (12.50), (12.51) and (12.52) above are suitable for the numerical evaluation of the integrals involving tangential differential operators which appear in the derivative integral equations.

Singular element integrals

Singular integration occurs when the collocation point $\boldsymbol{x}$ lies on the current element E. A typical singular element integral, arising from the discretization of eq. (12.45), is:

$$I^s = \int_E [\theta_r(\boldsymbol{y}) - \theta_r(\boldsymbol{x})]\,[\mu_q(y) - \mu_q(x)] G_{,ij}(\boldsymbol{x}, \boldsymbol{y}) D_{ij} u \,\mathrm{d}S_y \tag{12.53}$$

The polar coordinate mapping in the parent element introduced in Chapter 3 is applicable. The modified shape functions $\hat{N}(\rho, \alpha; \boldsymbol{p})$ are again introduced, so that:

$$\boldsymbol{\theta}(\boldsymbol{y}) - \boldsymbol{\theta}(\boldsymbol{x}) = \rho \sum_{k=1}^{N} \boldsymbol{y}^k_{,p_1} \hat{N}_k(\rho, \alpha; \boldsymbol{p})$$

$$\boldsymbol{\mu}(\boldsymbol{y}) - \boldsymbol{\mu}(\boldsymbol{x}) = \rho \sum_{k=1}^{N} \boldsymbol{y}^k_{,p_2} \hat{N}_k(\rho, \alpha; \boldsymbol{p})$$

In view of the $|\boldsymbol{y} - \boldsymbol{x}|^{-3}$ singularity of $G_{,ij}$, a nonsingular integral in (ρ, α) is eventually obtained.

12.8 DERIVATIVE BOUNDARY INTEGRAL EQUATIONS IN ELASTOSTATICS

The analysis conducted in Sections 12.4 and 12.5 in the framework of scalar potential problems applies to elastodynamics as well. Here only the main results pertaining to first-order and second-order derivative boundary integral equations are given; the reader is referred to [4].

The regularized displacement boundary integral equation (4.32) can be written (assuming for simplicity the absence of body forces and using the Kelvin elastic fundamental solution) as:

$$I_1(\boldsymbol{x}, \boldsymbol{u}) - I_2(\boldsymbol{x}, \boldsymbol{t}) = 0 \tag{12.54}$$

having put:

$$I_1(\boldsymbol{x}, \boldsymbol{u}) = \bar{\kappa} u_k(\boldsymbol{x}) + \int_{\partial\Omega} [u_i(\boldsymbol{y}) - u_i(\boldsymbol{x})] T^k_i(\boldsymbol{x}, \boldsymbol{y}) \,\mathrm{d}S_y$$

$$I_2(\boldsymbol{x}, \boldsymbol{t}) = \int_{\partial\Omega} t_i(\boldsymbol{y}) U^k_i(\boldsymbol{x}, \boldsymbol{y}) \,\mathrm{d}S_y$$

The first-order regularized derivative boundary integral equation reads:

$$I_1(\boldsymbol{x}, \overset{\star}{\boldsymbol{u}}) - I_2(\boldsymbol{x}, \overset{\star}{\boldsymbol{t}}) = -J_1(\boldsymbol{x}, \boldsymbol{u}; \boldsymbol{\theta}) + J_2(\boldsymbol{x}, \boldsymbol{t}; \boldsymbol{\theta}) \tag{12.55}$$

with

$$\begin{aligned} J_1(\boldsymbol{x}, \boldsymbol{u}; \boldsymbol{\theta}) &= \int_{\partial\Omega} [\theta_r(\boldsymbol{y}) - \theta_r(\boldsymbol{x})]\, D_{rj} u_i(\boldsymbol{y}) \Sigma_{ij}^k(\boldsymbol{x}, \boldsymbol{y})\, \mathrm{d}S_y \\ J_2(\boldsymbol{x}, \boldsymbol{t}; \boldsymbol{\theta}) &= \int_{\partial\Omega} t_i(\boldsymbol{y}) \left\{ \overset{\star}{U}{}_i^k(\boldsymbol{x}, \boldsymbol{y}) + U_i^k(\boldsymbol{x}, \boldsymbol{y}) D_r \theta_r(\boldsymbol{y}) \right\} \mathrm{d}S_y \end{aligned}$$

The second-order regularized derivative boundary integral equation subsequently reads:

$$\begin{aligned} I_1(\boldsymbol{x}, \overset{\star\vee}{\boldsymbol{u}}) - I_2(\boldsymbol{x}, \overset{\star\vee}{\boldsymbol{t}}) = {} & J_2(\boldsymbol{x}, \overset{\star}{\boldsymbol{t}}; \boldsymbol{\mu}) + J_2(\boldsymbol{x}, \overset{\vee}{\boldsymbol{t}}; \boldsymbol{\theta}) + J_2(\boldsymbol{x}, \boldsymbol{t}; \boldsymbol{\chi}) \\ & - J_1(\boldsymbol{x}, \overset{\star}{\boldsymbol{u}}; \boldsymbol{\mu}) - J_1(\boldsymbol{x}, \overset{\vee}{\boldsymbol{u}}; \boldsymbol{\theta}) - J_1(\boldsymbol{x}, \boldsymbol{u}; \boldsymbol{\chi}) \\ & + K_2(\boldsymbol{x}, \boldsymbol{t}; \boldsymbol{\theta}, \boldsymbol{\mu}) - K_1(\boldsymbol{x}, \boldsymbol{u}; \boldsymbol{\theta}, \boldsymbol{\mu}) \end{aligned} \tag{12.56}$$

where J_1, J_2 are as given above and K_1, K_2 are given by:

$$\begin{aligned} K_1(\boldsymbol{x}, \boldsymbol{u}; \boldsymbol{\theta}, \boldsymbol{\mu}) = {} & \int_{\partial\Omega} [\theta_r(\boldsymbol{y}) - \theta_r(\boldsymbol{x})]\, [\mu_q(y) - \mu_q(x)] \Sigma_{ij,q}^k(\boldsymbol{x}, \boldsymbol{y}) D_{rj} u_i(\boldsymbol{y})\, \mathrm{d}S_y \\ & + \int_{\partial\Omega} [\theta_r(\boldsymbol{y}) - \theta_r(\boldsymbol{x})]\, D_q \mu_q(\boldsymbol{y}) \Sigma_{ij}^k(\boldsymbol{x}, \boldsymbol{y}) D_{rj} u_i(\boldsymbol{y})\, \mathrm{d}S_y \\ & + \int_{\partial\Omega} [\theta_r(\boldsymbol{y}) - \theta_r(\boldsymbol{x})]\, D_j \mu_q(\boldsymbol{y}) D_{qr} u_i(\boldsymbol{y}) \Sigma_{ij}^k(\boldsymbol{x}, \boldsymbol{y})\, \mathrm{d}S_y \\ & + \int_{\partial\Omega} [\theta_r(\boldsymbol{y}) - \theta_r(\boldsymbol{x})]\, D_r \mu_q(\boldsymbol{y}) D_{jq} u_i(\boldsymbol{y}) \Sigma_{ij}^k(\boldsymbol{x}, \boldsymbol{y})\, \mathrm{d}S_y \\ K_2(\boldsymbol{x}, \boldsymbol{t}; \boldsymbol{\theta}, \boldsymbol{\mu}) = {} & \int_{\partial\Omega} t_i(\boldsymbol{y}) U_{i,qr}^k(\boldsymbol{x}, \boldsymbol{y})\, [\theta_r(\boldsymbol{y}) - \theta_r(\boldsymbol{x})]\, [\mu_q(y) - \mu_q(x)]\, \mathrm{d}S_y \\ & + \int_{\partial\Omega} t_i(\boldsymbol{y}) U_{i,r}^k(\boldsymbol{x}, \boldsymbol{y})\, [\theta_r(\boldsymbol{y}) - \theta_r(\boldsymbol{x})]\, D_q \mu_q(\boldsymbol{y})\, \mathrm{d}S_y \\ & + \int_{\partial\Omega} t_i(\boldsymbol{y}) U_{i,r}^k(\boldsymbol{x}, \boldsymbol{y}) [\mu_r(y) - \mu_r(x)] D_q \theta_q(\boldsymbol{y})\, \mathrm{d}S_y \\ & + \int_{\partial\Omega} t_i(\boldsymbol{y}) U_i^k(\boldsymbol{x}, \boldsymbol{y}) \{ n_p n_r D_q \mu_p D_q \theta_r \}(\boldsymbol{y})\, \mathrm{d}S_y \\ & + \int_{\partial\Omega} t_i(\boldsymbol{y}) U_i^k(\boldsymbol{x}, \boldsymbol{y})\, \{ D_r \theta_r D_q \mu_q - D_r \mu_q D_q \theta_r \}\, (\boldsymbol{y})\, \mathrm{d}S_y \end{aligned}$$

The symmetry in $(\boldsymbol{\theta}, \boldsymbol{\mu})$ of K_1 can be established following the same approach as in the scalar case.

12.9 THE ADJOINT STATE METHOD

Evaluating $\overset{\star}{J}$ by the direct differentiation approach relies on solving the derivative integral equation (12.32) or (12.55). This intermediary stage, which furnishes the boundary traces of the Lagrangian derivatives of the relevant field variables, must be performed for each variable geometrical parameter p_i considered, bearing in mind that each p_i is associated to a different transformation velocity field $\boldsymbol{\theta}_i$.

The adjoint state method, to be presented in this section, allows us to replace the computation of those intermediary computations with the solution of a well-chosen supplementary boundary-value problem: the *adjoint state*, which is unique whatever the number of geometrical parameters p_i considered.

12.9.1 Formulation of the adjoint state

The basic idea goes as follows: to evaluate the derivative of $\mathcal{J}(\Omega)$, the differentiation of $J(u, q, \Omega)$ is performed by treating (u, q, Ω) as *independent* variables, and enforcing the fact that u, q depend on p through problem (12.15)) explicitly as a *constraint*. This leads to the introduction of a Lagrangian $\mathcal{L}$:

$$\mathcal{L}(u,q,\Omega) = J(u,q,\Omega) + \int_{\Omega} \boldsymbol{\nabla} u.\boldsymbol{\nabla}\hat{u}\,\mathrm{d}\Omega - \int_{S_q} \bar{q}\hat{u}\,\mathrm{d}S - \int_{S_u} q\hat{u}\,\mathrm{d}S - \int_{S_u} (u-\bar{u})\hat{q}\,\mathrm{d}S \quad (12.57)$$

where the constraint term is the weak formulation of problem (12.15), the test functions $(\hat{u}, \hat{q})$ acting as Lagrange multipliers.

The derivative of the Lagrangian in a domain transformation is then expressed using formulas (12.9) and (12.11):

$$\begin{aligned}\overset{\star}{\mathcal{L}} = {} & \int_{S_q} \left[\varphi_{,u}\,\overset{\star}{u} + \varphi_{,y}.\boldsymbol{\theta} + \varphi \mathrm{div}_S\boldsymbol{\theta}\right] \mathrm{d}S_y + \int_{S_u} \left[\psi_{,q}\,\overset{\star}{q} + \psi_{,y}.\boldsymbol{\theta} + \psi \mathrm{div}_S\boldsymbol{\theta}\right] \mathrm{d}S_y \\ & + \int_{\Omega} \boldsymbol{\nabla}\overset{\star}{u}.\boldsymbol{\nabla}\hat{u}\,\mathrm{d}\Omega \\ & + \int_{\Omega} \left[(\boldsymbol{\nabla}u.\boldsymbol{\nabla}\hat{u})\mathrm{div}\boldsymbol{\theta} - \boldsymbol{\nabla}u.(\boldsymbol{\nabla}\hat{u}.\boldsymbol{\nabla}\boldsymbol{\theta}) - \boldsymbol{\nabla}\hat{u}.(\boldsymbol{\nabla}u.\boldsymbol{\nabla}\boldsymbol{\theta})\right] \mathrm{d}\Omega \\ & - \int_{S_q} \overset{\star}{\bar{q}}\,\hat{u}\,\mathrm{d}S - \int_{S_q} \bar{q}\hat{u}\mathrm{div}_S\boldsymbol{\theta}\,\mathrm{d}S - \int_{S_u} \overset{\star}{q}\,\hat{u}\,\mathrm{d}S - \int_{S_u} q\hat{u}\mathrm{div}_S\boldsymbol{\theta}\,\mathrm{d}S \\ & - \int_{S_u} (\overset{\star}{u} - \overset{\star}{\bar{u}})\hat{q}\,\mathrm{d}S - \int_{S_u} (u-\bar{u})\hat{q}\mathrm{div}_S\boldsymbol{\theta}\,\mathrm{d}S \end{aligned} \quad (12.58)$$

(the terms containing $(\hat{u})^{\star}$ and $(\hat{q})^{\star}$ vanish due to the constraint (12.15)).

At this point, the adjoint state is sought as a particular choice of the multipliers $(\hat{u}, \hat{q})$. Since the objective is to evaluate the effect of *geometrical* perturbations on the objective function, the derivative $\overset{\star}{\mathcal{L}}$ of the Lagrangian should be zero in the absence of any domain perturbation, i.e. for $\boldsymbol{\theta} = \boldsymbol{\nabla\theta} = \mathbf{0}$). Accordingly, the multipliers $(\hat{u}, \hat{q})$ are sought so that the conditions:

$$(\forall \overset{\star}{u}) \qquad \int_{\Omega} \boldsymbol{\nabla} \overset{\star}{u} . \boldsymbol{\nabla} \hat{u} \, d\Omega + \int_{S_q} \varphi_{,u} \overset{\star}{u} \, dS_y - \int_{S_u} \overset{\star}{u} \, \hat{q} \, dS = 0 \tag{12.59}$$

$$(\forall \overset{\star}{q}) \qquad \int_{S_u} (\psi_{,q} - \hat{u}) \overset{\star}{q} \, dS_y = 0 \tag{12.60}$$

hold. The latter are the statement, in weak form, of a boundary-value problem of the same general type as the initial problem (12.15) and whose local equations are:

$$\begin{cases} \Delta \hat{u} = 0 & (\text{in } \Omega(p)) \\ \hat{q} \equiv \dfrac{\partial \hat{u}}{\partial n} = -\varphi_{,u} & (\text{on } S_q(p)) \\ \hat{u} = \psi_{,q} & (\text{on } S_u(p)) \end{cases} \tag{12.61}$$

It is in particular important to note that the definition of the adjoint state depends on the objective function under consideration but does not depend on the transformation velocity $\boldsymbol{\theta}$, and hence is independent of the definition and the number of geometrical parameters considered.

Then, upon substituting the solutions (u_Ω, q_Ω) and $(\hat{u}_\Omega, \hat{q}_\Omega)$ of problems (12.15) and (12.61) into eq. (12.58), the derivative of $\mathcal{J}$ in *any* domain transformation is obtained as:

$$\begin{aligned} \overset{\star}{\mathcal{J}} = & \int_{S_q} \left[\varphi_{,y} . \boldsymbol{\theta} + \varphi \text{div}_S \boldsymbol{\theta} \right] dS_y + \int_{S_u} \left[\psi_{,y} . \boldsymbol{\theta} + \psi \text{div}_S \boldsymbol{\theta} \right] dS_y \\ & + \int_{\Omega} \left[(\boldsymbol{\nabla} u_\Omega . \boldsymbol{\nabla} \hat{u}_\Omega) \text{div} \boldsymbol{\theta} - \boldsymbol{\nabla} u_\Omega . (\boldsymbol{\nabla} \hat{u}_\Omega . \boldsymbol{\nabla\theta}) - \boldsymbol{\nabla} \hat{u}_\Omega . (\boldsymbol{\nabla} u_\Omega . \boldsymbol{\nabla\theta}) \right] d\Omega \\ & - \int_{S_q} \overset{\star}{q} \, \hat{u}_\Omega \, dS + \int_{S_u} \overset{\star}{\bar{u}} \, \hat{q}_\Omega \, dS - \int_{\partial\Omega} q_\Omega \hat{u}_\Omega \text{div}_S \boldsymbol{\theta} \, dS \end{aligned} \tag{12.62}$$

This formula involves a domain integral and is thus not well suited to a boundary element treatment. In fact, after some analytical manipulations using integration by parts, the domain integral in eq. (12.62) can be recast into:

$$\begin{aligned} & \int_{\Omega} \left[(\boldsymbol{\nabla} u_\Omega . \boldsymbol{\nabla} \hat{u}_\Omega) \text{div} \boldsymbol{\theta} - \boldsymbol{\nabla} u_\Omega . (\boldsymbol{\nabla} \hat{u}_\Omega . \boldsymbol{\nabla\theta}) - \boldsymbol{\nabla} \hat{u}_\Omega . (\boldsymbol{\nabla} u_\Omega . \boldsymbol{\nabla\theta}) \right] d\Omega \\ & \quad = \int_{\partial\Omega} \left[(\boldsymbol{\nabla} u_\Omega . \boldsymbol{\nabla} \hat{u}_\Omega)(\boldsymbol{\theta} . \boldsymbol{n}) - (\boldsymbol{\nabla} u_\Omega . \boldsymbol{\nabla\theta}) . \boldsymbol{\nabla} \hat{u}_\Omega - (\boldsymbol{\nabla} \hat{u}_\Omega . \boldsymbol{\nabla\theta}) . \boldsymbol{\nabla} u_\Omega \right] dS \end{aligned} \tag{12.63}$$

Substituting this identity into eq. (12.62) yields a boundary-only expression for $\overset{\star}{\mathcal{J}}$, which can thus be computed from the boundary traces of the primal and adjoint states.

Solution using the boundary element method

The foregoing development is not specifically made within a boundary integral equation framework; it is in fact independent of the particular technique being used to solve numerically the primal and adjoint problems (12.15) and (12.61). Both problems can be formulated as boundary integral equations; moreover, the same governing integral operator is involved in both integral equations. Introducing a boundary element discretization thus leads to linear systems of equations of the form:

$$[\boldsymbol{K}]\{\boldsymbol{v}\} = \{\boldsymbol{f}\}$$
$$[\boldsymbol{K}]\{\hat{\boldsymbol{v}}\} = \{\hat{\boldsymbol{f}}\}$$

so that, once the discrete primal state $\{\boldsymbol{v}\}$ is computed, the solution of the discrete adjoint problem relies on setting up *one* new right-hand side $\{\hat{\boldsymbol{f}}\}$ and performing one backsubstitution, since the factorization of the matrix $[\boldsymbol{K}]$ can be reused. The additional computational burden entailed by setting up and solving the adjoint problem is thus quite small, in relative terms (a fraction of the computational cost of the primal problem).

12.10 EXAMPLE: ENERGY RELEASE RATE IN FRACTURE MECHANICS

The *energy release rate* G, defined as the derivative of the potential energy at equilibrium W in extensions of a crack Γ (physically, G is the energy consumed in the creation of a unit crack area), plays an important role in fracture mechanics. Mathematically, G is the kernel function of the domain derivative of W in crack extensions (G is a function of the arc length s on the crack edge). The numerical evaluation of $G(s)$ thus constitutes one of the applications of derivative integral equations; this aspect is treated in more detail in Section 13.8.

Appendix 12.A DERIVATIVE OF THE DIFFERENTIAL AREA ELEMENT

Denote by S_p a surface material with respect to a transformation (12.1) and by $(\boldsymbol{a}, \boldsymbol{b})$ a pair of *material* vectors attached to a *material* point $\boldsymbol{y} = \boldsymbol{\Phi}(\boldsymbol{Y}; p)$. Moreover, $(\boldsymbol{a}, \boldsymbol{b})$ are supposed to be unitary and orthogonal for a given value p_0 of the shape parameter, and are further assumed to lie in the plane tangent to S_p at $\boldsymbol{y}(p)$ for any p in a neighbourhood of p_0. For any such p, the unit normal to S_p at $\boldsymbol{y}$ is thus given by:

$$\boldsymbol{n}(\boldsymbol{y}; p) = \frac{1}{|\boldsymbol{a} \wedge \boldsymbol{b}|}(\boldsymbol{a} \wedge \boldsymbol{b})$$

Moreover, the differential area element $\mathrm{d}S$ at $\boldsymbol{y}$ is proportional to the norm $|\boldsymbol{a} \wedge \boldsymbol{b}|$, so that:

$$\overset{\star}{\widehat{\mathrm{d}S}} = |\boldsymbol{a} \wedge \boldsymbol{b}|^{\star}\ \mathrm{d}S \tag{12.A.1}$$

The Lagrangian derivative of the vector product $\boldsymbol{a} \wedge \boldsymbol{b}$ is now evaluated at $p = p_0$, using eq. (12.6):

$$\begin{aligned}(\boldsymbol{a} \wedge \boldsymbol{b})^{\star} &= (\boldsymbol{\nabla\theta}.\boldsymbol{a}) \wedge \boldsymbol{b} + \boldsymbol{a} \wedge (\boldsymbol{\nabla\theta}.\boldsymbol{b}) \\ &= [(\boldsymbol{\nabla\theta})_{aa} + (\boldsymbol{\nabla\theta})_{bb}]\boldsymbol{n} - (\boldsymbol{\nabla\theta})_{na}\boldsymbol{a} - (\boldsymbol{\nabla\theta})_{nb}\boldsymbol{b} \\ &= (\mathrm{div}_S\boldsymbol{\theta})\boldsymbol{n} - \boldsymbol{n}.\boldsymbol{\nabla}_S\boldsymbol{\theta}\end{aligned}$$

where the fact that $(\boldsymbol{a}, \boldsymbol{b}, \boldsymbol{n})$ is an orthonormal frame for $p = p_0$ has been used (the surface gradient and divergence are defined in Appendix A). Moreover, one has:

$$|\boldsymbol{a} \wedge \boldsymbol{b}|^{\star} = \frac{\boldsymbol{a} \wedge \boldsymbol{b}}{|\boldsymbol{a} \wedge \boldsymbol{b}|}.(\boldsymbol{a} \wedge \boldsymbol{b})^{\star} = \boldsymbol{n}.(\boldsymbol{a} \wedge \boldsymbol{b})^{\star} = \mathrm{div}_S\boldsymbol{\theta}$$

As a result, the Lagrangian derivative of the unit normal $\boldsymbol{n}$ and the differential area element $\mathrm{d}S$ at $\boldsymbol{y} \in S_p$ are given by:

$$\overset{\star}{\mathrm{d}S} = \mathrm{div}_S\boldsymbol{\theta}\,\mathrm{d}S = D_r\theta_r\ \mathrm{d}S \qquad \overset{\star}{\boldsymbol{n}} = -\boldsymbol{n}.\boldsymbol{\nabla}_S\boldsymbol{\theta} = -n_r D_j\theta_r \boldsymbol{e}_j \tag{12.A.2}$$

Appendix 12.B PROOF OF IDENTITY (12.36)

The proof rests upon several integrations by parts. First, the classical divergence formula leads to:

$$\int_{\partial\Omega} A_i(\boldsymbol{y})n_i(\boldsymbol{y})G(\boldsymbol{x},\boldsymbol{y})\,\mathrm{d}S_y = \int_{\Omega} \{G(\boldsymbol{x},\boldsymbol{y})A_{i,i}(\boldsymbol{y}) + G_{,i}(\boldsymbol{x},\boldsymbol{y})A_i(\boldsymbol{y})\}\ \mathrm{d}V_y \tag{12.A.3}$$

Next, from the definition (5.24) of $\boldsymbol{A}$, the second integral in the above equation becomes:

$$I = \int_{\Omega} G_{,i}\ \{u_{,j}\theta_{j,i} - u_{,i}\theta_{j,j} + \theta_{j,i}u_{,j}\}\ \mathrm{d}V \tag{12.A.4}$$

(the arguments $\boldsymbol{x}, \boldsymbol{y}$ are omitted). Also, applying the general identity

$$\int_{\Omega} (fg_{,i}h_{,j} - fg_{,j}h_{,i})\,\mathrm{d}V = \int_{\Omega} (gh_{,i}f_{,j} - gh_{,j}f_{,i})\,\mathrm{d}V + \int_{\partial\Omega} fgD_{ij}h\,\mathrm{d}S$$

(which is established from the divergence formula), one shows that:

$$\begin{aligned}\int_{\Omega} \{u_{,j}\theta_{j,i} - u_{,i}\theta_{j,j}\} \, \mathrm{d}V &= \int_{\partial\Omega} D_{ij}u \, [\theta_j(\boldsymbol{y}) - \theta_j(\boldsymbol{x})] \, G_{,i} \, \mathrm{d}S \\ &\quad + \int_{\Omega} [\theta_j(\boldsymbol{y}) - \theta_j(\boldsymbol{x})] \, \{G_{,ij}u_{,i} - G_{,ii}u_{,j}\} \, \mathrm{d}V \\ &= \int_{\partial\Omega} D_{ij}u \, [\theta_j(\boldsymbol{y}) - \theta_j(\boldsymbol{x})] \, G_{,i} \, \mathrm{d}S \\ &\quad + \int_{\Omega} [\theta_j(\boldsymbol{y}) - \theta_j(\boldsymbol{x})] \, G_{,ij}u_{,i} \, \mathrm{d}V \end{aligned} \tag{12.A.5}$$

where the equality $\theta_{j,i}(\boldsymbol{y}) = [\theta_j(\boldsymbol{y}) - \theta_j(\boldsymbol{x})]_{,i}$ has been used for regularization purposes, as well as the equilibrium equation $G_{,ii} = 0$ $(\boldsymbol{y} \neq \boldsymbol{x})$. Moreover, one also has:

$$\begin{aligned}&\int_{\Omega} \theta_{j,i}u_{,j} \, \mathrm{d}V \\ &= \int_{\partial\Omega} G_{,j} \, [\theta_j(\boldsymbol{y}) - \theta_j(\boldsymbol{x})] \, q \, \mathrm{d}S - \int_{\Omega} \{G_{,ij}u_{,i} + G_{,j}u_{,ii}\} \, [\theta_j(\boldsymbol{y}) - \theta_j(\boldsymbol{x})] \, \mathrm{d}V \\ &= \int_{\partial\Omega} G_{,j} \, [\theta_j(\boldsymbol{y}) - \theta_j(\boldsymbol{x})] \, q \, \mathrm{d}S + \int_{\Omega} \{G_{,j}b - G_{,ij}u_{,i}\} \, [\theta_j(\boldsymbol{y}) - \theta_j(\boldsymbol{x})] \, \mathrm{d}V \\ &= \int_{\partial\Omega} \{q(\boldsymbol{y})G_{,j} \, [\theta_j(\boldsymbol{y}) - \theta_j(\boldsymbol{x})] + bG \, [\theta_j(\boldsymbol{y}) - \theta_j(\boldsymbol{x})] \, n_j\} \, \mathrm{d}S \\ &\quad - \int_{\Omega} \{b_{,j}G + G_{,ij}u_{,i}\} \, [\theta_j(\boldsymbol{y}) - \theta_j(\boldsymbol{x})] \, \mathrm{d}V \end{aligned} \tag{12.A.6}$$

(where the equilibrium equation $u_{,ii} + b = 0$ has been used). Finally, substituting equations (12.A.5–12.A.6) into the integral I of eq. (12.A.4) gives the result:

$$\begin{aligned}I = &\int_{\partial\Omega} q \, [\theta_j(\boldsymbol{y}) - \theta_j(\boldsymbol{x})] \, G_{,j} \, \mathrm{d}S - \int_{\partial\Omega} [\theta_i(\boldsymbol{y}) - \theta_i(\boldsymbol{x})] \, D_{ij}u \, \mathrm{d}S \\ &+ \int_{\partial\Omega} bG \, [\theta_j(\boldsymbol{y}) - \theta_j(\boldsymbol{x})] \, n_j \, \mathrm{d}S - \int_{\Omega} G(b \, [\theta_j(\boldsymbol{y}) - \theta_j(\boldsymbol{x})])_{,j} \, \mathrm{d}V\end{aligned}$$

Its substitution into (12.A.3) finally establishes identity (12.36).

Appendix 12.C EXPRESSION OF $(D_{ij}f)^{\star}$, $(D_i f)^{\star}$, $(D_i u_i)^{\star}$

Expression of $(D_{ij}f)^{\star}$

From the definition:

$$D_{ij}f = n_i f_{,j} - n_j f_{,i}$$

and rewriting equations (12.5), (12.8) in component notation:

$$(f_{,i})^\star = \overset{\star}{f}_{,i} - f_{,j}\theta_{j,i} \qquad \overset{\star}{n}_i = -n_j D_i\theta_j$$

one has:

$$\begin{aligned}(D_{ij}f)^\star &= n_i(\overset{\star}{f}_{,j} - f_{,k}\theta_{k,j}) - n_k D_i\theta_k f_{,j} - n_j(\overset{\star}{f}_{,i} - f_{,k}\theta_{k,i}) + n_k D_j\theta_k f_{,i} \\ &= D_{ij}\overset{\star}{f} - f_{,k}D_{ij}\theta_k + n_k D_j\theta_k f_{,i} - n_k D_i\theta_k f_{,j} \\ &= D_{ij}\overset{\star}{f} - f_{,k}D_{ij}\theta_k + D_j\theta_k(n_k f_{,i} - n_i f_{,k}) + n_i D_j\theta_k f_{,k} \\ &\quad - D_i\theta_k(n_k f_{,j} - n_j f_{,k}) - n_j D_i\theta_k f_{,k} \\ &= D_{ij}\overset{\star}{f} - f_{,k}D_{ij}\theta_k + f_{,k}D_{ij}\theta_k + D_j\theta_k D_{ki}f + D_i\theta_k D_{jk}f\end{aligned}$$

Hence the result:

$$(D_{ij}f)^\star = D_{ij}\overset{\star}{f} + D_j\theta_k D_{qr}f + D_r\theta_k D_{sq}f \qquad (12.A.7)$$

Expression of $(D_i f)^\star$

One notices that:

$$-n_i D_{ij}f = n_j(n_j D_i f - n_i D_i f) = D_i f$$

since $\nabla_S f.\boldsymbol{n} = 0$ from definition (A.4) of the tangential gradient. Consequently, equations (12.8) and (12.A.7) can be combined, giving:

$$(D_i f)^\star = -n_j(D_{ij}f)^\star - \overset{\star}{n}_j D_{ij}f = -n_j(D_{ij}f)^\star + n_k D_j\theta_k n_i D_j f$$

and thus the sought-after expression:

$$(D_i f)^\star = D_i\overset{\star}{f} - D_i\theta_k D_k f + n_k D_j\theta_k n_i D_j f \qquad (12.A.8)$$

Expression of $(D_i u_i)^\star$

Applying the previous result (12.A.8) in the special case $f = u_i$ and summing with respect to i, one obtains (in component notation):

$$(D_i u_i)^\star = D_i\overset{\star}{u}_i - D_i\theta_k D_k u_i + n_k D_j\theta_k n_i D_j u_i \qquad (12.A.9)$$

or, equivalently (in intrinsic notation):

$$(\text{div}_S\boldsymbol{u})^\star = \text{div}_S\overset{\star}{\boldsymbol{u}} - \nabla_S\boldsymbol{\theta} : \nabla_S\boldsymbol{u} - \overset{\star}{\boldsymbol{n}}.(\boldsymbol{n}.\nabla_S\boldsymbol{u}) \qquad (12.A.10)$$

Appendix 12.D SYMMETRIC EXPRESSION FOR $K_1(x, u; \theta, \mu)$

Recall that originally (i.e. before an integration by parts is performed), the quantity $J_1(\boldsymbol{x}, \boldsymbol{u}; \boldsymbol{\theta})$ is given by eq. (12.29):

$$J_1(\boldsymbol{x}, \boldsymbol{u}; \boldsymbol{\theta}) = \int_{\partial\Omega} [u(\boldsymbol{y}) - u(\boldsymbol{x})] \times \{[\theta_i(\boldsymbol{y}) - \theta_i(\boldsymbol{x})]\, n_j(\boldsymbol{y}) G_{,ij}(\boldsymbol{x}, \boldsymbol{y}) - D_{ij}\theta_i(\boldsymbol{y}) G_{,j}(\boldsymbol{x}, \boldsymbol{y})\} \, \mathrm{d}S_y$$

Lagrangian differentiation of this expression using formula (12.11) and equations (12.7), (12.28) then yields:

$$\begin{aligned}
\overset{\vee}{J}_1(\boldsymbol{x}, \boldsymbol{u}; \boldsymbol{\theta}) = & \int_{\partial\Omega} [\overset{\vee}{u}(\boldsymbol{y}) - \overset{\vee}{u}(\boldsymbol{x})]\, [\theta_i(\boldsymbol{y}) - \theta_i(\boldsymbol{x})]\, n_j(\boldsymbol{y}) (G_{,j})^\star(\boldsymbol{x}, \boldsymbol{y}) \, \mathrm{d}S_y \\
& - \int_{\partial\Omega} [\overset{\vee}{u}(\boldsymbol{y}) - \overset{\vee}{u}(\boldsymbol{x})] D_{ij}\theta_i(\boldsymbol{y}) G_{,j}(\boldsymbol{x}, \boldsymbol{y}) \, \mathrm{d}S_y \\
& + \int_{\partial\Omega} [u(\boldsymbol{y}) - u(\boldsymbol{x})][\chi_i(\boldsymbol{y}) - \chi_i(\boldsymbol{x})] n_j(\boldsymbol{y}) G_{,ij}(\boldsymbol{x}, \boldsymbol{y}) \, \mathrm{d}S_y \\
& - \int_{\partial\Omega} [u(\boldsymbol{y}) - u(\boldsymbol{x})](D_{ij}\theta_i)^\vee(\boldsymbol{y}) G_{,j}(\boldsymbol{x}, \boldsymbol{y}) \, \mathrm{d}S_y \\
& + \int_{\partial\Omega} [u(\boldsymbol{y}) - u(\boldsymbol{x})]\, [\theta_i(\boldsymbol{y}) - \theta_i(\boldsymbol{x})]\, n_j(\boldsymbol{y}) (G_{,j})^\vee(\boldsymbol{x}, \boldsymbol{y}) \, \mathrm{d}S_y \\
& + \int_{\partial\Omega} [u(\boldsymbol{y}) - u(\boldsymbol{x})]\, [\theta_i(\boldsymbol{y}) - \theta_i(\boldsymbol{x})]\, (G_{,j})^\star(\boldsymbol{x}, \boldsymbol{y}) D_{jk}\mu_k(\boldsymbol{y}) \, \mathrm{d}S_y \\
& - \int_{\partial\Omega} [u(\boldsymbol{y}) - u(\boldsymbol{x})] D_{ij}\theta_i(\boldsymbol{y}) G_{,j}(\boldsymbol{x}, \boldsymbol{y}) D_k\mu_k(\boldsymbol{y}) \, \mathrm{d}S_y \\
& - \int_{\partial\Omega} [u(\boldsymbol{y}) - u(\boldsymbol{x})] D_{ij}\theta_i(\boldsymbol{y}) (G_{,j})^\vee(\boldsymbol{x}, \boldsymbol{y}) \, \mathrm{d}S_y
\end{aligned}$$

Also, from eq. (12.A.7):

$$(D_{ij}\theta_i)^\vee = D_{ij}\chi_i + D_j\theta_k D_{ki}\theta_i + D_i\theta_k D_{jk}\theta_i$$

which, after suitable term rearrangement, leads to the result:

$$\overset{\vee}{J}_1(\boldsymbol{x}, \boldsymbol{u}; \boldsymbol{\theta}) = J_1(\boldsymbol{x}, \boldsymbol{u}; \boldsymbol{\chi}) + J_1(\boldsymbol{x}, \overset{\vee}{\boldsymbol{u}}; \boldsymbol{\theta}) + K_1(\boldsymbol{x}, \boldsymbol{u}; \boldsymbol{\theta}, \boldsymbol{\mu})$$

with:

$$\begin{aligned}
K_1(\boldsymbol{x}, u; \boldsymbol{\theta}, \boldsymbol{\mu}) = & \int_{\partial\Omega} [u(\boldsymbol{y}) - u(\boldsymbol{x})] n_j(\boldsymbol{y}) (G_{,j})^{\star\vee}(\boldsymbol{x}, \boldsymbol{y}) \, \mathrm{d}S_y \\
& + \int_{\partial\Omega} [u(\boldsymbol{y}) - u(\boldsymbol{x})] (G_{,j})^\star(\boldsymbol{x}, \boldsymbol{y}) D_{jk}\mu_j(\boldsymbol{y}) \, \mathrm{d}S_y \\
& + \int_{\partial\Omega} [u(\boldsymbol{y}) - u(\boldsymbol{x})] (G_{,j})^\vee(\boldsymbol{x}, \boldsymbol{y}) D_{jk}\theta_k(\boldsymbol{y}) \, \mathrm{d}S_y \\
& - \int_{\partial\Omega} [u(\boldsymbol{y}) - u(\boldsymbol{x})] G_{,j}(\boldsymbol{x}, \boldsymbol{y}) d_j(\boldsymbol{y}) \, \mathrm{d}S_y
\end{aligned}$$

and

$$d_j = n_i \left(D_j \theta_i D_k \mu_k + D_j \mu_i D_k \theta_k - D_j \mu_k D_k \theta_i - D_j \theta_k D_k \mu_i \right) + n_j \left(D_k \theta_i D_i \mu_k - D_i \theta_i D_k \mu_k \right)$$

The symmetry of K_1 in $(\boldsymbol{\theta}, \boldsymbol{\mu})$ is now apparent. It is however preferable in practice to use the more compact form (12.44) of K_1 in numerical applications.

REFERENCES

1. Barone, M. R., Yang, R. J. A boundary element approach for recovery of shape sensitivities in threedimensional elastic solids. *Comp. Meth. in Appl. Mech. Engng.*, **74**, 69–82 (1989).
2. Bonnet, M. A numerical approach for shape identification problems using BIE and shape differentiation. In F. Paris, C. A. Brebbia, J. Dominguez (eds), *Boundary elements XIV (vol. 2: stress analysis, computational aspects)*, pp. 541–553. Comp. Mech. Publ., Southampton (1992).
3. Bonnet, M. Equations intégrales variationnelles pour le problème en vitesse de propagation de fissures en élasticité linéaire. *C.R. Acad. Sci. Paris, série II,*, **318**, 429–434 (1994).
4. Bonnet, M. Regularized BIE formulations for first- and second-order shape sensitivity of elastic fields. *Computers and Structures*, **56**, 799–811 (1995). (Invited paper, special issue, S. Saigal, guest editor).
5. Burczynski, T., Fedelinski, P. Boundary elements in shape design sensitivity analysis and optimal design of vibrating structures. *Engng. Anal. with Bound. Elem.*, **9**, 195–201 (1992).
6. Choi, J. O., Kwak, B. M. Boundary integral equation method for shape optimization of elastic structures. *Int. J. Num. Meth. in Eng.*, **26**, 1579–1595 (1988).
7. Haug, E. J., Choi, K. K., Komkov, V. *Design sensitivity analysis of structural systems*. Academic Press (1986).
8. Kane, J. H., Zhao, G., Wang, H., Guru Prasad, K. Boundary formulations for three-dimensional continuum structural shape sensitivity analysis. *ASME J. Appl. Mech.*, **59**, 827–834 (1992).
9. Meric, R. A. Shape optimization of thermoelastic solids. *J. Therm. Sc.*, **11**, 187–206 (1988).
10. Mialon, P. Calcul de la dérivée d'une grandeur par rapport à un fond de fissure par la méthode θ. Bulletin EDF/DER série c vol. 3, Electricité de France (1987).
11. Mukherjee, S., Chandra, A. A boundary element formulation for design sensitivities in problems involving both geometric and material nonlinearities. *Mathl. Comput. Modelling*, **15**, 245–255 (1991).

12. Nishimura, N, Kobayashi, S. A boundary integral equation method for an inverse problem related to crack detection. *Int. J. Num. Meth. in Eng.*, **32**, 1371–1387 (1991).

13. Petryk, H., Mroz, Z. Time derivatives of integrals and functionals defined on varying volume and surface domains. *Arch. Mech.*, **38**, 694–724 (1986).

14. Salençon, J. *Mécanique des milieux continus (vols. 1,2,3)*. Presses de l'Ecole Polytechnique, Palaiseau, France (1992).

15. Simon, J. Second variations for domain optimization problems. In F. Kappel, K. Kunisch, W. Schappacher (eds), *Control theory of distributed parameter systems and applications*, vol. 91 of *International Series of Numerical Mathematics*, pp. 361–378. Birkhäuser Verlag, Basel (1989).

16. Sokolowski, J., Zolesio, J. P. *Introduction to shape optimization. Shape sensitivity analysis*, vol. 16 of *Springer series in Computational Mathematics*. Springer-Verlag (1992).

17. Wei, X., Leu, L. J., Chandra, A., Mukher jee, S. Shape optimization in elasticity and elastoplasticity. *Int. J. Solids Struct.*, **31**, 533–550 (1994).

18. Zeng, X., Saigal, S. An inverse formulation with boundary elements. *ASME J. Appl. Mech.*, **59**, 835–840 (1992).

19. Zhang, Q., Mukherjee, S. Second-order design sensitivity analysis for linear elastic problems by the derivative boundary element method. *Comp. Meth. in Appl. Mech. Engng.*, **86**, 321–335 (1991).

Part IV

Solid mechanics

13

Linear fracture mechanics

Boundary element methods are frequently applied to fracture mechanics analyses. In addition to dimensionality considerations, boundary element methods lead to accurate evaluations of local quantities (e.g. stress intensity factors) in the vicinity of the crack front. Also, numerical simulations of crack propagation is made much easier by the fact that only the crack surface mesh, which for internal cracks is completely disconnected from the external boundary, needs to be updated. This advantage is made even greater by the fact that the mesh must be refined around the crack front because of the singular behaviour of the mechanical fields.

13.1 INAPPLICABILITY OF THE DISPLACEMENT BOUNDARY INTEGRAL EQUATION

Consider an elastic body Ω of external boundary S, containing a crack Γ (Figure 13.1). The two crack faces Γ^+, Γ^- are geometrically identical surfaces having opposite unit normal vectors $\boldsymbol{n}^+$, $\boldsymbol{n}^-$ chosen so that $\boldsymbol{n}^-$ points from S^- to S^+ (this choice is consistent with the usual convention whereby the unit normal points towards the *exterior of the medium*). The surface $\Gamma = \Gamma^+ = \Gamma^-$ is assumed to be twice piecewise continuously differentiable.

Assume for the sake of argument that a given load is applied to the whole external boundary: $S_T = S$. The displacement $\boldsymbol{u}(\boldsymbol{x})$ at an observation point

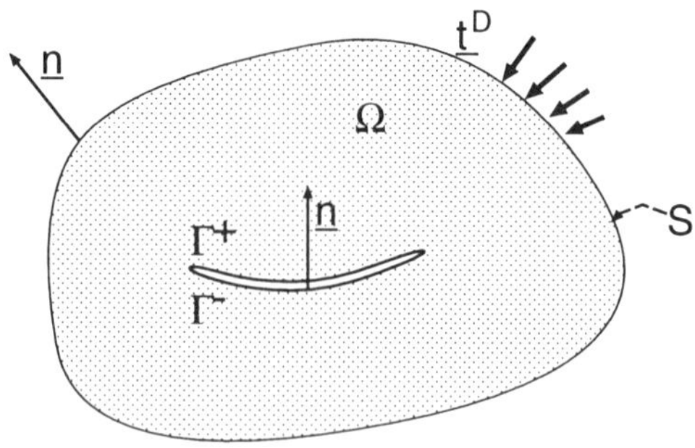

Figure 13.1 Cracked body.

$\boldsymbol{x}$ interior to the body Ω reads:

$$u_k(\boldsymbol{x}) = \int_S \{u_i(\boldsymbol{y})T_i^k(\boldsymbol{x},\boldsymbol{y}) - t_i^D(\boldsymbol{y})U_i^k(\boldsymbol{x},\boldsymbol{y})\} \, \mathrm{d}S_y + \int_\Gamma \phi_i(\boldsymbol{y})T_i^k(\boldsymbol{x},\boldsymbol{y}) \, \mathrm{d}S_y \quad (13.1)$$

($\boldsymbol{\phi}(\boldsymbol{y}) = \boldsymbol{u}^+(\boldsymbol{y}) - \boldsymbol{u}^-(\boldsymbol{y})$: displacement jump across the crack surface). This expression is obtained by applying the representation formula (4.21) for a body containing an infinitely thin cavity limited by the (slightly apart) surfaces $\Gamma^\pm$; the Kelvin traction vector T_i^k refers implicitly to the normal $\boldsymbol{n} = \boldsymbol{n}^-$.

Then, consider the displacement $\hat{\boldsymbol{u}}$ generated *in the uncracked solid* by the loading $\boldsymbol{t}^D$ applied on S_T. It is expressed at any point $\boldsymbol{x}$ of Ω by the integral representation:

$$\hat{u}_k(\boldsymbol{x}) = \int_S \{\hat{u}_i(\boldsymbol{y})T_i^k(\boldsymbol{x},\boldsymbol{y}) - t_i^D(\boldsymbol{y})U_i^k(\boldsymbol{x},\boldsymbol{y})\} \, \mathrm{d}S_y \quad (13.2)$$

Also, $\hat{\boldsymbol{u}}(\boldsymbol{y})$ is continuous across the surface Γ. Upon subtracting eq. (13.2) from eq. (13.1), one obtains:

$$\Delta u_k(\boldsymbol{x}) = \int_S \Delta u_i(\boldsymbol{y})T_i^k(\boldsymbol{x},\boldsymbol{y}) \, \mathrm{d}S_y + \int_\Gamma \phi_i(\boldsymbol{y})T_i^k(\boldsymbol{x},\boldsymbol{y}) \, \mathrm{d}S_y \quad (13.3)$$

where $\Delta\boldsymbol{u} = \boldsymbol{u} - \hat{\boldsymbol{u}}$ is the displacement variation caused by the presence of the crack, the external loading $\boldsymbol{t}^D$ being kept fixed. Equation (13.3) is linear in $\Delta\boldsymbol{u}$ and homogeneous, and thus admits infinitely many solutions $\Delta\boldsymbol{u}$. Equation (13.1) thus also admits infinitely many solutions $\boldsymbol{u} = \hat{\boldsymbol{u}} + \Delta\boldsymbol{u}$ and is therefore not exploitable. The consideration of other types of boundary data on the boundary S would have led to essentially the same discussion (which is an extension to bounded bodies of that made in the introductory section of Chapter 5) for any boundary conditions on the external boundary S.

Two distinct approaches for formulating boundary integral equations for cracked bodies thus present themselves. The *multiregion approach* consists in splitting Ω into two fictitious subdomains chosen such that Γ lies in their common interface. The usual displacement boundary integral equation is then invoked separately for each subdomain. This results in solving two coupled elastic problems, in the fashion of Section 3.7. The *displacement discontinuity approach*, on the other hand, avoids the coupling but relies on traction boundary integral equations, which were introduced in Chapter 5.

It is useful to point out that fundamental solutions whose definition incorporates the presence of a crack (with stress-free faces) are known for a few special geometrical configurations (e.g. an infinite plane with a straight crack, see Snyder and Cruse [18]). In such cases, the above discussion is meaningless.

13.2 THE MULTIREGION APPROACH

It consists in splitting the cracked body Ω into two complementary subdomains Ω^+ and Ω^- chosen so that the crack lies in the interface, which is thus of the form $\Gamma \cup \tilde{\Gamma}$, where $\tilde{\Gamma}$ is an arbitrarily chosen complementary surface (Figure 13.2). The cracked body problem is then recast into two coupled elasticity subproblems for non-cracked bodies, defined by the following boundary conditions for each subdomain $\Omega^\pm$:

$$\begin{aligned} \boldsymbol{u} &= \boldsymbol{u}^D && \text{on } S_u^\pm \\ \boldsymbol{t} &= \boldsymbol{t}^D && \text{on } S_T^\pm \\ \boldsymbol{t} &= 0 && \text{on } \Gamma^\pm \end{aligned} \tag{13.4}$$

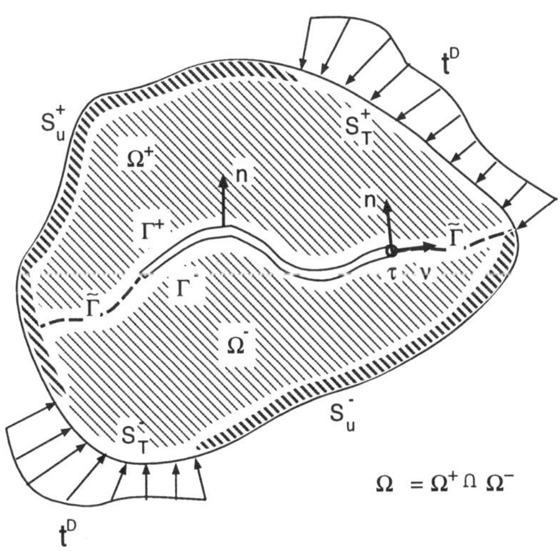

Figure 13.2 Cracked body Ω: the multiregion approach.

and interface conditions (continuity between Ω^+ and Ω^-):

$$\begin{aligned} \boldsymbol{u}^+ &= \boldsymbol{u}^- \\ \boldsymbol{t}^+ &= \boldsymbol{t}^- \end{aligned} \qquad \text{on } \tilde{\Gamma} \tag{13.5}$$

Each subproblem can be formulated in terms of the displacement boundary integral equation (4.32), which is simpler to implement than the traction integral equation. On the other hand, one must solve two coupled integral equations, and the introduction of the fictitious complementary interface $\tilde{\Gamma}$ results in additional nodal unknowns. Moreover, special modelling of the traction vector on $\tilde{\Gamma}$ must be introduced in the crack front region.

Crack in a plane of symmetry

For particular situations where Ω admits a plane of symmetry Π, the external loading $\boldsymbol{t}^D$ is symmetrical with respect to Π and Γ is planar and lies in Π, the analysis reduces to solving the subproblem on Ω^+ (or Ω^-) in which the interface conditions are replaced by the boundary conditions:

$$\begin{aligned} \boldsymbol{u}^+ . \boldsymbol{n} &= 0 \\ \boldsymbol{t}^+ - (\boldsymbol{t}^+ . \boldsymbol{n})\boldsymbol{n} &= 0 \end{aligned} \qquad \text{on } (\Pi \cap \Omega) - \Gamma \tag{13.6}$$

When the external loading is not symmetric with respect to Π, the concepts introduced in Chapter 11 are applicable: the loading is decomposed into symmetric and antisymmetric parts, and suitable boundary conditions on $\tilde{\Gamma}$ must be introduced for the antisymmetric subproblem.

13.3 DISPLACEMENT DISCONTINUITY APPROACH: CRACKS OF ARBITRARY SHAPE IN INFINITE MEDIA

In order to introduce the main features of the displacement discontinuity approach, this section is restricted to infinite media, the extension to bounded bodies being discussed in the next section. The crack faces $\Gamma^{\pm}$, which constitute the only boundary, are loaded with prescribed tractions $\boldsymbol{t}^{D\pm}$. For such boundary conditions, which generally result from the superposition principle (the given load being initially applied at infinity), the mechanical fields are expected to decay at infinity and thus admit representation formulas expressed as integrals over Γ.

Let the surface $\tilde{\Gamma}$ be chosen so that $\Gamma \cup \tilde{\Gamma}$ is a *closed* smooth surface of exterior unit normal $\boldsymbol{n}$. Denote by Ω^- the bounded domain enclosed by $\Gamma \cup \tilde{\Gamma}$ and let Ω^+ be the unbounded complement: $\Omega^+ = \mathbb{R}^3 - \Omega^-$. The respective unit normals are $\boldsymbol{n}^- = \boldsymbol{n}$ and $\boldsymbol{n}^+ = -\boldsymbol{n}$. The prescribed loading on the crack faces and the continuity conditions on the complementary surface impose that the elastostatic variables $(\boldsymbol{u}^-, \boldsymbol{\sigma}^-)$ on Ω^- and $(\boldsymbol{u}^+, \boldsymbol{\sigma}^+)$ on Ω^+ are linked on

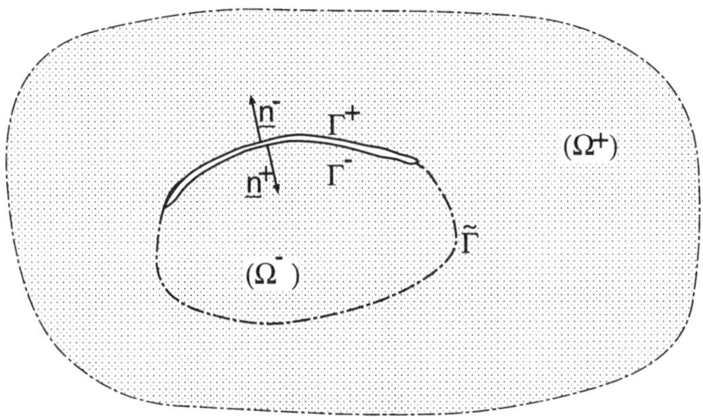

Figure 13.3 Crack in an infinite elastic medium and auxiliary subdomains: geometrical notation.

$\Gamma \cup \tilde{\Gamma}$, satisfy the following conditions:

$$\begin{cases} (\boldsymbol{\sigma}.\boldsymbol{n})^+ = \boldsymbol{t}^{D+} & (\text{on } \Gamma^+) \\ (\boldsymbol{\sigma}.\boldsymbol{n})^- = \boldsymbol{t}^{D-} & (\text{on } \Gamma^-) \end{cases} \qquad \begin{cases} \boldsymbol{u}^+ = \boldsymbol{u}^- \\ (\boldsymbol{\sigma}.\boldsymbol{n})^+ = -(\boldsymbol{\sigma}.\boldsymbol{n})^- \end{cases} (\text{on } \tilde{\Gamma}) \quad (13.7)$$

Then, the regularized integral representation formula (5.34) for the stress tensor at a point $\boldsymbol{x}$ located on the crack surface and relative to each subdomain $\Omega^\pm$ reads:

$$\begin{aligned} \frac{1}{2}\sigma_{ij}^{\pm}(\boldsymbol{x}) = C_{ijk\ell} \int_\Gamma \left[D_{\ell b}^{\pm} u_a^{\pm}(\boldsymbol{y}) - D_{\ell b}^{\pm} u_a^{\pm}(\boldsymbol{x})\right] \Sigma_{ab}^k(\boldsymbol{x}, \boldsymbol{y}) \, \mathrm{d}S_y \\ - \int_\Gamma [t_k^{\pm}(\boldsymbol{y}) - t_k^{\pm}(\boldsymbol{x})] \Sigma_{ij}^k(\boldsymbol{x}, \boldsymbol{y}) \, \mathrm{d}S_y \\ + C_{ijk\ell} A_{ab}^k(\boldsymbol{x}, \Gamma \cup \tilde{\Gamma}) D_{\ell b}^{\pm} u_a^{\pm}(\boldsymbol{x}) - t_k^{\pm}(\boldsymbol{x}) A_{ij}^k(\boldsymbol{x}, \Gamma \cup \tilde{\Gamma}) \end{aligned} \quad (13.8)$$

where the choice of unit normal in the operator $D_{\ell b}$ is emphasized by the notation $D_{\ell b}^{\pm}$. All integrals are convergent. The $A_{ab}^k(\boldsymbol{x}, \cdot)$ are given by eq. (5.A.1).

Since the two normals $\boldsymbol{n}^\pm$ have opposite signs and the $A_{pq}^k(\boldsymbol{x}, \Gamma \cup \tilde{\Gamma})$ do not involve normals, taking the previous equation for Ω^- and Ω^+ and adding the resulting identities results into the identity:

$$\begin{aligned} \frac{1}{2}[\sigma_{ij}^+ + \sigma_{ij}^-](\boldsymbol{x}) = -C_{ijk\ell} \int_{\Gamma \cup \tilde{\Gamma}} \left[D_{\ell b}\phi_a(\boldsymbol{y}) - D_{\ell b}\phi_a(\boldsymbol{x})\right] \Sigma_{ab}^k(\boldsymbol{x}, \boldsymbol{y}) \, \mathrm{d}S_y \\ - \int_{\Gamma \cup \tilde{\Gamma}} [\psi_k(\boldsymbol{y}) - \psi_k(\boldsymbol{x})] \Sigma_{ij}^k(\boldsymbol{x}, \boldsymbol{y}) \, \mathrm{d}S_y \\ - C_{ijk\ell} A_{ab}^k(\boldsymbol{x}, \Gamma \cup \tilde{\Gamma}) D_{\ell b}\phi_a(\boldsymbol{x}) - \psi_k(\boldsymbol{x}) A_{ij}^k(\boldsymbol{x}, \Gamma \cup \tilde{\Gamma}) \end{aligned}$$

having put $\phi_i = u_i^+ - u_i^-$ (displacement discontinuity across $\Gamma \cup \tilde{\Gamma}$) and $\psi_i = t_i^+ + t_i^-$ (traction discontinuity across $\Gamma \cup \tilde{\Gamma}$), while $D_{\ell b}$ refers to the normal

$\boldsymbol{n} = \boldsymbol{n}^-$. Then, upon incorporating conditions (13.7), separating integrations over Γ and $\tilde{\Gamma}$ and noticing that:

$$A^k_{ab}(\boldsymbol{x}, \Gamma \cup \tilde{\Gamma}) = A^k_{ab}(\boldsymbol{x}, \Gamma) + A^k_{ab}(\boldsymbol{x}, \tilde{\Gamma})$$

one obtains:

$$\begin{aligned}
\frac{1}{2}[\sigma^+_{ij} + \sigma^-_{ij}](\boldsymbol{x}) = &-C_{ijk\ell} \int_\Gamma [D_{\ell b}\phi_a(\boldsymbol{y}) - D_{\ell b}\phi_a(\boldsymbol{x})]\, \Sigma^k_{ab}(\boldsymbol{x}, \boldsymbol{y})\, \mathrm{d}S_y \\
&- \int_\Gamma [\psi_k(\boldsymbol{y}) - \psi_k(\boldsymbol{x})] \Sigma^k_{ij}(\boldsymbol{x}, \boldsymbol{y})\, \mathrm{d}S_y \\
&- C_{ijk\ell} A^k_{ab}(\boldsymbol{x}, \Gamma) D_{\ell b}\phi_a(\boldsymbol{x}) - \psi_k(\boldsymbol{x}) A^k_{ij}(\boldsymbol{x}, \Gamma)
\end{aligned}$$

Then, the inner product of this equation with a unit normal to Γ should be taken in order to make the traction vector appear. It is easy to see that choosing $\boldsymbol{n}^+$ or $\boldsymbol{n}^-$ leads to the same equality (with sign reversed). Putting $\boldsymbol{n} = \boldsymbol{n}^-$ and $\boldsymbol{t}^D = \boldsymbol{t}^{D-}$, one has:

$$\frac{1}{2}[\sigma^+_{ij} + \sigma^-_{ij}](\boldsymbol{x}) n_j(\boldsymbol{x}) = t^D_i(\boldsymbol{x}) - \frac{1}{2}\psi_i(\boldsymbol{x})$$

and this manipulation results in the following *traction boundary integral equation*:, which relates the tractions $\boldsymbol{t}^D, \boldsymbol{t}^{D+} = \boldsymbol{\psi} - \boldsymbol{t}^D$ on the crack faces to the displacement discontinuity $\boldsymbol{\phi}$:

$$\begin{aligned}
t^D_i(\boldsymbol{x}) = &-n_j(\boldsymbol{x}) C_{ijk\ell} A^k_{ab}(\boldsymbol{x}, \Gamma) D_\ell \phi_a(\boldsymbol{x}) - \psi_k(\boldsymbol{x}) \Big(A^k_{ij}(\boldsymbol{x}, \Gamma) n_j(\boldsymbol{x}) - \frac{1}{2}\delta_{ik} \Big) \\
&- n_j(\boldsymbol{x}) C_{ijk\ell} \int_\Gamma [D_{\ell b}\phi_a(\boldsymbol{y}) - D_{\ell b}\phi_a(\boldsymbol{x})]\, \Sigma^k_{ab}(\boldsymbol{x}, \boldsymbol{y})\, \mathrm{d}S_y \\
&- n_j(\boldsymbol{x}) \int_\Gamma [\psi_k(\boldsymbol{y}) - \psi_k(\boldsymbol{x})] \Sigma^k_{ij}(\boldsymbol{x}, \boldsymbol{y})\, \mathrm{d}S_y
\end{aligned} \tag{13.9}$$

Once $\boldsymbol{\phi}, \boldsymbol{\psi}$ are known, the displacement at points lying outside the crack surface is given by the integral representation formula:

$$u_k(\boldsymbol{x}) = \int_\Gamma \{\phi_i(\boldsymbol{y}) T^k_i(\boldsymbol{x}, \boldsymbol{y}) + \psi_i(\boldsymbol{y}) U^k_i(\boldsymbol{x}, \boldsymbol{y})\}\ \mathrm{d}S_y$$

from which the other mechanical variables (strain, stress) at $\boldsymbol{x}$ can be expressed as well.

Equation (13.9) is in fact the representation on Γ of the stress generated by an elastic double-layer potential (Section 5.6), whose density $\boldsymbol{\phi}$ here possesses a clear physical meaning, and has thus a direct link with eq. (5.38).

It is important to point out that, since the displacement must be continuous outside the crack surface, the displacement discontinuity $\boldsymbol{\phi}$ necessarily vanishes on the crack front $\partial\Gamma$.

For symmetrically loaded crack faces ($\boldsymbol{t}^{D+} + \boldsymbol{t}^{D-} = 0$, i.e. $\boldsymbol{\psi} = 0$), the integral equation (13.9) reduces to:

$$t_i^D(\boldsymbol{x}) = -C_{ijk\ell}n_j(\boldsymbol{x})A_{ab}^k(\boldsymbol{x}, G)D_{\ell b}\phi_a \\ - C_{ijk\ell}n_j(\boldsymbol{x}) \int_\Gamma [D_{\ell b}\phi_a(\boldsymbol{y}) - D_{\ell b}\phi_a(\boldsymbol{x})] \, \Sigma_{ab}^k(\boldsymbol{x}, \boldsymbol{y}) \, \mathrm{d}S_y \quad (13.10)$$

Extension to elastodynamics

The displacement discontinuity approach is also applicable for the integral equation of elastodynamics (Chapters 7, 8), in either time or frequency domain. The resulting regularized traction boundary integral equations are given here for infinite media and symmetrical loadings ($\boldsymbol{\psi} = 0$). Extensions to other situations are left to the reader. In the time domain, the integral equation relating the loading $\boldsymbol{t}^D(\boldsymbol{x}, t)$ (t: collocation time instant) to the displacement discontinuity history $\boldsymbol{\phi}(\boldsymbol{y}, \tau)$ ($0 \leq \tau \leq t$) is:

$$t_i^D(\boldsymbol{x}, t) = -C_{ijk\ell}n_j(\boldsymbol{x})D_{\ell b}\phi_a(\boldsymbol{x}, t)A_{ab}^k(\boldsymbol{x}, \Gamma) \\ - C_{ijk\ell}n_j(\boldsymbol{x}) \int_\Gamma \left[\Sigma_{ab}^k[\boldsymbol{x}, t, \boldsymbol{y}|D_{\ell b}\phi_a(\boldsymbol{y}, t)] - D_{\ell b}\phi_a(\boldsymbol{x}, t)\Sigma_{ab}^k(\boldsymbol{x}, \boldsymbol{y})\right] \, \mathrm{d}S_y \\ - C_{ijk\ell}n_j(\boldsymbol{x})\rho \int_\Gamma U_a^k[\boldsymbol{x}, t, \boldsymbol{y}|\ddot{\phi}_a(\boldsymbol{y}, t)]n_\ell(\boldsymbol{y}) \, \mathrm{d}S_y \quad (13.11)$$

In the frequency domain, the integral equation relating the loading $\boldsymbol{t}^D(\boldsymbol{x})$, applied with a given angular frequency ω, to the displacement discontinuity $\boldsymbol{\phi}(\boldsymbol{y})$ is:

$$t_i^D(\boldsymbol{x}) = -C_{ijk\ell}n_j(\boldsymbol{x})D_{\ell b}\phi_a(\boldsymbol{x})A_{ab}^k(\boldsymbol{x}, \Gamma) \\ - C_{ijk\ell}n_j(\boldsymbol{x}) \int_\Gamma \left[D_{\ell b}\phi_a(\boldsymbol{y})\Sigma_{ab}^k(\boldsymbol{x}, \boldsymbol{y}; \omega) - D_{\ell b}\phi_a(\boldsymbol{x})\Sigma_{ab}^k(\boldsymbol{x}, \boldsymbol{y})\right] \, \mathrm{d}S_y \\ + \rho\omega^2 C_{ijk\ell}n_j(\boldsymbol{x}) \int_\Gamma U_a^k(\boldsymbol{x}, \boldsymbol{y}; \omega)\phi_a(\boldsymbol{y})n_\ell(\boldsymbol{y}) \, \mathrm{d}S_y \quad (13.12)$$

13.4 DISPLACEMENT DISCONTINUITY APPROACH: PLANAR CRACKS IN INFINITE MEDIA

Assume that the crack surface Γ lies in the $(\boldsymbol{e}_1, \boldsymbol{e}_2)$ plane. For any two points $\boldsymbol{x}, \boldsymbol{y}$ of Γ, one then has:

$$\boldsymbol{n}(\boldsymbol{x}) = \boldsymbol{n}(\boldsymbol{y}) = \boldsymbol{e}_3 \\ r_{,3} = r_{,n} = 0 \quad (13.13) \\ D_{\ell b}f = \delta_{\ell 3}f_{,b} - \delta_{b3}f_{,\ell}$$

The elastostatic traction integral equation (13.10) then becomes:

$$\begin{aligned} t_i^D(\boldsymbol{x}) = &-C_{i3k3}\int_\Gamma [\phi_{a,\alpha}(\boldsymbol{y}) - \phi_{a,\alpha}(\boldsymbol{x})]\,\Sigma^k_{a\alpha}(\boldsymbol{x},\boldsymbol{y})\,\mathrm{d}S_y \\ &+ C_{i3k\alpha}\int_\Gamma [\phi_{a,\alpha}(\boldsymbol{y}) - \phi_{a,\alpha}(\boldsymbol{x})]\,\Sigma^k_{a3}(\boldsymbol{x},\boldsymbol{y})\,\mathrm{d}S_y \\ &- C_{i3k3}\phi_{a,\alpha}(\boldsymbol{x})A^k_{a\alpha}(\boldsymbol{x},\Gamma) + C_{i3k\alpha}\phi_{a,\alpha}(\boldsymbol{x})A^k_{a3}(\boldsymbol{x},\Gamma) \end{aligned} \tag{13.14}$$

where the index α range is $\{1,2\}$. When the elastic properties of the medium are isotropic (the Hooke tensor being thus given by eq. (1.18)), it is interesting to establish the explicit forms taken by eq. (13.14) for tensile or shear loadings.

Tensile loading

It is the so-called mode I situation ($i = 3$) in fracture mechanics terminology. In that case, one has:

$$C_{33k3} = (\lambda + 2\mu)\delta_{k3} \qquad C_{33k\alpha} = \lambda\delta_{k\alpha}$$

so that one obtains:

$$\begin{aligned} t_3^D(\boldsymbol{x}) = &-(\lambda+2\mu)\int_S [\phi_{a,\alpha}(\boldsymbol{y}) - \phi_{a,\alpha}(\boldsymbol{x})]\,\Sigma^3_{a\alpha}(\boldsymbol{x},\boldsymbol{y})\,\mathrm{d}S_y \\ &\mp \lambda\int_S [\phi_{a,\alpha}(\boldsymbol{y}) - \phi_{a,\alpha}(\boldsymbol{x})]\,\Sigma^\alpha_{a3}(\boldsymbol{x},\boldsymbol{y})\,\mathrm{d}S_y \\ &- (\lambda+2\mu)\phi_{a,\alpha}(\boldsymbol{x})A^3_{a\alpha}(\boldsymbol{x},\Gamma) + \lambda\phi_{a,\alpha}(\boldsymbol{x})A^\alpha_{a3}(\boldsymbol{x},\Gamma) \end{aligned}$$

Also, expression (4.9) of the Kelvin solution implies that:

$$\Sigma^3_{a\alpha}(\boldsymbol{x},\boldsymbol{y}) = -\Sigma^\alpha_{a3}(\boldsymbol{x},\boldsymbol{y}) = -\frac{1-2\nu}{8\pi(1-\nu)}\frac{r_{,\alpha}}{r^2}\delta_{a3}$$

and hence:

$$A^a_{\alpha 3}(\boldsymbol{x},\Gamma) = -A^\alpha_{a3}(\boldsymbol{x},\Gamma) = -\frac{1-2\nu}{8\pi(1-\nu)}\delta_{a3}I_\alpha(\boldsymbol{x},\Gamma)$$

The regularized traction boundary integral equation relating the applied tensile load t_3^D to the normal displacement discontinuity $\phi \equiv \phi_3$ across Γ is readily obtained (using $\lambda + \mu = \mu/(1-2\nu)$) as:

$$t_3^D(\boldsymbol{x}) = \frac{\mu}{4\pi(1-\nu)}\left\{\int_S [\phi_{,\alpha}(\boldsymbol{y}) - \phi_{,\alpha}(\boldsymbol{x})]\,r_{,\alpha}\frac{\mathrm{d}S_y}{r^2} + \phi_{,\alpha}(\boldsymbol{x})I_\alpha(\boldsymbol{x},\Gamma)\right\} \tag{13.15}$$

Shear loading

It is the mode II-III situation ($i = 1, 2$). In that case, one has:

$$C_{\beta 3k3} = \mu\delta_{k\beta} \qquad C_{\beta 3k\alpha} = \mu\delta_{\alpha\beta}\delta_{k3}$$

so that one obtains:

$$\begin{aligned} t^D_\beta(\boldsymbol{x}) = &-\mu\int_S [\phi_{a,\alpha}(\boldsymbol{y}) - \phi_{a,\alpha}(\boldsymbol{x})]\,\Sigma^\beta_{a\alpha}(\boldsymbol{x},\boldsymbol{y})\,\mathrm{d}S_y - \mu\phi_{a,\alpha}(\boldsymbol{x})A^\beta_{a\alpha}(\boldsymbol{x},\Gamma) \\ &+\mu\int_S [\phi_{a,\beta}(\boldsymbol{y}) - \phi_{a,\beta}(\boldsymbol{x})]\,\Sigma^3_{a3}(\boldsymbol{x},\boldsymbol{y})\,\mathrm{d}S_y + \mu\phi_{a,\beta}(\boldsymbol{x})A^3_{a3}(\boldsymbol{x},\Gamma) \end{aligned}$$

Also, expression (4.9) of the Kelvin solution implies that:

$$\begin{aligned} \Sigma^3_{3\alpha}(\boldsymbol{x},\boldsymbol{y}) &= -\frac{1-2\nu}{8\pi(1-\nu)}\frac{r_{,\alpha}}{r^2} \\ \Sigma^\beta_{a\alpha}(\boldsymbol{x},\boldsymbol{y}) &= -\frac{1}{8\pi(1-\nu)r^2}\left\{3r_{,a}r_{,\alpha}r_{,\beta} + (1-2\nu)(\delta_{a\beta}r_{,\alpha} + \delta_{\alpha\beta}r_{,a} - \delta_{a\alpha}r_{,\beta})\right\} \end{aligned}$$

In particular, eq. $(13.13)_2$ implies $\Sigma^\beta_{3\alpha}(\boldsymbol{x},\boldsymbol{y}) = 0$ for $\boldsymbol{x},\boldsymbol{y} \in \Gamma$. One then has:

$$\begin{aligned} A^3_{3\alpha}(\boldsymbol{x},\Gamma) &= -\frac{1-2\nu}{8\pi(1-\nu)}I_\alpha(\boldsymbol{x},\Gamma) \\ A^\beta_{\gamma\alpha}(\boldsymbol{x},\boldsymbol{y}) &= -\frac{3}{8\pi(1-\nu)}J_{\alpha\beta\gamma}(\boldsymbol{x},\Gamma) \\ &\quad - \frac{1-2\nu}{8\pi(1-\nu)}[\delta_{\gamma\beta}I_\alpha(\boldsymbol{x},\Gamma) + \delta_{\alpha\beta}I_\gamma(\boldsymbol{x},\Gamma) - \delta_{\gamma\alpha}I_\beta(\boldsymbol{x},\Gamma)] \end{aligned}$$

The regularized traction boundary integral equation obtained as a result relates the applied shear load t^D_β to the tangential displacement discontinuity ϕ_α across Γ:

$$\begin{aligned} t^D_\beta(\boldsymbol{x}) - &\frac{3\mu}{8\pi(1-\nu)}\left\{\int_S r_{,\alpha}r_{,\beta}r_{,\gamma}[\phi_{\alpha,\gamma}(\boldsymbol{y}) \quad \phi_{\alpha,\gamma}(\boldsymbol{x})]\frac{\mathrm{d}S_y}{r^2}\right. \\ &+\frac{\mu(1-2\nu)}{8\pi(1-\nu)}\int_S r_{,\beta}[\phi_{\alpha,\alpha}(\boldsymbol{y}) - \phi_{\alpha,\alpha}(\boldsymbol{x})]\frac{\mathrm{d}S_y}{r^2} \\ &-\frac{\mu(1-2\nu)}{8\pi(1-\nu)}\int_S r_{,\alpha}[\phi_{\beta,\alpha}(\boldsymbol{y}) - \phi_{\beta,\alpha}(\boldsymbol{x})]\frac{\mathrm{d}S_y}{r^2} \\ &+\frac{3\mu}{8\pi(1-\nu)}\phi_{\alpha,\gamma}(\boldsymbol{x})J_{\alpha\beta\gamma}(\boldsymbol{x},\Gamma) \\ &+\frac{\mu(1-2\nu)}{8\pi(1-\nu)}[\phi_{\alpha,\alpha}(\boldsymbol{x})I_\beta(\boldsymbol{x},\Gamma) - \phi_{\beta,\alpha}(\boldsymbol{x})I_\alpha(\boldsymbol{x},\Gamma)] \end{aligned} \qquad (13.16)$$

Comments

The integral equations (13.15), (13.16) are the regularized versions of CPV-singular integral equations proposed by Weaver [23] and Bui [2]. They govern the normal and tangential parts of the displacement discontinuity, respectively, and are uncoupled.

The auxiliary integrals $I_\alpha(\boldsymbol{x},\Gamma)$ and $J_{\alpha\beta\gamma}(\boldsymbol{x},\Gamma)$ appearing in eqs (13.15), (13.16) are expressed in terms of contour integrals only (see Appendix 5.A):

$$I_\alpha(\boldsymbol{x},\Gamma) = -\int_{\partial S} \nu_\alpha \frac{\mathrm{d}s_y}{r} \tag{13.17}$$

$$3J_{\alpha\beta\gamma}(\boldsymbol{x},\Gamma) = \delta_{\alpha\beta} I_\gamma(\boldsymbol{x},\Gamma) + \delta_{\beta\gamma} I_\alpha(\boldsymbol{x},\Gamma) - \int_{\partial S} r_{,\alpha} r_{,\beta} \nu_{,\gamma} \frac{\mathrm{d}s_y}{r} \tag{13.18}$$

This makes the implementation of the regularization much easier than in the general case of non-planar cracks (compare with the general expressions (5.31), (5.32) in Appendix 5.A).

Frequency-domain elastodynamics

The traction boundary integral equation for the normal opening (mode I) case reads:

$$\begin{aligned} t_3^D(\boldsymbol{x}) = &-\frac{\mu}{2\pi}\int_S \phi_{3,\alpha}(\boldsymbol{y})\left[\frac{1-\nu}{1-2\nu}D - \frac{\nu}{1-2\nu}E + \frac{1}{1-\nu}\right] r_{,\alpha}\frac{\mathrm{d}S_y}{r^2} \\ &+ \frac{\mu}{4\pi(1-\nu)}\left\{\int_S [\phi_{3,\alpha}(\boldsymbol{y}) - \phi_{3,\alpha}(\boldsymbol{x})]\, r_{,\alpha}\frac{\mathrm{d}S_y}{r^2} + \phi_{3,\alpha}(\boldsymbol{x}) I_\alpha(\boldsymbol{x},\Gamma)\right\} \\ &+ \frac{\mu(1-\nu)}{2\pi(1-2\nu)} k_T^2 \int_S \phi_3(\boldsymbol{y})[A + \mathrm{e}^{\mathrm{i}k_T r}]\frac{\mathrm{d}S_y}{r} \end{aligned} \tag{13.19}$$

where A, D, E are given by eq. (8.37). The combination:

$$\frac{1-\nu}{1-2\nu}D - \frac{\nu}{1-2\nu}E + \frac{1}{1-\nu}$$

is weakly singular and should be expressed by means of the expansions (8.38) in order to evaluate its integral in the singular case.

13.5 DISPLACEMENT DISCONTINUITY APPROACH FOR A BOUNDED BODY

Consider now a body Ω externally bounded by a closed surface S and containing a crack. The integral formulation in this case consists of a juxtaposition of the traction integral equation (for collocation points on Γ) and the displacement integral equation (for collocation points on S).

In order to do this, introduce an auxiliary surface $\tilde{\Gamma}$ lying inside the domain Ω and chosen again so that $\Gamma \cup \tilde{\Gamma}$ is closed and smooth, and let $\Omega = \Omega^- \cup \Omega^+$, where the subdomain Ω^- is bounded by $\Gamma \cup \tilde{\Gamma}$ (Figure 13.4).

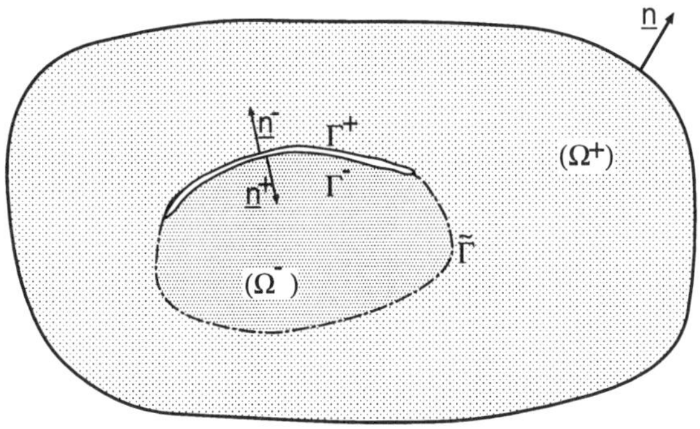

Figure 13.4 Crack in a bounded elastic body and auxiliary subdomains: geometrical notation.

First, consider the displacement representation formula (4.21) for the domain Ω^- (of boundary $\partial\Omega^- = \Gamma \cup \tilde{\Gamma}$) and a collocation point $\boldsymbol{x}$ lying on S, and thus in particular exterior to Ω^-:

$$0 = \int_{\Gamma \cup \tilde{\Gamma}} \left\{ u_i^-(\boldsymbol{y}) T_i^k(\boldsymbol{x}, \boldsymbol{y}) - t_i^-(\boldsymbol{y}) U_i^k(\boldsymbol{x}, \boldsymbol{y}) \right\} \, \mathrm{d}S_y$$

and the displacement integral equation (4.32) for the domain Ω^+ (of boundary $\partial\Omega^+ = S \cup \Gamma \cup \tilde{\Gamma}$) and the same collocation point $\boldsymbol{x} \in S$:

$$\int_{\Gamma \cup \tilde{\Gamma}} \left\{ u_i^+(\boldsymbol{y}) T_i^k(\boldsymbol{x}, \boldsymbol{y}) - t_i^+(\boldsymbol{y}) U_i^k(\boldsymbol{x}, \boldsymbol{y}) \right\} \, \mathrm{d}S_y \\ + \int_S \left\{ [u_i^+(\boldsymbol{y}) - u_i^+(\boldsymbol{x})] T_i^k(\boldsymbol{x}, \boldsymbol{y}) - t_i^+(\boldsymbol{y}) U_i^k(\boldsymbol{x}, \boldsymbol{y}) \right\} \, \mathrm{d}S_y = 0$$

Adding these two relations gives:

$$\int_S \left\{ [u_i(\boldsymbol{y}) - u_i(\boldsymbol{x})] T_i^k(\boldsymbol{x}, \boldsymbol{y}) - t_i(\boldsymbol{y}) U_i^k(\boldsymbol{x}, \boldsymbol{y}) \right\} \, \mathrm{d}S_y \\ - \int_\Gamma \left\{ \phi_i(\boldsymbol{y}) T_i^k(\boldsymbol{x}, \boldsymbol{y}) + \psi_i^+(\boldsymbol{y}) U_i^k(\boldsymbol{x}, \boldsymbol{y}) \right\} \, \mathrm{d}S_y = 0 \quad (13.20)$$

The traction boundary integral equation for $\boldsymbol{x} \in \Gamma$ is established following the steps of Section 13.3, the only difference being that the term

$$- \int_S \left\{ D_{\ell b} u_a(\boldsymbol{y}) C_{ijk\ell} \Sigma_{ab}^k(\boldsymbol{x}, \boldsymbol{y}) - t_k(\boldsymbol{y}) \Sigma_{ij}^k(\boldsymbol{x}, \boldsymbol{y}) \right\} \, \mathrm{d}S_y$$

should be added to the right-hand side of eq. (13.8) when applied for the subdomain Ω^+. The resulting regularized integral equation is:

$$\begin{aligned} t_i^D(\boldsymbol{x}) = -n_j(\boldsymbol{x}) C_{ijk\ell} \Big\{ & A_{ab}^k(\boldsymbol{x}, \Gamma) D_{\ell b}\phi_a(\boldsymbol{x}) \\ & + \int_\Gamma [D_{\ell b}\phi_a(\boldsymbol{y}) - D_{\ell b}\phi_a(\boldsymbol{x})] \, \Sigma_{ab}^k(\boldsymbol{x}, \boldsymbol{y}) \, \mathrm{d}S_y \Big\} \\ - n_j(\boldsymbol{x}) & \Big\{ A_{ij}^k(\boldsymbol{x}, \Gamma)\psi_k(\boldsymbol{x}) + \int_\Gamma [\psi_k(\boldsymbol{y}) - \psi_k(\boldsymbol{x})] \, \Sigma_{ij}^k(\boldsymbol{x}, \boldsymbol{y}) \, \mathrm{d}S_y \Big\} \\ + n_j(\boldsymbol{x}) & \int_S \{ D_{\ell b} u_a(\boldsymbol{y}) C_{ijk\ell} \Sigma_{ab}^k(\boldsymbol{x}, \boldsymbol{y}) - t_k(\boldsymbol{y}) \Sigma_{ij}^k(\boldsymbol{x}, \boldsymbol{y}) \} \, \mathrm{d}S_y \end{aligned} \tag{13.21}$$

The displacement integral equation (13.20) at collocation points $\boldsymbol{x} \in S$ and the traction integral equation (13.21) at collocation points $\boldsymbol{x} \in \Gamma$ together constitute the elastostatic integral formulation for the cracked body. The traction integral equation, necessary to avoid degeneracy, is introduced only where necessary, i.e. for collocation points on the crack surface; the displacement integral equation, which is simpler, is invoked otherwise.

The 'dual' formulation on the crack

One notes that the kinematical unknown on Γ in the system (13.20)–(13.21) is the displacement discontinuity $\boldsymbol{\phi} = \boldsymbol{u}^+ - \boldsymbol{u}^-$, not the individual displacements $\boldsymbol{u}^+, \boldsymbol{u}^-$ on the crack faces. This is usually of no consequence, as the usual fracture mechanics parameters, e.g. stress intensity factors (see next section) or energy release rate depend on $\boldsymbol{\phi}$, not on $\boldsymbol{u}^+, \boldsymbol{u}^-$ separately.

Nevertheless, in a variant to the approach described here called the 'dual' boundary element method for fracture mechanics problems (see e.g. Mi and Aliabadi [11, 12]), the individual displacements $\boldsymbol{u}^+, \boldsymbol{u}^-$, not the crack opening displacement $\boldsymbol{\phi}$, are taken as unknowns on the crack surface. Since in this case *two* vector fields are unknown on Γ, the system (13.20)–(13.21) becomes undetermined and is supplemented for collocation points $\boldsymbol{x} \in \Gamma$ with the displacement integral equation

$$\begin{aligned} u_k^+(\boldsymbol{x}) + \int_S \Big\{ u_i(\boldsymbol{y}) T_i^k(\boldsymbol{x}, \boldsymbol{y}) - t_i(\boldsymbol{y}) U_i^k(\boldsymbol{x}, \boldsymbol{y}) \Big\} \, \mathrm{d}S_y \\ - \int_\Gamma \Big\{ [\phi_i(\boldsymbol{y}) - \phi_i(\boldsymbol{x})] T_i^k(\boldsymbol{x}, \boldsymbol{y}) - \psi_i(\boldsymbol{y}) U_i^k(\boldsymbol{x}, \boldsymbol{y}) \Big\} \, \mathrm{d}S_y = 0 \end{aligned}$$

13.6 EVALUATION OF STRESS INTENSITY FACTORS

The boundary element discretization process uses the techniques introduced in Chapters 3 and 5. The crack surface Γ is the geometrical support of the crack opening displacement $\boldsymbol{\phi}$. The discretization of Γ and $\boldsymbol{\phi}$ must enforce $\boldsymbol{\phi} = \mathbf{0}$ on the crack front $\partial\Gamma$ and take into account the restrictions concerning the location of the collocation points or the choice of interpolation functions in connection with the traction integral equation (Chapter 5).

Also, the singular behaviour of strains and elastic stresses at the crack front is well-known (Bui [3]). Let the point $\boldsymbol{y}_0$ lie on the crack front $\partial\Gamma$ and introduce the right-handed orthonormal frame $(\boldsymbol{\tau}, \boldsymbol{n}, \boldsymbol{\nu})$ ($\boldsymbol{n}$: normal to Γ, $\boldsymbol{\tau}$: tangent to $\partial\Gamma$, $\boldsymbol{\nu} = \boldsymbol{\tau} \wedge \boldsymbol{n}$: normal to $\partial\Gamma$ exterior to Γ and lying in the tangent plane to Γ at $\boldsymbol{y}_0$), see Figure 13.5. The asymptotic behaviour of the displacement at a point $\boldsymbol{y} = \boldsymbol{y}_0 + d(\cos\alpha\boldsymbol{\nu} + \sin\alpha\boldsymbol{n})$ located at a small distance d away from $\boldsymbol{y}_0$ is:

$$
\begin{aligned}
u_\nu(\boldsymbol{y}) &= \frac{1}{2\mu}\sqrt{\frac{d}{2\pi}}\Big\{ K_I(\boldsymbol{y}_0)\cos\frac{\alpha}{2}(k - \cos\alpha) \\
&\qquad + K_{II}(\boldsymbol{y}_0)\sin\frac{\alpha}{2}(k + \cos\alpha + 2)\Big\} + O(d) \\
u_n(\boldsymbol{y}) &= \frac{1}{2\mu}\sqrt{\frac{d}{2\pi}}\Big\{ K_I(\boldsymbol{y}_0)\sin\frac{\alpha}{2}(k - \cos\alpha) \\
&\qquad - K_{II}(\boldsymbol{y}_0)\cos\frac{\alpha}{2}(k + \cos\alpha - 2)\Big\} + O(d) \qquad (13.22) \\
u_\tau(\boldsymbol{y}) &= \frac{2}{\mu}\sqrt{\frac{d}{2\pi}} K_{III}(\boldsymbol{y}_0)\sin\frac{\alpha}{2} + O(d)
\end{aligned}
$$

with $k = 3 - 4\nu$. The strains and stresses are therefore $d^{-1/2}$-singular on the crack front. The *stress intensity factors* K_I, K_{II}, K_{III} depend on the geometry and loading and vary along the crack front. The crack may propagate when some function of the K's reaches a threshold which is a physical characteristic of the constitutive material; e.g. when the stress intensity factors reach the *toughness* K_c. Moreover, fatigue (crack advance under a large number of loading cycles) is also often described by criteria in terms of the stress intensity factors.

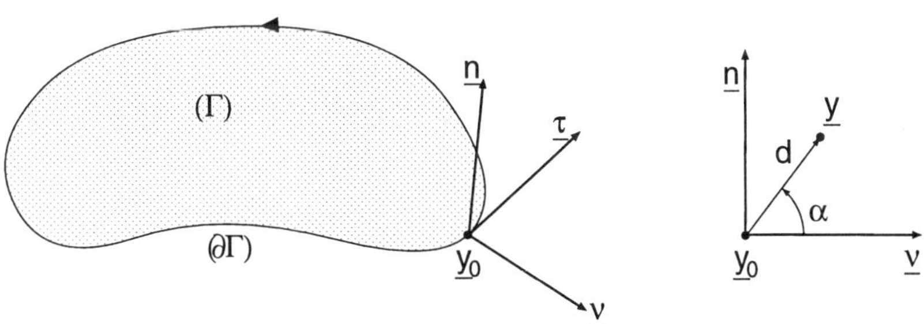

Figure 13.5 Geometrical notations for points in the vicinity of the crack front.

A correct evaluation of K_I, K_{II}, K_{III} at any point of the crack front is thus often of great practical importance. Analytical and semi-analytical solutions have been extensively investigated, see e.g. Tada et al. [19]. In the context of boundary element methods, they are generally computed using *kinematical extrapolation*: formulas (13.22) for $\alpha = \pm\pi$ yield the displacement $\boldsymbol{u}^{\pm}$ on the upper and lower crack faces in the vicinity of the crack front, and one obtains:

$$\begin{aligned}
\phi_n &= K_I \frac{4(1-\nu)}{\mu}\sqrt{\frac{d}{2\pi}} + O(d) \\
\phi_\nu &= K_{II} \frac{4(1-\nu)}{\mu}\sqrt{\frac{d}{2\pi}} + O(d) \\
\phi_\tau &= K_{III} \frac{4}{\mu}\sqrt{\frac{d}{2\pi}} + O(d)
\end{aligned} \tag{13.23}$$

The stress intensity factors can thus be expressed using the crack opening displacement:

$$\begin{aligned}
K_I &= \lim_{d\to 0} \frac{\mu}{4(1-\nu)}\sqrt{\frac{2\pi}{d}}\phi_n \\
K_{II} &= \lim_{d\to 0} \frac{\mu}{4(1-\nu)}\sqrt{\frac{2\pi}{d}}\phi_\nu \\
K_{III} &= \lim_{d\to 0} \frac{\mu}{4}\sqrt{\frac{2\pi}{d}}\phi_\tau
\end{aligned} \tag{13.24}$$

The simplest numerical evaluation method thus consists in applying the above formulas at the nodes closest to (but not on) the crack front; using notations of Figure 13.6, approximate values of K_I at nodes 1,2,3 are thus:

$$K_I^{7,8,1} \approx \frac{\mu}{4(1-\nu)}\sqrt{\frac{2\pi}{d_{6,9,2}}}\phi_n^{6,9,2}$$

The accuracy of this technique is not always satisfactory, due to the fact that the usual interpolations of $\boldsymbol{\phi}$ do not reproduce the $\sqrt{d}$ variation near $\partial\Gamma$. For this reason, *quarter-node* elements have been proposed (initially for finite element methods). To fix our ideas, consider a plane, nine-noded boundary element adjacent to $\partial\Gamma$, and let the nodes $\boldsymbol{y}^1, \boldsymbol{y}^2, \boldsymbol{y}^3$ be aligned on a line perpendicular to $\partial\Gamma$ and placed so that $\boldsymbol{y}^2$ is the quarter-point of the interval $[\boldsymbol{y}^1, \boldsymbol{y}^3]$ of length $4a$:

$$\boldsymbol{y}^2 - \boldsymbol{y}^1 = -a\boldsymbol{\nu} \qquad \boldsymbol{y}^3 - \boldsymbol{y}^1 = -4a\boldsymbol{\nu} \tag{13.25}$$

Using such an element, the geometrical interpolation along $[\boldsymbol{y}^1, \boldsymbol{y}^3]$ is quadratic:

$$\boldsymbol{y} = N_1(\xi_1)\boldsymbol{y}^1 + N_2(\xi_1)\boldsymbol{y}^2 + N_3(\xi_1)\boldsymbol{y}^3$$

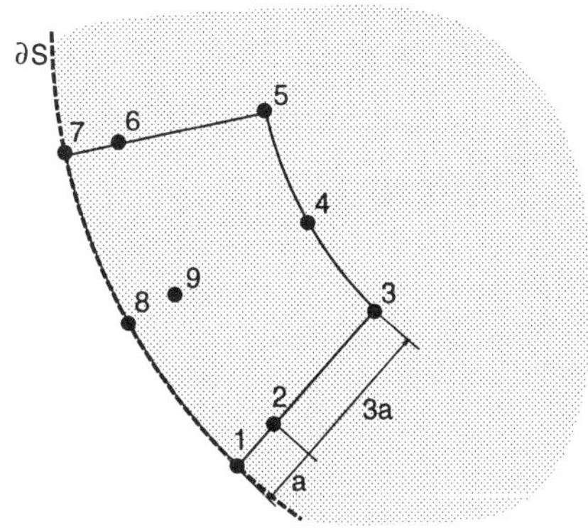

Figure 13.6 Nine-noded element with quarter nodes 2,6,9.

with the interpolation functions (B.1). Taken into account the specific placing (13.25) of $\boldsymbol{y}^2$, one gets:

$$\boldsymbol{y} - \boldsymbol{y}^1 = -(1+\xi)^2 a\boldsymbol{\nu}$$

and the distance $d = \left|\boldsymbol{y} - \boldsymbol{y}^1\right|$ evaluated in terms of element parameters is:

$$d = a(1+\xi)^2 \tag{13.26}$$

Also, the interpolation of a component ϕ of the crack-opening displacement vector is:

$$\phi(\boldsymbol{y}) = N_2(\xi)\phi^2 + N_3(\xi)\phi^3 = (1+\xi)[\phi^2 + (\frac{1}{2}\phi^3 - \phi^2)\xi] \tag{13.27}$$

having used that $\phi^1 = 0$ (ϕ^1 being associated with a node on the crack front). Eliminating ξ_1 between eqs (13.26) and (13.26), one finds:

$$\phi(\boldsymbol{y}) = \sqrt{\frac{d}{a}}\left[2\phi^2 - \frac{1}{2}\phi^3 + \sqrt{\frac{d}{a}}\left(\frac{1}{2}\phi^3 - \phi^2\right)\right] \tag{13.28}$$

The $d^{1/2}$ behaviour of ϕ near $\partial\Gamma$ is thus reproduced using standard quadratic interpolation functions. Then, the factor K_I is computed from the values of ϕ_n at nodes 2 and 3 using the formula:

$$K_I = \frac{\mu}{4(1-\nu)}\sqrt{\frac{2\pi}{a}}\left(2\phi_n^2 - \frac{1}{2}\phi_n^3\right) \tag{13.29}$$

and similar formulas are obtained for K_{II}, K_{III}.

If the subregion approach (Section 13.2) is used, the $d_{-1/2}$ singular behaviour of the traction vector on $\tilde{\Gamma}$ and in the vicinity of $\partial\Gamma$ calls for a similar interpolation technique. It is then also possible to evaluate the stress intensity factors from the computed nodal tractions.

13.7 NUMERICAL EXAMPLES

Semi-elliptical edge crack (Le Van [10])

A cylindrical bar (radius R) with a semi-elliptical edge crack (semi-axes a, b) is considered (Figure 13.7). The crack faces are subjected to either tensile (uniform or linear distributions) or shear loadings, allowing us to simulate tension, flexion or torsion.

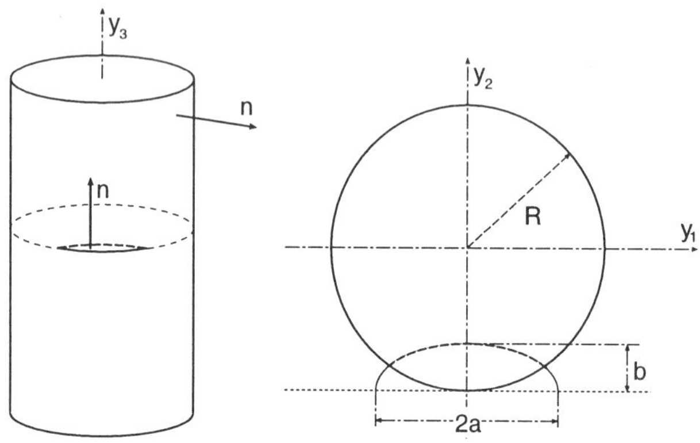

Figure 13.7 Cylindrical bar with edge crack: geometrical notation.

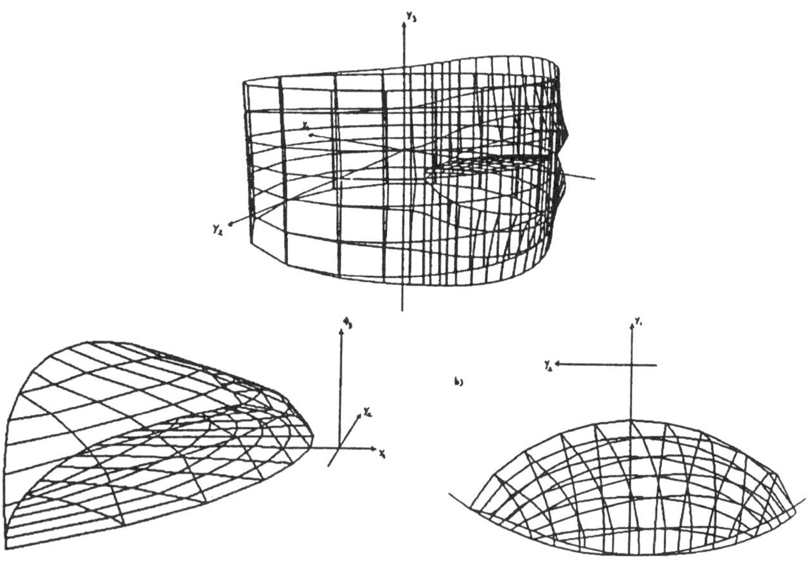

Figure 13.8 Deformed bar (top) and crack opening (linear pressure, bottom left; torsion, bottom right), from [10]).

Le Van [10] developed a boundary element discretization of the traction integral equation using four-noded elements (58 elements on the crack surface). Results for the case $b/a = 5/7$, $b/R = 0.4$ are displayed in Figure 13.8 (deformed mesh). For the pressure loading, the stress intensity factor K_I at the rear of the crack front ($y_1 = b - R, y_2 = 0$) was found to be within a few percentage points of published values for geometrically similar configurations.

Numerical simulation of dynamical crack propagation (Koller et al. [8])

The spontaneous propagation of a straight crack C extending along the x_1-axis in an infinite elastic space under antiplane shear (tractions $\boldsymbol{t}^{n\pm}(\boldsymbol{y}) = \pm\tau H(t)\boldsymbol{e}_3$ applied on the crack faces, where τ is a constant and $H(t)$ is the Heaviside step function) is considered. The propagation of the right crack tip is governed by $K_{III} = K_{III}^c$ (K_{III}^c: material toughness), the left tip being kept fixed. Let $\ell(t)$ denote the abscissa of the propagating crack tip time t.

The crack is discretized into J elements of equal length Δx and the time interval $[0, T]$ of interest into I equal time intervals Δt. The ratio $c\Delta t/\Delta x$ has been set to the value $1/2$, in order to be able to evaluate most element integrals analytically and thus circumvent inaccuracy problems arising when dealing with numerical integration of time-domain kernels, see Chapter 7. The only nonzero component $\phi_3(\boldsymbol{y}, t) = \phi(\boldsymbol{y}, t)$ is interpolated linearly in space and time. The regularized traction boundary integral equation is collocated at the midpoint of each boundary element and at $t = i\Delta t$, $i = 1, 2, \ldots I$. This results in a $(j \times J - 1)$ linear system of equations, hence solved in a least squares sense using the Householder factorization (LINPACK software library [6]).

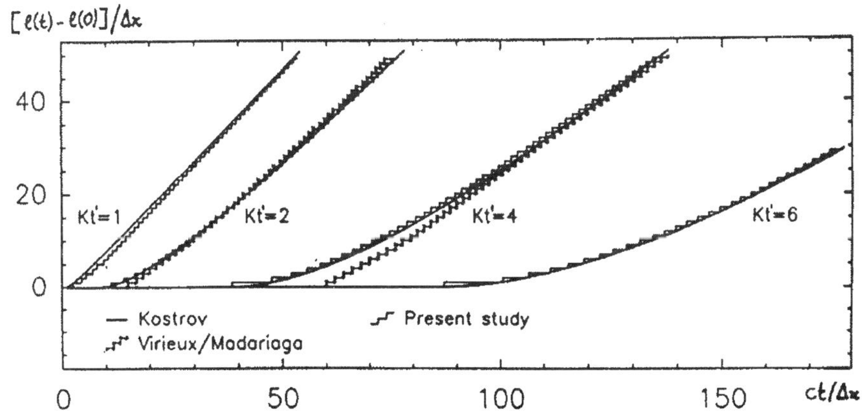

Figure 13.9 Crack-tip location during the spontaneous rupture propagation of a semi-infinite crack. Results from [8] are compared to Kostrov's [9] exact solution and Virieux and Madariaga's [21] numerical results (obtained using FDM). The results are normalized, with $K' = K_{III}^c/(\mu\sqrt{\pi\Delta x})$.

The propagation is simulated by adding a new element if the propagation criterion is met, so that I increases during the time-stepping scheme. The numerical results for the spontaneously propagating crack, displayed in Figure 13.9, are seen to compare very favourably with the analytical solution for the semi-infinite spontaneously propagating crack in antiplane strain (Kostrov [9]) and to be more accurate than other numerical results obtained using a finite-difference method (Virieux and Madariaga [21]), especially for the initiation phase, which is delayed due to the poor stress resolution of the FD technique.

13.8 BOUNDARY ELEMENT FORMULATION FOR THE ENERGY APPROACH IN FRACTURE MECHANICS

Consider a body Ω containing a (possibly multiple) crack Γ. Many fracture analyses can be performed using energy considerations. In particular, the first-order derivative of the potential energy at equilibrium W with respect to perturbations of the crack front $\partial\Gamma$ (see Chapter 12 for the concepts and notations used in this section) defines the energy release rate $G(s)$:

$$\int_{\Gamma} G\theta_{\nu}\,\mathrm{d}s + \overset{\star}{W} = 0 \qquad \forall \boldsymbol{\theta} \in \Theta \tag{13.30}$$

where Θ denotes the set of all transformation velocities $\boldsymbol{\theta}$ describing a crack front advance in the otherwise unperturbed body. G characterizes the potential or actual energy dissipation in a virtual or real crack advance. In elasticity, G is linked to the stress intensity factors by Irwin's formula:

$$G(s) = \frac{1+\nu}{2\mu}\left[K_I^2(s) + K_{II}^2(s)\right] + \frac{1}{2\mu}K_{III}(s) \tag{13.31}$$

Then, analysis of actual crack propagation (computation of the crack front velocity distribution, investigation of the uniqueness of this solution and of the stability of the crack front perturbation) in the framework of a Griffith-type criterion $g(G) \leq 0$ involve the perturbation of G, and hence the second-order domain derivative of W.

Energy approach in fracture mechanics

Early formulations of the energy approach, like the so-called virtual crack extension method (Hellen [7] and Parks [16]), consisted in evaluating approximate values of the first-order derivative of the potential energy at equilibrium in crack extensions using finite element models and numerical differentiation, i.e. invoking small but finite crack extensions. In two-dimensional situations, the crack tip is advanced by a length $\Delta\ell$, and the approximate value of G is:

$$G \simeq -\frac{W(\ell + \Delta\ell)}{\Delta\ell}$$

such a numerical differentiation technique is an ill-posed mathematical operation, whose accuracy is sensitive to the choice of $\Delta \ell$.

It is thus conceptually better (and also more economical from a computational viewpoint) to differentiate W analytically and discretize the resulting formulation. This viewpoint is at the heart of the so-called 'θ method' (see Destuynder, Djaoua and Lescure [5], Wadier and Malak [22] and also Delorenzi [4], Nguyen, Stolz and Debruyne [15]) whereby the potential energy at equilibrium W, which (for linear elasticity and in the absence of body forces) is given by:

$$W = \frac{1}{2} \int_{\Omega} \boldsymbol{\sigma}(\boldsymbol{u}) : \boldsymbol{\nabla u} \, \mathrm{d}V - \int_{S_T} \boldsymbol{t}^D . \boldsymbol{u} \, \mathrm{d}S$$

($\boldsymbol{u}$ being the displacement field solving the elastic equilibrium problem with boundary data $\boldsymbol{u}^D = \boldsymbol{0}, \boldsymbol{t}^D$), is analytically differentiated in a domain transformation. The derivative $\overset{\star}{W}$ is found to be:

$$\overset{\star}{W} = \int_{\Omega} \left\{ \frac{1}{2} \boldsymbol{\sigma}(\boldsymbol{u}) : \boldsymbol{\nabla u} \operatorname{div} \boldsymbol{\theta} - \boldsymbol{\sigma}(\boldsymbol{u}) : (\boldsymbol{\nabla u} . \boldsymbol{\nabla \theta}) \right\} \mathrm{d}V \qquad (13.32)$$

and is seen to depend on the current elastic solution and the transformation velocity $\boldsymbol{\theta}$. As expected from basic fracture mechanics considerations and like the J-integral, the domain derivative $\overset{\star}{u}$ of the elastostatic state does not appear in the result (13.32). Mathematically, $\overset{\star}{u}$ could be eliminated by invoking the virtual work principle and the symmetry of the strain energy bilinear form.

Equations (13.32) and (13.30) are the basis of finite element implementations of the energy approach. Upon defining a basis $\boldsymbol{\theta}^1, \dots, \boldsymbol{\theta}^{\mathrm{NC}}$, i.e. a finite-dimensional subspace of the space Θ of admissible transformation velocities, and evaluating $\overset{\star}{W}$ by means of eq. (13.32) for each $\boldsymbol{\theta}_k$, eq. (13.30) leads to a linear system of equations whose unknowns are the nodal values of G at the NC nodes of the crack front.

Note that eq. (13.32) is derived by by means of differentiation formulas in Lagrangian form, i.e. eqs (12.9-12.11), which ensure that all singularities at the crack front resulting from this process are integrable, whereas Eulerian differentiation would generate non-integrable quantities. Conversely, eq. (13.32) hides the fact (established by (Mialon [13] and Pradeilles-Duval [17]) that the value of $\overset{\star}{W}$ depends only on the normal trace $\theta_\nu \mid_{\partial\Gamma}$ on the crack front of the velocity $\theta_\nu \mid_{\partial\Gamma}$ (in other words, two velocity fields $\boldsymbol{\theta}^a, \boldsymbol{\theta}^b$ such that $\theta^a_\nu = \theta^b_\nu$ on $\partial\Gamma$ produce the same value of $\overset{\star}{W}$).

The 'θ-integral method'.

It consists in transposing the 'θ method' to integral equation formulations and boundary element techniques. The main idea is to start from a boundary-only expression of W:

$$W = \frac{1}{2}\int_{S_u} \boldsymbol{u}^D.\boldsymbol{t}\,\mathrm{d}S - \frac{1}{2}\int_{S_T} \boldsymbol{u}.\boldsymbol{t}^D\,\mathrm{d}S \tag{13.33}$$

which is applicable only under restrictive conditions: geometrical and material linearity, no body forces. The first-order derivative of W in a virtual crack extension (i.e. such that $\boldsymbol{\theta} = \mathbf{0}$ on the external boundary) and for fixed boundary data $(\boldsymbol{u}^D, \boldsymbol{t}^D)$ is then simply given by:

$$W_{,\Gamma}.\boldsymbol{\theta} = \frac{1}{2}\int_{S_u} \overset{\star}{\boldsymbol{t}}\,.\boldsymbol{u}^D\,\mathrm{d}S - \frac{1}{2}\int_{S_T} \boldsymbol{t}^D.\,\overset{\star}{\boldsymbol{u}}\;\mathrm{d}S \tag{13.34}$$

This expression involves the boundary traces of the first-order domain derivatives $(\overset{\star}{\boldsymbol{u}}, \overset{\star}{\boldsymbol{t}})$ in a given transformation velocity $\boldsymbol{\theta}$. Assuming the multidomain approach (Section 13.2) is being used, $(\overset{\star}{\boldsymbol{u}}, \overset{\star}{\boldsymbol{t}})$ are governed on each subdomain by the derivative integral equation (12.32), which must then be solved for each $\boldsymbol{\theta}^i$. Then $\overset{\star}{W}$ given by eq. (13.34) is evaluated for each $\boldsymbol{\theta}^i$ and the result is substituted into the variational equation (13.30). This operation again produces a linear system of equations over the nodal values of G.

One key ingredient of the numerical implementation of this 'θ-integral method' (Bonnet and Xiao [1]) is the definition of a basis of vector functions $\boldsymbol{\theta}^k(\boldsymbol{\xi})$ such that:

$$\boldsymbol{\theta}(\boldsymbol{y}) = \sum_{k=1}^{\mathrm{NC}} \boldsymbol{B}^k(\boldsymbol{\xi})\theta_k$$

where the θ_k are the nodal values of θ_ν at the crack front nodes. In practice, these functions are constructed so that (i) their geometrical support is restricted to the boundary elements adjacent to the crack front and (ii) the above interpolation formula reduces to a standard one-dimensional interpolation when $\boldsymbol{y} \in \partial\Gamma$.

Numerical example: semi-elliptical surface crack

A semi-elliptical surface crack (see Figure 13.10 for the geometrical notations) is situated in the symmetry plane $y_3 = 0$ of a rectangular parallelepiped subjected to a uniform tension p along the axial direction (mode I). Owing to geometrical symmetry (see Chapter 11), only one-quarter of the boundary is discretized (136 9-noded boundary elements); the crack front itself supports 13 nodal values of G and $\boldsymbol{\theta}$. Two variant meshes M_1 and M_2 (Figure 13.10) were used, with respectively uniform ($\Delta\theta = \pi/24$, mesh M_1) and non-uniform ($\Delta\theta = \pi/32$ (resp. $\pi/16$) for $\theta \in [0, \pi/4]$ (resp. $\theta \in [\pi/4, \pi/2]$), mesh M_2) angular spacing between crack front nodes, the crack edge being located at $\theta = 0$.

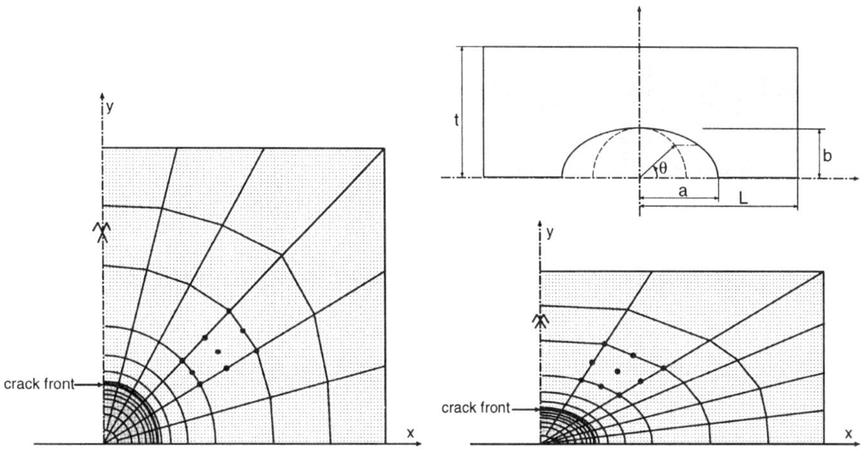

Figure 13.10 Semi-elliptical surface crack: geometrical notation for the crack plane (top right), boundary element meshes M_1 with $b = a$ (bottom left) and M_2 with $b = 0.6a$ (bottom right) of the crack plane.

Numerical values of the non-dimensional SIF $K_I^\star Q/(p\sqrt{\pi a})$, where

$$K_I^e = \frac{1}{Q} p\sqrt{\pi a} \qquad Q = E(k) = \int_0^{\pi/2} \sqrt{1 - k^2 \sin^2\theta}\, \mathrm{d}\theta$$

and

$$k^2 = \begin{cases} 1 - (b/a)^2 & (b \le a) \\ 1 - (a/b)^2 & (b \ge a) \end{cases}$$

were obtained from the values of G computed with the present method and using Irwin's formula (13.31). They are compared (Figure 13.11) to other numerical results from Newman and Raju [14] and Tanaka and Itoh [20]. The latter were obtained using a sophisticated special crack-front element which allows for the modelling of both the square-root crack-front singularity and the crack-edge singularity (whose exponent differs from $-1/2$ except for $\nu = 0$), and are thus expected to provide the better reference solution near the crack edge. Our results show an improvement in reproducing the small peak near the crack edge when mesh M_2, which is finer than M_1 near the crack edge, is used; generally speaking, they agree reasonably well with the reference ones. For comparison purposes, Figure 13.11 also shows the values of $K_I^\star$ obtained by extrapolation of the crack opening displacement, which tend to be somewhat poorer than those obtained using the present approach, despite the use of quarter-node elements along the crack front.

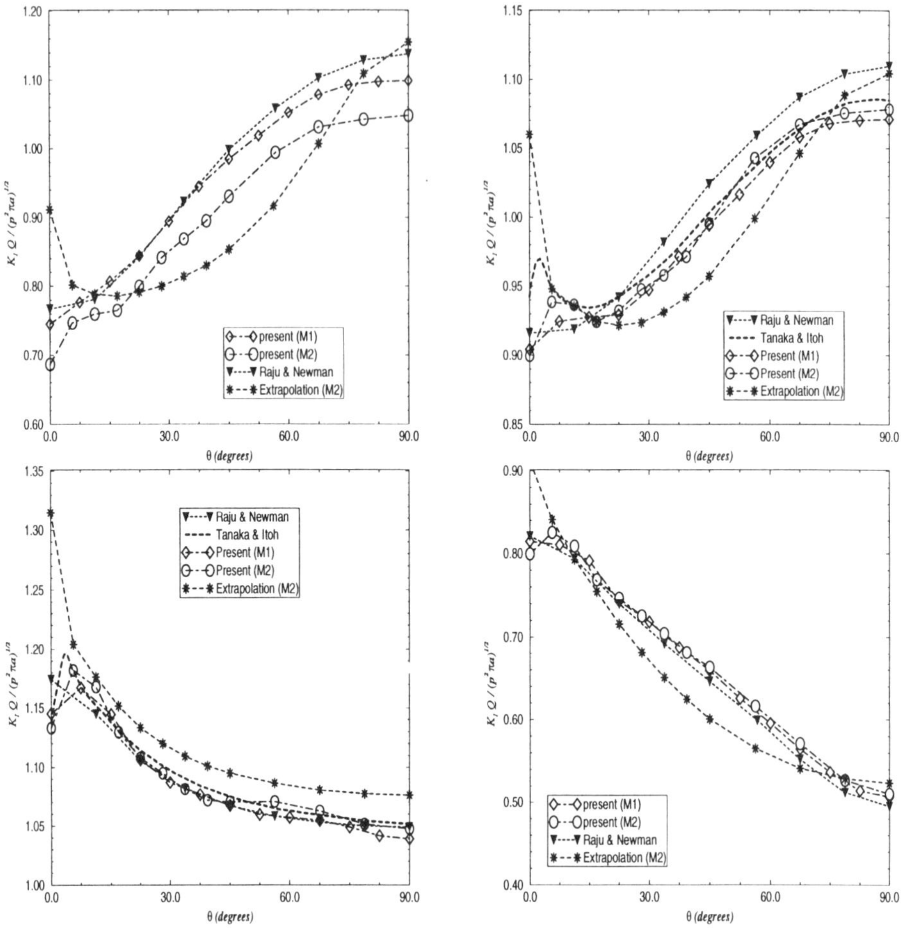

Figure 13.11 Semi-elliptical surface crack (top left: $b = 0.4a$, top right: $b = 0.6a$, bottom left: $b = a$, bottom right: $b = 2a$): numerical results for K_I.

REFERENCES

1. BONNET, M., XIAO, H. Computation of energy release rate along a crack front using material differentiation of elastic BIE. *Engng. Anal. with Bound. Elem.*, **15**, 137–150 (1995).

2. BUI, H. D. An integral equation method for solving the problem of a plane crack of arbitrary shape. *J. Mech. Phys. Solids*, **25**, 29–39 (1977).

3. BUI, H. D. *Mécanique de la rupture fragile.*. Masson (1978).

4. DELORENZI, H. G. On the energy release rate and the J-integral for 3-D crack configurations. *Int. J. Fract.*, **19**, 183–194 (1982).

5. DESTUYNDER, PH., DJAOUA, M., LESCURE, S. Quelques remarques sur la mécanique de la rupture élastique. *J. Mécan. Théor. Appl.*, **2**, 113–135 (1983).

6. DONGARRA, J. J., BUNCH, J. R., MOLER, C. B., STEWART, G. B. LINPACK *users' guide*. Society for Industrial and Applied Mathematics (SIAM), Philadelphia, USA (1979).

7. HELLEN, T. K. On the method of virtual crack extension. *Int. J. Num. Meth. in Eng.*, **9**, 187–207 (1975).

8. KOLLER, M. G., BONNET, M., MADARIAGA, R. Modelling of dynamical crack propagation using time-domain boundary integral equations. *Wave Motion*, **16**, 339–366 (1992).

9. KOSTROV, B. V. Unsteady propagation of longitudinal shear cracks. *Int. J. Appl. Math. Mech. (PMM)*, **30**, 1241–1248 (1966).

10. LE VAN, A. *Equations intégrales pour fissures 3D. Contribution à la régularisation*. Ph.D. thesis, Ecole Nationale Supérieure de Mécanique de Nantes, France (1988).

11. MI, Y., ALIABADI, M. H. Dual boundary element method for analysis of 3-D crack problems. In J. Dominguez, C. A. Brebbia, F. Paris (eds), *Boundary elements XIV*, pp. 315–329. Comp. Mech. Publ., Southampton (1992).

12. MI, Y., ALIABADI, M. H. Dual boundary element method for three-dimensional fracture mechanics analysis. *Engng. Anal. with Bound. Elem.*, **10**, 161–171 (1992).

13. MIALON, P. Calcul de la dérivée d'une grandeur par rapport à un fond de fissure par la méthode θ. Bulletin EDF/DER série c vol. 3, Electricité de France (1987).

14. NEWMAN, J. C., RAJU, I. S. An empirical stress-intensity factor equation for the surface crack. *Engng. Frac. Mech.*, **15**, 185–192 (1981).

15. NGUYEN, Q. S., STOLZ, C., DEBRUYNE, G. Energy methods in fracture mechanics: stability, bifurcation and second variations. *Engng. Structures*, **9**, 157–173 (1990).

16. PARKS, D. M. A stiffness derivative finite element technique for determination of crack tip stress intensity factors. *Int. J. Fract.*, **10**, 487–501 (1974).

17. PRADEILLES-DUVAL, R. M. *Evolution de systèmes avec surfaces de discontinuité mobiles: application au délaminage*. Ph.D. thesis, Ecole Polytechnique, Palaiseau, France (1992).

18. SNYDER, M. D., CRUSE, T. A. Boundary integral equation analysis of cracked anisotropic plates. *Int. J. Fract.*, **11**, 315–328 (1975).

19. TADA, H., PARIS, P., IRWIN, G. *The stress analysis of cracks handbook*. Technical report, Del. Research Corporation, Hellertown, Pennsylvania, USA (1973).

20. TANAKA, M., ITOH, H. New crack elements for boundary element analysis of elastostatics considering arbitrary stress singularities. *Appl. Math. Modelling*, **11**, 357–363 (1987).

21. VIRIEUX, J., MADARIAGA, R. Dynamic faulting studied by a finite difference method. *Bull. Seism. Soc. Am.*, **72**, 345–369 (1982).

22. WADIER, Y., MALAK, O. The theta method applied to the analysis of 3D elastic–plastic cracked bodies. In *SMIRT proceedings* (1989).

23. WEAVER, J. Three-dimensional crack analysis. *Int. J. Solids Struct.*, **13**, 321–330 (1977).

14

Initial strains and stresses – elastoplasticity

Distributions of initial strain $\boldsymbol{\varepsilon}^I$ or stress $\boldsymbol{\sigma}^I$ occur in thermoelastic problems (for an isotropic material: $\boldsymbol{\varepsilon}^I = \alpha T \boldsymbol{I}$, where T is a temperature distribution and α denotes the thermal dilatation coefficient) or in micromechanics (inclusions, i.e. geometrical support of residual plastic strains, or inhomogeneities embedded in an elastic matrix). Also, the numerical procedures for elastic–plastic analyses are based on incremental and iterative algorithms formulated in terms of sequences of elastic problems with nonlinear effects gathered into 'initial' strains or stresses distributions.

14.1 ELASTOSTATIC PROBLEMS WITH INITIAL STRAIN OR STRESS DISTRIBUTIONS

Initial strains and stresses

The total strain tensor $\boldsymbol{\varepsilon}$ in an elastic body Ω can be decomposed additively into initial and elastic strains $\boldsymbol{\varepsilon}^I$ and $\boldsymbol{\varepsilon}^E$ according to:

$$\boldsymbol{\varepsilon} = \boldsymbol{\varepsilon}^E + \boldsymbol{\varepsilon}^I \qquad \text{where} \quad \boldsymbol{\sigma} = \boldsymbol{C} : \boldsymbol{\varepsilon}^E$$

In similar fashion, the stress tensor $\boldsymbol{\sigma}$ is decomposable into initial and elastic stresses $\boldsymbol{\sigma}^I$ and $\boldsymbol{\sigma}^E$ such that:

$$\boldsymbol{\sigma} = \boldsymbol{\sigma}^E - \boldsymbol{\sigma}^I \qquad \text{where} \quad \boldsymbol{\sigma}^E = \boldsymbol{C} : \boldsymbol{\varepsilon}$$

These definitions imply that $\boldsymbol{\sigma}^I = \boldsymbol{C} : \boldsymbol{\varepsilon}^I$; however $\boldsymbol{\sigma}^E \neq \boldsymbol{C} : \boldsymbol{\varepsilon}^E$ in general.

Field equations and boundary conditions

Assume that the body Ω is loaded by body forces ρF, prescribed tractions $\boldsymbol{t}^D$ over S_T and prescribed displacement $\bar{\boldsymbol{u}}$ over S_u). The equilibrium equation and boundary conditions take the form

$$\text{div}\boldsymbol{\sigma} = \text{div}(\boldsymbol{C} : \boldsymbol{\varepsilon}^I) - \rho \boldsymbol{F}$$
$$\boldsymbol{T}^n(\boldsymbol{u}) = \boldsymbol{t}^D + (\boldsymbol{C} : \boldsymbol{\varepsilon}^I).\boldsymbol{n}$$

in terms of initial strain, and

$$\text{div}\boldsymbol{\sigma} = \text{div}(\boldsymbol{\sigma}^I) - \rho \boldsymbol{F}$$
$$\boldsymbol{T}^n(\boldsymbol{u}) = \boldsymbol{t}^D + (\boldsymbol{\sigma}^I).\boldsymbol{n}$$

in terms of initial stresses.

Displacement boundary integral equation

The Kelvin fundamental solution is used for simplicity throughout this chapter. The displacement boundary integral equation (4.32) is applicable to the elasticity problem with initial strains or stresses, the body force distribution in the right-hand side being taken equal respectively to $\rho\boldsymbol{F} - \text{div}(\boldsymbol{C} : \boldsymbol{\varepsilon}^I)$ or $\rho\boldsymbol{F} - \text{div}(\boldsymbol{\sigma}^I)$. With initial strains, one obtains:

$$\bar{\kappa} u_k(\boldsymbol{x}) + \int_{\partial\Omega} \{[u_i(\boldsymbol{y}) - u_i(\boldsymbol{x})]T_i^k(\boldsymbol{x}, \boldsymbol{y}) - f_i(\boldsymbol{y})U_i^k(\boldsymbol{x}, \boldsymbol{y})\} \, \mathrm{d}S_y$$
$$= \int_{\Omega} [\rho F_i(\boldsymbol{y}) - C_{ijab}\varepsilon^I_{ab,j}(\boldsymbol{y})]U_i^k(\boldsymbol{x}, \boldsymbol{y}) \, \mathrm{d}V_y \quad (14.1)$$

with $\bar{\kappa} = 0$ (Ω bounded) or $\bar{\kappa} = 1$ ($\mathbb{R}^3 - \Omega$ bounded); the notation $\boldsymbol{f}(\boldsymbol{y})$ is a shorthand for $\boldsymbol{T}^n(\boldsymbol{u})(\boldsymbol{y}) = \boldsymbol{f} = (\boldsymbol{\sigma} + \boldsymbol{C} : \boldsymbol{\varepsilon}^I).\boldsymbol{n}$, which differs from the actual traction vector $\boldsymbol{t} = \boldsymbol{\sigma}.\boldsymbol{n}$. Integrating by parts the domain integral containing $\boldsymbol{\varepsilon}^I$ by means of the usual divergence formula, one obtains:

$$\int_{\Omega} C_{ijab}\varepsilon^I_{ab,j}(\boldsymbol{y})U_i^k(\boldsymbol{x}, \boldsymbol{y}) \, \mathrm{d}V_y$$
$$= \int_{\partial\Omega} C_{ijab}\varepsilon^I_{ab}(\boldsymbol{y})n_j(\boldsymbol{y})U_i^k(\boldsymbol{x}, \boldsymbol{y}) \, \mathrm{d}S_y - \int_{\Omega} \varepsilon^I_{ij}(\boldsymbol{y})\Sigma^k_{ij}(\boldsymbol{x}, \boldsymbol{y}) \, \mathrm{d}V_y$$

where the last domain integral is convergent since $\Sigma^k_{ij}(\boldsymbol{x}, \boldsymbol{y}) \sim |\boldsymbol{y} - \boldsymbol{x}|^{-2}$). Substitution of this result into eq. (14.1) produces an integral equation for the values of $\boldsymbol{u}$ and $\boldsymbol{t} = \boldsymbol{\sigma}.\boldsymbol{n}$ on the boundary of Ω :

$$\bar{\kappa} u_k(\boldsymbol{x}) + \int_{\partial\Omega} \{[u_i(\boldsymbol{y}) - u_i(\boldsymbol{x})]T_i^k(\boldsymbol{x}, \boldsymbol{y}) - t_i(\boldsymbol{y})U_i^k(\boldsymbol{x}, \boldsymbol{y})\} \, \mathrm{d}S_y$$
$$= \int_{\Omega} [\rho F_i(\boldsymbol{y})U_i^k(\boldsymbol{x}, \boldsymbol{y}) + \varepsilon^I_{ij}(\boldsymbol{y})\Sigma^k_{ij}(\boldsymbol{x}, \boldsymbol{y})] \, \mathrm{d}V_y \quad (14.2)$$

where use has been made of

$$\boldsymbol{t} = \boldsymbol{T}^n(\boldsymbol{u}) - (\boldsymbol{C} : \boldsymbol{\varepsilon}^I).\boldsymbol{n} = \boldsymbol{f} - (\boldsymbol{C} : \boldsymbol{\varepsilon}^I).\boldsymbol{n}$$

Similarly, the displacement in an interior point $\boldsymbol{x} \in \Omega, \boldsymbol{x} \notin \partial\Omega$ admits the integral representation:

$$u_k(\boldsymbol{x}) = \int_{\partial\Omega} \left\{ t_i(\boldsymbol{y}) U_i^k(\boldsymbol{x}, \boldsymbol{y}) - u_i(\boldsymbol{y}) T_i^k(\boldsymbol{x}, \boldsymbol{y}) \right\} \, \mathrm{d}S_y + \int_{\Omega} \left[\rho F_i(\boldsymbol{y}) U_i^k(\boldsymbol{x}, \boldsymbol{y}) + \varepsilon_{ij}^I(\boldsymbol{y}) \Sigma_{ij}^k(\boldsymbol{x}, \boldsymbol{y}) \right] \, \mathrm{d}V_y \tag{14.3}$$

If the distribution $\boldsymbol{\varepsilon}^I$ of initial strain is known, the integral equation (14.2) can be solved using the concepts developed in Chapters 3 and 4, including analytical transformations (conversion into boundary integrals) or numerical methods (integration cells) for the evaluation of known domain integrals (Section 3.6).

Integral representation of strain or stress at interior points

The evaluation at points interior to the body of strains or stresses associated to the integral representation (14.3) is useful for various applications, e.g. in micromechanics, and crucial for elastic–plastic analyses. This capability rests upon differentiating the integral representation (14.3) with respect to coordinates of the observation point $\boldsymbol{x}$. A technical difficulty appears at this stage, namely that the differentiation of eq. (14.3) with respect to x_ℓ necessitates the evaluation of:

$$I_{k\ell}(\boldsymbol{x}) = \frac{\partial}{\partial x_\ell} \int_{\Omega} \varepsilon_{ij}^I(\boldsymbol{y}) \Sigma_{ij}^k(\boldsymbol{x}, \boldsymbol{y}) \, \mathrm{d}V_y$$

Indeed, the derivatives $\Sigma_{ij,\bar{\ell}}^k(\boldsymbol{x}, \boldsymbol{y})$ exhibit a $|\boldsymbol{y} - \boldsymbol{x}|^{-3}$ singularity which is not integrable over Ω. Thus the differentiation with respect to $\boldsymbol{x}$ in $I_{k\ell}(\boldsymbol{x})$ cannot be transferred under the integral sign. This difficulty, which arises only in the presence of initial strains or stresses, is serious, as witness the fact that the issue was incorrectly solved in the literature until Bui [3] established the correct expression of $I_{k\ell}(\boldsymbol{x})$, which involves a free term and a domain integral in the Cauchy principal value sense:

$$I_{k\ell}(\boldsymbol{x}) = \int_{\Omega}^{\star} \varepsilon_{ij}^I(\boldsymbol{y}) \Sigma_{ij,\bar{\ell}}^k(\boldsymbol{x}, \boldsymbol{y}) \, \mathrm{d}V_y - 2\mu \left[\frac{8 - 10\nu}{15(1-\nu)} \varepsilon_{k\ell}^I(\boldsymbol{x}) - \frac{1 - 5\nu}{15(1-\nu)} \varepsilon_{aa}^I(\boldsymbol{x}) \delta_{k\ell} \right] \tag{14.4}$$

Bui's derivation proceeds by excluding the singular point $\boldsymbol{x}$ by removing a spherical neighbourhood $v_\varepsilon(\boldsymbol{x})$ and *then* differentiating the result with respect to x_ℓ. The integration domain $\Omega - v_\varepsilon(\boldsymbol{x})$ thus depends on $\boldsymbol{x}$ through $v_\varepsilon(\boldsymbol{x})$, so that the process involves a *convective derivative*, which was overlooked in previous attempts at solving this issue.

Regularized integral representation of strain or stress at interior points

Application of formula (14.4) entails the evaluation of a Cauchy principal value domain integral. The direct approach of Guiggiani and Gigante [6] can be adapted to this task. It is also possible, with the help of an auxiliary integral identity, to perform an indirect regularization approach in order to reformulate the integral representation in an integrable form.

Consider a *constant* initial strain tensor $\boldsymbol{\varepsilon}^0$. Since $\boldsymbol{\varepsilon}^0$ is then geometrically compatible, it induces in the absence of body forces in Ω and kinematical constraints on $\partial\Omega$ the following elastic state in Ω, defined up to additive rigid-body motions:

$$\boldsymbol{u}(\boldsymbol{y}) = \boldsymbol{\varepsilon}^0.\boldsymbol{y} \qquad \boldsymbol{F}(\boldsymbol{y}) = \boldsymbol{0} \qquad \boldsymbol{\varepsilon}^E(\boldsymbol{y}) = \boldsymbol{\sigma}(\boldsymbol{y}) = \boldsymbol{0} \tag{14.5}$$

Substituting this simple solution into the representation formula (14.3), one obtains an integral identity that holds for any constant strain tensor $\boldsymbol{\varepsilon}^0$:

$$\left\{ \delta_{ki}x_j + \int_{\partial\Omega} y_j T_i^k(\boldsymbol{x},\boldsymbol{y})\,\mathrm{d}S_y - \int_{\Omega} \Sigma_{ij}^k(\boldsymbol{x},\boldsymbol{y})\,\mathrm{d}V_y \right\} \varepsilon_{ij}^0 = 0 \tag{14.6}$$

In particular, putting $\boldsymbol{\varepsilon}^0 = \boldsymbol{\varepsilon}^I(\boldsymbol{x})$ and subtracting the resulting identity from eq. (14.3), one obtains the integral representation of displacement in regularized form:

$$\begin{aligned} u_k(\boldsymbol{x}) = &\int_{\partial\Omega} \{ t_i(\boldsymbol{y}) U_i^k(\boldsymbol{x},\boldsymbol{y}) - u_i(\boldsymbol{y}) T_i^k(\boldsymbol{x},\boldsymbol{y}) \}\,\mathrm{d}S_y \\ &+ \int_{\Omega} \left[\rho F_i(\boldsymbol{y}) U_i^k(\boldsymbol{x},\boldsymbol{y}) + [\varepsilon_{ij}^I(\boldsymbol{y}) - \varepsilon_{ij}^I(\boldsymbol{x})]\Sigma_{ij}^k(\boldsymbol{x},\boldsymbol{y}) \right] \mathrm{d}V_y \\ &+ \varepsilon_{ij}^I(\boldsymbol{x}) \left\{ \delta_{ki}x_j + \int_{\partial\Omega} y_j T_i^k(\boldsymbol{x},\boldsymbol{y})\,\mathrm{d}S_y \right\} \end{aligned} \tag{14.7}$$

Then, thanks to the term $[\boldsymbol{\varepsilon}^I(\boldsymbol{y}) - \boldsymbol{\varepsilon}^I(\boldsymbol{x})]$, a differentiation of the domain equation under the integral sign produces a convergent domain integral and is thus mathematically valid. This requires the initial strain $\boldsymbol{\varepsilon}^I$ to possess $C^{0,\alpha}$ smoothness at $\boldsymbol{y} = \boldsymbol{x}$. Indeed, under this assumption, one has:

$$[\varepsilon_{ij}^I(\boldsymbol{y}) - \varepsilon_{ij}^I(\boldsymbol{x})]\Sigma_{ij,\ell}^k(\boldsymbol{x},\boldsymbol{y}) \sim |\boldsymbol{y} - \boldsymbol{x}|^{-3+\alpha}$$

so that this quantity is integrable in the usual sense over Ω. A differentiation of eq. (14.7) with respect to $\boldsymbol{x}$ then yields the relation:

$$\begin{aligned} u_{k,\ell}(\boldsymbol{x}) = &\int_{\partial\Omega} \{ u_i(\boldsymbol{y}) n_j(\boldsymbol{y}) \Sigma_{ij,\ell}^k(\boldsymbol{x},\boldsymbol{y}) - t_i(\boldsymbol{y}) U_{i,\ell}^k(\boldsymbol{x},\boldsymbol{y}) \}\,\mathrm{d}S_y \\ &- \int_{\Omega} \left[\rho F_i(\boldsymbol{y}) U_{i,\ell}^k(\boldsymbol{x},\boldsymbol{y}) + [\varepsilon_{ij}^I(\boldsymbol{y}) - \varepsilon_{ij}^I(\boldsymbol{x})]\Sigma_{ij,\ell}^k(\boldsymbol{x},\boldsymbol{y}) \right] \mathrm{d}V_y \\ &+ \varepsilon_{ij,\ell}^I(\boldsymbol{x}) \left\{ \delta_{ki}x_j + \int_{\partial\Omega} y_j T_i^k(\boldsymbol{x},\boldsymbol{y})\,\mathrm{d}S_y - \int_{\Omega} \Sigma_{ij}^k(\boldsymbol{x},\boldsymbol{y})\,\mathrm{d}V_y \right\} \\ &+ \varepsilon_{ij}^I(\boldsymbol{x}) \left\{ \delta_{ki}\delta_{\ell j} - \int_{\partial\Omega} y_j \Sigma_{aj,\ell}^k(\boldsymbol{x},\boldsymbol{y}) n_a(\boldsymbol{y})\,\mathrm{d}S_y \right\} \end{aligned}$$

having used property (4.13) of the Kelvin solution. The factor of $\varepsilon^I_{ij,\ell}(\boldsymbol{x})$ vanishes identically as a consequence of eq. (14.6), while that of $\varepsilon^I_{ij}(\boldsymbol{x})$ can be transformed into:

$$\int_{\partial\Omega} y_j \Sigma^k_{aj,\ell}(\boldsymbol{x},\boldsymbol{y}) n_a(\boldsymbol{y})\,\mathrm{d}S_y = \int_{\partial\Omega} D_{\ell a} y_j \Sigma^k_{aj}(\boldsymbol{x},\boldsymbol{y})\,\mathrm{d}S_y$$
$$= (1-\bar{\kappa})\delta_{ki}\delta_{\ell j} + \int_{\partial\Omega} n_\ell(\boldsymbol{y})\Sigma^k_{ij}(\boldsymbol{x},\boldsymbol{y})\,\mathrm{d}S_y$$

using the integration by parts identity (A.8). As a result, the regularized integral representation formula for the displacement gradient at an interior point $\boldsymbol{x}$ reads:

$$u_{k,\ell}(\boldsymbol{x}) = \int_{\partial\Omega} \{u_i(\boldsymbol{y})n_j(\boldsymbol{y})\Sigma^k_{ij,\ell}(\boldsymbol{x},\boldsymbol{y}) - t_i(\boldsymbol{y})U^k_{i,\ell}(\boldsymbol{x},\boldsymbol{y})\}\,\mathrm{d}S_y$$
$$- \int_{\Omega} [\rho F_i(\boldsymbol{y})U^k_{i,\ell}(\boldsymbol{x},\boldsymbol{y}) + [\varepsilon^I_{ij}(\boldsymbol{y}) - \varepsilon^I_{ij}(\boldsymbol{x})]\Sigma^k_{ij,\ell}(\boldsymbol{x},\boldsymbol{y})]\,\mathrm{d}V_y$$
$$+ \varepsilon^I_{ij}(\boldsymbol{x})\left\{\bar{\kappa}\delta_{ki}\delta_{\ell j} - \int_{\partial\Omega} n_\ell(\boldsymbol{y})\Sigma^k_{ij}(\boldsymbol{x},\boldsymbol{y})\,\mathrm{d}S_y\right\} \tag{14.8}$$

Comments. Like other regularized formulations given in previous chapters, the result (14.8) is well suited to applications in that the numerical evaluation of the Cauchy principal value integral (14.4) is avoided. Since $\boldsymbol{x}$-derivatives have been changed into $\boldsymbol{y}$-derivatives according to (4.13), it applies only for the Kelvin fundamental solution. The representation formula (14.8) can be at once expressed in terms of initial stresses through the substitution $\boldsymbol{\varepsilon}^I = \boldsymbol{C}^{-1} : \boldsymbol{\sigma}^I$.

14.2 ELASTIC FIELDS GENERATED BY INCLUSIONS

Fundamental problems in micromechanics (see Mura [13]) include finding elastic states for bodies Ω containing *inclusions* (initial strain $\boldsymbol{\varepsilon}^I$ distributed over a subset $D \subset \Omega$) or *inhomogeneities* (region $D \subset \Omega$ where elastic constants $\boldsymbol{C}^\star$ differ from those of the *matrix* $\Omega - D$, denoted $\boldsymbol{C}$). Of particular interest in this area of research are the strain or stress fields generated by the presence of an inclusion or inhomogeneity D, ignoring body forces ($\boldsymbol{F} = \boldsymbol{0}$); they can be expressed by integral representation formulas like (14.8). The case $\Omega = \mathbb{R}^3$ (infinite matrix), for which integral representations lead to exact analytical solutions for simple shapes of D, has been the subject of numerous studies (see [13] and the many references therein).

Integral representation of strain tensor

The regularized integral representation for displacement gradient, eq. (14.8), is applicable to this class of problems. Considering an inclusion embedded in an infinite matrix, eq. (14.8) can be applied separately to the inclusion D and the matrix $\mathbb{R}^3 - D$. Summing the two relations and taking into account the continuity conditions (equal displacements and opposite tractions on ∂D), one obtains the result:

$$u_{k,\ell}(\boldsymbol{x}) = -\varepsilon^I_{ij}(\boldsymbol{x}) \int_{\partial D} n_\ell(\boldsymbol{y}) \Sigma^k_{ij}(\boldsymbol{x}, \boldsymbol{y}) \, \mathrm{d}S_y - \int_D \left[\varepsilon^I_{ij}(\boldsymbol{y}) - \varepsilon^I_{ij}(\boldsymbol{x})\right] \Sigma^k_{ij,\ell}(\boldsymbol{x}, \boldsymbol{y}) \, \mathrm{d}V_y \quad (14.9)$$

It is implicitly assumed that the resultant of the body forces $C_{ijab}\varepsilon^I_{ab,j}$ is zero, so that the fields $\boldsymbol{u}, \boldsymbol{\varepsilon}, \boldsymbol{\sigma}$ decay at infinity according to (4.35). The above result is valid for any observation point $\boldsymbol{x}$ in $\mathbb{R}^3$, the unit normal $\boldsymbol{n}$ being taken exterior to the subdomain (D or $\mathbb{R}^3 - D$) containing the point $\boldsymbol{x}$. The total strain $\boldsymbol{\varepsilon}$ thus apears to depend linearly on the initial strain distribution $\boldsymbol{\varepsilon}^I$ prescribed in the inclusion.

For the particular case of initial strain $\boldsymbol{\varepsilon}^I$ *uniform* in D ($\varepsilon^I_{ij}(\boldsymbol{x}) = \varepsilon^I_{ij}$), eq. (14.9) gives for a point $\boldsymbol{x}$ interior to D:

$$\begin{aligned} \varepsilon_{k\ell}(\boldsymbol{x}) &= -\frac{1}{2}\varepsilon^I_{ij} \int_{\partial\Omega} \left[n_\ell(\boldsymbol{y}) \Sigma^k_{ij}(\boldsymbol{x}, \boldsymbol{y}) + n_k(\boldsymbol{y}) \Sigma^\ell_{ij}(\boldsymbol{x}, \boldsymbol{y})\right] \, \mathrm{d}S_y \\ &= S_{k\ell ij}(\boldsymbol{x})\varepsilon^I_{ij} \end{aligned} \quad (14.10)$$

This proves that the total strain tensor $\boldsymbol{\varepsilon}(\boldsymbol{x})$ inside the inclusion depends linearly on $\boldsymbol{\varepsilon}^I$. Moreover, for inclusions D of spheroidal shape, an analytical exact expression can be obtained for the integral in eq. (14.10). This calculation further establishes a classical result of micromechanics, namely that $\boldsymbol{S}(\boldsymbol{x})$ is then uniform in D: $\boldsymbol{S}(\boldsymbol{x}) = \boldsymbol{S}$ is known as the *Eshelby tensor*.

The equivalent inclusion method for inhomogeneities. Assume now that D is an elastic inhomogeneity. Assuming the strain $\boldsymbol{\varepsilon}^0$ and stress $\boldsymbol{\sigma}^0 = \boldsymbol{C} : \boldsymbol{\varepsilon}^0$ distributions are known in Ω in the absence of inhomogeneity (i.e. if $\boldsymbol{C}^\star = \boldsymbol{C}$), it is possible to define an *equivalent initial strain* $\boldsymbol{\varepsilon}^\star$, allowing to replace the inhomogeneity by a fictitious inclusion giving rise to the same mechanical effects in Ω. Indeed, one has for both cases in D:

$$\begin{aligned} \boldsymbol{\sigma} + \boldsymbol{\sigma}^0 &= \boldsymbol{C} : (\boldsymbol{\varepsilon} + \boldsymbol{\varepsilon}^0 - \boldsymbol{\varepsilon}^\star) \qquad \text{(inclusion)} \\ &= \boldsymbol{C}^\star : (\boldsymbol{\varepsilon} + \boldsymbol{\varepsilon}^0) \qquad \text{(inhomogeneity)} \end{aligned}$$

On the other hand, $\boldsymbol{\varepsilon}$ is known to depend linearly on $\boldsymbol{\varepsilon}^\star$ through the integral representation (14.9), so that the previous equalities relate $\boldsymbol{\varepsilon}^\star$ to $\boldsymbol{\varepsilon}^0$. For instance, if $\boldsymbol{\varepsilon}^0, \boldsymbol{\sigma}^0$ are uniform in D, $\boldsymbol{\varepsilon}^\star$ is also uniform in D. Then, using eq. (14.10), one obtains the following relation:

$$\boldsymbol{C}^\star : (\boldsymbol{\varepsilon}^0 + \boldsymbol{S} : \boldsymbol{\varepsilon}^\star - \boldsymbol{\varepsilon}^\star) = \boldsymbol{C} : (\boldsymbol{S} : \boldsymbol{\varepsilon}^\star + \boldsymbol{\varepsilon}^0)$$

which expresses $\boldsymbol{\varepsilon}^\star$ as a function of $\boldsymbol{S}, \boldsymbol{C}, \boldsymbol{C}^\star, \boldsymbol{\varepsilon}^0$. Then, an inhomogeneity problem can be reduced to a mathematically equivalent inclusion problem.

Inclusions or inhomogeneities in a bounded matrix

The previous ideas can be extended to the determination of elastic states generated by inclusions in bounded matrices. In this case, it is of course necessary to prescribe boundary conditions on the external boundary $\partial\Omega$. The corresponding elasticity problem then must in general be solved numerically, by means of a boundary element discretization of eq. (14.2). The strain and stress fields can then be computed using the representation formula (14.8).

14.3 SMALL-STRAIN ELASTOPLASTICITY

The boundary integral equation method is applicable to situations involving nonlinearities, both constitutive (e.g. elastoplasticity, viscoplasticity) and geometrical (finite deformations and strains). Such applications have been developed by e.g. Mukherjee [11], Mukherjee and Chandra [12] for quasi-static evolution problems. Ahmad and Banerjee [1], Kontoni and Beskos [8] have proposed boundary element formulations for transient elastic–plastic analyses.

This section is devoted to the boundary integral equation approach to small-strain elastoplasticity problems. As in other approaches, any such algorithm consists of treating the inelastic part of the strain tensor as an 'initial' strain distribution. Contrarily to other situations (thermal strains, inclusion problems ...), the initial strains are not known in advance but must be determined by means of an iterative procedure for each given load step.

Incremental formulation

Elastic–plastic constitutive behaviour gives rise to history-dependent evolution problems. Quasi-static situations are characterized by the absence of a physical time scale: the evolution is then described by a kinematical parameter t (e.g. a proportionality factor of the loading), called 'time' for simplicity. Elastic–plastic analyses generally consist of prescribing loading increments and finding the corresponding increments of mechanical variables (displacements, strains, stresses ...). Many solution strategies exist in the literature. They are often formulated in terms of rates (i.e. 'time' derivatives) of mechanical variables. During the last fifteen years, implicit algorithms, based on the concept of *consistent tangent operator* (CTO) and formulated in terms of finite increments instead of rates, have been proposed in the finite element framework (see Simo and Taylor [15]) and later adapted to boundary element methods (Bonnet and Mukherjee [2]).

Let $\boldsymbol{\varepsilon}^P$ denote the (unknown) plastic strain, such that

$$\boldsymbol{\sigma} = \boldsymbol{C} : (\boldsymbol{\varepsilon} - \boldsymbol{\varepsilon}^P)$$

Basically, the boundary element approach to elastic–plastic analysis is based on the boundary integral equation (14.2) and the integral representation (14.8) in which the inelastic strain ε^P plays the role of initial strain. These equations are discretized using boundary elements for the boundary unknowns $\boldsymbol{u}, \boldsymbol{t}$ and integration cells for the domain integrals containing ε^P. Since the plastic region, i.e. the geometrical support of ε^P, is not known *a priori*, a *potentially plastic* subset Ω_p of Ω, outside of which no plastic strain is expected, must be defined *a priori*. Interpolation functions for the plastic strain ε^P or the total strain ε are defined on the cells. One then obtains matrix equations associated to the integral equation (14.2) and the integral representation (14.8) which have the form:

$$[\boldsymbol{B}]\{\boldsymbol{u}\} - [\boldsymbol{A}]\{\boldsymbol{t}\} = [\boldsymbol{Q}]\{\boldsymbol{C} : \varepsilon^p\}$$
$$\{\varepsilon\} = [\boldsymbol{A}']\{\boldsymbol{t}\} - [\boldsymbol{B}']\{\boldsymbol{u}\} + [\boldsymbol{Q}']\{\boldsymbol{C} : \varepsilon^p\}$$

or, after separation of known and unknown degrees of freedom in $\{\boldsymbol{u}\}, \{\boldsymbol{t}\}$ (the latter being gathered into the vector $\{\boldsymbol{y}\}$) and appropriate column-switching in the matrices $[\boldsymbol{A}], [\boldsymbol{B}]$ and $[\boldsymbol{A}'], [\boldsymbol{B}']$:

$$[\boldsymbol{K}]\{\boldsymbol{y}\} = \{\boldsymbol{f}\} + [\boldsymbol{Q}]\{\boldsymbol{C} : \varepsilon^p\} \tag{14.11}$$
$$\{\varepsilon\} = \{\boldsymbol{f}'\} - [\boldsymbol{K}']\{\boldsymbol{y}\} + [\boldsymbol{Q}']\{\boldsymbol{C} : \varepsilon^p\} \tag{14.12}$$

where $\{\boldsymbol{f}\}, \{\boldsymbol{f}'\}$ incorporate the boundary data. Solving eq. (14.11) for $\{\boldsymbol{y}\}$ and substituting the result into eq. (14.12), one obtains:

$$\{\varepsilon\} = \{\varepsilon^0\} + [\boldsymbol{S}]\{\boldsymbol{C} : \varepsilon^p\} \tag{14.13}$$

with:

$$\{\varepsilon^0\} = \{\boldsymbol{f}'\} - [\boldsymbol{K}'][\boldsymbol{K}]^{-1}\{\boldsymbol{f}\} \qquad [\boldsymbol{S}] = [\boldsymbol{Q}'] - [\boldsymbol{K}'][\boldsymbol{K}]^{-1}[\boldsymbol{Q}]$$

Thus $\{\varepsilon^0\}$ is the total strain under the given loading and in the purely elastic case, i.e. assuming that no plastic strain is present.[1] Then, the elastic constitutive equation $\boldsymbol{C} : \varepsilon^p = \boldsymbol{C} : \varepsilon - \boldsymbol{\sigma}$ is introduced in eq. (14.13) in order to eliminate ε^P. This yields a relation between the total strain and the stress which takes into account the equilibrium and the elastic part of the constitutive equation:

$$[\boldsymbol{S}]\{\boldsymbol{\sigma} - \boldsymbol{C}\varepsilon\} - \{\varepsilon^0\} + [\boldsymbol{I}]\{\varepsilon\} = \{0\} \tag{14.14}$$

To take into account the incremental nature of the problem, the evolution between the time instants t_n and $t_{n+1} = t_n + \Delta t$ is considered. Writing eq. (14.14) for both instants and taking the difference of the two resulting relations gives:

$$[\boldsymbol{S}]\{\Delta\boldsymbol{\sigma}_n - \boldsymbol{C}\Delta\varepsilon_n\} - \{\Delta n_n\} + [\boldsymbol{I}]\{\Delta\varepsilon_n\} = \{0\} \tag{14.15}$$

[1] This should not be mistaken for the elastic strain ε^E in the additive decomposition of the total strain ε.

using the notation $\Delta()_n = ()_{n+1} - ()_n$. The plastic part of the constitutive behaviour still remains to be accounted for. Following the approach of Simo and Taylor [15], the *radial return algorithm* (RRA), symbolically denoted by $\bar{\boldsymbol{\sigma}}$, outputs the stress $\boldsymbol{\sigma}_{n+1}$ given a strain increment $\Delta\varepsilon_n$ and the mechanical variables $(\boldsymbol{\varepsilon}_n, \boldsymbol{\sigma}_n, \bar{e}^p_n)$ at $t = t_n$:

$$\boldsymbol{\sigma}_{n+1} = \bar{\boldsymbol{\sigma}}(\boldsymbol{\varepsilon}_n, \boldsymbol{\sigma}_n, \bar{e}^p_n, \Delta\varepsilon_n) \tag{14.16}$$

where $\bar{e}^p$ denotes the cumulated plastic equivalent strain:

$$\bar{e}^p = \int_0^t \sqrt{\tfrac{2}{3}}(\boldsymbol{d}^p : \boldsymbol{d}^p)\,\mathrm{d}\tau$$

Since the RRA is directly formulated in terms of finite time and strain increments, a rate formulation of the evolution problem is not necessary. After substituting the RRA into eq. (14.15), one obtains a nonlinear system of equations:

$$\begin{aligned}\{G(\Delta\varepsilon_n)\} &\equiv [\boldsymbol{S}]\,\{\bar{\boldsymbol{\sigma}}(\boldsymbol{\varepsilon}_n, \boldsymbol{\sigma}_n, \bar{e}^p_n, \Delta\varepsilon_n) - \boldsymbol{\sigma}_n - \boldsymbol{C}\Delta\varepsilon_n\} - \{\Delta n_n\} + [\boldsymbol{I}]\{\Delta\varepsilon_n\} \\ &= \{0\}\end{aligned} \tag{14.17}$$

whose primary unknown is $\{\Delta\varepsilon_n\}$, the strain increment compatible with both the elastic equilibrium and the plastic constitutive behaviour (left unspecified at this point). Equation (14.17) is then solved by means of an iterative procedure. For example, using the Newton method, the additive correction $\delta\varepsilon^i_n = \Delta\varepsilon^{i+1}_n - \Delta\varepsilon^i_n$ to $\Delta\varepsilon^i_n$ solves the linear system of equations:

$$([\boldsymbol{S}][\boldsymbol{C} - \boldsymbol{C}^i_{n+1}] - [\boldsymbol{I}])\{\delta\varepsilon^i_n\} = \{G(\Delta\varepsilon^i_n)\} \tag{14.18}$$

where $\boldsymbol{C}_{n+1} \equiv \frac{\partial\bar{\sigma}}{\partial\Delta\varepsilon_n}(\boldsymbol{\varepsilon}_n, \boldsymbol{\sigma}_n, \bar{e}^p_n, \Delta\varepsilon_n)$ and $([\boldsymbol{S}][\boldsymbol{C} - \boldsymbol{C}^i_{n+1}] - [\boldsymbol{I}])$ are respectively the local and global consistent tangent operators. The iterative procedure is carried out until convergence, and the value of $\{\Delta\varepsilon_n\}$ thus obtained is used to update the mechanical variables:

$$\boldsymbol{\varepsilon}_{n+1} = \boldsymbol{\varepsilon}_n + \Delta\varepsilon_n \qquad \boldsymbol{\sigma}_{n+1} = \bar{\boldsymbol{\sigma}}(\boldsymbol{\varepsilon}_n, \boldsymbol{\sigma}_n, \bar{e}^p_n, \Delta\varepsilon_n)$$

It is interesting to note that the Newton step (14.18) involves the difference $[\boldsymbol{C} - \boldsymbol{C}^i_{n+1}]$ between the elastic constitutive law and the local CTO, rather than the local CTO itself; this is entirely consistent with the fact that eq. (14.15) accounts for equilibrium as well as the elastic constitutive law.

Observe that at those material points where the incremental deformation is purely elastic (i.e. trial stress lies within the yield surface), the local CTO coincides with $\boldsymbol{C}$. The implication is that the quasi-diagonal matrix $[\boldsymbol{C} - \boldsymbol{C}^i_{n+1}]$ can have zero blocks on the diagonal, leading to corresponding zero block columns in the matrix $[\boldsymbol{D}^i_{n+1}]$. Hence, the Newton step (14.18) admits the following block decomposition:

$$([\boldsymbol{D}^i_{n+1}] - [\boldsymbol{I}])_{PP}\{\delta\varepsilon^i_n\}_P = \{G(\Delta\varepsilon^i_n)\}_P \tag{14.19}$$

$$\{\delta\varepsilon^i_n\}_E = ([\boldsymbol{D}^i_{n+1}])_{EP}\{\delta\varepsilon^i_n\}_P - \{G(\Delta\varepsilon^i_n)\}_E \tag{14.20}$$

where the subscripts E, P indicate partitionings based on currently elastic (E) or plastic (P) nodes. Also, use is made of the conglomeration of zero block columns:

$$[\boldsymbol{D}^i_{n+1}]_{PE} = [\boldsymbol{D}^i_{n+1}]_{EE} = [0]$$

In effect, (14.19) says that the size of the linear system ties in with the size of the plastically deforming zone, representing significant savings in computer time.

For completeness, the RRA is now explicitly given for a specific constitutive behaviour, namely isotropic elasticity and a von Mises yield criterion ($\boldsymbol{s} = \boldsymbol{\sigma} - \frac{1}{3}\mathrm{Tr}(\boldsymbol{\sigma})$: deviatoric stress, $\boldsymbol{\alpha}$: kinematical hardening variable):

$$f(\boldsymbol{\xi}, \kappa) \equiv \sqrt{\boldsymbol{\xi} : \boldsymbol{\xi}} - \sqrt{\tfrac{2}{3}}\kappa(\bar{e}^p) \leq 0 \qquad (\boldsymbol{\xi} = \boldsymbol{s} - \boldsymbol{\alpha}) \tag{14.21}$$

(the threshold function $\kappa(\bar{e}^p)$ being given). Incompressibility of plastic strain is assumed.

First, the (elastic) trial deviatoric stress $\boldsymbol{\xi}^T_{n+1}$ is defined by:

$$\boldsymbol{\xi}^T_{n+1} = \boldsymbol{s}_n + 2\mu\Delta \boldsymbol{e}_n - \boldsymbol{\alpha}_n$$

($\boldsymbol{e}$: deviatoric strain). If $f(\boldsymbol{\xi}^T_{n+1}, \kappa_n) \leq 0$, i.e. $\boldsymbol{\xi}^T_{n+1}$ is elastic, one has

$$\bar{\boldsymbol{\sigma}} = K\Delta\varepsilon_n : (\mathbf{1} \otimes \mathbf{1}) + 2G\Delta \boldsymbol{e}_n + \boldsymbol{\sigma}_n$$

On the other hand, if $f(\boldsymbol{\xi}^T_{n+1}, \kappa_n) > 0$, $\bar{\boldsymbol{\sigma}}$ is given by the following equations, which constitute the RRA [15]:

$$\bar{\boldsymbol{\sigma}} = K\Delta\varepsilon_n : (\boldsymbol{\delta} \otimes \boldsymbol{\delta}) + \boldsymbol{s}_{n+1} \tag{14.22}$$

$$\boldsymbol{s}_{n+1} = \boldsymbol{\alpha}_n + \sqrt{\tfrac{2}{3}}(\kappa_{n+1} + \Delta H_n)\hat{\boldsymbol{n}} \tag{14.23}$$

$$\bar{e}^p_{n+1} = \bar{e}^p_n + \sqrt{\tfrac{2}{3}}\gamma\Delta t$$

$$\hat{\boldsymbol{n}} = \boldsymbol{\xi}^T_{n+1} / \left|\boldsymbol{\xi}^T_{n+1}\right|$$

$$\boldsymbol{\alpha}_{n+1} = \boldsymbol{\alpha}_n + \sqrt{\tfrac{2}{3}}\Delta H_n \hat{\boldsymbol{n}} \tag{14.24}$$

where $K = (3\lambda + 2\mu)/3$ is the compressibility modulus and with $\Delta H_n = H_{n+1} - H_n$, $H_n = H(\bar{e}^p_n)$, the function H being given. The scalar $\gamma\Delta t$ solves a nonlinear equation expressing that the yield criterion is verified at $t = t_{n+1}$:

$$\left|\boldsymbol{\xi}^T_{n+1}\right| - \sqrt{\tfrac{2}{3}}\kappa(\bar{e}^p_n + \sqrt{\tfrac{2}{3}}\gamma\Delta t) - 2\mu\gamma\Delta t - \sqrt{\tfrac{2}{3}}\Delta H_n = 0 \tag{14.25}$$

Finally, the local CTO is given in this case by [15]:

$$\boldsymbol{C}_{n+1} = \boldsymbol{C} - 2\mu\left[(\beta - \delta)\hat{\boldsymbol{n}} \otimes \hat{\boldsymbol{n}} + (1 - \beta)(\boldsymbol{I} - \frac{1}{3}\boldsymbol{\delta} \otimes \boldsymbol{\delta})\right] \tag{14.26}$$

where $\boldsymbol{\delta}$ is the second-order identity tensor, $2I_{abcd} = \delta_{ac}\delta_{bd} + \delta_{ad}\delta_{bc}$ and:

$$\beta = \frac{1}{\left|\xi^T_{n+1}\right|}\sqrt{\tfrac{2}{3}}(\Delta H_n + \kappa_{n+1}) \qquad \delta = \frac{\kappa'_{n+1} + H'_{n+1}}{\kappa'_{n+1} + H'_{n+1} + 3\mu}$$

Computational algorithm

The following pseudo-code outlines the overall solution procedure, from initial time t_0 (which corresponds to the first yield load) to final time t_{N_T}.

For $0 \leq n \leq (N_T - 1)$:

1. – *Compute* $\{\Delta n_n\}$ *(purely elastic internal strain)*
2. – *Initialize* $\{\Delta\varepsilon^0_n\}$ *(e.g. to the elastic value)*
   $i = 0$

   Iterative solution of (14.17):

   (a) $i = 0$
   (b) Compute the residual $\{G(\Delta\varepsilon^i_n)\}$ *from (14.17)*
   (this requires iterative solution of $\gamma\Delta t$ *from the RRA)*
   (c) Convergence test: if $\{G(\Delta\varepsilon^i_n)\} \leq$ EPS, GOTO 3.
   (d) $i := i + 1$
   (e) Compute the local CTOs $\boldsymbol{C}^i_{n+1}$ *at all nodes;*
   classify into currently elastic (E) and plastic (P) sets.
   (f) Set up and factor the global CTO $[\boldsymbol{S}(\boldsymbol{C} - \boldsymbol{C}^i_{n+1}) - \boldsymbol{I}]_{PP}$,
   set up $[\boldsymbol{S}(\boldsymbol{C} - \boldsymbol{C}^i_{n+1}) - \boldsymbol{I}]_{EP}$.
   (g) Solve (14.19) for $\{\delta\varepsilon^i_n\}_P$ *and compute* $\{\delta\varepsilon^i_n\}_E$ *using (14.20).*
   (h) Update: $\{\Delta\varepsilon^i_n\} := \{\Delta\varepsilon^i_n\} + \{\delta\varepsilon^i_n\}$.
   (i) GOTO *(b) (start new iteration).*

3. – *Update:* $\{\bar{e}^p\}_{n+1} = \{\bar{e}^p\}_n + \sqrt{\tfrac{2}{3}}\{\gamma\Delta t\}$,
   $\{\sigma_{n+1}\} = \{\bar{\sigma}(\Delta\varepsilon^i_{n+1})\}$,
   $\{\varepsilon_{n+1}\} = \{\varepsilon_n\} + \{\Delta\varepsilon_n\}$

End for

Discussion

The boundary element/domain cell treatment is also applicable to e.g. elasto-viscoplasticity, and also to geometrically nonlinear problems (finite strains), see Mukherjee and Chandra [12]. Galerkin boundary element techniques (see Chapter 10) have also been proposed for elastoplasticity by e.g. Maier et al. [10]).

Most numerical algorithms devoted to nonlinear mechanical analyses are based on finite elements rather than boundary elements. This may result from several factors:

- The more advanced state of the art in finite element methods.
- The domain cells are discretization entities very close to finite elements, the main difference being that they are not necessarily used for the

whole body. Their use tends to weaken the advantages usually provided by the boundary element techniques.

- The rather complex character of integral formulations, especially concerning the evaluation of strains or stresses at points interior to the potentially plastic region. Also, when many numerical evaluations of the integral representation formula of strain or stress are necessary, the computational costs become substantial. The evaluation of singular domain integrals in the regularized formula (14.8) requires careful implementation.
- The boundary integral approach is more or less limited to piecewise homogeneous elastic properties, because otherwise the availability of a fundamental solution for the linear part of the constitutive behaviour is problematic.

However, it is also important to point out that:

- The domain discretization into integration cells is restricted to the geometrical support Ω_p of plastic strains. The boundary element treatment is best suited for nonlinear analyses when the size of Ω_p is small compared to the overall size of Ω. Recall that in practice the exact plastic region is unknown, so that Ω_p is a *potentially plastic* region, outside of which no plastic strain is expected, which must be guessed *a priori*.
- In particular, a boundary element approach is worthwhile for unbounded domains containing a plastic region of finite extension. Such configurations arise for example in geotechnical analyses (tunnels, foundations ...).
- Contrarily to finite element methods, the stress fields evaluated using eq. (14.8) satisfy for any distribution ε^I or $\boldsymbol{\sigma}^I$ the local equilibrium equation. The iterative procedure is meant to ensure verification of the constitutive equation and yield criterion.

The understanding of key computational issues, like the correct treatment of three-dimensional singular domain integrals, has seen substantial advances in the recent past. The range of applications of boundary element techniques is accordingly expected to expand, notably to three-dimensional situations like elastic–plastic fracture mechanics where the plastic region is expected to be confined in a neighbourhood of the crack front.

14.4 EXAMPLES

Plane-strain implementation of the implicit CTO-based approach

The results presented here (Poon, Mukherjee and Bonnet [14]) relate to internal pressure loading of an elastoplastic hollow cylinder, in plane strain.

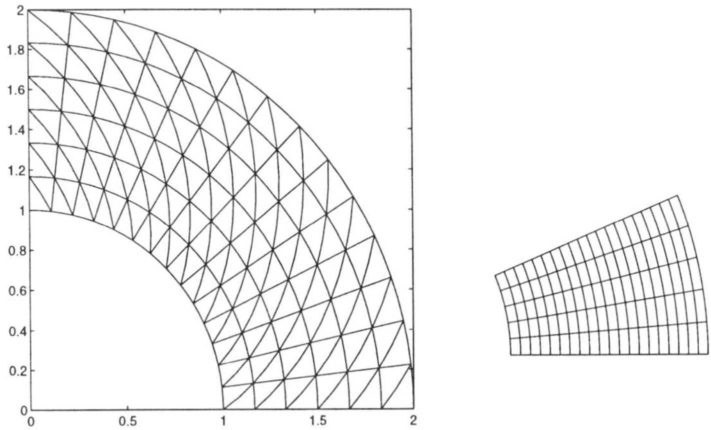

Figure 14.1 BEM mesh (with integration cells) and FEM meshes used for the hollow cylinder example (from Poon et al. [14]).

Even though the physics of this problem is really one-dimensional, a general 2D implementation of the present implicit formulation is employed here. Despite the apparent physical simplicity of the problem at hand, closed-form solutions, describing the evolution of mechanical variables as the plastic front moves outward, are not available for the von Mises yield function case. Here, the commercial finite-element code ABAQUS provides a reference solution.

The hollow cylinder has inner radius 1 and outer radius 2. The elastic constants are: $\mu = 1$, $\nu = 0.3$. The material deforms plastically according to the classical J_2 theory, with isotropic strain-hardening (i.e. $\alpha = \mathbf{0}$) defined by the Mises threshold (see eq. (14.21)):

$$\kappa(\bar{e}^p) = 2G(0.001 + 0.001(\bar{e}^p)^m)$$

where m takes on values of 1 or 0.2 depending on the example. The usual practice of non-dimensionalization (e.g. dividing stresses by the shear modulus, radial distances by the inner radius, etc.) becomes unnecessary with the current choice of material parameters.

The BEM implementation uses 3-noded boundary elements and 6-noded, triangular internal cells. The FEM setup (ABAQUS) employs 8-noded quadrilateral elements. To exploit symmetry, only one quarter of the body is modelled in the BEM implementation (meshes are shown in Figure 14.1).

Example 1. This example traces the loading history of the elastoplastic cylinder under increasing internal pressure loading. Linear hardening is considered here, so that the Mises threshold $\kappa(\bar{e}^p)$ of eq. (14.21) takes the form:

$$\kappa(\bar{e}^p) = 2\mu(0.001 + 0.001\bar{e}^p)$$

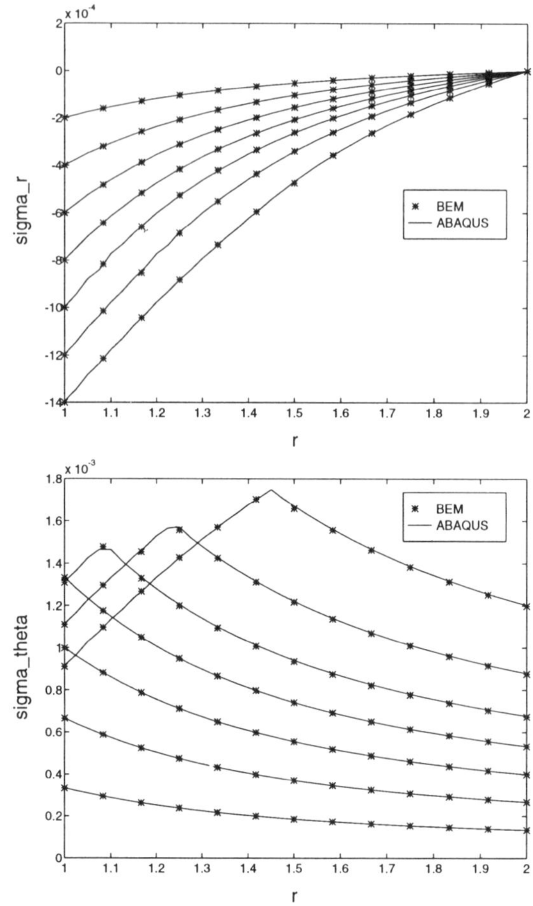

Figure 14.2 Pressurized hollow cylinder: radial (top) and hoop (bottom) stress (linear hardening), comparison of BEM and FEM results (from Poon et al. [14]).

The pressure at first yield is $p_y \approx 8.7 \times 10^{-4}$. The maximum pressure applied here, 14×10^{-4}, is such that a significant amount of plastic flow has occurred, yet the plastic front remains inside the body. The total load is divided into 7 equal increments, the first 4 of which are elastic.

Figure 14.2 shows the evolution of radial and hoop stress profiles. The agreement between BEM and FEM results is excellent.

Example 2. This example concerns nonlinear isotropic hardening of the form:

$$\kappa = 2G(0.001 + 0.001(\bar{e}^p)^m)$$

with $m = 0.2$. The internal pressure of 19×10^{-4} is chosen so that a good portion (but not all) of the cylinder has yielded. This time, only one increment

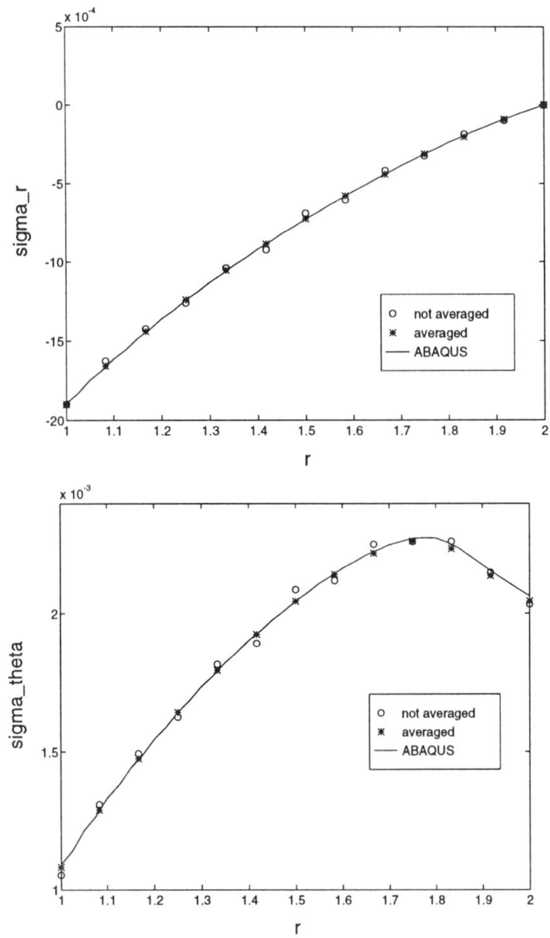

Figure 14.3 Pressurized hollow cylinder: radial (top) and hoop (bottom) stress (nonlinear hardening), comparison of BEM and FEM results (from Poon et al. [14]).

is used (re-analysis using multiple increments confirms the validity of this choice). The focus here is on the oscillatory behaviour exhibited by the BEM results in the absence of averaging.

Figure 14.3 shows the radial and hoop stress. The circles, which mark the BEM results without averaging (along a radial edge in the quarter hollow cylinder), follow a visible pattern of small oscillations around the solid line (ABAQUS reference solution). The oscillation is largely reduced upon averaging (see the crosses), implying a higher level of accuracy globally than on the boundary. The origin of the oscillation appears to be the use of traction and tangential derivatives of displacements (as explained in Section 4.7) to evaluate stresses. As a matter of fact, Guiggiani [5] reported the same situation in linear elasticity and showed that evaluating the hypersingular representation formula (in his case using the direct approach) provides much better results.

Sliding/rolling contact on an elastoplastic body

This work (Lederer, Bonnet and Maitournam [9]) is concerned with a boundary integral approach to the determination of mechanical fields during rolling or sliding of an object on an elastic–plastic semi-infinite medium, using the steady-state assumption ([4], Figure 14.4): the equations are formulated in a frame moving (with constant velocity) with the load, with a considerable computational gain over procedures solving the evolution problem in a fixed frame.

If a purely elastic constitutive behaviour is assumed, analytical solutions are obtained using the fundamental solution of Mindlin (in three dimensions) or Melan (in plane strain), under the classical idealization of contacting half-spaces (see Johnson [7]). In the elastic-plastic case and under the small-strain assumption, a potentially plastic region, discretized using integration cells, is assumed beneath the contact zone. In addition, the contact pressure, when unknown, must be determined by an iterative procedure and a potential contact

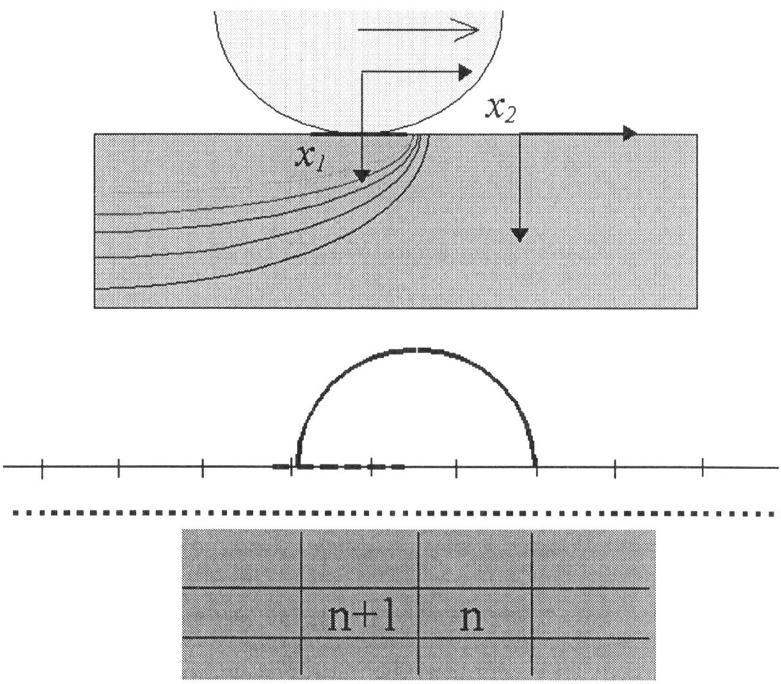

Figure 14.4 Contact and steady-state rolling/sliding: principle (top) and discretisation using boundary elements and integration cells (bottom), from Lederer et al. [9]).

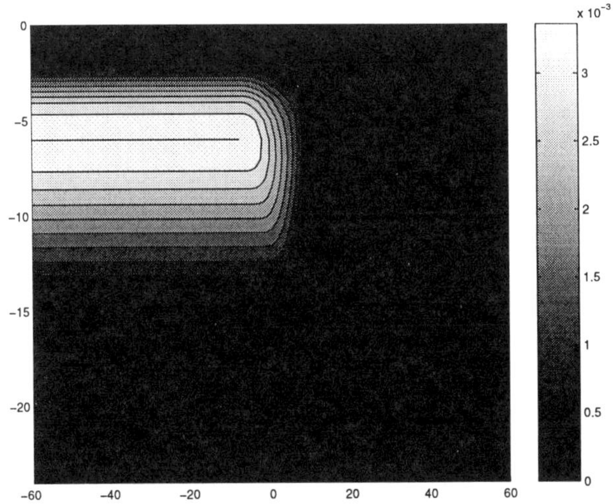

Figure 14.5 Isovalues of $\bar{e}^p$ for first load passage (from Lederer et al. [9]).

area, discretized using boundary elements, must be introduced (Figure 14.4). In any case, the discretized area is of finite extent, and the behaviour of the solution at infinity is accounted for by the integral formulation. In contrast, a finite element treatment requires a computational region of size much larger than the contact area and the plastic zone, as well as artificial additional boundary conditions (zero displacement).

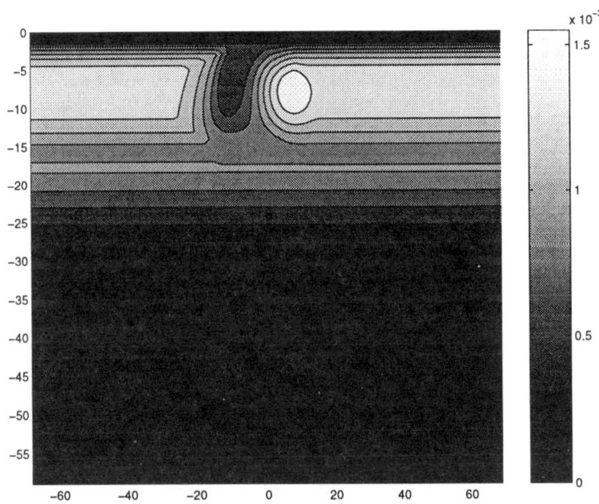

Figure 14.6 Isovalues of $\bar{e}^p$ in stabilized regime (from Lederer et al. [9]).

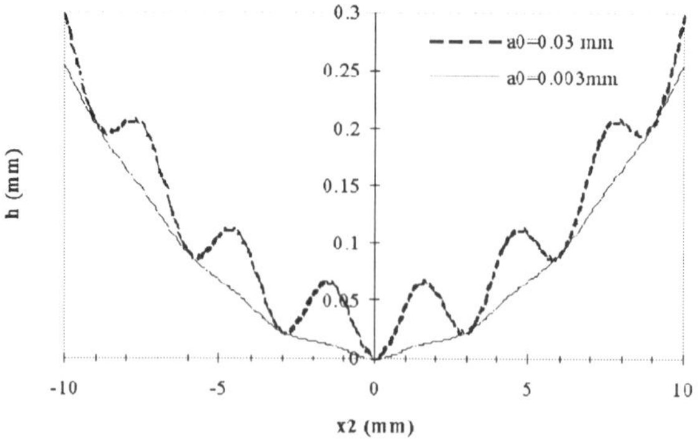

Figure 14.7 Contact and steady-state rolling/sliding: 'wavy' contact profiles (amplitudes $a_0 = 0.003$ mm and $a_0 = 0.03$ mm), from Lederer et al. [9].

It is important to emphasize that, due to the steady-state character of the problem, the horizontal spatial coordinate x_1 (parallel to the free surface) plays the role of physical time. The strain ε has a nonzero limiting value (at least within a region of finite depth) as $x_1 \to -\infty$ (permanent strain left after the first passage of the moving load) and even as $x_1 \to +\infty$ (permanent strain after the previous pass of the moving load, in the case of repeated passes).

The constitutive integration in time is performed by means of a variant of the implicit radial return algorithm (Section 14.3) which incorporates the steady-state character of the problem.

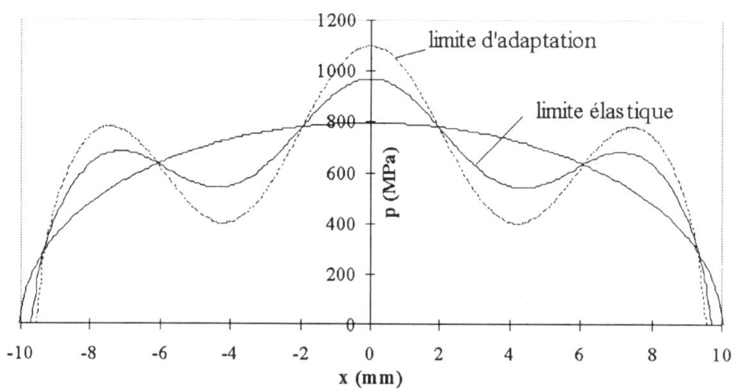

Figure 14.8 Contact and steady-state rolling/sliding: contact pressure distribution corresponding to the elasticity limit ($a_0 = 0.004$ mm) and the adaptation limit ($a_0 = 0.007$ mm) for frictionless 'wavy' contact (from Lederer et al. [9]).

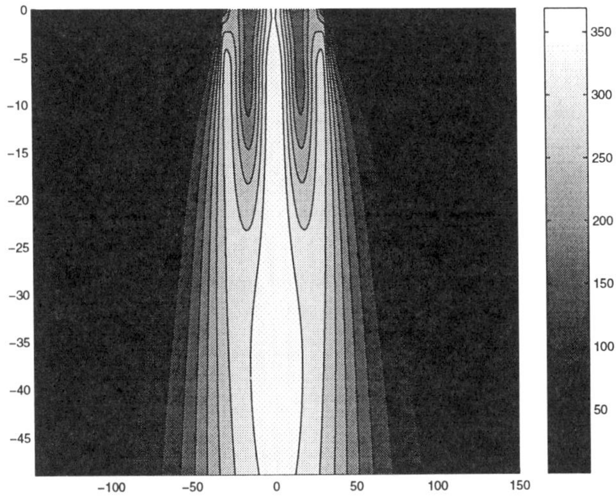

Figure 14.9 Von Mises equivalent stress in stabilized regime for 'wavy' pressure distribution (from Lederer et al. [9]).

Figures 14.5, 14.6 correspond to an example of stabilized cycle for steady-state rolling on a semi-infinite elastic–plastic medium ($E = 210000$ MPa, $\nu = 0.3$, von Mises yield function with kinematic linear hardening: elasticity limit $k = 159$, hardening modulus $C = 6900$ MPa). The contact is assumed Hertzian (contact pressure known *a priori*); the maximum contact pressure on the contact zone (centred at $x_1 = 0$, half-width 10mm) is $p_0 = 5k$. Figure 14.5 depicts the distribution of equivalent plastic strain for the first loading pass. When a stabilized state is reached (after several passes), the plastic strains are equal at $x_1 = \pm\infty$ but the plastic strain distribution is not uniform along the x_1-direction (see Figure 14.6).

Figure 14.9 depicts the distribution of von Mises equivalent stresses under a 'wavy' punch (Figure 14.7) (irregularity of amplitude 0.007 millimetres); this time the contact pressure has been determined by means of an iterative process (Figure 14.8). The stress distribution differs strongly from the Hertzian case, with substantial interaction between the three pressure 'peaks'.

REFERENCES

1. Ahmad, S., Banerjee, P. K. Inelastic Transient elastodynamic analysis of three-dimensional problems by BEM. *Int. J. Num. Meth. in Eng.*, **29**, 371–390 (1990).

2. Bonnet, M., Mukherjee, S. The consistent tangent operator and BEM formulations of regular and sensitivity problems in elasto-plasticity. In Shan-Fu

Shen, Paul Dawson (eds), *Simulation of materials processing : theory, methods and applications*, pp. 33–38. A. A. Balkema (1995). (Numiform'95, USA).

3. BUI, H. D. Some remarks about the formulation of three-dimensional thermoelastoplastic problems by integral equations. *Int. J. Solids Struct.*, **14**, 935–939 (1978).

4. DANG VAN, K., MAITOURNAM, M. H. Steady-state flow in classical elastoplasticity: applications to repeated rolling and sliding contact. *J. Mech. Phys. Solids*, **41**, 1691–1710 (1993).

5. GUIGGIANI, M. Hypersingular formulation for boundary stress evaluation. *Engng. Anal. with Bound. Elem.*, **14**, 169–179 (1994). (invited paper, special issue).

6. GUIGGIANI, M., GIGANTE, A. A general algorithm for multidimensional Cauchy principal value integrals in the boundary element method. *ASME J. Appl. Mech.*, **57**, 906–915 (1990).

7. JOHNSON, K. L. *Contact mechanics*. Cambridge University Press (1985).

8. KONTONI, D. P. N., BESKOS, D. E. Application of the DR-BEM to inelastic dynamic problems. In J. Dominguez, C. A. Brebbia, F. Paris (eds), *Boundary elements XIV*. Comp. Mech. Publ., Southampton (1992).

9. LEDERER, G., BONNET, M., MAITOURNAM, H. M. Modélisation par équations intégrales du frottement sur un demi-espace élasto-plastique. In *Troisième Colloque National de Calcul des Structures*, pp. 435–440. Presses Académiques de l'Ouest (1997).

10. MAIER, G., MICCOLI, S., NOVATI, G., SIRTORI, S.. A Galerkin symmetric boundary-element method in plasticity: formulation and implementation. In J. H. Kane, G. Maier, N. Tosaka, S. N. Atluri (eds), *Advances in boundary element techniques*, pp. 288–328. Springer-Verlag (1993).

11. MUKHERJEE, S. *Boundary element methods in creep and fracture*. Elsevier (1982).

12. MUKHERJEE, S., CHANDRA, A. Nonlinear solid mechanics. In D. E. Beskos (ed.), *Boundary element methods in mechanics.*, vol. 1, pp. 285–332. North-Holland (1987).

13. MURA, T. *Micromechanics of defects in solids*. Martinus Nijhoff (1982).

14. POON, H., MUKHERJEE, S., BONNET, M. Numerical implementation of a CTO-based implicit approach for the BEM solution of usual and sensitivity problems in elasto-plasticity. *Engng. Anal. with Bound. Elem.*, **22**, 257–269 (1998).

15. SIMO, J. C., TAYLOR, R. L. Consistent tangent operators for rate-independent elastoplasticity. *Comp. Meth. in Appl. Mech. Engng.*, **48**, 101–118 (1985).

Appendices

Appendix A
Differential operators and integration by parts on surfaces

Tangential differential operators

Consider a twice continuously differentiable (C^2) surface S of unit normal $\boldsymbol{n}$. Denote by $\boldsymbol{\nu}$ the unit normal exterior to the edge ∂S and lying in the tangent plane to S and let $\boldsymbol{\tau} = \boldsymbol{n} \wedge \boldsymbol{\nu}$ be the unit tangent to ∂S.

Consider a scalar function $f(\boldsymbol{y})$, $\boldsymbol{y} \in S$. In some cases, f is not defined outside of S (e.g. $f = n_i$ or $f = \phi_i$, the i-component of the crack opening displacement) and the Cartesian derivatives $f_{,i}$ are not defined. One can nevertheless define a continuation $\hat{f}$ of f in an open neighbourhood V of S. The simplest way to do so is:

$$(\boldsymbol{y} \in V) \qquad \hat{f}(\boldsymbol{y}) = f(\mathcal{P}(\boldsymbol{y})) \tag{A.1}$$

where $\mathcal{P}(\boldsymbol{y})$ is the orthogonal projection of $\boldsymbol{y} \in V$ onto S. The trace of $\hat{f}$ on S is clearly f. Moreover, the normal derivative of $\hat{f}$ vanishes, i.e. the gradient of $\hat{f}$ is tangent to S. The *tangential gradient* of f, denoted by $\boldsymbol{\nabla}_S$, is then defined by:

$$\boldsymbol{\nabla}_S f = \boldsymbol{\nabla}_S \hat{f} = \boldsymbol{\nabla} \hat{f} \tag{A.2}$$

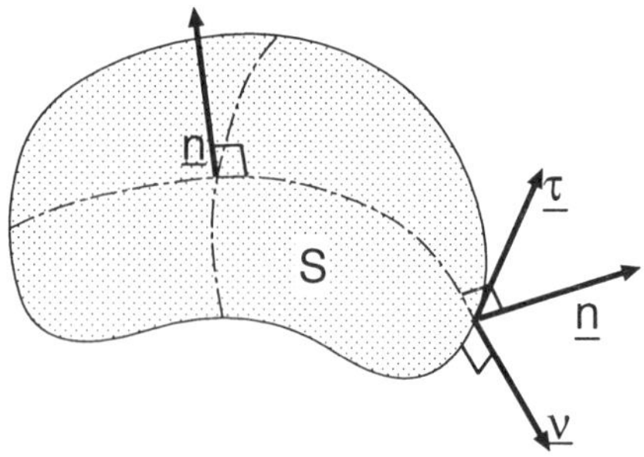

Figure A.1 Open surface: notations.

and can be expressed in terms of tangential partial derivatives D_r:

$$\nabla_S f = \hat{D}_r f \boldsymbol{e}_r \qquad \hat{D}_r f = (\nabla_S f).\boldsymbol{e}_r \tag{A.3}$$

If f is an arbitrary scalar function defined in V, the tangential gradient $\nabla_S f$ is given by:

$$\nabla_S f = \nabla f - \boldsymbol{n} f_{,n} \qquad D_r f = f_{,r} - n_r f_{,n} \tag{A.4}$$

Integration by parts of combinations of Cartesian partial derivatives

The classical Stokes identity for a vector field $\boldsymbol{u}$ defined in V, expressed in Cartesian components, reads:

$$\int_S e_{abc} n_a u_{c,b} \,\mathrm{d}S = \int_{\partial S} u_a \tau_a \,\mathrm{d}s \tag{A.5}$$

Substituting a vector field of the $\boldsymbol{u} = \boldsymbol{e}_j f$, or $u_c = \delta_{cj} f$ into the above equation yields the following identity:

$$\int_S e_{jab} f_{,b} n_{,a} \,\mathrm{d}S = \int_{\partial S} f \tau_j \,\mathrm{d}s \tag{A.6}$$

Further, the inner product of this equation by the permutation symbol e_{jrs} ((r, s) being any fixed pair of indices) is taken. Since the permutation symbol verifies the algebraic identity:

$$e_{jrs} e_{jab} = \delta_{ra}\delta_{sb} - \delta_{rb}\delta_{sa} \tag{A.7}$$

a variant of eq. (A.6) is obtained:

$$\int_S D_{rs} f \,\mathrm{d}S = e_{jrs} \int_{\partial S} f \tau_j \,\mathrm{d}s \tag{A.8}$$

where the differential operator D_{rs} is defined by:

$$D_{rs} f = e_{jrs} e_{jab} n_a f_{,b} = n_r f_{,s} - n_s f_{,r} \tag{A.9}$$

In view of eq. (A.4), the definition (A.9) becomes:

$$D_{rs}f = n_r D_s f - n_s D_r f$$

which shows clearly that $D_{rs}f$ is a tangential differential operator, i.e. involves only derivatives along the surface S.

Integration by parts of a partial tangential derivative

Applying the Stokes identity (A.5) to the vector field $\boldsymbol{u} = \boldsymbol{n} \wedge \boldsymbol{e_r} f$, i.e. $u_c = e_{cdr} n_d f$ for some fixed index r leads to the identity:

$$\int_S (-n_r K f + D_r f)\,\mathrm{d}S = \int_{\partial S} f \nu_r\,\mathrm{d}s \tag{A.10}$$

where $K(\boldsymbol{y})$ is (minus twice) the mean curvature on S:

$$K(\boldsymbol{y}) = \mathrm{div}_S \boldsymbol{n} = D_r n_r(\boldsymbol{y}) = -(R_1^{-1} + R_2^{-1})$$

- Formulas (A.6), (A.8) and (A.10) still hold true when the surface S is *piecewise smooth*, provided $f(\boldsymbol{y})$ is continuous across the edges of S. For example, if S is made of two smooth surfaces S_1, S_2 connected to one another along the line Γ, eq. (A.8) leads to:

$$\int_{S_1 \cup S_2} D_{rj} f\,\mathrm{d}S = \varepsilon_{irj} \left\{ \int_{\partial S} f\tau_i\,\mathrm{d}s + \int_\Gamma f(\tau_i^1 + \tau_i^2)\,\mathrm{d}s \right\} \tag{A.11}$$

  and the integral over Γ vanishes since, using consistent orientation conventions on S_1 and S_2, $\boldsymbol{\tau}^1 + \boldsymbol{\tau}^2 = \boldsymbol{0}$ on Γ.

- If S is piecewise smooth and closed, the contour integrals appearing in the right-hand sides of eqs (A.6), (A.8) and (A.10) vanish.

Formulas expressed in terms of a parametrization of S

Let S be defined by a parametrization $\boldsymbol{\xi} \in \Delta \rightarrow \boldsymbol{y} \in S$. The gradient of f defined in V is given by:

$$\boldsymbol{\nabla} f = f_{,\alpha}\boldsymbol{a}^\alpha + f_{,n}\boldsymbol{n} = f_{,\alpha} g^{\alpha\beta} \boldsymbol{a}_\beta + f_{,n}\boldsymbol{n} = f_{,\alpha} g^{\alpha\beta} (\boldsymbol{a}_\beta . \boldsymbol{e}_r) \boldsymbol{e}_r + f_{,n}\boldsymbol{n} \tag{A.12}$$

where $\boldsymbol{a}_\alpha$ $\boldsymbol{a}^\alpha = g^{\alpha\beta}\boldsymbol{a}_\beta$ $(\alpha = 1,2)$ are the natural basis and its dual, the $g^{\alpha\beta}$ being the contravariant components of the metric tensor. The expression of the *surface gradient* $\boldsymbol{\nabla}_S f$, which also applies when F is defined only on S, follows:

$$\boldsymbol{\nabla}_S f = D_r f \boldsymbol{e}_r = \boldsymbol{\nabla} f - \boldsymbol{n} f_{,n} = f_{,\alpha} g^{\alpha\beta} (\boldsymbol{a}_\beta . \boldsymbol{e}_r) \boldsymbol{e}_r \tag{A.13}$$

Now consider a vector field $\boldsymbol{u}(\boldsymbol{\xi}) = u_r(\boldsymbol{\xi})\boldsymbol{e}_r$ defined on S, whose Cartesian components u_r are functions of $\boldsymbol{\xi} \in \Delta$. In that case, identity (A.13) is applicable to each component u_r, which yields the following expression for the surface divergence $\text{div}_S \boldsymbol{u}$:

$$\text{div}_S \boldsymbol{u} = D_r u_r = \boldsymbol{\nabla}_S \tilde{f}.\boldsymbol{e}_r = g^{\alpha\beta} \boldsymbol{u}_{,\alpha}.\boldsymbol{a}_\beta \tag{A.14}$$

Taking into account the algebraic identity $g_{\alpha\beta}g^{\beta\gamma} = \delta_\alpha^\gamma$, eq. (A.14) takes the more explicit form:

$$\text{div}_S \boldsymbol{u} = \frac{1}{J}\Big\{ \boldsymbol{a}_1.(g_{22}\boldsymbol{u}_{,1} - g_{12}\boldsymbol{u}_{,2}) + \boldsymbol{a}_2.(g_{11}\boldsymbol{u}_{,2} - g_{12}\boldsymbol{u}_{,1}) \Big\} \tag{A.15}$$

where:

$$J(\boldsymbol{\xi}) = |\boldsymbol{a_1} \wedge \boldsymbol{a_2}|$$

In particular, one has:

$$\text{div}_S \boldsymbol{n} = -\boldsymbol{n}.(g^{\alpha 1}\boldsymbol{a}_{1,\alpha} + g^{\alpha 2}\boldsymbol{a}_{2,\alpha}) \tag{A.16}$$

Finally, notice that eq. (A.9) can be written as a vector product:

$$D_{rs} f = e_{ars}(\boldsymbol{n} \wedge \boldsymbol{\nabla}_S f).\boldsymbol{e}_a$$

and recall (chapter 3, eq. (3.6)) that:

$$\boldsymbol{n} = \frac{1}{J}(\boldsymbol{a_1} \wedge \boldsymbol{a_2})$$

Applying the double vector product formula, one then obtains:

$$\begin{aligned} J(\boldsymbol{n} \wedge \boldsymbol{\nabla}_S \boldsymbol{f}) &= f_{,\gamma} g^{\gamma\beta} (\boldsymbol{a_1} \wedge \boldsymbol{a_2}) \wedge \boldsymbol{a_\beta} \\ &= f_{,\gamma} g^{\gamma\beta} [g_{1\beta}\boldsymbol{a}_2 - g_{2\beta}\boldsymbol{a}_1] = f_{,1}\boldsymbol{a}_2 - f_{,2}\boldsymbol{a}_1 \end{aligned}$$

and the expression of D_{rs} in terms of the surface parametrization follows:

$$J D_{rs} f \,\mathrm{d}S_y = e_{ars}\Big[(\boldsymbol{a}_2.\boldsymbol{e}_a) f_{,1} - (\boldsymbol{a}_1.\boldsymbol{e}_a) f_{,2} \Big] \,\mathrm{d}\boldsymbol{\xi} \tag{A.17}$$

Appendix B
Interpolation functions and numerical integration

This appendix provides some details relative to the boundary element method as presented in Chapter 3:

- Usual interpolation functions in one and two dimensions;
- Modified interpolation functions used for singular integration on elements;
- Standard Gauss points and weights used for nonsingular numerical integration over segments[1] and triangles;
- Modified Gauss points and weights used for integrating logarithmic singularities over segments.

The modified interpolation functions are specific to boundary element methods. By materializing the singularity weakening brought by regularization, they contribute to correct, stable and accurate singular integration and thus have great practical importance. The other information given in this appendix, mostly for self-sufficiency purposes, is standard.

[1]Recall that numerical integrations over a quadrilateral region are performed using products of one-dimensional Gaussian rules, see Section 3.4.

B.1 ONE-DIMENSIONAL INTERPOLATION

One-dimensional interpolation is used in boundary element methods for two-dimensional geometrical configurations. In this case, the modified interpolation functions $\hat{N}$ are defined in terms of the standard interpolation functions N by:

$$N(\xi) - N(\eta) = \rho \hat{N}(\rho, \eta) \qquad \rho \equiv \xi - \eta$$

Two-noded linear element

The standard linear interpolation functions are given by:

$$N_1(\xi) = \frac{1}{2}(1 - \xi) \qquad N_2(\xi) = \frac{1}{2}(1 + \xi)$$

and the modified interpolation functions are then:

$$\hat{N}_1(\rho; \eta) = -\frac{1}{2} \qquad \hat{N}_2(\rho; \eta) = \frac{1}{2}$$

This element results in a piecewise straight approximate boundary.

Three-noded quadratic element

The standard quadratic interpolation functions are given by:

$$N_1(\xi) = \frac{1}{2}\xi(\xi - 1) \quad N_2(\xi) = 1 - \xi^2 \quad N_3(\xi) = \frac{1}{2}\xi(\xi + 1) \tag{B.1}$$

and the modified interpolation functions are then:

$$\begin{aligned} \hat{N}_1(\rho; \eta) &= \eta + \frac{1}{2}(\rho - 1) \\ \hat{N}_2(\rho; \eta) &= -2\eta - \rho \\ \hat{N}_3(\rho; \eta) &= \eta + \frac{1}{2}(\rho + 1) \end{aligned} \tag{B.2}$$

Four-noded cubic element

The standard quadratic interpolation functions are given by:

$$\begin{aligned} N_1(\xi) &= -\tfrac{1}{16}(1 - \xi)(1 - 9\xi^2) \qquad & N_2(\xi) &= \tfrac{9}{16}(1 - 3\xi)(1 - \xi^2) \\ N_3(\xi) &= \tfrac{9}{16}(1 + 3\xi)(1 - \xi^2) \qquad & N_4(\xi) &= -\tfrac{1}{16}(1 + \xi)(1 - 9\xi^2) \end{aligned}$$

and the modified interpolation functions are then:

$$\begin{aligned} \hat{N}_1(\xi) &= \frac{1}{16}\left[1 + 18\eta - 27\eta^2 + \rho(9 - 27\eta) - 9\rho^2\right] \\ \hat{N}_2(\xi) &= \frac{9}{16}\left[-3 - 2\eta + 9\eta^2 + \rho(-1 + 9\eta) + 3\rho^2\right] \end{aligned}$$

$$\hat{N}_3(\xi) = \frac{9}{16}\left[3 - 2\eta - 9\eta^2 + \rho(-1 - 9\eta) - 3\rho^2\right]$$
$$\hat{N}_4(\xi) = \frac{1}{16}\left[-1 + 18\eta + 27\eta^2 + \rho(9 + 27\eta) + 9\rho^2\right]$$

B.2 TWO-DIMENSIONAL INTERPOLATION

The standard interpolation functions, which originate from finite element methods, are polynomial functions of degree ≥ 1 in (ξ_1, ξ_2) defined so as to verify requirements (3.4). Again, both the standard functions $N(\boldsymbol{\xi})$ and the modified functions $\hat{N}(\rho, \alpha; \boldsymbol{\eta})$ are given in each case.

Three-noded triangular element

The standard interpolation functions are given (Figure B.1) by:

$$N_1(\xi_1, \xi_2) = \xi_3 \qquad N_2(\xi_1, \xi_2) = \xi_1 \qquad N_3(\xi_1, \xi_2) = \xi_2$$

where ξ_1, ξ_2 are the coordinates of the point $\boldsymbol{\xi}$ in the reference triangle $\xi_1 \geq 0, \xi_2 \geq 0, \xi_3 \geq 0)$ and

$$\xi_3 = 1 - \xi_1 - \xi_2$$

Since N_1, N_2, N_3 are linear, the boundary element is a flat triangular panel with straight edges. The modified interpolation functions are then:

$$\hat{N}^1(\rho, \alpha; \boldsymbol{\eta}) = -(\cos\alpha + \sin\alpha)$$
$$\hat{N}^2(\rho, \alpha; \boldsymbol{\eta}) = \cos\alpha$$
$$\hat{N}^3(\rho, \alpha; \boldsymbol{\eta}) = \sin\alpha$$

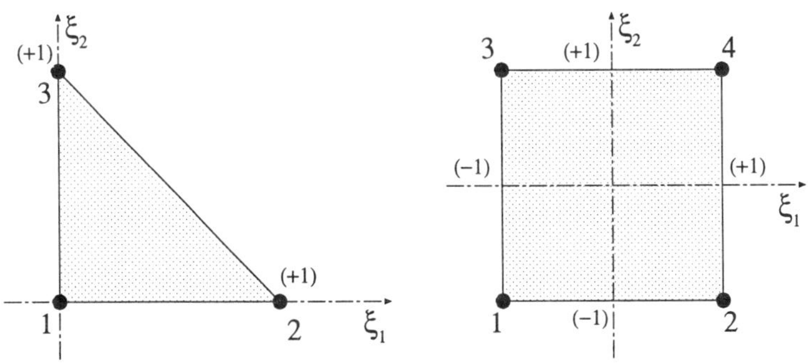

Figure B.1 Three-noded triangle and four-noded quadrilateral.

Four-noded quadrilateral element

The standard interpolation functions are given (Figure B.1) by:

$$N_1(\xi_1,\xi_2) = \frac{1}{4}(\xi_1 - 1)(\xi_2 - 1)$$
$$N_3(\xi_1,\xi_2) = \frac{1}{4}(\xi_1 + 1)(\xi_2 + 1)$$
$$N_2(\xi_1,\xi_2) = \frac{1}{4}(\xi_1 + 1)(\xi_2 - 1)$$
$$N_4(\xi_1,\xi_2) = \frac{1}{4}(\xi_1 - 1)(\xi_2 + 1)$$

This element is often termed 'bilinear' because the interpolation functions are linear with each coordinate ξ_α and have a total degree of 2. The boundary element in the physical space has four straight edges but is nonplanar (unless of course all four nodes are coplanar). The modified interpolation functions are then:

$$\hat{N}^1(\rho,\alpha;\boldsymbol{\eta}) = \frac{1}{4}\left[-(1-\eta_1)\sin\alpha - (1-\eta_2)\cos\alpha + \rho\cos\alpha\sin\alpha\right]$$
$$\hat{N}^2(\rho,\alpha;\boldsymbol{\eta}) = \frac{1}{4}\left[-(1+\eta_1)\sin\alpha + (1-\eta_2)\cos\alpha + \rho\cos\alpha\sin\alpha\right]$$
$$\hat{N}^3(\rho,\alpha;\boldsymbol{\eta}) = \frac{1}{4}\left[(1+\eta_1)\sin\alpha + (1+\eta_2)\cos\alpha + \rho\cos\alpha\sin\alpha\right]$$
$$\hat{N}^4(\rho,\alpha;\boldsymbol{\eta}) = \frac{1}{4}\left[(1-\eta_1)\sin\alpha - (1+\eta_2)\cos\alpha + \rho\cos\alpha\sin\alpha\right]$$

Six-noded triangular element

The standard interpolation functions are given (Figure B.2)by:

$$N_1(\xi_1,\xi_2) = -\xi_3(1-2\xi_3) \qquad N_2(\xi_1,\xi_2) = 4\xi_1\xi_3$$
$$N_3(\xi_1,\xi_2) = -\xi_1(1-2\xi_1) \qquad N_4(\xi_1,\xi_2) = 4\xi_1\xi_2$$
$$N_5(\xi_1,\xi_2) = -\xi_2(1-2\xi_2) \qquad N_6(\xi_1,\xi_2) = 4\xi_2\xi_3$$

In contrast with the three-noded triangular element, the quadratic variation in ξ_1, ξ_2 of the functions $N_1, \ldots, N_6$ allows the creation of curved surfaces with curved edges.

The first-order derivatives $N_{m,\alpha}$ ($m = 1, \ldots, 6$, $\alpha = 1, 2$) are given by:

$$N_{1,1}(\boldsymbol{\xi}) = 1 - 4\xi_3 \qquad N_{1,2}(\boldsymbol{\xi}) = 1 - 4\xi_3$$
$$N_{2,1}(\boldsymbol{\xi}) = 4(\xi_3 - \xi_1) \qquad N_{2,2}(\boldsymbol{\xi}) = -4\xi_1$$
$$N_{3,1}(\boldsymbol{\xi}) = -1 + 4\xi_1 \qquad N_{3,2}(\boldsymbol{\xi}) = 0$$
$$N_{4,1}(\boldsymbol{\xi}) = 4\xi_2 \qquad N_{4,2}(\boldsymbol{\xi}) = 4\xi_1$$
$$N_{5,1}(\boldsymbol{\xi}) = 0 \qquad N_{5,2}(\boldsymbol{\xi}) = -1 + 4\xi_2$$
$$N_{6,1}(\boldsymbol{\xi}) = -4\xi_2 \qquad N_{6,2}(\boldsymbol{\xi}) = 4(\xi_3 - \xi_2)$$

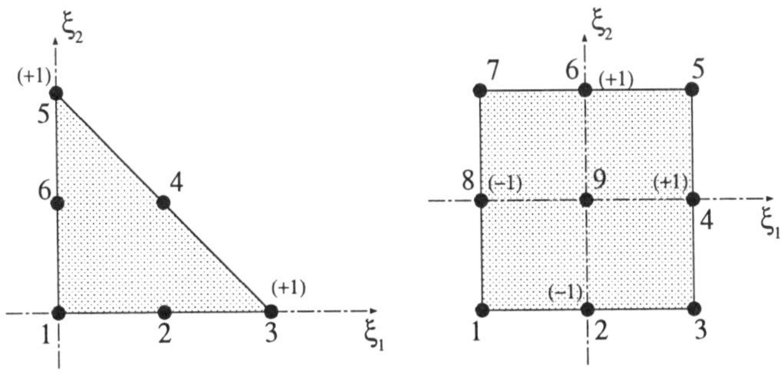

Figure B.2 Six-noded triangle and nine-noded quadrilateral.

and the second-order derivatives $N_{m,\alpha\beta}$ $(m = 1, \ldots, 6,\ \alpha, \beta = 1, 2)$ by:

$$
\begin{array}{lll}
N_{1,11}(\boldsymbol{\xi}) = 4 & N_{1,12}(\boldsymbol{\xi}) = 4 & N_{1,22}(\boldsymbol{\xi}) = 4 \\
N_{2,11}(\boldsymbol{\xi}) = -8 & N_{2,12}(\boldsymbol{\xi}) = -4 & N_{2,22}(\boldsymbol{\xi}) = 0 \\
N_{3,11}(\boldsymbol{\xi}) = 4 & N_{3,12}(\boldsymbol{\xi}) = 0 & N_{3,22}(\boldsymbol{\xi}) = 0 \\
N_{4,11}(\boldsymbol{\xi}) = 0 & N_{4,12}(\boldsymbol{\xi}) = 4 & N_{4,22}(\boldsymbol{\xi}) = 0 \\
N_{5,11}(\boldsymbol{\xi}) = 0 & N_{5,12}(\boldsymbol{\xi}) = 0 & N_{5,22}(\boldsymbol{\xi}) = 4 \\
N_{6,11}(\boldsymbol{\xi}) = 0 & N_{6,12}(\boldsymbol{\xi}) = 4 & N_{6,22}(\boldsymbol{\xi}) = -8
\end{array}
$$

The modified interpolation functions are then:

$$
\begin{aligned}
\hat{N}^1(\rho, \alpha; \boldsymbol{\eta}) &= -(4\eta_3 - 1)(\cos\alpha + \sin\alpha) + 2\rho(\cos\alpha + \sin\alpha)^2 \\
\hat{N}^2(\rho, \alpha; \boldsymbol{\eta}) &= 4(\eta_3 \cos\alpha - \eta_1(\cos\alpha + \sin\alpha)) + 4\rho\cos\alpha(\cos\alpha + \sin\alpha) \\
\hat{N}^3(\rho, \alpha; \boldsymbol{\eta}) &= (4\eta_1 - 1)\cos\alpha + 2\rho\cos^2\alpha \\
\hat{N}^4(\rho, \alpha; \boldsymbol{\eta}) &= 4(\eta_2 \cos\alpha + \eta_1 \sin\alpha) + 4\rho\cos\alpha\sin\alpha \\
\hat{N}^5(\rho, \alpha; \boldsymbol{\eta}) &= (4\eta_2 - 1)\sin\alpha + 2\rho\sin^2\alpha \\
\hat{N}^6(\rho, \alpha; \boldsymbol{\eta}) &= 4(\eta_3 \sin\alpha - \eta_2(\cos\alpha + \sin\alpha)) - 4\rho\sin\alpha(\cos\alpha + \sin\alpha)
\end{aligned}
$$

Eight-noded quadrilateral element

The standard interpolation functions are given by:

$$
\begin{array}{ll}
N_1(\boldsymbol{\xi}) = \frac{1}{4}(1-\xi_1)(1-\xi_2)(-1-\xi_1-\xi_2) & N_2(\boldsymbol{\xi}) = \frac{1}{2}(1-\xi_1^2)(1-\xi_2) \\
N_3(\boldsymbol{\xi}) = \frac{1}{4}(1+\xi_1)(1-\xi_2)(-1+\xi_1-\xi_2) & N_4(\boldsymbol{\xi}) = \frac{1}{2}(1-\xi_2^2)(1+\xi_1) \\
N_5(\boldsymbol{\xi}) = \frac{1}{4}(1+\xi_1)(1+\xi_2)(-1+\xi_1+\xi_2) & N_6(\boldsymbol{\xi}) = \frac{1}{2}(1-\xi_1^2)(1+\xi_2) \\
N_7(\boldsymbol{\xi}) = \frac{1}{4}(1-\xi_1)(1+\xi_2)(-1-\xi_1+\xi_2) & N_8(\boldsymbol{\xi}) = \frac{1}{2}(1-\xi_2^2)(1-\xi_1)
\end{array}
$$

The first-order derivatives $N_{m,\alpha}$ ($m = 1, \ldots, 8$, $\alpha = 1, 2$) are given by:

$$
\begin{aligned}
N_{1,1}(\boldsymbol{\xi}) &= \tfrac{1}{4}(1-\xi_2)(2\xi_1+\xi_2) & N_{2,1}(\boldsymbol{\xi}) &= -\xi_1(1-\xi_2)\\
N_{3,1}(\boldsymbol{\xi}) &= \tfrac{1}{4}(1-\xi_2)(2\xi_1-\xi_2) & N_{4,1}(\boldsymbol{\xi}) &= \tfrac{1}{2}(1-\xi_2^2)\\
N_{5,1}(\boldsymbol{\xi}) &= \tfrac{1}{4}(1+\xi_2)(2\xi_1+\xi_2) & N_{6,1}(\boldsymbol{\xi}) &= -\xi_1(1+\xi_2)\\
N_{7,1}(\boldsymbol{\xi}) &= \tfrac{1}{4}(1+\xi_2)(2\xi_1-\xi_2) & N_{8,1}(\boldsymbol{\xi}) &= -\tfrac{1}{2}(1-\xi_2^2)
\end{aligned}
$$

$$
\begin{aligned}
N_{1,2}(\boldsymbol{\xi}) &= \tfrac{1}{4}(1-\xi_1)(\xi_1+2\xi_2) & N_{2,2}(\boldsymbol{\xi}) &= -\tfrac{1}{2}(1-\xi_1^2)\\
N_{3,2}(\boldsymbol{\xi}) &= \tfrac{1}{4}(1+\xi_1)(-\xi_1+2\xi_2) & N_{4,2}(\boldsymbol{\xi}) &= -\xi_2(1+\xi_1)\\
N_{5,2}(\boldsymbol{\xi}) &= \tfrac{1}{4}(1+\xi_1)(\xi_1+2\xi_2) & N_{6,2}(\boldsymbol{\xi}) &= \tfrac{1}{2}(1-\xi_1^2)\\
N_{7,2}(\boldsymbol{\xi}) &= \tfrac{1}{4}(1-\xi_1)(-\xi_1+2\xi_2) & N_{8,2}(\boldsymbol{\xi}) &= -\xi_2(1-\xi_1)
\end{aligned}
$$

and the second-order derivatives $N_{m,\alpha\beta}$ ($m = 1, \ldots, 8$, $\alpha, \beta = 1, 2$) by:

$$
\begin{aligned}
N_{1,11} &= \tfrac{1}{2}(1-\xi_2) & N_{1,22} &= \tfrac{1}{2}(1-\xi_1) & N_{1,12} &= \tfrac{1}{4}(1-2\xi_1-2\xi_2)\\
N_{2,11} &= -(1-\xi_2) & N_{2,22} &= 0 & N_{2,12} &= \xi_1\\
N_{3,11} &= N_{1,11} & N_{3,22} &= \tfrac{1}{2}(1+\xi_1) & N_{3,12} &= \tfrac{1}{4}(-1-2\xi_1+2\xi_2)\\
N_{4,11} &= 0 & N_{4,22} &= -(1+\xi_1) & N_{4,12} &= -\xi_2\\
N_{5,11} &= \tfrac{1}{2}(1+\xi_2) & N_{5,22} &= N_{3,22} & N_{5,12} &= \tfrac{1}{4}(1+2\xi_1+2\xi_2)\\
N_{6,11} &= -(1+\xi_2) & N_{6,22} &= 0 & N_{6,12} &= -\xi_1\\
N_{7,11} &= N_{5,11} & N_{7,22} &= N_{1,22} & N_{7,12} &= \tfrac{1}{4}(-1+2\xi_1-2\xi_2)\\
N_{8,11} &= 0 & N_{8,22} &= -(1-\xi_1) & N_{8,12} &= \xi_2
\end{aligned}
$$

The modified interpolation functions are then:

$$
\begin{aligned}
\hat{N}_1(\rho,\alpha;\boldsymbol{\eta}) = \frac{1}{4}\Big\{&\cos\alpha(c_1-a_1)a_2+\sin\alpha(c_1-a_2)a_1\\
&+\rho(a_2\cos^2\alpha+3\cos\alpha\sin\alpha+a_1\sin^2\alpha)\\
&-\rho^2\cos\alpha\sin\alpha(\cos\alpha+\sin\alpha)\Big\}
\end{aligned}
$$

$$
\begin{aligned}
\hat{N}_2(\rho,\alpha;\boldsymbol{\eta}) = \frac{1}{2}\Big\{&a_2(a_1-a_3)\cos\alpha-a_1a_3\sin\alpha\\
&+\rho(\cos\alpha\sin\alpha(a_3-a_1)-a_2\cos^2\alpha)+\rho^2\cos^2\alpha\sin\alpha\Big\}
\end{aligned}
$$

$$
\begin{aligned}
\hat{N}_3(\rho,\alpha;\boldsymbol{\eta}) = \frac{1}{4}\Big\{&\sin\alpha(c_2-a_2)a_3-\cos\alpha(c_2-a_3)a_2\\
&+\rho(a_3\sin^2\alpha-3\cos\alpha\sin\alpha+a_2\cos^2\alpha)\\
&+\rho^2\cos\alpha\sin\alpha(\sin\alpha-\cos\alpha)\Big\}
\end{aligned}
$$

$$
\begin{aligned}
\hat{N}_4(\rho,\alpha;\boldsymbol{\eta}) = \frac{1}{2}\Big\{&a_3(a_2-a_4)\sin\alpha+a_2a_4\cos\alpha\\
&-\rho(\cos\alpha\sin\alpha(a_4-a_2)+a_3\sin^2\alpha)-\rho^2\sin^2\alpha\cos\alpha\Big\}
\end{aligned}
$$

$$\hat{N}_5(\rho,\alpha;\boldsymbol{\eta}) = \frac{1}{4}\Big\{-\cos\alpha(c_3-a_3)a_4 - \sin\alpha(c_3-a_4)a_3 + \rho(a_4\cos^2\alpha + 3\cos\alpha\sin\alpha + a_3\sin^2\alpha) + \rho^2\cos\alpha\sin\alpha(\sin\alpha+\cos\alpha)\Big\}$$

$$\hat{N}_6(\rho,\alpha;\boldsymbol{\eta}) = \frac{1}{2}\Big\{-a_4(a_3-a_1)\cos\alpha + a_3a_1\sin\alpha + \rho(\cos\alpha\sin\alpha(a_1-a_3) - a_4\cos^2\alpha) - \rho^2\cos^2\alpha\sin\alpha\Big\}$$

$$\hat{N}_7(\rho,\alpha;\boldsymbol{\eta}) = \frac{1}{4}\Big\{-\cos\alpha(c_4-a_4)a_1 - \sin\alpha(c_4-a_1)a_4 + \rho(a_1\cos^2\alpha + 3\cos\alpha\sin\alpha + a_4\sin^2\alpha) + \rho^2\cos\alpha\sin\alpha(\sin\alpha+\cos\alpha)\Big\}$$

$$\hat{N}_8(\rho,\alpha;\boldsymbol{\eta}) = \frac{1}{2}\Big\{-a_1(a_4-a_2)\cos\alpha + a_4a_2\sin\alpha + \rho(\cos\alpha\sin\alpha(a_2-a_4) - a_1\cos^2\alpha) - \rho^2\cos^2\alpha\sin\alpha\Big\}$$

with:

$a_1 = 1-\eta_1$	$a_5 = \eta_2$	$b_1 = 2\eta_1+\eta_2$	$c_1 = -1-\eta_1-\eta_2$
$a_2 = 1-\eta_2$	$a_6 = \eta_1$	$b_2 = 2\eta_1-\eta_2$	$c_2 = -1+\eta_1-\eta_2$
$a_3 = 1+\eta_1$		$b_3 = 2\eta_2+\eta_1$	$c_3 = -1+\eta_1+\eta_2$
$a_4 = 1+\eta_2$		$b_4 = 2\eta_2-\eta_1$	$c_4 = -1-\eta_1+\eta_2$

Nine-noded quadrilateral element

The standard interpolation functions are given (Figure B.2) by:

$$N_1(\boldsymbol{\xi}) = S_1(\xi_1)S_1(\xi_2) \quad N_4(\boldsymbol{\xi}) = S_3(\xi_1)S_2(\xi_2) \quad N_7(\boldsymbol{\xi}) = S_1(\xi_1)S_3(\xi_2)$$
$$N_2(\boldsymbol{\xi}) = S_2(\xi_1)S_1(\xi_2) \quad N_5(\boldsymbol{\xi}) = S_3(\xi_1)S_3(\xi_2) \quad N_8(\boldsymbol{\xi}) = S_1(\xi_1)S_2(\xi_2)$$
$$N_3(\boldsymbol{\xi}) = S_3(\xi_1)S_1(\xi_2) \quad N_6(\boldsymbol{\xi}) = S_2(\xi_1)S_3(\xi_2) \quad N_9(\boldsymbol{\xi}) = S_2(\xi_1)S_2(\xi_2)$$

where S_1, S_2, S_3 are the one-dimensional quadratic interpolation functions (B.1). The modified interpolation functions are then given in terms of the S_i and the one-dimensional modified interpolation functions $\hat{S}_i$ (B.2):

$$\hat{N}_1(\rho,\alpha;\boldsymbol{\eta}) = S_1(\xi_1)\hat{S}_1(\rho\sin\alpha;\eta_2) + \hat{S}_1(\rho\cos\alpha;\eta_1)S_1(\eta_2)$$
$$\hat{N}_2(\rho,\alpha;\boldsymbol{\eta}) = S_2(\xi_1)\hat{S}_2(\rho\sin\alpha;\eta_2) + \hat{S}_1(\rho\cos\alpha;\eta_1)S_1(\eta_2)$$
$$\hat{N}_3(\rho,\alpha;\boldsymbol{\eta}) = S_3(\xi_1)\hat{S}_3(\rho\sin\alpha;\eta_2) + \hat{S}_1(\rho\cos\alpha;\eta_1)S_1(\eta_2)$$
$$\hat{N}_4(\rho,\alpha;\boldsymbol{\eta}) = S_3(\xi_1)\hat{S}_3(\rho\sin\alpha;\eta_2) + \hat{S}_2(\rho\cos\alpha;\eta_2)S_1(\eta_2)$$
$$\hat{N}_5(\rho,\alpha;\boldsymbol{\eta}) = S_3(\xi_1)\hat{S}_3(\rho\sin\alpha;\eta_2) + \hat{S}_3(\rho\cos\alpha;\eta_3)S_1(\eta_2)$$
$$\hat{N}_6(\rho,\alpha;\boldsymbol{\eta}) = S_2(\xi_1)\hat{S}_2(\rho\sin\alpha;\eta_2) + \hat{S}_3(\rho\cos\alpha;\eta_3)S_1(\eta_2)$$

$$\hat{N}_7(\rho,\alpha;\boldsymbol{\eta}) = S_1(\xi_1)\hat{S}_1(\rho\sin\alpha;\eta_2) + \hat{S}_3(\rho\cos\alpha;\eta_3)S_1(\eta_2)$$
$$\hat{N}_8(\rho,\alpha;\boldsymbol{\eta}) = S_1(\xi_1)\hat{S}_1(\rho\sin\alpha;\eta_2) + \hat{S}_2(\rho\cos\alpha;\eta_2)S_1(\eta_2)$$
$$\hat{N}_9(\rho,\alpha;\boldsymbol{\eta}) = S_2(\xi_1)\hat{S}_2(\rho\sin\alpha;\eta_2) + \hat{S}_2(\rho\cos\alpha;\eta_2)S_1(\eta_2)$$

B.3 GAUSSIAN NUMERICAL INTEGRATION RULES

Throughout this section, N denotes the number of Gauss points for the integration rule being considered.

Gaussian rule for integration of logarithmic singularities

Fundamental solutions for two-dimensional problems have a logarithmic singularity, for which a specific integration rule is needed. Gaussian type integration rules are constructible for any integrable singularity. Table B.1 gives the abscissae $x_i \in [0,1]$ and weights w_i corresponding to the logarithmic case with $N = 10$, i.e.:

$$\int_0^1 \ln\frac{1}{x} f(x)\,\mathrm{d}x \approx \sum_{i=1}^{10} w_i f(x_i)$$

Gaussian rules for one-dimensional nonsingular integration

The Gauss points x_i and weights w_i for integration over the line interval $S = [-1,1]$ are displayed for $2 \le N \le 7$ in Table B.2. The Gauss points are always interior to S; the weights are always strictly positive. Points and

Table B.1 Modified Gauss points and weights for integration of logarithmic singularities over $-1 \le \xi \le 1$ ($N = 10$).

ξ_i	w_i
0.0090426309	0.1209551319
0.0539712662	0.1863635425
0.1353118246	0.1956608732
0.2470524162	0.1735771421
0.3802125396	0.1356956729
0.5237923179	0.0936467585
0.6657752055	0.0557877273
0.7941904160	0.0271598109
0.8981610912	0.0095151826
0.9688479887	0.0016381576

Table B.2 Gauss points and weights for nonsingular integration over $-1 \leq \xi \leq 1$ [2].

N	ξ_i	w_i
1	0.	2.
2	±0.57735026918962576450	1.
3	±0.	0.88888888888888888889
	±0.77459666924148337703	0.55555555555555555556
4	±0.33998104358485626480	0.65214515486254614262
	±0.86113631159405257522	0.34785484513745385737
5	0.	0.56888888888888888889
	±0.53846931010568309103	0.47862867049936646804
	±0.90617984593866399279	0.23692688505618908751
6	±0.23861918608319690863	0.46791393457269104738
	±0.66120938646626451366	0.36076157304813860756
	±0.93246951420315202781	0.17132449237917034504
7	0	0.41795918367346938775
	±0.40584515137739716690	0.38183005050511894495
	±0.74153118559939443986	0.27970539148927666790
	±0.94910791234275852452	0.12948496616886969327

weights are symmetrically distributed about $x = 0$. The book by Stroud and Secrest [2] is a classical reference on Gaussian quadrature.

Gaussian rules for integration over triangles

The Gauss points (x_i^T, y_i^T) and weights w_i for integration over the reference triangle $\mathcal{T} = x \geq 0, y \geq 0, 1-x-y \geq 0$ (see Lyness and Jespersen [1]) are given in Table B.3. The Gauss points are always interior to $\mathcal{T}$; the weights are always strictly positive. In addition, these rules respect the ternary symmetry of the triangle: all permutations of $x_i^T, y_i^T, 1 - x_i^T - y_i^T$ define Gauss points, with identical weights. Accordingly, the information in Table B.3 is compressed: to each given point a 'multiplicity' M ($M = 1, 3, 6$) indicates the number of distinct Gauss points obtained by permutation.

REFERENCES

1. LYNESS, J. N., JESPERSEN, D. Moderate degree symmetric quadrature rules for the triangle. *J. Inst. Math. Appl.*, **15**, 19–32 (1975).
2. STROUD, A. H., SECREST, D. *Gaussian quadrature formulas.* Prentice-Hall (1966).

Table B.3 Gauss points and weights for integration over the triangle $x \geq 0, y \geq 0, 1 - x - y \geq 0$, from [1].

N	x_i^T	y_i^T	w_i	M
3	0.166666666666667	0.166666666666667	0.166666666666667	3
6	0.445948490915965	0.445948490915965	0.111690794839005	3
	0.091576213509771	0.091576213509771	0.054975871827661	3
7	0.333333333333333	0.333333333333333	0.112500000000000	1
	0.470142064105115	0.470142064105115	0.066197076394253	3
	0.101286507323456	0.101286507323456	0.062969590272414	3
16	0.333333333333333	0.333333333333333	0.072157803838893	1
	0.081414823414554	0.459292588292722	0.047545817133642	3
	0.898905543365937	0.050547228317031	0.016229248811599	3
	0.658861384496479	0.170569307751760	0.051608685267359	3
	0.008394777409957	0.728492392955404	0.013615157087217	6
19	0.333333333333333	0.333333333333333	0.048567898141398	1
	0.020634961602525	0.489682519198737	0.015667350113570	3
	0.125820817014129	0.437089591492935	0.038913770502388	6
	0.910540973211094	0.044729513394452	0.012788837829349	3
	0.036838412054736	0.741198598784498	0.021641769688645	6

Appendix C
Bibliography

Note. *Each chapter of this book contains a separate bibliography list, to which citations (appearing as numbers in brackets) in that chapter are referred. This bibliographical appendix provides a more global survey of the boundary element literature. Entries are sorted by type of publication. Journal articles have been further (loosely) sorted by topics (a paper addressing more than one topic is listed for each topic). No attempt at completeness is claimed, and many excellent publications are certainly missing despite the author's efforts.*

C.1 BOOKS ON BIE/BEM

ALIABADI, M. H., ROOKE, D. P. *Numerical fracture mechanics*. Comp. Mech. Publ., Southampton and Kluwer Academic Press (1991).

ANTES, H., PANAGIOTOPOULOS, P. D. *The boundary integral approach to static and dynamic contact problems*. Birkhäuser, Basel (1992).

BALAS, J., SLADEK, J., SLADEK, V. *Stress analysis by boundary element methods*. Elsevier (1989).

BANERJEE, P. K. *The boundary element method in engineering (2nd edition)*. McGraw Hill, London (1994).

BANERJEE, P. K., BUTTERFIELD, R. *Boundary element methods in engineering sciences*. McGraw Hill, London (1981).

BESKOS, D. E. (ed.) *Boundary element methods in mechanics*, vol. 3 of *Computational methods in mechanics*. North-Holland (1987).

BESKOS, D. E. (ed.) *Boundary element analysis of plates and shells*. Springer-Verlag (1991).

BONNET, M. *Equations intégrales et éléments de frontière*. CNRS Editions/Eyrolles, Paris, France (1995).

BREBBIA, C. A. (ed.) *Topics in boundary element research (vols. 1, 2)*. Springer-Verlag (1984).

BREBBIA, C. A., TELLES, J. C. F., WROBEL, L. C. *Boundary element techniques*. Springer-Verlag (1984).

CHANDRA, A., MUKHERJEE, S. *Boundary element methods in manufacturing*. Oxford University Press, Oxford (1997).

CROUCH, S. L., STARFIELD, A. M. *Boundary element methods in solid mechanics*. George Allen and Unwin (1983).

CRUSE, T. A. *Boundary element analysis in computational fracture mechanics*. Kluwer Academic Publishers (1988).

DOMINGUEZ, J. *Boundary elements in dynamics*. Comp. Mech. Publ., Southampton (1993).

KANE, J. H. *Boundary element analysis in engineering continuum mechanics*. Prentice-Hall (1994).

KITAHARA, M. *Boundary integral equation methods to eigenvalue problems of elastodynamics and thin plates*. Elsevier (1985).

LEITAO, V. M. A. *Boundary elements in nonlinear fracture mechanics*. Comp. Mech. Publ., Southampton (1994).

MANOLIS, G., BESKOS, D. E. *Boundary element methods in elastodynamics*. Unwin Hyman, London (1988).

MUKHERJEE, S. *Boundary element methods in creep and fracture*. Elsevier (1982).

POZRIKIDIS, C. *Boundary integral and singularity methods for linearized viscous flow*. Cambridge University Press (1992).

Expositions of the potential theory and other mathematical issues can be found in the following books:

COLTON, D., KRESS, R. *Integral equation method in scattering theory*. John Wiley (1983).

GUNTHER, N. M. *La théorie du potentiel et ses applications aux problèmes fondamentaux de la physique mathématique*. Gauthier-Villars, Paris (1934).

HACKBUSCH, W., WITTUM, G. (eds). *Boundary elements: implementation and analysis of advanced algorithms*. Wieweg, Braunschweig (1996).

KELLOGG, O. D. *Fundations of potential theory*. Springer-Verlag, Berlin, Heidelberg (1967).

KUPRADZE, V. D. *Dynamical problems in elasticity*, vol. 3 of *Progress in solids mechanics*. North-Holland (1963).

KUPRADZE, V. D. (ed.) *Three-dimensional problems of the mathematical theory of elasticity and thermoelasticity*. North-Holland (1979).

MIKHLIN, S. G. *Multi-dimensional singular integrals and singular integral equations*. Pergamon Press (1965).

PARTON, V. Z., PERLIN, P. I. *Integral equations in elasticity*. Mir Publishers, Moscow (1982).

PRÖSSDORF, S., SILBERMANN, B. *Numerical analysis for integral and related operator equations*. Birkhäuser, Basel (1991).

TRICOMI, F. G. *Integral equations*. Interscience Publishers, New York (1957).

WENDLAND, W. L. (ed.) *Boundary element topics*. Springer-Verlag (1997).

WING, G. M. *A primer on integral equations of the first kind*. SIAM Publications, Philadelphia, USA (1991).

C.2 ARTICLES

Fundamental solutions

BENITEZ, F. G., ROSAKIS, A. J. Three-dimensional elastostatics of a layer and a layered medium. *J. Elast.*, **18**, 3–50 (1987).

BERGER, J. R. Boundary element analysis of anisotropic bimaterials with special Green's functions. *Engng. Anal. with Bound. Elem.*, **14**, 123–131 (1994).

CHAN, K. S., KARASUDHI, P., LEE, S. L. Force at a point in the interior of a layered elastic half-space. *Int. J. Solids Struct.*, **10**, 1179–1199 (1974).

DUNDURS, J., HETENYI, M. The elastic plane with a circular insert, loaded by a radial force. *ASME J. Appl. Mech.*, **28**, 103–110 (1961).

GEBBIA, M. Le deformazioni tipiche dei corpi solidi elastici. *Annali Mat. Pura et Applicata*, **7**, 141–230 (1902).

HETENYI, M., DUNDURS, J. The elastic plane with a circular insert, loaded by a radial force. *ASME J. Appl. Mech.*, **29**, 362–368 (1962).

KENNETT, B. L. N. Seismic wave propagation in stratified media. In *Advances in Applied Mechanics*, vol. 21. Academic Press (1981).

MINDLIN, R. D. Force at a point in the interior of a semi-infinite solid. *Physics*, **7**, 195–201 (1936).

PAN, Y. C., CHOU, T. W. Point force solution for an infinite transversely isotropic solid. *ASME J. Appl. Mech.*, **43**, 608–612 (1976).

RONGVED, L. Force interior to one of two joined semi-infinite solids. In *Proc. 2nd. Midwestern Conf. Solid Mechanics.*, pp. 1–13 (1955).

TELLES, J. C. F., BREBBIA, C. A. Boundary element solution for half-plane problems. *Int. J. Solids Struct.*, **17**, 1149–1158 (1981).

Heat transfer

CARINI, A., DILIGENTI, M., MAIER, G.. Variational formulation of the boundary element method in transient heat conduction. *Mathl. Comput. Modelling.*, **15**, 71–80 (1991).

CHANG, Y. P., KANG, C. S., CHEN, D. J. The use of fundamental Green's functions for the solutions of problems of heat conduction in anisotropic media. *Int. J. Heat Mass Transfer*, **16**, 1905–1918 (1973).

CHAUDOUET, A. Three-dimensional transient thermo-elastic analyses by the BIE method. *Int. J. Num. Meth. in Eng.*, **24**, 25–45 (1987).

KIKUTA, M., TOGOH, H., TANAKA, M. Boundary element analysis of nonlinear transient heat conduction problems. *Comp. Meth. in Appl. Mech. Engng.*, **62**, 321–329 (1987).

SHARP, S., CROUCH, S. L. Boundary integral methods for thermoelasticity problems. *ASME J. Appl. Mech.*, **53**, 298–302 (1986).

Elasticity

CHAUDOUET, A. Three-dimensional transient thermo-elastic analyses by the BIE method. *Int. J. Num. Meth. in Eng.*, **24**, 25–45 (1987).

CRUSE, T. A. Numerical solutions in three-dimensional elastostatics. *Int. J. Solids Struct.*, **5**, 1259–1274 (1969).

GHOSH, N., RAJIYAH, H., GHOSH, S., MUKHERJEE, S. A new boundary element method formulation for linear elasticity. *ASME J. Appl. Mech.*, **53**, 69–76 (1986).

LACHAT, J. C., WATSON, J. O. Effective numerical treatment of boundary integral equations: a formulation for three-dimensional elastostatics. *Int. J. Num. Meth. in Eng.*, **10**, 991–1005 (1976).

PHAN, A.-V., MUKHERJEE, S., RENÉ MAYER, J. R.. A boundary contour formulation for design sensitivity analysis in two-dimensional linear elasticity. *Int. J. Solids Struct.*, **35**, 1981–1999 (1997).

RIZZO, F. J. An integral equation approach to boundary-value problems of classical elastostatics. *Quart. Appl. Math.*, **25**, 83–95 (1967).

RIZZO, F. J., SHIPPY, D. J. An advanced boundary integral equation method for three-dimensional elasticity. *Int. J. Num. Meth. in Eng.*, **11**, 1753–1768 (1977).

SHARP, S., CROUCH, S. L. Boundary integral methods for thermoelasticity problems. *ASME J. Appl. Mech.*, **53**, 298–302 (1986).

SOMIGLIANA, C. Sopra l'equilibrio di un corpo elastico isotrope. *Il Nuovo Ciemento, serie III*, **20**, 181–185 (1886).

Some references dealing specifically with boundary element methods for plate problems are:

ALTIERO, N. J., SIKARSKIE, D. L. A boundary integral method applied to plates of arbitrary plane form. *Computers and Structures*, **9**, 163–168 (1978).

ANTES, H. Static and dynamic analysis of Reissner-Mindlin plates. In D. E. Beskos (ed.), *Boundary element analysis of plates and shells*, pp. 312–340. Springer-Verlag (1991).

BESKOS, D. E. Dynamic analysis of plates. In D. E. Beskos (ed.), *Boundary element analysis of plates and shells*, pp. 35–92. Springer-Verlag (1991).

BEZINE, G. Boundary integral formulation for plate flexure with arbitrary boundary conditions. *Mech. Res. Commun.*, **5**, 197–206 (1978).

BEZINE, G. A new boundary element method for bending of plates on elastic foundations. *Int. J. Solids Struct.*, **24**, 557–565 (1988).

BREBBIA, C. A., TELLES, J. C. F., WROBEL, L. C. *Boundary element techniques*, chap. 9, pp. 324–337. Springer-Verlag (1984).

CASTRO, L. M., PINHO, J., PARREIRA, P. Solving plate problems using boundary element method. In C. A. Brebbia (ed.), *Boundary elements technology VI*, pp. 215–225. Comp. Mech. Publ., Southampton (1991).

EL-ZAFRANY, A., FADHIL, S. A new fundamental solution for boundary element analysis of thin plates on Winkler foundation. *Int. J. Num. Meth. in Eng.*, **38**, 887–903 (1995).

FRANGI, A., BONNET, M. A Galerkin symmetric and direct BIE method for Kirchhoff elastic plates: formulation and implementation. *Int. J. Num. Meth. in Eng.*, **41**, 337–369 (1998).

FRANGI, A., NOVATI, G. Symmetric BEM for fracture mechanics of thin plates. In B. Karihaloo, et al. (eds), *Advances in fracture research*, vol. 4, pp. 2055–2062. Pergamon, Oxford (1997).

GIROIRE, J., NEDELEC, J. C. A new system of boundary integral equations for plates with free edges. *Math. Meth. in Appl. Sc.*, **8**, 755–772 (1995).

GUOSHU, S., MUKHERJEE, S. Boundary element method analysis of bending of elastic plates of arbitrary shape with general boundary conditions. *Engng. Anal. with Bound. Elem.*, **3**, 2049–2069 (1986).

HANSEN, E. B. Numerical solution of integro-differential and singular integral equations for plate bending problems. *J. Elast.*, **6**, 39–56 (1976).

HARTMANN, F., ZOTEMANTEL, R. The direct boundary element method in plate bending. *Int. J. Num. Meth. in Eng.*, **23**, 2049–2069 (1986).

JASWON, M. A., MAITI, M. An integral equation formulation for plate bending problems. *J. Engng. Math.*, **2**, 83–93 (1968).

PROVIDAKIS, D. E., BESKOS, D. E., SOTIROPOULOS, D. A. Dynamic analysis of inelastic plates by the D/BEM. *Comp. Mech.*, **13**, 276–284 (1994).

PUTTONEN, J., VARPASUO, P. Boundary element analysis of a plate on elastic foundations. *Int. J. Num. Meth. in Eng.*, **23**, 287–303 (1986).

SLADEK, V., SLADEK, J. Multiple reciprocity method in BEM formulations for solution of plate bending problems. *Engng. Anal. with Bound. Elem.*, **17**, 161–173 (1996).

STERN, M. A general boundary integral formulation for the numerical solution of plate bending problems. *Int. J. Solids Struct.*, **15**, 769–782 (1979).

TOTTENHAM, H. The boundary element method for plates and shells. In P. K. Banerjee, R. Butterfield (eds), *Developments in boundary element methods.*, vol. 1. Elsevier (1979).

Nonlinear solid mechanics

AHMAD, S., BANERJEE, P. K. Time-domain Transient elastodynamic analysis of 3-d solids by BEM. *Int. J. Num. Meth. in Eng.*, **26**, 1709–1728 (1988).

AHMAD, S., BANERJEE, P. K. Inelastic Transient elastodynamic analysis of three-dimensional problems by BEM. *Int. J. Num. Meth. in Eng.*, **29**, 371–390 (1990).

BANERJEE, P. K., HENRY, D. P. (JR), RAVEENDRA, S. T. Advanced inelastic analysis of solids by the boundary element method. *Int. J. Mech. Sci.*, **31**, 309–322 (1989).

BANERJEE, P. K., RAVEENDRA, S. T. Advanced boundary element analysis of two- and three-dimensional problems of elasto-plasticity. *Int. J. Num. Meth. in Eng.*, **23**, 985–1002 (1986).

BONNET, M., MUKHERJEE, S. Implicit BEM formulations for usual and sensitivity problems in elasto-plasticity using the consistent tangent operator concept. *Int. J. Solids Struct.*, **33**, 4461–4480 (1996).

BONNET, M., POON, H., MUKHERJEE, S. Hypersingular formulation for boundary strain evaluation in the context of a CTO-based implicit BEM scheme for small strain elasto-plasticity. *Int. J. Plasticity*, **14**, 1033–1058 (1998).

BUI, H. D. Some remarks about the formulation of three-dimensional thermo-elastoplastic problems by integral equations. *Int. J. Solids Struct.*, **14**, 935–939 (1978).

COMI, C., MAIER, G. Extremum, convergence and stability properties of the finite-increment problem in elastic–plastic boundary element analysis. *Int. J. Solids Struct.*, **29**, 249–270 (1992).

HUBER, O., DALLNER, R., PARTHEYMULLER, P., KUHN, G. Evaluation of the stress tensor in 3D elastoplasticity by direct solving of hypersingular integrals. *Int. J. Num. Meth. in Eng.*, **39**, 2555–2573 (1996).

HUBER, O., KUHN, G. Evaluation of the J-integral for 3D crack problems using the hypersingular identity. In J. Dominguez, C. A. Brebbia, F. Paris (eds), *Boundary elements XIV (vol. 2)*, pp. 367–381. Comp. Mech. Publ., Southampton (1992).

ISRAIL, A. S. M., BANERJEE, P. K. Advanced development of boundary element method for two-dimensional dynamic elastoplasticity. *Int. J. Solids Struct.*, **29**, 1433–1451 (1992).

KONTONI, D. P. N., BESKOS, D. E. Application of the DR-BEM to inelastic dynamic problems. In J. Dominguez, C. A. Brebbia, F. Paris (eds), *Boundary elements XIV*, Comp. Mech. Publ., Southampton (1992).

LEU, L. J., MUKHERJEE, S. Sensitivity analysis and shape optimization in nonlinear solid mechanics. *Engng. Anal. with Bound. Elem.*, **12**, 251–260 (1993).

LIAO, S. J. High-order BEM formulations for strongly non-linear problems governed by quite general non-linear differential operators. *Int. J. Num. Meth. in Fluids*, **23**, 739–751 (1996).

LIAO, S. J., CHWANG, A. T. General boundary element methods for non-linear problems. *Int. J. Num. Meth. in Fluids*, **23**, 467–483 (1996).

LIU, Q., CEN, Z., XU, B. Plastic analysis by quasi-high order Galerkin BEM. *Engng. Structures*, **19**, 1050–1056 (1997).

MAIER, G. On elastoplastic analysis by boundary elements. *Mech. Res. Commun.*, **10**, 45–52 (1983).

MAIER, G., FRANGI, A. Dynamic elastoplasticity analysis by Galerkin discretization of symmetric integral equations with elastostatic kernels. In D. R. G. Owen, E. Oñate, E. Hinton (eds), *Computational plasticity '97*, vol. 2, pp. 1003–1038. CIMNE, Barcelona (1997).

MAIER, G., MICCOLI, S., NOVATI, G., PEREGO, U. Symmetric Galerkin boundary element method in plasticity and gradient-plasticity. *Comp. Mech.*, **17**, 115–129 (1995).

MAIER, G., MICCOLI, S., NOVATI, G., SIRTORI, S.. A Galerkin symmetric boundary-element method in plasticity: formulation and implementation. In J. H. Kane, G. Maier, N. Tosaka, S. N. Atluri (eds), *Advances in boundary element techniques*, pp. 288–328. Springer-Verlag (1993).

MAIER, G., POLIZZOTTO, C.. A Galerkin approach to boundary element elastoplastic analysis. *Comp. Meth. in Appl. Mech. Engng.*, **60**, 175–194 (1987).

MUKHERJEE, S., CHANDRA, A. Nonlinear solid mechanics. In D. E. Beskos (ed.), *Boundary element methods in mechanics.*, vol. 1, pp. 285–332. North-Holland (1987).

MUKHERJEE, S., CHANDRA, A. A boundary element formulation for design sensitivities in problems involving both geometric and material nonlinearities. *Mathl. Comput. Modelling*, **15**, 245–255 (1991).

POLIZZOTTO, C. An energy approach to the boundary element method; Part II: Elastic–plastic solids. *Comp. Meth. in Appl. Mech. Engng.*, **69**, 263–276 (1988).

POON, H., MUKHERJEE, S., BONNET, M. Numerical implementation of a CTO-based implicit approach for the BEM solution of usual and sensitivity problems in elasto-plasticity. *Engng. Anal. with Bound. Elem.*, **22**, 257–269 (1998).

PROVIDAKIS, D. E., BESKOS, D. E., SOTIROPOULOS, D. A. Dynamic analysis of inelastic plates by the D/BEM. *Comp. Mech.*, **13**, 276–284 (1994).

TELLES, J. C. F., BREBBIA, C. A. Boundary elements: new developments in elastoplastic analyses. *Appl. Math. Modelling*, **5**, 376–382 (1981).

TELLES, J. C. F., CARRER, J. A. M. Implicit procedures for the solution of elastoplastic problems by the boundary element method. *Mathl. Comput. Modelling*, **15**, 303–311 (1991).

TELLES, J. C. F., CARRER, J. A. M. Static and transient dynamic nonlinear stress analysis by the boundary element method with implicit techniques. *Engng. Anal. with Bound. Elem.*, **14**, 65–4 (1994).

WEI, X., LEU, L. J., CHANDRA, A., MUKHER JEE, S. Shape optimization in elasticity and elastoplasticity. *Int. J. Solids Struct.*, **31**, 533–550 (1994).

ZHANG, Q., MUKHERJEE, S., CHANDRA, A. Design sensitivity coefficients for elasto-visco-plastic problems by boundary element methods. *Int. J. Num. Meth. in Eng.*, **34**, 947–966 (1992).

ZHANG, Q., MUKHERJEE, S., CHANDRA, A. Shape design sensitivity analysis for geometrically and materially nonlinear problems by the boundary element method. *Int. J. Solids Struct.*, **29**, 2503–2525 (1992).

Contact mechanics

ALESSANDRI, C., TRALLI, A. Sensitivity analysis for unilateral contact problems: 'Boundary' variational formulations and BEM discretisations. *Comp. Mech.*, **15**, 287–300 (1995).

ALONSO, M. P., GARRIDO, J. A., FOCES, A. Application of BEM to solve two-dimensional thermoelastic contact problems with convection and radiation conditions. *Computers and Structures*, **66**, 115–125 (1998).

BLAZQUEZ, A., PARIS, F., MANTIC, V. BEM solution of two-dimensional contact problems by weak application of contact conditions with non-conforming discretizations. *Int. J. Solids Struct.*, **35**, 3259–3278 (1998).

DOSTAL, Z., FRIEDLANDER, A., SANTOS, S. A, MALIK, J. Analysis of semicoercive contact problems using symmetric BEM and augmented Lagrangians. *Engng. Anal. with Bound. Elem.*, **18**, 195–201 (1996).

ECK, C., WENDLAND, W. L. A boundary integral formulation of contact problems with friction. In L. Morino, W. L. Wendland (eds), *Boundary integral methods for nonlinear problems*, pp. 47–55. Kluwer Academic Publishers (1997).

GARRIDO, J. A., ALONSO, P., DEL CAÑO, J. C., PARIS, F. A boundary integral formulation for frictionless contact problems involving steady heat conduction. *Eur. J. Mech. A/Solids*, **14**, 119–136 (1995).

HUESMANN, A., KUHN, G. Automatic load incrementation technique for plane elastoplastic frictional contact problems using boundary element method. *Computers and Structures*, **56**, 733–744 (1995).

LEITAO, V., ALIABADI, M.H., ROOKE, D.P. Elastoplastic dual boundary elements: application to crack-face contact. *Computers and Structures*, **54**, 443–454 (1995).

OLUKOKO, O. A., BECKER, A. A., FENNER, R. T. A new boundary element approach for contact problems with friction. *Int. J. Num. Meth. in Eng.*, **36**, 2625–2642 (1993).

SIMUNOVIC, S., SAIGAL, S. Contact surface optimization using boundary element method. *Computers and Structures*, **56**, 745–750 (1995).

Fracture mechanics

ALIABADI, M. H. Boundary element formulations in fracture mechanics. *Appl. Mech. Rev.*, **50**, 83–96 (1997).

ALTIERO, N. J., GIODA, G. An integral equation approach to fracture propagation in rocks. *Riv. Ital. Geotecnica*, **16**, 55–69 (1982).

BONNET, M. Equations intégrales variationnelles pour le problème en vitesse de propagation de fissures en élasticité linéaire. *C.R. Acad. Sci. Paris, série II*, **318**, 429–434 (1994).

BONNET, M. Stability of crack fronts under Griffith criterion: a computational approach using integral equations and domain derivatives of potential energy. *Comp. Meth. in Appl. Mech. Engng.* (1999, to appear).

BONNET, M., XIAO, H. Computation of energy release rate along a crack front using material differentiation of elastic BIE. *Engng. Anal. with Bound. Elem.*, **15**, 137–150 (1995).

BUI, H. D. An integral equation method for solving the problem of a plane crack of arbitrary shape. *J. Mech. Phys. Solids*, **25**, 29–39 (1977).

CRUSE, T. A., BESUNER, P. M. Residual life prediction for surface cracks in complex structural details. *Journal of Aircraft*, **12**, 369–375 (1975).

CRUSE, T. A., RICHARDSON, J. D. Nonsingular Somigliana stress identities in elasticity. *Int. J. Num. Meth. in Eng.*, **39**, 3273–3304 (1996).

FUKUYAMA, E., MADARIAGA, R. Integral equation method for plane crack with arbitrary shape in 3D elastic medium. *Bull. Seism. Soc. Am.*, **85**, 614–628 (1995).

GAUTESEN, A. K., DUNDURS, J. On the solution to a Cauchy principal value integral equation which arises in fracture mechanics. *SIAM J. Appl. Math.*, **47**, 109–116 (1987).

GU, H., YEW, C. H. Finite element solution of a boundary integral equation for mode I embedded three-dimensional fractures. *Int. J. Num. Meth. in Eng.*, **26**, 1525–1540 (1988).

HAYASHI, K., ABÉ, H. Stress intensity factors for a semi-elliptical crack in the surface of a semi-infinite solid. *Int. J. Fract.*, **16**, 275–285 (1980).

HUBER, O., NICKEL, J., KUHN, G. On the decomposition of the J-integral for 3D problems. *Int. J. Fract.*, **64**, 339–348 (1993).

LEBLOND, J. B. Equations intégrales régularisées pour un corps bidimensionnel contenant une fissure de forme quelconque. *C.R. Acad. Sci. Paris, série II*, **303**, 1521–1524 (1986).

LE VAN, A., ROYER, J. Integral equations for three-dimensional problems. *Int. J. Fract.*, **31**, 125–142 (1986).

LI, S., MEAR, M.E., XIAO, L. Symmetric weak-form integral equation method for three-dimensional fracture analysis. *Comp. Meth. in Appl. Mech. Engng.*, **151**, 435–459 (1998).

MI, Y., ALIABADI, M. H. Dual boundary element method for analysis of 3-D crack problems. In J. Dominguez, C. A. Brebbia, F. Paris (eds), *Boundary elements XIV*, pp. 315–329. Comp. Mech. Publ., Southampton (1992).

MI, Y., ALIABADI, M. H. Dual boundary element method for three-dimensional fracture mechanics analysis. *Engng. Anal. with Bound. Elem.*, **10**, 161–171 (1992).

POLCH, E. Z., CRUSE, T. A., HUANG, C. J. Traction BIE solutions for flat cracks. *Comp. Mech.*, **2**, 253–267 (1987).

PORTELA, A., ALIABADI, M. H., ROOKE, D. P. Dual boundary element incremental analysis of crack propagation. In C. A. Brebbia, M. H. Aliabadi (eds), *Advances in BEM for fracture mechanics*. Comp. Mech. Publ., Southampton (1992).

PORTELA, A., ALIABADI, M. H., ROOKE, D. P. Dual boundary element method: efficient implementation for cracked problems. *Int. J. Num. Meth. in Eng.*, **33**, 1269–1287 (1992).

SAEZ, A., GALLEGO, R., DOMINGUEZ, J. Hypersingular quarter-point boundary elements for crack problems. *Int. J. Num. Meth. in Eng.*, **38**, 1681–1701 (1995).

SEELIG, TH., GROSS, D. Analysis of dynamic crack propagation using a time-domain boundary integral equation method. *Int. J. Solids Struct.*, **34**, 2087–2103 (1997).

SLADEK, J., SLADEK, V. Three-dimensional crack analysis for an anisotropic body. *Appl. Math. Modelling*, **6**, 374–380 (1982).

SLADEK, J., SLADEK, V. Three-dimensional curved crack in an elastic body. *Int. J. Solids Struct.*, **19**, 425–436 (1983).

SLADEK, V., SLADEK, J. Regularization of hypersingular integrals in BEM formulations using various kinds of continuous elements. *Engng. Anal. with Bound. Elem.*, **17**, 5–18 (1996).

SNYDER, M. D., CRUSE, T. A. Boundary integral equation analysis of cracked anisotropic plates. *Int. J. Fract.*, **11**, 315–328 (1975).

TANAKA, M., ITOH, H. New crack elements for boundary element analysis of elastostatics considering arbitrary stress singularities. *Appl. Math. Modelling*, **11**, 357–363 (1987).

WATSON, J. O. Hermitian cubic boundary elements for plane problems of fracture mechanics. *Res Mechanica*, **4**, 23–42 (1982).

WEAVER, J. Three-dimensional crack analysis. *Int. J. Solids Struct.*, **13**, 321–330 (1977).

XU, G., ARGON, A. S., ORTIZ, M., BOWER, A. Development of a variational boundary integral method for the analysis of fully three dimensional crack advance problems. In S. N. Atluri, G. Yagawa, T. A. Cruse (eds), *Computational mechanics '95*, vol. 2, pp. 2874–2889. Springer-Verlag (1995).

XU, G., ORTIZ, M. A variational boundary integral method for the analysis of 3D cracks of arbitrary geometry modelled as continuous distributions of dislocation loops. *Int. J. Num. Meth. in Eng.*, **36**, 3675–3701 (1993).

YIN, H. P., EHRLACHER, A. Variational approach of displacement discontinuity method and application to crack problems. *Int. J. Fract.*, **63**, 135–153 (1993).

Dynamics, acoustics, wave propagation

AHMAD, S., BANERJEE, P. K. Time-domain Transient elastodynamic analysis of 3-d solids by BEM. *Int. J. Num. Meth. in Eng.*, **26**, 1709–1728 (1988).

AHMAD, S., BANERJEE, P. K. Inelastic Transient elastodynamic analysis of three-dimensional problems by BEM. *Int. J. Num. Meth. in Eng.*, **29**, 371–390 (1990).

ALVES, C., DUONG, T. H. Numerical resolution of the boundary integral equations for elastic scattering by a plane crack. *Int. J. Num. Meth. in Eng.*, **38**, 2347–2371 (1995).

ANTES, H. Static and dynamic analysis of Reissner-Mindlin plates. In D. E. Beskos (ed.), *Boundary element analysis of plates and shells*, pp. 312–340. Springer-Verlag (1991).

BAMBERGER, A., HA DUONG, T. Formulation variationnelle espace-temps pour le calcul par potentiel retardé de la diffraction d'une onde acoustique (I). *Math. Meth. in Appl. Sc.*, **8**, 405–435 (1986).

BAMBERGER, A, HA DUONG, T. Formulation variationnelle pour le calcul de la diffraction d'une onde acoustique par une surface rigide. *Math. Meth. in Appl. Sc.*, **8**, 598–608 (1986).

BANERJEE, P. K., AHMAD, S., MANOLIS, G. D. Advanced elastodynamic analysis. In D. E. Beskos (ed.), *Boundary element methods in mechanics*. North-Holland (1987).

BECACHE, E. A variational boundary integral equation method for an elastodynamic antiplane crack. *Int. J. Num. Meth. in Eng.*, **36**, 969–993 (1993).

BECACHE, E. *Equations intégrales pour l'équation des ondes*, chap. 8. INRIA, Collection Didactique (1994).

BENDALI, A. Numerical analysis of the exterior boundary-value problem for the time-harmonic Maxwell equations by a boundary finite element method. *Math. Comp.*, **43**, 29–46 and 47–68 (1984).

BEN MARIEM, J., HAMDI, M. A. A new boundary finite element method for fluid-structure interaction problems. *Int. J. Num. Meth. in Eng.*, **24**, 1251–1267 (1987).

BESKOS, D. E. Boundary element methods in dynamic analysis. *Appl. Mech. Rev.*, **40**, 1–23 (1987).

BESKOS, D. E. . Boundary element methods in dynamic analysis, part. II (1986–1996). *Appl. Mech. Rev.*, **50**, 149–197 (1997).

BIELAK, J., MACCAMY, R. C., ZENG, X. Stable coupling method for interface scattering problems by combined integral equations and finite elements. *J. Comp. Math.*, **119**, 374–384 (1995).

BLUCK, M. J., WALKER, S. P. Analysis of three-dimensional transient acoustic wave propagation using the boundary integral equation method. *Int. J. Num. Meth. in Eng.*, **39**, 1419–1431 (1996).

BUDRECK, D. E., ACHENBACH, J. D. Scattering from three-dimensional planar cracks by the boundary integral equation method. *ASME J. Appl. Mech.*, **55**, 405–412 (1988).

BURRIDGE, R. The numerical solution of certain integral equations with non-integrable kernels arising in the theory of crack propagation and elastic wave diffraction. *Phil. Trans. Roy. Soc. London*, **A-265**, 353–381 (1969).

BURTON, A. J., MILLER, G. F. The application of integral equations to the numerical solution of some exterior boundary-value problems. *Phil. Trans. Roy. Soc. London*, **A-323**, 201–220 (1971).

CAPUANI, D., WILLIS, J. R. Wave propagation in elastic media with cracks: I. transient nonlinear response of a single crack. *Eur. J. Mech. A/Solids*, **16**, 377–408 (1997).

CARRER, J. A. M., MANSUR, W. J. Time-domain BEM analysis for the 2D scalar wave equation: initial conditions contributions to space and time derivatives. *Int. J. Num. Meth. in Eng.*, **39**, 2169–2188 (1996).

COLE, D. M., KOSLOFF, D. D., MINSTER, J. B. A numerical boundary integral equation method for elastodynamics I. *Bull. Seism. Soc. Am.*, **68**, 1331–1357 (1978).

DE HOOP, A. T., STAM, H. J. Time-domain reciprocity theorems for elastodynamic wave fields with relaxation and their application to inverse problems. *Wave Motion*, **10**, 479–489 (1988).

DEMIREL, V., WANG, S. An efficient boundary-element method for two-dimensional transient wave propagation problems. *Appl. Math. Modelling*, **11**, 411–416 (1987).

DOHNER, J. L., RAHMATALLAH, S., BERNHARD, R. J. Transient analysis of three-dimensional wave propagation using the boundary element method. *Int. J. Num. Meth. in Eng.*, **24**, 621–634 (1987).

GALLEGO, R., DOMINGUEZ, J. Dynamic crack propagation analysis by moving singular boundary elements. *ASME J. Appl. Mech.*, **59**, S158–S162 (1992).

GALLEGO, R., DOMINGUEZ, J. Hypersingular BEM for transient elastodynamics. *Int. J. Num. Meth. in Eng.*, **39**, 1681–1705 (1996).

HAMDI, M. A. Une formulation variationnelle par équations intégrales pour la résolution de l'équation de Helmholtz avec des conditions aux limites mixtes. *C.R. Acad. Sci. Paris, série II,*, **292**, 17–20 (1981).

HAMDI, M. A., JEAN, PH. Résolution numérique des problèmes couplés par une méthode d'éléments de frontière. In *Tendances actuelles en calcul des structures*, pp. 449–464. Pluralis (1985).

HIROSE, S., ACHENBACH, J. D. Time-domain boundary element analysis of elastic wave interaction with a crack. *Int. J. Num. Meth. in Eng.*, **28**, 629–644 (1989).

HUBER, O., KUHN, G. Evaluation of the J-integral for 3D crack problems using the hypersingular identity. In J. Dominguez, C. A. Brebbia, F. Paris (eds), *Boundary elements XIV (vol. 2)*, pp. 367–381. Comp. Mech. Publ., Southampton (1992).

ISRAIL, A. S. M., BANERJEE, P. K. Advanced development of boundary element method for two-dimensional dynamic elastoplasticity. *Int. J. Solids Struct.*, **29**, 1433–1451 (1992).

KARABALIS, D. L. Dynamic soil-structure interaction by BEM. In C. A. Brebbia (ed.), *Boundary elements X*, pp. 351–374. Springer-Verlag (1988).

KARABALIS, D. L. A simplified 3-D time-domain BEM for dynamic soil-structure interaction problems. *Engng. Anal. with Bound. Elem.*, **8**, 139–145 (1991).

KOLLER, M. G., BONNET, M., MADARIAGA, R. Modelling of dynamical crack propagation using time-domain boundary integral equations. *Wave Motion*, **16**, 339–366 (1992).

KONTONI, D. P. N., BESKOS, D. E. Application of the DR-BEM to inelastic dynamic problems. In J. Dominguez, C. A. Brebbia, F. Paris (eds), *Boundary elements XIV*. Comp. Mech. Publ., Southampton (1992).

KRISHNASAMY, G., SCHMERR, L. W., RUDOLPHI, T. J., RIZZO, F. J. Hypersingular boundary integral equations: some applications in acoustic and elastic wave scattering. *ASME J. Appl. Mech.*, **57**, 404–414 (1990).

LEE, C. H., NEWMAN, J. N., ZHU, X. An extended boundary integral equation method for the removal of irregular frequency effects. *Int. J. Num. Meth. in Fluids*, **23**, 637–660 (1996).

LIU, Y., RIZZO, F. J. A weakly singular form of the hypersingular boundary integral equation applied to 3D acoustic wave problems. *Comp. Meth. in Appl. Mech. Engng.*, **96**, 271–287 (1992).

MAIER, G., DILIGENTI, M., CARINI, A.. A variational formulation of the boundary element method in elastodynamics. In K. Tseng, B. S. Annigeri (eds), *Boundary element methods in engineering*, pp. 268–275. Springer-Verlag, Berlin, Heidelberg (1989).

MAIER, G, DILIGENTI, M, CARINI, A. A variational approach to boundary element elastodynamic analysis and extension to multidomain problems. *Comp. Meth. in Appl. Mech. Engng.*, **92**, 193–213 (1991).

MANOLIS, G. D. A comparative study on three boundary element method approaches to problems in elastodynamics. *Int. J. Num. Meth. in Eng.*, **19**, 73–91 (1983).

MANOLIS, G. D., BESKOS, D. E. Dynamic stress concentration studies by boundary integrals and Laplace transform. *Int. J. Num. Meth. in Eng.*, **17**, 573–599 (1981).

MANSUR, W. J., CASTRO, A. Numerical solution of Kirchhoff integral equation by boundary elements. In *Boundary elements VIII*. Springer-Verlag (1986).

MITZNER, K. M. Numerical solution for transient scattering from a hard surface of arbitrary shape – retarded potential technique. *J. Acoust. Soc. Am.*, **42**, 391–397 (1967).

NISHIMURA, N., GUO, Q. C., KOBAYASHI, S. Boundary integral equation method for elastodynamic crack problems. In G. Kuhn, C.A. Brebbia, W. L. Wendland (eds), *Boundary elements IX (vol. 2)*, pp. 279–291. Springer-Verlag (1987).

NISHIMURA, N., KOBAYASHI, S. A regularized boundary integral equation method for elastodynamic crack problems. *Comp. Mech.*, **4**, 319–328 (1989).

PEIRCE, A., SIEBRITS, E. Stability analysis and design of time-stepping schemes for general elastodynamic boundary element models. *Int. J. Num. Meth. in Eng.*, **40**, 319–342 (1997).

REZAYAT, M., SHIPPY, D. J., RIZZO, F. J. On time-harmonic elastic wave analysis by the boundary element method for moderate to high frequencies. *Comp. Meth. in Appl. Mech. Engng.*, **55**, 349–367 (1986).

RIZZO, F. J., SHIPPY, D. J., REZAYAT, M. A boundary integral equation method for radiation and scattering of elastic waves in three dimensions. *Int. J. Num. Meth. in Eng.*, **21**, 115–129 (1985).

ROKHLIN, V. Solution of acoustic scattering problems by means of second kind integral equations. *Wave Motion*, **5**, 257–272 (1983).

RYNNE, B. P. Stability and convergence of time marching methods in scattering problems. *IMA J. Appl. Math.*, **35**, 297–310 (1985).

SCHANZ, M., ANTES, H. Application of 'operational quadrature methods' in time domain boundary element methods. *Meccanica* ().

SCHANZ, M., ANTES, H. A new visco- and elastodynamic time domain boundary element formulation. *Comp. Mech.*, **20**, 452–459 (1997).

SCHENCK, H. A. Improved integral equation formulation for acoustic radiation problem. *J. Acoust. Soc. Am.*, **44**, 41–58 (1968).

SEELIG, TH., GROSS, D. Analysis of dynamic crack propagation using a time-domain boundary integral equation method. *Int. J. Solids Struct.*, **34**, 2087–2103 (1997).

SHAW, R. P. Diffraction of acoustic pulses by obstacles of arbitrary shape with a Robin condition – part A. *J. Acoust. Soc. Am.*, **41**, 855–859 (1962).

SLADEK, J., SLADEK, V. Dynamic stress-intensity factors studied by boundary integro-differential equations. *Int. J. Num. Meth. in Eng.*, **23**, 919–928 (1986).

SONG, C., WOLF, J. P. The scaled boundary-finite element method – alias consistent infinitesimal finite-element cell method – for elastodynamics. *Comp. Meth. in Appl. Mech. Engng.*, **147**, 329–355 (1997).

TELLES, J. C. F., CARRER, J. A. M. Static and transient dynamic nonlinear stress analysis by the boundary element method with implicit techniques. *Engng. Anal. with Bound. Elem.*, **14**, 65–4 (1994).

VON ESTORFF, O., PAIS, A. L., KAUSEL, E. Some observations on time-domain and frequency domain boundary elements. *Int. J. Num. Meth. in Eng.*, **29**, 785–800 (1990).

WANG, C. Y., ACHENBACH, J. D. Elastodynamic fundamental solutions for anisotropic solids. *Geophys. J. Int.*, **118**, 384–392 (1994).

WANG, C. Y., ACHENBACH, J. D., HIROSE, S. Two-dimensional time-domain BEM for scattering of elastic waves in solids of general anisotropy. *Int. J. Solids Struct.*, **33**, 3843–3864 (1996).

WANG, C. Y., TAKEMIYA, H. Analytical elements of time-domain BEM for two-dimensional scalar wave problems. *Int. J. Num. Meth. in Eng.*, **33**, 1737–1754 (1992).

WANG, H. C., BANERJEE, P. K. Axisymmetric free-vibration problems by boundary element method. *ASME J. Appl. Mech.*, **55**, 437–442 (1988).

Geophysics and geodynamics

AUBRY, D., CLOUTEAU, D. Seismic site effect and soil structure interaction. In W. B. Krätzig et al. (eds), *Structural dynamics, vol. 2*, pp. 731–738. A. A. Balkema, Rotterdam (1991).

COCHARD, R., MADARIAGA, R. Dynamic faulting under rate-dependent friction. PAGEOPH, **142**, 419–445 (1994).

FUKUYAMA, E., MADARIAGA, R. Integral equation method for plane crack with arbitrary shape in 3D elastic medium. *Bull. Seism. Soc. Am.*, **85**, 614–628 (1995).

FUKUYAMA, E., MADARIAGA, R. Rupture dynamics of a planar fault in a 3D elastic medium: rate- and slip-weakening friction. *Bull. Seism. Soc. Am.*, **88**, 1–17 (1998).

KAWASE, H., AKI, K. A study of the response of a soft basin for incident *S*, *P* and Rayleigh waves with special reference to the long duration observed in Mexico City. *Bull. Seism. Soc. Am.*, **79**, 1361–1382 (1989).

KOLLER, M. G., BONNET, M., MADARIAGA, R. Modelling of dynamical crack propagation using time-domain boundary integral equations. *Wave Motion*, **16**, 339–366 (1992).

SANCHEZ-SESMA, F. J., CAMPILLO, M. Diffraction of P, SV and Rayleigh waves by topographic features: a boundary integral formulation. *Bull. Seism. Soc. Am.*, **81**, 2234–2253 (1991).

Inverse problems, variable domains

AITHAL, R., SAIGAL, S. Shape sensitivity analysis in thermal problems using BEM. *Engng. Anal. with Bound. Elem.*, **15**, 115–120 (1995).

BANKS, H. T., KOJIMA, F. Boundary shape identification problems in two-dimensional domains related to thermal testing of materials. *Quart. Appl. Math.*, **47**, 273–293 (1989).

BARONE, M. R., YANG, R. J. A boundary element approach for recovery of shape sensitivities in threedimensional elastic solids. *Comp. Meth. in Appl. Mech. Engng.*, **74**, 69–82 (1989).

BONNET, M. Equations intégrales variationnelles pour le problème en vitesse de propagation de fissures en élasticité linéaire. *C.R. Acad. Sci. Paris, série II*, **318**, 429–434 (1994).

BONNET, M. BIE and material differentiation applied to the formulation of obstacle inverse problems. *Engng. Anal. with Bound. Elem.*, **15**, 121–136 (1995).

BONNET, M. Regularized BIE formulations for first- and second-order shape sensitivity of elastic fields. *Computers and Structures*, **56**, 799–811 (1995). (Invited paper, special issue, S. Saigal, guest editor).

BONNET, M. Differentiability of strongly singular and hypersingular boundary integral formulations with respect to boundary perturbations. *Comp. Mech.*, **19**, 240–246 (1997).

BONNET, M. A formulation for crack shape sensitivity analysis based on Galerkin BIE, domain differentiation and adjoint variable. In *Mathematical aspects of boundary element methods (volume dedicated to the 60th birthday of Prof. V.G. Maz'ya)*. CRC Press (1999, to appear).

BUI, H. D., BONNET, M. (eds). *Unknown or variable domains, inverse problems.*, vol. 15(2) of *Engng. Anal. with Bound. Elem., special issue* (1995).

BURCZYŃSKI, T. Application of BEM in sensitivity analysis and optimization. *Comp. Mech.*, **13**, 29–44 (1993).

BURCZYŃSKI, T., FEDELINSKI, P. Shape sensitivity analysis of natural frequencies of structures using boundary elements. *Struct. Optim.*, **2**, 47–54 (1990).

BURCZYŃSKI, T., FEDELINSKI, P. Boundary elements in shape design sensitivity analysis and optimal design of vibrating structures. *Engng. Anal. with Bound. Elem.*, **9**, 195–201 (1992).

BURCZYŃSKI, T., KANE, J. H., BALAKRISHNA, C. Shape design sensitivity analysis via material derivative – adjoint variable approach. *Int. J. Num. Meth. in Eng.*, **38**, 2839–2866 (1995).

BURCZYŃSKI, T., KANE, J.H., BALAKRISHNA, C. Comparison of shape design sensitivity analysis formulations via material derivative-adjoint variable and implicit differentiation techniques for 3-D and 2-D curved boundary element. *Comp. Meth. in Appl. Mech. Engng.*, **142**, 89–109 (1997).

BURCZYŃSKI, T., KUHN, G., ANTES, H., NOWAKOWSKI, M. Boundary element formulation of shape sensitivity analysis for defect identification in free vibration problem. *Engng. Anal. with Bound. Elem.*, **19**, 167–175 (1997).

BURCZYŃSKI, T., POLCH, E. Z. Path-independent and boundary integral approach to sensitivity analysis and identification of cracks. In M. Tanaka, H. D. Bui et al. (eds), *Inverse problems in engineering mechanics*, pp. 355–361. A. A. Balkema, Rotterdam (1994).

CHOI, J. O., KWAK, B. M. Boundary integral equation method for shape optimization of elastic structures. *Int. J. Num. Meth. in Eng.*, **26**, 1579–1595 (1988).

DELFOUR, M. C., P., ZOLESIO J. Velocity method and lagrangian formulation for the computation of the shape hessian. *SIAM J. Contr. Opt.*, **29**, 1414–1442 (1991).

ERMAN, Z., FENNER, R. T. Three-dimensional design sensitivity analysis using a boundary integral approach. *Int. J. Num. Meth. in Eng.*, **40**, 637–654 (1997).

GAO, Z., MURA, T. On the inversion of residual stresses from surface measurements. *ASME J. Appl. Mech.*, **56**, 508–513 (1989).

HUANG, C. H., CHEN, C. W. A boundary element-based inverse-problem in estimating transient boundary conditions with conjugate gradient method. *Int. J. Num. Meth. in Eng.*, **42**, 943–965 (1998).

KANE, J. H., ZHAO, G., WANG, H., GURU PRASAD, K. Boundary formulations for three-dimensional continuum structural shape sensitivity analysis. *ASME J. Appl. Mech.*, **59**, 827–834 (1992).

KASSAB, A., MOSLEHY, F. A., DARYAPURKAR, A. B. Nondestructive evaluation of cavities by an inverse elastostatics boundary element method. *Engng. Anal. with Bound. Elem.*, **13**, 45–55 (1994).

KIRSCH, A. The domain derivative and two applications in inverse scattering theory. *Inverse Problems*, **9**, 81–96 (1993).

KOO, B. U. Shape design sensitivity analysis of acoustic problems using a boundary element method. *Computers and Structures*, **65**, 713–719 (1997).

KOYA, T., YEIH, W., MURA, T. An inverse problem in elasticity with partially overprescribed boundary conditions, part I: numerical details. *ASME J. Appl. Mech.*, **60**, 601–606 (1993).

KRESS, R. Integral equation methods in inverse obstacle scattering. *Engng. Anal. with Bound. Elem.*, **15**, 171–180 (1995).

KUBO, S., KAWAGUCHI, M. A mathematical and numerical study on regularization of an inverse boundary-value problem in elasto-static field. In G. S. Dulikravich, M. Tanaka (eds), *Inverse problems in engineering mechanics (proceedings of the ISIP'98 conference, Nagano, Japan, March 1998)*, pp. 353–361. Elsevier (1998).

KUBO, S., OHJI, K. Adjoint boundary integral method applied to the determination of concentrated forces acting in a body. In M. Tanaka, H. D. Bui et al. (eds), *Inverse problems in engineering mechanics*, pp. 289–293. A. A. Balkema, Rotterdam (1994).

LEU, L. J., MUKHERJEE, S. Sensitivity analysis and shape optimization in nonlinear solid mechanics. *Engng. Anal. with Bound. Elem.*, **12**, 251–260 (1993).

MATSUMOTO, T., TANAKA, M., MIYAGAWA, M., ISHII, N. Optimum design of cooling lines in injection moulds by using boundary element design sensitivity analysis. *Finite Elements in Analysis and Design*, **14**, 177–185 (1993).

MELLINGS, S. C., ALIABADI, M. H. Flaw identification using the boundary element method. *Int. J. Num. Meth. in Eng.*, **38**, 399–419 (1995).

MERIC, R. A. Shape optimization of thermoelastic solids. *J. Therm. Sc.*, **11**, 187–206 (1988).

MERIC, R. A. Differential and integral sensitivity formulations and shape optimization by BEM. *Engng. Anal. with Bound. Elem.*, **15**, 181–188 (1995).

MERIC, R. A. Shape design sensitivity analysis and optimization for the nonhomogeneous Helmholtz equation by the BEM. *Comm. Num. Meth. Engng.*, **12**, 95–105 (1996).

NISHIMURA, N. Application of boundary integral equation method to various crack determination problems. In M. H. Aliabadi (ed.), *Dynamic fracture mechanics*, chap. 7. Comp. Mech. Publ., Southampton (1995).

NISHIMURA, N, KOBAYASHI, S. A boundary integral equation method for an inverse problem related to crack detection. *Int. J. Num. Meth. in Eng.*, **32**, 1371–1387 (1991).

NISHIMURA, N, KOBAYASHI, S. Determination of cracks having arbitrary shape with boundary integral method. *Engng. Anal. with Bound. Elem.*, **15**, 189–197 (1995).

PETRYK, H., MROZ, Z. Time derivatives of integrals and functionals defined on varying volume and surface domains. *Arch. Mech.*, **38**, 694–724 (1986).

ROCHE, J. R., SOKOLOWSKI, J. Numerical methods for shape identification problems. *Control and Cybernetics*, **25**, 867–894 (1996).

SIMON, J. Second variations for domain optimization problems. In F. Kappel, K. Kunisch, W. Schappacher (eds), *Control theory of distributed parameter systems and applications.*, vol. 91 of *International Series of Numerical Mathematics*, pp. 361–378. Birkhäuser Verlag, Basel (1989).

TANAKA, M., YAMAGIWA, K. A boundary element method for some inverse problems in elastodynamics. *Appl. Math. Modelling*, **13**, 307–312 (1989).

WEI, X., LEU, L. J., CHANDRA, A., MUKHERJEE, S. Shape optimization in elasticity and elastoplasticity. *Int. J. Solids Struct.*, **31**, 533–550 (1994).

YEIH, W., KOYA, T., MURA, T. An inverse problem in elasticity with partially overprescribed boundary conditions, part I: theoretical approach. *ASME J. Appl. Mech.*, **60**, 595–600 (1993).

ZENG, X., SAIGAL, S. An inverse formulation with boundary elements. *ASME J. Appl. Mech.*, **59**, 835–840 (1992).

ZHANG, Q., MUKHERJEE, S. Second-order design sensitivity analysis for linear elastic problems by the derivative boundary element method. *Comp. Meth. in Appl. Mech. Engng.*, **86**, 321–335 (1991).

Fluid mechanics

ABOUSLEIMAN, Y., CHENG, A. H. D. Boundary element solution for steady and unsteady Stokes flow. *Comp. Meth. in Appl. Mech. Engng.*, **117**, 1–13 (1994).

BUSH, M. B., TANNER, R. I. Numerical solution of viscous flow using integral equation method. *Int. J. Num. Meth. in Fluids*, **3**, 71–92 (1983).

DARGUSH, G. F., BANERJEE, P. K. A boundary element method for steady incompressible thermoviscous flow. *Int. J. Num. Meth. in Eng.*, **31**, 1605–1626 (1991).

DARGUSH, G. F., BANERJEE, P. K. A time-dependent incompressible viscous BEM for moderate Reynolds numbers. *Int. J. Num. Meth. in Eng.*, **31**, 1627–1648 (1991).

HRIBERŠEK, M., ŠKERGET, L. Iterative methods in solving Navier–Stokes equations by the boundary element method. *Int. J. Num. Meth. in Eng.*, **39**, 115–139 (1996).

HU, H. Study of integral equation methods for transonic flow calculations. *Engng. Anal. with Bound. Elem.*, **11**, 101–107 (1993).

LEE, C. H., NEWMAN, J. N., ZHU, X. An extended boundary integral equation method for the removal of irregular frequency effects. *Int. J. Num. Meth. in Fluids*, **23**, 637–660 (1996).

LIGGETT, J. A. Fluid mechanics. In D. E. Beskos (ed.), *Boundary element methods in mechanics.*, vol. 1, pp. 367–416. North-Holland (1987).

ONISHI, K., KUROKI, T., TANAKA, M. Boundary element method for laminar viscous flow and convective diffusion problems. In C. A. Brebbia (ed.), *Topics in boundary element research (vol. 2)*, pp. 209–229. Springer-Verlag (1984).

POWER, H., PARTRIDGE, P. W. The use of Stokes' fundamental solution for the boundary-only element formulation of the three-dimensional Navier–Stokes equations for moderate Reynolds numbers. *Int. J. Num. Meth. in Eng.*, **37**, 1825–1840 (1994).

SU, J., TABARROK, B., DOST, S. A time marching boundary element method for two-dimensional incompressible thermally coupled flows. *Comp. Meth. in Appl. Mech. Engng.*, **128**, 11–23 (1995).

TANZOSH, J., MANGA, M., STONE, H. A. Boundary integral methods for viscous free-boundary problems: deformation of single and multiple fluid-fluid interfaces. In C. A. Brebbia and M. Ingber (eds), *Boundary element techniques VII*, pp. 19–39. Comp. Mech. Publ., Southampton (1992).

VAN DER WEIDEN, R., DE HOOP, A. T. Boundary integral equations for the computational modeling of three-dimensional groundwater flow problems. *Comp. Mech.*, **4**, 283–291 (1989).

WU, J. C., WAHBAH, M. *Numerical solution of viscous flow equations using integral representations.*, vol. 59 of *Lecture notes in Physics*, pp. 448–453. Springer-Verlag (1976).

Mathematical and numerical methods

ALLGOWER, E. L., BÖHMER, K., GEORG, K., MIRANDA, R. Exploiting symmetry in boundary element methods. *SIAM J. Numer. Anal.*, **29**, 534–552 (1992).

BALAKRISHNA, C., GRAY, L. J., KANE, J. H.. Efficient analytical integration of symmetric Galerkin boundary integrals over curved elements: thermal conduction formulation. *Comp. Meth. in Appl. Mech. Engng.*, **111**, 335–355 (1994).

BALAKRISHNA, C., GRAY, L. J., KANE, J. H.. Efficient analytical integration of symmetric Galerkin boundary integrals over curved elements: elasticity formulation. *Comp. Meth. in Appl. Mech. Engng.*, **117**, 157–179 (1994).

BIALECKI, R. A., DALLNER, R., KUHN, G. Minimum distance calculation between a source point and a boundary point. *Engng. Anal. with Bound. Elem.*, **12**, 211–218 (1993).

BIALECKI, R. A., MERKEL, M., MEWS, H., KUHN, G. In- and out-of-core BEM equations solver with parallel and non-linear options. *Int. J. Num. Meth. in Eng.*, **39**, 4215–4242 (1996).

BOSSAVIT, A. Symmetry, groups and boundary-value problems : a progressive introduction to noncommutative harmonic analysis of partial differential equations in domains with geometrical symmetry. *Comp. Meth. in Appl. Mech. Engng.*, **56**, 167–215 (1986).

BULGAKOV, V. E., BIALECKI, R. A., KUHN, G. Coarse division transform based preconditioner for boundary element problems. *Int. J. Num. Meth. in Eng.*, **38**, 2115–2142 (1995).

CAMP, C. V., GIPSON, G. S. Overhauser elements in boundary element analysis. *Mathl. Comput. Modelling.*, **15**, 59–69 (1991).

CARINI, A., DILIGENTI, M., MARANESI, P., ZANELLA, M. Analytical integrations for two-dimensional elastic analysis by the symmetric Galerkin boundary element method. *Comp. Mech.* (submitted).

GUIGGIANI, M., GIGANTE, A. A general algorithm for multidimensional Cauchy principal value integrals in the boundary element method. *ASME J. Appl. Mech.*, **57**, 906–915 (1990).

GUIGGIANI, M., KRISHNASAMY, G., RUDOLPHI, T. J., RIZZO, F. J. A general algorithm for the numerical solution of hypersingular boundary integral equations. *ASME J. Appl. Mech.*, **59**, 604–614 (1992).

HAYAMI, K., MATSUMOTO, H. A numerical quadrature for nearly singular boundary element integrals. *Engng. Anal. with Bound. Elem.*, **13**, 143–154 (1994).

KANE, J. H. *Equation solving and multizone BEA*, chap. 9. Prentice-Hall (1994).

KANE, J. H., KEYES, D. E., GURU PRASAD, K. Iterative solution techniques in boundary element analysis. *Int. J. Num. Meth. in Eng.*, **31**, 1511–1536 (1991).

KUTT, H. R. The numerical evaluation of principal value integrals by finite-part integration. *Numerische Mathematik*, **24**, 205–210 (1975).

LACHAT, J. C., WATSON, J. O. Effective numerical treatment of boundary integral equations: a formulation for three-dimensional elastostatics. *Int. J. Num. Meth. in Eng.*, **10**, 991–1005 (1976).

LUTZ, E. Exact gaussian quadrature methods for near-singular integrals in the boundary element method. *Engng. Anal. with Bound. Elem.*, **9**, 233–245 (1992).

LYNESS, J. N., JESPERSEN, D. Moderate degree symmetric quadrature rules for the triangle. *J. Inst. Math. Appl.*, **15**, 19–32 (1975).

MACKERLE, J. Contact mechanics – finite element and boundary element approaches – A bibliography (1995–1997). *Finite Elements in Analysis and Design*, **29**, 275–285 (1998).

MANOLIS, G. D., BANERJEE, P. K. Conforming versus non-conforming boundary elements in three-dimensional elastostatics. *Int. J. Num. Meth. in Eng.*, **23**, 1885–1904 (1986).

MICHAEL, O., BARBONE, P. E. Galerkin formulation and singularity subtraction for spectral solutions of boundary integral equations. *Int. J. Num. Meth. in Eng.*, **41**, 95–111 (1998).

MILROY, J., HINDUJA, S., DAVEY, K. The elastostatic three-dimensional boundary element method: analytical integration for linear isoparametric triangular elements. *Appl. Math. Modelling*, **21**, 763–782 (1997).

MONEGATO, G. Numerical evaluation of hypersingular integrals. *J. Comp. Appl. Math.*, **50**, 9–31 (1994).

MULLEN, R. L., RENCIS, J. J. Iterative methods for solving boundary element equations. *Computers and Structures*, **25**, 713–723 (1987).

NOWAK, A. J., BREBBIA, C. A. The multiple-reciprocity method. A new approach for transforming BEM domain integrals to the boundary. *Engng. Anal. with Bound. Elem.*, **6**, 164–168 (1989).

PATTERSON, C., SHEIKH, M. A. Interelement continuity in the boundary element method. In C. A. Brebbia (ed.), *Topics in boundary element research.*, vol. 1. Springer-Verlag (1984).

REZAYAT, M., SHIPPY, D. J., RIZZO, F. J. On time-harmonic elastic wave analysis by the boundary element method for moderate to high frequencies. *Comp. Meth. in Appl. Mech. Engng.*, **55**, 349–367 (1986).

SAEZ, A., GALLEGO, R., DOMINGUEZ, J. Hypersingular quarter-point boundary elements for crack problems. *Int. J. Num. Meth. in Eng.*, **38**, 1681–1701 (1995).

SLADEK, V., SLADEK, J. Why use double nodes in BEM? *Engng. Anal. with Bound. Elem.*, **8**, 109–112 (1991).

SLADEK, V., SLADEK, J. Multiple reciprocity method in BEM formulations for solution of plate bending problems. *Engng. Anal. with Bound. Elem.*, **17**, 161–173 (1996).

TANAKA, M., ITOH, H. New crack elements for boundary element analysis of elastostatics considering arbitrary stress singularities. *Appl. Math. Modelling*, **11**, 357–363 (1987).

TELLES, J. C. F. A self-adaptive coordinate transformation for efficient numerical evaluation of general boundary element integrals. *Int. J. Num. Meth. in Eng.*, **24**, 959–973 (1987).

WATSON, J. O. Hermitian cubic boundary elements for plane problems of fracture mechanics. *Res Mechanica*, **4**, 23–42 (1982).

Galerkin boundary element methods

ALVES, C., DUONG, T. H. Numerical resolution of the boundary integral equations for elastic scattering by a plane crack. *Int. J. Num. Meth. in Eng.*, **38**, 2347–2371 (1995).

ANDRÄ, H., SCHNACK, E. Integration of singular Galerkin-type boundary element integrals for 3D elasticity problems. *Numerische Mathematik*, **76**, 143–165 (1997).

BALAKRISHNA, C., GRAY, L. J., KANE, J. H.. Efficient analytical integration of symmetric Galerkin boundary integrals over curved elements: thermal conduction formulation. *Comp. Meth. in Appl. Mech. Engng.*, **111**, 335–355 (1994).

BALAKRISHNA, C., GRAY, L. J., KANE, J. H.. Efficient analytical integration of symmetric Galerkin boundary integrals over curved elements: elasticity formulation. *Comp. Meth. in Appl. Mech. Engng.*, **117**, 157–179 (1994).

BAMBERGER, A., HA DUONG, T. Formulation variationnelle espace-temps pour le calcul par potentiel retardé de la diffraction d'une onde acoustique (I). *Math. Meth. in Appl. Sc.*, **8**, 405–435 (1986).

BAMBERGER, A., HA DUONG, T. Formulation variationnelle pour le calcul de la diffraction d'une onde acoustique par une surface rigide. *Math. Meth. in Appl. Sc.*, **8**, 598–608 (1986).

BÉCACHE, E. A variational boundary integral equation method for an elastodynamic antiplane crack. *Int. J. Num. Meth. in Eng.*, **36**, 969–993 (1993).

BÉCACHE, E., NEDELEC, J.-C., NISHIMURA, N.. Regularization in 3D for anisotropic elastodynamic crack and obstacle problems. *J. Elast.*, **31**, 25–46 (1993).

BENDALI, A. Numerical analysis of the exterior boundary-value problem for the time-harmonic Maxwell equations by a boundary finite element method. *Math. Comp.*, **43**, 29–46 and 47–68 (1984).

BEN MARIEM, J., HAMDI, M. A. A new boundary finite element method for fluid-structure interaction problems. *Int. J. Num. Meth. in Eng.*, **24**, 1251–1267 (1987).

BONNET, M. Regularized direct and indirect symmetric variational BIE formulations for three-dimensional elasticity. *Engng. Anal. with Bound. Elem.*, **15**, 93–102 (1995).

BONNET, M. A formulation for crack shape sensitivity analysis based on Galerkin BIE, domain differentiation and adjoint variable. In *Mathematical aspects of boundary element methods (volume dedicated to the 60th birthday of Prof. V. G. Maz'ya)*. CRC Press (1999, to appear).

BONNET, M. Stability of crack fronts under Griffith criterion: a computational approach using integral equations and domain derivatives of potential energy. *Comp. Meth. in Appl. Mech. Engng.* (1999, to appear).

BONNET, M., MAIER, G., POLIZZOTTO, C. On symmetric galerkin boundary element method. *Appl. Mech. Rev.*, **51**, 669–704 (1998).

BUI, H. D. Symétrie, dualité et variables conjuguées dans les équations intégrales de frontière. *C.R. Acad. Sci. Paris, série II*, **311**, 7–13 (1990).

CARINI, A., DILIGENTI, M., MAIER, G. Boundary integral equation analysis in linear viscoelasticity: variational and saddle point formulations. *Comp. Mech.*, **8**, 87–98 (1991).

CARINI, A., DILIGENTI, M., MAIER, G. Symmetric boundary integral formulations of transient heat conduction: saddle-point theorems for BE analysis and BE-FE coupling. *Arch. Mech.*, **49**, 253–283 (1997).

CARINI, A., DILIGENTI, M., MARANESI, P., ZANELLA, M. Analytical integrations for two-dimensional elastic analysis by the symmetric Galerkin boundary element method. *Comp. Mech.* (1997). (submitted).

DE FIGUEIREDO, T. B. G., BREBBIA, C. A. A new hybrid displacement variational formulation of BEM for elastostatics. In J. J. Connor, C. A. Brebbia (eds), *Advances in boundary elements*, pp. 47–58. Springer-Verlag, Berlin, Heidelberg (1989).

DUMONT, N. A. The hybrid boundary element method. In C. A. Brebbia, W. L. Wendland, G. Kuhn (eds), *Boundary elements IX*, pp. 117–130. Springer-Verlag, Berlin, Heidelberg (1989).

FRANGI, A., NOVATI, G. Symmetric BE method in two-dimensional elasticity: evaluation of double integrals for curved elements. *Comp. Mech.*, **19**, 58–68 (1996).

FRANGI, A., NOVATI, G. Symmetric BEM for fracture mechanics of thin plates. In B. Karihaloo et al. (eds), *Advances in fracture research*, vol. 4, pp. 2055–2062. Pergamon, Oxford (1997).

GRAY, L. J., CHINTA, B., KANE, J. H. Symmetric Galerkin fracture analysis. *Engng. Anal. with Bound. Elem.*, **15**, 103–109 (1995).

GRAY, L. J., PAULINO, G. H. Symmetric Galerkin boundary integral formulation for interface and multi-zone problems. *Int. J. Num. Meth. in Eng.*, **40**, 3085–3101 (1997).

GRAY, L. J., PAULINO, G. H. Symmetric Galerkin boundary integral fracture analysis for plane orthotropic elasticity. *Comp. Mech.*, **20**, 26–33 (1997).

GU, H., YEW, C. H. Finite element solution of a boundary integral equation for mode I embedded three-dimensional fractures. *Int. J. Num. Meth. in Eng.*, **26**, 1525–1540 (1988).

HAMDI, M. A. Une formulation variationnelle par équations intégrales pour la résolution de l'équation de Helmholtz avec des conditions aux limites mixtes. *C.R. Acad. Sci. Paris, série II*, **292**, 17–20 (1981).

HAMDI, M. A., JEAN, PH. Résolution numérique des problèmes couplés par une méthode d'éléments de frontière. In *Tendances actuelles en calcul des structures*, pp. 449–464. Pluralis (1985).

HOLZER, S.M. How to deal with hypersingular integrals in the symmetric BEM. *Comm. Num. Meth. Engng.*, **9**, 219–232 (1993).

KANE, J. H., BALAKRISHNA, C. Symmetric Galerkin boundary formulations employing curved elements. *Int. J. Num. Meth. in Eng.*, **36**, 2157–2187 (1993).

LI, S., MEAR, M. E., XIAO, L. Symmetric weak-form integral equation method for three-dimensional fracture analysis. *Comp. Meth. in Appl. Mech. Engng.*, **151**, 435–459 (1998).

NEDELEC, J-C. Curved finite element methods for the solution of singular integral equations on surfaces in $\mathbb{R}^3$. *Comp. Meth. in Appl. Mech. Engng.*, **8**, 61–80 (1976).

NEDELEC, J. C. Integral equations with non integrable kernels. *Integral equations and operator theory*, **5**, 562–572 (1982).

NEDELEC, J. C. New trends in the use and analysis of integral equations. *Math. Comp.*, **48**, 151–175 (1993).

PAULINO, G. H., GRAY, L. J. Galerkin residuals for error estimation and adaptivity in the symmetric Galerkin boundary element method. *Advances in Comp. Math.* (1999, to appear).

POLIZZOTTO, C. An energy approach to the boundary element method; Part I: Elastic solids,. *Comp. Meth. in Appl. Mech. Engng.*, **69**, 167–184 (1988).

POLIZZOTTO, C. An energy approach to the boundary element method; Part II: Elastic–plastic solids. *Comp. Meth. in Appl. Mech. Engng.*, **69**, 263–276 (1988).

SIRTORI, S., MAIER, G., NOVATI, G., MICCOLI, S. A Galerkin symmetric boundary element method in elasticity: formulation and implementation. *Int. J. Num. Meth. in Eng.*, **35**, 255–282 (1992).

TERESHCHENKO, V. YA. An algorithm and error estimates for the variational boundary-element method in elasticity theory. *J. Appl. Math. Mech.*, **56**, 365–372 (1992).

XU, G., ORTIZ, M. A variational boundary integral method for the analysis of 3D cracks of arbitrary geometry modelled as continuous distributions of dislocation loops. *Int. J. Num. Meth. in Eng.*, **36**, 3675–3701 (1993).

BEM-FEM coupling

BELYTSCHKO, T., CHANG, H. S., LU, Y. Y.. A variationally coupled finite element-boundary-element method. *Computers and Structures*, **33**, 21–29 (1989).

BELYTSCHKO, T., LU, Y. Y.. Singular integration in variationally coupled FE-BE method. *J. Engng. Math.*, **117**, 820–835 (1991).

BELYTSCHKO, T., LU, Y. Y.. A variationally coupled FE-BE method for transient problem. *Int. J. Num. Meth. in Eng.*, **37**, 91–104 (1994).

HSIAO, G. C.. The coupling of boundary element and finite element methods. *Z. Angew. Math. Mech.*, **70**, 493–503 (1990).

ZENG, X., BIELAK, J.. Stable symmetric finite element - boundary integral coupling methods for fluid-structure interface problems. *Engng. Anal. with Bound. Elem.*, **15**, 79–91 (1995).

ZENG, X., BIELAK, J., MACCAMY, R. C.. Unified symmetric finite element and boundary integral variational coupling methods for linear fluid-structure interation. *Num. Meth. Partial Diff. Eq.*, **8**, 451–467 (1992).

Singular integrals, regularization

BÉCACHE, E., NEDELEC, J.-C., NISHIMURA, N.. Regularization in 3D for anisotropic elastodynamic crack and obstacle problems. *J. Elast.*, **31**, 25–46 (1993).

BONNET, M., BUI, H. D. Regularization of the displacement and traction BIE for 3D elastodynamics using indirect methods. In J. H. Kane, G. Maier, N. Tosaka, S. N. Atluri (eds), *Advances in boundary element techniques*, pp. 1–29. Springer-Verlag (1993).

BUI, H. D. New trends in BIE methods. In *Computational mechanics '86 (vol. 2)*, pp. XI. 3–8. Springer-Verlag (1986).

BUI, H. D., LORET, B., BONNET, M. Régularisation des équations intégrales de l'élastostatique et de l'élastodynamique. *C.R. Acad. Sci. Paris, série II*, **300**, 633–636 (1985).

CRUSE, T. A., RICHARDSON, J. D. Nonsingular Somigliana stress identities in elasticity. *Int. J. Num. Meth. in Eng.*, **39**, 3273–3304 (1996).

DILIGENTI, M., MONEGATO, G. Finite part integrals: their occurrence and computation. *Rendiconti del Circolo Matematico di Palermo*, **33**, 39–61 (1993).

FIEDLER, C. On the calculation of boundary stresses with the Somigliana stress identity. *Int. J. Num. Meth. in Eng.*, **38**, 3275–3295 (1995).

FRANGI, A. A new regularized be formulation for kirchhoff plates. *Eur. J. Mech. A/Solids*, **15**, 915–931 (1996).

GUIGGIANI, M. Hypersingular formulation for boundary stress evaluation. *Engng. Anal. with Bound. Elem.*, **14**, 169–179 (1994). (invited paper, special issue).

GUIGGIANI, M., GIGANTE, A. A general algorithm for multidimensional Cauchy principal value integrals in the boundary element method. *ASME J. Appl. Mech.*, **57**, 906–915 (1990).

GUIGGIANI, M., KRISHNASAMY, G., RUDOLPHI, T. J., RIZZO, F. J. A general algorithm for the numerical solution of hypersingular boundary integral equations. *ASME J. Appl. Mech.*, **59**, 604–614 (1992).

HILDEBRAND, J., KUHN, G. Numerical computation of hypersingular integrals and application to the boundary integral equation for the stress tensor. *Engng. Anal. with Bound. Elem.*, **10**, 209–218 (1992).

KRISHNASAMY, G., RIZZO, F. J., RUDOLPHI, T. J. Hypersingular boundary integral equations: their occurrence, interpretation, regularization and computation. In P. K. Banerjee, S. Kobayashi (eds), *Developments in boundary element methods*, vol. 7: Advanced dynamic analysis, chap. 7. Elsevier (1992).

KRISHNASAMY, G., SCHMERR, L. W., RUDOLPHI, T. J., RIZZO, F. J. Hypersingular boundary integral equations: some applications in acoustic and elastic wave scattering. *ASME J. Appl. Mech.*, **57**, 404–414 (1990).

KUTT, H. R. The numerical evaluation of principal value integrals by finite-part integration. *Numerische Mathematik*, **24**, 205–210 (1975).

LE VAN, A., ROYER, J. Integral equations for three-dimensional problems. *Int. J. Fract.*, **31**, 125–142 (1986).

MARTIN, P. A., RIZZO, F. J. Hypersingular integrals: how smooth must the density be? *Int. J. Num. Meth. in Eng.*, **39**, 687–704 (1996).

MARTIN, P. A., RIZZO, F. J., CRUSE, T. A. Smoothness relaxation strategies for singular and hypersingular integral equations. *Int. J. Num. Meth. in Eng.*, **42**, 885–906 (1998).

NISHIMURA, N., KOBAYASHI, S. A regularized boundary integral equation method for elastodynamic crack problems. *Comp. Mech.*, **4**, 319–328 (1989).

RIZZO, F. J., SHIPPY, D. J. An advanced boundary integral equation method for three-dimensional elasticity. *Int. J. Num. Meth. in Eng.*, **11**, 1753–1768 (1977).

RIZZO, F. J., SHIPPY, D. J., REZAYAT, M. A boundary integral equation method for radiation and scattering of elastic waves in three dimensions. *Int. J. Num. Meth. in Eng.*, **21**, 115–129 (1985).

SLADEK, J., SLADEK, V. Three-dimensional curved crack in an elastic body. *Int. J. Solids Struct.*, **19**, 425–436 (1983).

SLADEK, V., SLADEK, J. Nonsingular boundary integral representation of stresses. *Int. J. Num. Meth. in Eng.*, **33**, 1481–1499 (1992).

TANAKA, M., SLADEK, V., SLADEK, J. Regularization techniques applied to boundary element methods. *Appl. Mech. Rev.*, **47**, 457–499 (1994).

TOH, K. C., MUKHERJEE, S. Hypersingular and finite part integrals in the boundary element method. *Int. J. Solids Struct.*, **31**, 2289–2312 (1994).

WEAVER, J. Three-dimensional crack analysis. *Int. J. Solids Struct.*, **13**, 321–330 (1977).

A posteriori error evaluation and adaptive techniques

GUIGGIANI, M. Non-residual type accuracy estimates and adaptive refinement in boundary element analysis. In C. A. Brebbia and N. G. Zamani (ed.), *Boundary element techniques : applications in engineering*, pp. 323–340. Comp. Mech. Publ., Southampton (1989).

GUIGGIANI, M. Error indicators for adaptive mesh refinement in the boundary element method, a new approach. *Int. J. Num. Meth. in Eng.*, **29**, 1247–1269 (1990).

GUIGGIANI, M. Sensitivity analysis for boundary element error estimation and mesh refinement. *Int. J. Num. Meth. in Eng.*, **39**, 2907–2920 (1996).

HOLZER, S. M. A *p*-version of the symmetric BEM. *Comp. Meth. in Appl. Mech. Engng.*, **115**, 339–357 (1994).

HOLZER, S. M. The *h*-, *p*- and *hp*- version of the BEM in elasticity: numerical results. *Comm. Num. Meth. Engng.*, **11**, 255–266 (1995).

INGBER, M. S., MITRA, A. K. Grid redistribution based on measurable error indicators for the direct boundary element method. *Engng. Anal. with Bound. Elem.*, **9**, 13–19 (1992).

MENON, G., PAULINO, G. H., MUKHERJEE, S. Analysis of hypersingular error estimates in boundary element methods for potential problems. *SIAM J. Numer. Anal.* (1997). (submitted).

PAULINO, G. H., GRAY, L. J. Galerkin residuals for error estimation and adaptivity in the symmetric Galerkin boundary element method. *Advances in Comp. Math.* (1998). (to appear).

PAULINO, G. H., GRAY, L. J., ZARIKIAN, V. Hypersingular residuals – a new approach for error estimation in the boundary element method. *Int. J. Num. Meth. in Eng.*, **39**, 2005–2029 (1996).

PAULINO, G. H., SHI, F., MUKHERJEE, S., RAMESH, P. Nodal sensitivities as error estimates in computational mechanics. *Arch. Mech.*, **121**, 191–213 (1997).

RANK, E. Adaptive *h*-, *p*-, and *hp*- versions for boundary integral element methods. *Int. J. Num. Meth. in Eng.*, **28**, 1335–1349 (1989).

STEPHAN, E. P. The *h*–*p* boundary element method for solving 2- and 3-dimensional problems. *Comp. Meth. in Appl. Mech. Engng.*, **133**, 183–208 (1996).

C.3 GENERAL PURPOSE BOOKS

This section gathers some references devoted to general subjects which are often associated with boundary element methods or useful in their investigation and development.

ABRAMOWITZ, M., STEGUN, I. *Handbook of mathematical functions.* Dover Publications, New York (1972).

CIARLET, P. G. *Analyse numérique matricielle et optimisation.* Masson, Paris (1982).

CIARLET, P. G., LIONS, J. L. (eds). *Finite element methods (part 1).* Handbook of numerical analysis. North-Holland (1991).

CIARLET, P. G., LIONS, J. L. (eds). *Finite element methods (part 2), numerical methods for solids (part 2).* Handbook of numerical analysis. North Holland (1996).

COURANT, R., HILBERT, D. *Methods of mathematical physics (2 volumes).* Interscience (1953).

DAUBECHIES, I. *Ten lectures on wavelets.* SIAM Publications, Philadelphia, USA (1992).

DAVIS, J. J., RABINOWITZ, P. *Methods of numerical integration.* Academic Press (1975).

DONGARRA, J. J., BUNCH, J. R., MOLER, C. B., STEWART, G. B. LINPACK *users' guide.* Society for Industrial and Applied Mathematics (SIAM), Philadelphia, USA (1979).

DOUBROVINE, B., NOVIKOV, S., FOMENKO, A. *Géométrie contemporaine, méthodes et applications (première partie : géométrie des surfaces, des groupes de transformations et des champs).* Editions de Moscou (1979).

ERINGEN, A. C., SUHUBI, E. S. *Elastodynamics (vol. II – linear theory).* Academic Press (1975).

FLETCHER, R. *Practical methods of optimization.* John Wiley (1987).

FOGEL, D. B. *Evolutionary computation: towards a new philosophy of machine intelligence.* IEEE Press (1995).

GILL, P. E., MURRAY, W., WRIGHT, M. H. *Practical optimization.* Academic Press (1981).

GOLDBERG, D. E. *Genetic algorithms in search, optimization and machine learning.* Addison-Wesley (1989).

GOLUB, G. H., VAN LOAN, C. F. *Matrix computations (second edition).* Johns Hopkins University Press, Baltimore (1989).

GRADSHTEYN, I. S., RYZHIK, I. M. *Tables of integrals, series and products.* Academic Press (1980).

GREENBAUM, A. *Iterative methods for solving linear systems.* SIAM, Philadelphia (1997).

HAUG, E. J., CHOI, K. K., KOMKOV, V. *Design sensitivity analysis of structural systems.* Academic Press (1986).

LASCAUX, P., THEODOR, R. *Analyse numérique matricielle appliquée à l'art de l'ingénieur*. Masson (Paris) (1986).

LUBLINER, J. *Plasticity theory*. Macmillan, New York (1992).

MURA, T. *Micromechanics of defects in solids*. Martinus Nijhoff (1982).

MUSHKELISHVILI, N. I. *Some basic problems of the mathematical theory of elasticity*. Noordhoff (1953).

PARTON, V. Z., PERLIN, P. I. *Mathematical methods of the theory of elasticity (vols. 1, 2)*. Mir Publishers, Moscow (1984).

PIRONNEAU, O. *Optimal shape design for elliptic systems*. Springer-Verlag (1984).

PRESS, W. H., FLANNERY, B. P., TEUKOLSKY, S. A., VETTERLING, W. T. *Numerical recipes: the art of scientific computing (second edition)*. Cambridge Academic Press (1992).

REDDY, J. N. *Applied functional analysis and variational methods in engineering*. McGraw-Hill (1986).

SALENCON, J. *Mécanique des milieux continus (vols. 1,2,3)*. Presses de l'Ecole Polytechnique, Palaiseau, France (1992).

SCHWARZ, L. *Analyse*. Hermann (1981).

SOKOLOWSKI, J., ZOLESIO, J. P. *Introduction to shape optimization. Shape sensitivity analysis*, vol. 16 of *Springer series in Computational Mathematics*. Springer-Verlag (1992).

STOER, J., BULIRSCH, R. *Introduction to numerical analysis*. Springer-Verlag (1980).

STROUD, A. H., SECREST, D. *Gaussian quadrature formulas*. Prentice-Hall (1966).

WASHIZU, K. *Variational methods in elasticity and plasticity*. Pergamon Press (1975).

Author index

Subject index